Integrated Land Use Planning

Integrated Land Use Planning

Editor: John Lacey

Callisto Reference
www.callistoreference.com

Callisto Reference,
118-35 Queens Blvd., Suite 400,
Forest Hills, NY 11375, USA

Visit us on the World Wide Web at:
www.callistoreference.com

ISBN: 978-1-63239-949-6 (Hardback)

Printed in China.

Cataloging-in-Publication Data

Integrated land use planning / edited by John Lacey.
p. cm.
Includes bibliographical references and index.
ISBN 978-1-63239-949-6
1. Land use--Planning. 2. Land use--Environmental aspects.
3. Environmental management. I. Lacey, John.
HD108.3 .I58 2018
333.73--dc23

Table of Contents

Preface ..... **IX**

Chapter 1 **Evaluation of Dissolved Organic Carbon as a Soil Quality Indicator in National Monitoring Schemes** **1**
David L. Jones, Paul Simfukwe, Paul W. Hill, Robert T. E. Mills and Bridget A. Emmett

Chapter 2 **Effects of Land Use, Topography and Socio-Economic Factors on River Water Quality in a Mountainous Watershed with Intensive Agricultural Production in East China** **7**
Jiabo Chen and Jun Lu

Chapter 3 **Stream Macroinvertebrate Response Models for Bioassessment Metrics: Addressing the Issue of Spatial Scale** **19**
Ian R. Waite, Jonathan G. Kennen, Jason T. May, Larry R. Brown, Thomas F. Cuffney, Kimberly A. Jones and James L. Orlando

Chapter 4 **Low-Intensity Agricultural Landscapes in Transylvania Support High Butterfly Diversity: Implications for Conservation** **40**
Jacqueline Loos, Ine Dorresteijn, Jan Hanspach, Pascal Fust, László Rakosy and Joern Fischer

Chapter 5 **Political Systems Affect Mobile and Sessile Species Diversity – A Legacy from the Post-WWII Period** **51**
Sara A. O. Cousins, Mitja Kaligarič , Branko Bakan and Regina Lindborg

Chapter 6 **Conservation of Sandy Calcareous Grassland: What can be Learned from the Land Use History?** **58**
Anja Madelen Ödman and Pål Axel Olsson

Chapter 7 **Whither the Rangeland?: Protection and Conversion in California's Rangeland Ecosystems** **68**
D. Richard Cameron, Jaymee Marty and Robert F. Holland

Chapter 8 **Temporal Variation of Streamflow, Sediment Load and their Relationship in the Yellow River Basin, China** **80**
Guangju Zhao, Xingmin Mu, Alex Strehmel, Peng Tian

Chapter 9 **Spatial and Temporal Habitat Segregation of Mosquitoes in Urban Florida** **93**
Paul T. Leisnham, Shannon L. LaDeau, Steven A. Juliano

Chapter 10 **The Push and Pull of Land Use Policy: Reconstructing 150 Years of Development and Conservation Land Acquisition** **103**
Maria João Santos, Terry Watt, Stephanie Pincetl

Chapter 11 **Estimating Demand for Industrial and Commercial Land Use given Economic Forecasts** 112
Filipe Batista e Silva, Eric Koomen, Vasco Diogo and Carlo Lavalle

Chapter 12 **Land Use and Land Cover Change Dynamics Across the Brazilian Amazon: Insights from Extensive Time-Series Analysis of Remote Sensing Data** 126
João M. B. Carreiras, Joshua Jones, Richard M. Lucas and Cristina Gabriel

Chapter 13 **Soil Bacterial Community Response to Differences in Agricultural Management along with Seasonal Changes in a Mediterranean Region** 150
Annamaria Bevivino, Patrizia Paganin, Giovanni Bacci, Alessandro Florio, Maite Sampedro Pellicer, Maria Cristiana Papaleo, Alessio Mengoni, Luigi Ledda, Renato Fani, Anna Benedetti and Claudia Dalmastri

Chapter 14 **Development and Applications of a Comprehensive Land Use Classification and Map for the US** 164
David M. Theobald

Chapter 15 **Digital Mapping of Soil Organic Carbon Contents and Stocks in Denmark** 178
Kabindra Adhikari, Alfred E. Hartemink, Budiman Minasny, Rania Bou Kheir, Mette B. Greve and Mogens H. Greve

Chapter 16 **Communities of Endophytic Sebacinales Associated with Roots of Herbaceous Plants in Agricultural and Grassland Ecosystems are Dominated by *Serendipita herbamans* sp. nov.** 191
Kai Riess, Franz Oberwinkler, Robert Bauer and Sigisfredo Garnica

Chapter 17 **Climate and Land Use Controls on Soil Organic Carbon in the Loess Plateau Region of China** 201
Yaai Dang, Wei Ren, Bo Tao, Guangsheng Chen, Chaoqun Lu, Jia Yang, Shufen Pan, Guodong Wang, Shiqing Li and Hanqin Tian

Chapter 18 **Local Scale Comparisons of Biodiversity as a Test for Global Protected Area Ecological Performance** 212
Bernard W. T. Coetzee, Kevin J. Gaston and Steven L. Chown

Chapter 19 **A Mixed Methods Approach to Exploring the Relationship between Norway Rat (*Rattus norvegicus*) Abundance and Features of the Urban Environment in an Inner-City Neighborhood of Vancouver, Canada** 223
Chelsea G. Himsworth , Kirbee L. Parsons, Alice Y. T. Feng, Thomas Kerr, Claire M. Jardine and David M. Patrick

Chapter 20 **Uncertainties around the Implementation of a Clearing-Control Policy in a Unique Catchment in Northern Australia: Exploring Equity Issues and Balancing Competing Objectives** 230
Vanessa M. Adams, Robert L. Pressey

Chapter 21 **Deforestation and Benthic Indicators: How much Vegetation Cover is Needed to Sustain Healthy Andean Streams?** **241**
Carlos Iñiguez-Armijos , Adrián Leiva, Hans-Georg Frede, Henrietta Hampel and Lutz Breuer

Permissions

List of Contributors

Index

Preface

This book has been an outcome of determined endeavour from a group of educationists in the field. The primary objective was to involve a broad spectrum of professionals from diverse cultural background involved in the field for developing new researches. The book not only targets students but also scholars pursuing higher research for further enhancement of the theoretical and practical applications of the subject.

The process of transforming a natural environment into a built habitat for living beings is referred to as land use. This involves the management and efficient utilization of resources such as soil, water, etc. Modern land use is considered a branch of urban planning and has a major focus on aesthetics. This book explores the varied aspects of land use planning and management. Some of the diverse topics covered in this book address the varied branches that fall under this category. The extensive content of this book provides the readers with a thorough understanding of the subject.

It was an honour to edit such a profound book and also a challenging task to compile and examine all the relevant data for accuracy and originality. I wish to acknowledge the efforts of the contributors for submitting such brilliant and diverse chapters in the field and for endlessly working for the completion of the book. Last, but not the least; I thank my family for being a constant source of support in all my research endeavours.

Editor

1

Evaluation of Dissolved Organic Carbon as a Soil Quality Indicator in National Monitoring Schemes

David L. Jones[1]*, Paul Simfukwe[1], Paul W. Hill[1], Robert T. E. Mills[2], Bridget A. Emmett[3]

1 School of the Environment, Natural Resources & Geography, Bangor University, Bangor, Gwynedd, United Kingdom, **2** Ecole Polytechnique Fédérale de Lausanne (EPFL), School of Architecture, Civil and Environmental Engineering (ENAC), Laboratory of Ecological Systems (ECOS), Lausanne, Switzerland, **3** Centre for Ecology and Hydrology, Environment Centre Wales, Bangor, Gwynedd, United Kingdom

Abstract

Background: Monitoring the properties of dissolved organic carbon (DOC) in soil water is frequently used to evaluate changes in soil quality and to explain shifts in freshwater ecosystem functioning.

Methods: Using >700 individual soils (0–15 cm) collected from a 209,331 km^2 area we evaluated the relationship between soil classification (7 major soil types) or vegetation cover (8 dominant classes, e.g. cropland, grassland, forest) and the absorbance properties (254 and 400 nm), DOC quantity and quality (SUVA, total soluble phenolics) of soil water.

Results: Overall, a good correlation ($r^2 = 0.58$) was apparent between soil water absorbance and DOC concentration across the diverse range of soil types tested. In contrast, both DOC and the absorbance properties of soil water provided a poor predictor of SUVA or soluble phenolics which we used as a measure of humic substance concentration. Significant overlap in the measured ranges for UV absorbance, DOC, phenolic content and especially SUVA of soil water were apparent between the 8 vegetation and 7 soil classes. A number of significant differences, however, were apparent within these populations with total soluble phenolics giving the greatest statistical separation between both soil and vegetation groups.

Conclusions: We conclude that the quality of DOC rather than its quantity provides a more useful measure of soil quality in large scale surveys.

Editor: Dafeng Hui, Tennessee State University, United States of America

Funding: This work was funded by the United Kingdom Department for Environment, Food and Rural Affairs. The funders had no role in study design, data collection and analysis, decision to publish, or preparation of the manuscript.

Competing Interests: The authors have declared that no competing interests exist.

* E-mail: d.jones@bangor.ac.uk

Introduction

Measurement of dissolved organic matter (DOM) concentrations have provided a sensitive indicator for assessing large scale geographical and temporal (e.g. decadal) changes in water quality [1]. In contrast, while DOM is frequently measured in soil studies, it has rarely been used for evaluating changes in soil quality in large scale (e.g. national) monitoring programmes [2]. To some extent, the DOM concentration of freshwaters should reflect that of the surrounding catchment area suggesting that an assessment of soil DOM might provide a sensitive soil quality indicator. This would also match with studies investigating the progressive increase in DOM within rivers, a widespread trend observed in both Europe and North America [3–4]. An investigation of the spatial or temporal changes in soil DOM within different soil types could therefore prove useful for explaining these trends in catchments containing a complex mixture of soils. One potential drawback, however, is that determination of DOM in soil can be very time consuming and expensive when large numbers of samples need to be processed. One potential solution is to use the spectral properties of soil water which provides a rapid proxy for estimating dissolved organic C (DOC) concentration [5–10]. For a single geographical location, strong relationships are often apparent between DOC concentration and UV absorbance [11–13]. However, these relationships are often non-linear and the mathematical function describing the relationship can vary with season or changes in hydrological flow patterns [14–16]. Further, whilst differences in the quality and quantity of DOC are frequently reported for contrasting soil and vegetation types, whether these dissimilarities remain valid over large geographical scales remains uncertain. In this context, we undertook a national large scale survey of the quality and quantity of DOM across Great Britain to evaluate its potential for evaluating changes to soil quality in national soil quality monitoring programmes. Our first aim was to critically evaluate the relationship between direct and indirect methods for DOC quantification. Secondly, we aimed to evaluate whether different soil classes and vegetation cover types possessed unique DOM signatures in terms of both quantity and quality and whether these might be useful for explaining changes in DOM occurring at a national scale.

Methods

Soil Sampling

To encompass all the major soil and land use types, a total of 702 soil samples were collected throughout Great Britain in June-

July, 2007 (area 209,331 km^2; ca. 300 km^2 $sample^{-1}$) as part of the Centre for Ecology and Hydrology Countryside Survey (CS) [17]. Samples were selected, based on a stratified random sample of 1 km squares at gridpoints on a 15 km grid using the Institute of Terrestrial Ecology (ITE) Land Classification as the basis of the stratification (Fig. S1). At each grid intersection, a 1 km^2 sample area was selected. Within the 1 km^2 sample area, a 5×5 m^2 plot was randomly located and replicate 15 cm long ×4 cm diameter soil cores were collected. Topsoils (0–15 cm) were only sampled to reflect past and current (1978-present) standard practice in UK national monitoring [18–20]. This depth was originally selected by the UK Government's Department for Environment, Food and Rural Affairs as it was hypothesized that it would show the greatest change in response to environmental perturbation (including changes in land use, climate, agronomic management or atmospheric deposition). The soil horizons sampled included H, O and A horizons with *a, e, i, h, g, k* and *p* sub-designations [21]. Across all land use categories, the dominant eight soil groupings (% of total) were: Brown soils (31%), Podzolic soils (15%), Surface water (SW) gley soils (18%), Peat soils (13%), Groundwater (GW) gley soils (12%), Lithomorphic soils (8%), and Pelosol soils (3%) [22]. The FAO World Reference Base Classification equivalent categories for these soil groups are presented in Table S1 and their major chemical, physical and biological properties presented in Table 1. Vegetation cover at each sampling point was classified into eight aggregated vegetation classes (AVC) with the following groupings (% of total): Cropland (15%), Tall grass and herbs (4%), Fertile grassland (19%), Infertile grassland (21%), Lowland woodland (3%), Upland woodland (8%), Moorland grassland mosaic (11%), and Heathland and bog (19%) [17]. Aggregate vegetation classes were derived by cluster analysis of the mean DECORANA scores for 100 smaller classes obtained by TWINSPAN analysis of plant species data in each sample plots [37]. Further descriptions of the vegetation types can be found in Table S2.

Soil Water Collection and Analysis

Soil water was obtained by adding artificial rainwater to the intact soil columns and collecting the leachate as detailed in [23–24]. The absorbance of the soil leachate water was measured at 254 and 400 nm on a Synergy 96 well plate spectrophotometer (BioTek Instruments Inc., Winooski, VT) using Falcon flat-bottom UV well plates (BD Biosciences, Franklin Lakes, NJ). DOC concentrations were measured with a TOC-V analyser (Shimadzu Corp., Kyoto, Japan). Total dissolved phenolics and tannins were assayed colorimetrically using the Folin-Ciocalteu reagent (F9252; Sigma-Aldrich Inc.) according to [25] using gallic acid as a standard. Specific ultraviolet absorbance (SUVA) was calculated by dividing the absorbance at 254 nm (cm^{-1}) by the DOC concentration (mg l^{-1}).

Background Soil Analysis

Background soil characteristics were measured on replicate cores matching those used in the analysis above. After collection from the field, the soil was extruded from the core, roots and stones removed and the soil homogenised and dried (105 °C). Total soil C and N were analysed on an Elementar Vario-EL elemental analyser (Elementar Analysensysteme GmbH, Hanau, Germany) using the UKAS accredited method SOP3102 [23]. Bulk density was calculated as mass/volume after the removal of stones (>2 mm) and accounting for their volume [23]. Soil organic matter was determined by loss-on-ignition (LOI) by first drying soil (10 g) at 105°C and then measuring the mass loss after further heating at 375 °C for 16 h. Available soil P was measured using

Table 1. Properties of the major soil groups.

Soil group	Soil pH	Soil organic matter (%)	C-to-N ratio	Olsen P (mg kg^{-1})	Bulk density (g cm^{-3})	Exchangeable Ca (mmol kg^{-1})	Exchangeable Al (mmol kg^{-1})	Soil respiration (mg C cm^{-3} h^{-1})	Soil respiration (g C g SOC^{-1} h^{-1})
Pelosol	6.8 ± 1.4^{a}	7.3 ± 3.3^{cd}	10.7 ± 0.8^{de}	28 ± 22^{abc}	1.10 ± 0.21^{a}	62 ± 40^{a}	$1.3\pm4.8a^{b}$	10.2 ± 26.4^{bc}	4.7 ± 3.4^{a}
Brown	6.5 ± 1.2^{a}	10.0 ± 12.5^{d}	12.2 ± 4.1^{e}	32 ± 27^{abc}	1.04 ± 0.33^{a}	39 ± 27^{a}	2.1 ± 4.8^{b}	8.6 ± 11.0^{c}	7.3 ± 5.6^{a}
Groundwater gley	6.6 ± 1.2^{a}	10.9 ± 13.5^{d}	13.4 ± 4.4^{cde}	32 ± 34^{abc}	0.98 ± 0.36^{a}	42 ± 31^{a}	2.3 ± 5.4^{ab}	13.2 ± 24.1^{bc}	8.1 ± 6.4^{a}
Surface water gley	5.9 ± 1.3^{a}	24.9 ± 30.0^{bc}	14.2 ± 5.5^{cd}	25 ± 28^{bc}	0.78 ± 0.43^{b}	54 ± 49^{a}	6.3 ± 14.6^{ab}	14.9 ± 19.2^{bc}	6.6 ± 5.0^{ab}
Podzolic	4.9 ± 0.9^{b}	28.8 ± 27.8^{b}	17.9 ± 7.0^{b}	21 ± 20^{c}	0.57 ± 0.33^{c}	20 ± 29^{b}	$7.1\pm8.9a^{b}$	17.9 ± 32.9^{bc}	5.1 ± 3.9^{b}
Lithomorphic	6.0 ± 1.4^{a}	35.1 ± 32.8^{b}	16.2 ± 6.4^{bc}	41 ± 69^{ab}	0.56 ± 0.38^{c}	42 ± 39^{a}	3.4 ± 5.6^{ab}	20.8 ± 33.0^{b}	6.4 ± 5.6^{ab}
Peat	4.6 ± 0.6^{b}	75.9 ± 27.9^{a}	24.1 ± 8.5^{a}	44 ± 51^{a}	0.20 ± 0.21^{d}	46 ± 50^{a}	9.6 ± 38.0^{a}	43.1 ± 37.1^{a}	4.5 ± 3.8^{b}
ANOVA	***	***	***	**	***	***	**	***	***

Values are expressed on a dry weight basis and represent mean ± SD.
** and *** indicate significant differences between soil groups at the $P<0.01$ and $P<0.001$ levels respectively. Superscript letters indicate significant between soil groups at the $P<0.05$ level.

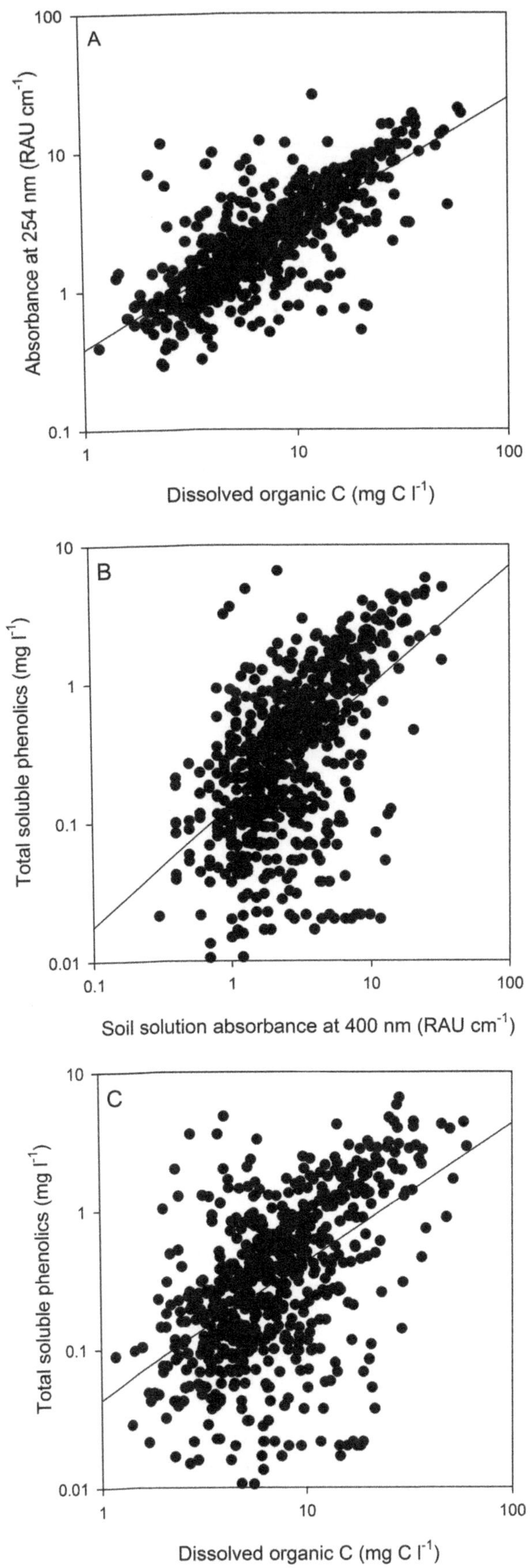

Figure 1. Relationship between the absorbance of soil water from 702 individual soils measured in the UV (254 nm) range and DOC concentration (Panel A), the relationship between total soluble phenolics and either solution absorbance at 400 nm (Panel B) or DOC concentration (Panel C). The lines are linear regression for all the data in the plots (Panel A, $r^2 = 0.579$; $y = 0.387 + 0.301x$; Panel B, $r^2 = 0.432$; $y = 0.12 + 0.139x$; Panel C, $r^2 = 0.375$; $y = 0.025 + 0.068x$).

the Olsen method whereby 5 g of soil was extracted with 100 ml of 0.5 M sodium bicarbonate (pH 8.5). The P in the extract was determined colorimetrically using the molybdate blue method (880 nm) using a continuous flow analyser [23]. Soil pH was measured by equilibrating 10 g of field-moist soil with 25 ml of deionised water. Exchangeable Ca and Al were determined by shaking 5 g of soil with 25 ml of 1.0 M NH_4Cl (250 rev min^{-1}, 60 min). Subsequently, the extracts were centrifuged (5000 *g*, 10 min) and the supernatant recovered for analysis. Ca in the extracts was determined by atomic absorption spectrometry on a Perkin Elmer Analyst 400 Atomic Absorption Spectrometer (PerkinElmer, Waltham, MA, USA). The extracts were diluted with $LaCl_3$ (0.5% w/v) prior to Ca determination. Al concentration in the extracts was determined using the modified catechol violet method [35]. The absorbance of the solution was measured at 580 nm using a PowerWave XS scanning microplate spectrophotometer (BioTek Instrument, Winooski, VT). Basal soil respiration was determined on one replicate core. The cores were wet to field capacity as described previously, placed in a sealed chamber (1250 cm^3 head space). The soils were then incubated at 10°C (average UK air temperature) for 1 h (at which linearity was known to be established following testing on selected cores which covered the range of soil types sampled). Subsequently, the head space gas was analysed for CO_2 concentration using a Clarus 500 Gas Chromatograph (PerkinElmer). SR was determined as the change in CO_2 concentration before and after incubation corrected for soil dry weight and soil organic matter content.

Statistical Analysis

Linear and stepwise regression analysis was undertaken using Minitab v16 (Minitab Inc., State College, PA). When the solution DOC, absorbance, soluble phenolic and SUVA values were grouped according to soil and vegetation type the data failed normality testing (Shapiro-Wilk). Consequently, the data were log transformed, normality verified and an ANOVA performed with Tukey-pairwise comparisons ($P<0.05$ cut-off) using Minitab v16.

Results

Soil Water Properties

The relationship between the absorbance of soil water in the UV (254 nm) and visible (400 nm) range across 702 individual sites showed a strong linear correlation ($r^2 = 0.931$; $P<0.001$; Fig. S2). A strong positive correlation was also observed between the DOC concentration in soil water and both absorbance at 254 nm ($r^2 = 0.58$; $P<0.001$; Fig. 1) and 400 nm ($r^2 = 0.47$; $P<0.001$). The prediction of DOC concentration using both absorbance values in a stepwise regression model did not result in a significantly better fit ($r^2 = 0.61$). In contrast, the absorbance characteristics of soil water proved to be a less reliable predictor of total soluble phenolics (254 nm $r^2 = 0.41$; $P<0.001$; 400 nm $r^2 = 0.43$; $P<0.001$; Fig. 1). Similarly, DOC concentration either with or without inclusion of spectral properties into the regression model, proved to be the least reliable predictor of total soluble phenolics ($r^2 = 0.38$; $P<0.001$). SUVA proved to be the worst predictor of

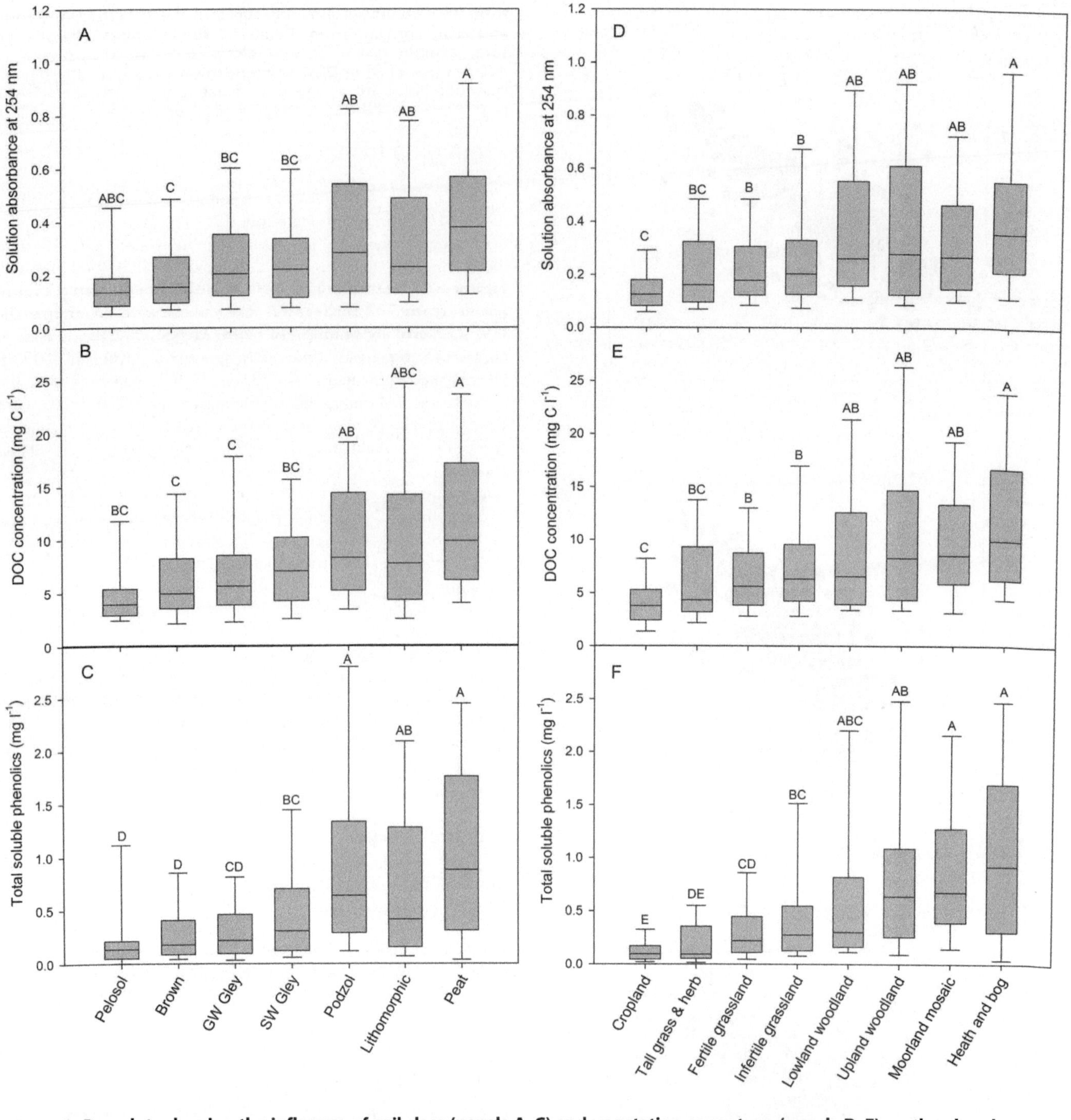

Figure 2. Box plots showing the influence of soil class (panels A–C) and vegetation cover type (panels D–F) on the absorbance of soil solutions at 254 nm, their DOC and total soluble phenolic concentration from 702 individual soils sampled as part of a nationwide soil quality assessment. The boundary of the box closest to zero indicates the 25th percentile, the line within the box marks the median, and the boundary of the box farthest from zero indicates the 75th percentile. Whiskers (error bars) above and below the box indicate the 90th and 10th percentiles respectively. Different letters indicate significant difference between individual groups ($P<0.05$). Soils are ranked in organic matter (OM) content from left (low OM) to right (high OM) while vegetation is broadly ranked according to agricultural productivity.

total soluble phenolics with either linear ($r^2 = 0.13$) or non-linear models (data not presented).

Is Soil or Vegetation a Better Predictor of DOC Quantity and Quality?

The quantity and quality of DOC as a function of different soil class and vegetation types is shown in Figure 2. Overall, there was a large similarity in the measured ranges of DOC, absorbance at 254 nm and soluble phenolics between the different groupings for both soil and vegetation. Despite this, ANOVA revealed significant differences between groups (all $P<0.001$) with response gradients apparent when the groups were ranked according to either organic matter content (soils) or an intensification/altitude scale (vegetation)(Fig. 2). Of the parameters measured, soluble phenolic content gave the greatest separation between groups. In contrast, SUVA revealed no significant differences between groupings for both soil ($P = 0.483$) and vegetation ($P = 0.819$; Fig. S3).

Discussion

Method of Soil Sampling

The sampling undertaken here formed part of a nationwide monitoring programme which is carried out each decade to assess changes in soil and water quality, landscape features and aquatic and terrestrial biodiversity [19,26]. Whilst the sampling depth (0–15 cm) is highly suited to agricultural soils which show a high degree of vertical homogeneity (due to cultivation), we acknowledge that the sampling regime may be less suited to highly stratified soils [27]. In these horizonated soils, differential amounts of O, E, A and possibly B horizons could be sampled by coring from 0–15 cm, depending on their thickness. This has led to many studies, particularly in forests, where soil quality is measured on organic and mineral horizons separately [27–28]. In our case, the historical legacy is such that the sampling protocols will not be changed for the foreseeable future (on scientific, socio-political and economic grounds) and therefore our results should be viewed in the context of this.

DOC and Soil Quality Assessment

Overall, our results show that in large geographical scale soil quality assessments across a broad range of vegetation and soil types, soil water DOC concentrations can be rapidly and cheaply estimated from their spectral properties. This is particularly relevant as DOC in itself can be used to predict the movement of both organic and inorganic pollutants in soil and is often seen as a pollutant itself when entering freshwaters [9,29–30]. Disappointingly, however, we found that all soil and vegetation types had a wide variability in their DOC concentrations. This indicates that the quantity of DOC may not represent a sensitive indicator for monitoring the stability of ecosystems when faced with anthropogenic perturbation (e.g. land use change). This is in contrast to [24] who found different rates of N mineralization ($NH_4^+ + NO_3^-$) in the same soils alongside other soil quality indicators as shown in Table 1. While SUVA is frequently used for assessing the quality of DOC in freshwaters [30], our results indicated that it was incapable of separating between land uses and soil classes over a wide geographical range and is therefore probably unsuited to large scale soil quality assessments. In contrast, total soluble phenols gave the best separation between ecosystem types. This was a surprising result as we had assumed that SUVA and soluble phenols would be highly correlated. This disagreement could be due to interference in both the determination of total phenolics (e.g. SO_2, DOC; [31–32]) or SUVA (e.g. pH, Fe^{3+}; [12]). However, more likely it is due to the natural variability in the total phenol content of humic substances originating from different soil classes and vegetation cover types [33–34]. As expected, the quality and quantity of DOC were dependent on both vegetation and soil type. This is to some extent expected considering that certain soil types favour certain vegetation covers, however, it does indicate that 2-way stratification by soil and vegetation type may prove more useful for revealing unique DOC signatures.

With respect to the measurement of DOC quality, we acknowledge that recent methodological advancements are increasing our potential to characterise the many thousands of compounds that comprise DOC [36]. It is highly likely that these new analytical approaches will reveal compounds that are specific to different functional soil or plant types (e.g. keystone compounds) improving our capacity to use DOC as a soil quality indicator. In addition, this analysis may also directly support the interpretation of temporal changes in other soil quality indicators (e.g. soil organic matter quality and quantity, soil biodiversity).

Supporting Information

Figure S1 Map of the UK showing the individual soil sampling locations used in the study. The total land area is 209,331 km2.

Figure S2 Relationship between the absorbance of soil water from 702 individual soils measured in either the UV (254 nm) or visible (400 nm) range. The line is a linear regression for all the data in the plot ($r^2 = 0.931$; $y = -0.01 + 0.123x$).

Figure S3 Box plots showing the influence of soil type (panel A) and vegetation cover (panel B) on the specific UV absorbance (SUVA) values from 702 individual soils sampled as part of a nationwide soil quality assessment. The boundary of the box closest to zero indicates the 25th percentile, the line within the box marks the median, and the boundary of the box farthest from zero indicates the 75th percentile. Whiskers (error bars) above and below the box indicate the 90th and 10th percentiles respectively. No significant differences were apparent between treatments.

Table S1 Comparable classification of the UK soil groups with those in the FAO World Reference Base Classification (WRB, 2006).

Table S2 Summary descriptions of the eight aggregate vegetation classes represented. Table adapted from Smart et al. (2003).

Acknowledgments

We would like to thank Ed Rowe at the Centre for Ecology and Hydrology for help with the experiments and data handling. Countryside Survey (CS) is funded by a consortium of UK government funded bodies led by the Natural Environment Research Council (NERC) and the Department for Environment, Food and Rural Affairs (Defra), which includes the Centre for Ecology and Hydrology, Countryside Council for Wales, Forestry Commission, Natural England, the Northern Ireland Environment Agency, the Scottish Government, Scottish Natural Heritage, and the Welsh Assembly Government. The completion of CS was only possible with the support and advice of many dedicated individuals from these and other organizations. The authors would like to thank the land owners, farmers, and other land managers who gave permission for the field surveyors to collect data and samples from their land; without such cooperation scientific field studies like CS would not be possible. Additionally, the authors would especially like to acknowledge the major contribution made by the team of laboratory staff who processed all the soils and undertook the chemical analyses.

Author Contributions

Conceived and designed the experiments: DLJ BAE. Performed the experiments: PS RTEM. Analyzed the data: PS PWH DLJ RTEM. Wrote the paper: DLJ PWH PS BAE.

References

1. Camarero L, Rogora M, Mosello T, Anderson NJ, Barbieri A, et al. (2009) Regionalisation of chemical variability in European mountain lakes. Freshwater Biol 54: 2452–2469.
2. Filep T, Rekasi M, Kincses I (2008) Effect of soil quality on dissolved organic matter (DOM) content of soils. Cereal Res Commun 36: 667–670.
3. Evans CD, Monteith DT, Cooper DM (2005) Long-term increases in surface water dissolved organic carbon: Observations, possible causes and environmental impacts. Environ Pollut 137: 55–7.
4. Clutterbuck B, Yallop AR (2010) Land management as a factor controlling dissolved organic carbon release from upland peat soils. 2 Changes in DOC productivity over four decades. Sci Total Environ 408: 6179–6191.
5. Mattson JS, Smith CA, Jones TT, Gerchako SM, Epstein BD (1974) Continuous monitoring of dissolved organic-matter by UV-visible photometry. Limnol Oceanogr 19: 530–535.
6. Dittman JA, Shanley JB, Driscoll CT, Aiken GR, Chalmers AT, et al. (2009) Ultraviolet absorbance as a proxy for total dissolved mercury in streams. Environ Pollut 157: 1953–1956.
7. Brandstetter A, Sletten RS, Mentler A, Wenzel WW (1996) Estimating dissolved organic carbon in natural waters by UV absorbance (254 nm). J Soil Sci Plant Nutr 159: 605–607.
8. Jaffrain J, Gerard F, Meyer M, Ranger J (2007) Assessing the quality of dissolved organic matter in forest soils using ultraviolet absorption spectrophotometry. Soil Sci Soc Am J 71: 1851–1858.
9. Amery F, Degryse F, Cheyns K, De Troyer I, Mertens J, et al. (2008) The UV-absorbance of dissolved organic matter predicts the fivefold variation in its affinity for mobilizing Cu in an agricultural soil horizon. Eur J Soil Sci 59: 1087–1095.
10. Hansson K, Kleja DB, Kalbitz K, Larsson H (2010) Amounts of carbon mineralised and leached as DOC during decomposition of Norway spruce needles and fine roots. Soil Biol Biochem 42: 178–185.
11. Dilling J, Kaiser K (2002) Estimation of the hydrophobic fraction of dissolved organic matter in water samples using UV photometry. Water Res 36: 5037–5044.
12. Weishaar JL, Aiken GR, Bergamaschi BA, Fram MS, Fujii R, et al. (2003) Evaluation of specific ultraviolet absorbance as an indicator of the chemical composition and reactivity of dissolved organic carbon. Environ Sci Technol 37: 4702–4708.
13. Tipping E, Corbishley HT, Koprivnjak JF, Lapworth DJ, Miller MP, et al. (2009) Quantification of natural DOM from UV absorption at two wavelengths. Environ Chem 6: 472–476.
14. Wheeler JR (1977) Dissolved organic-carbon - spectral relationships in coastal waters. Limnol Oceanogr 22: 573–575.
15. Edwards AC, Cresser MS (1987) Relationships between ultraviolet absorbency and total organic-carbon in 2 upland catchments. Water Res 21: 49–56.
16. Izbicki JA, Pimentel IM, Johnson R, Aiken GR, Leenheer J (2007) Concentration, UV-spectroscopic characteristics and fractionation of DOC in stormflow from an urban stream, Southern California, USA. Environ Chem 4: 35–48.
17. Emmett BA, Reynolds B, Chamberlain PM, Rowe E, Spurgeon D, et al. (2010) Countryside Survey: Soils Report from 2007. CS Technical Report No. 9/07, Centre for Ecology and Hydrology, Wallingford, UK.
18. Bellamy PH, Loveland PJ, Bradley RI, Lark RM, Kirk GJD (2005) Carbon losses from all soils across England and Wales 1978–2003. Nature 437: 245–248.
19. Norton LR, Maskell LC, Smart SS, Dunbar MJ, Emmett BA, et al. (2012) Measuring stock and change in the GB countryside for policy - Key findings and developments from the Countryside Survey 2007 field survey. J Environ Manage 113: 117–127.
20. Reynolds B, Chamberlain PM, Poskitt J, Woods C, Scott WA, et al. (2013) Countryside Survey: National "soil change" 1978-2007 for topsoils in Great Britain-acidity, carbon, and total nitrogen status. Vadose Zone J 12: vzj2012.0114
21. FAO (2006) Guidelines for Soil Description. Fourth Edition.Food and Agriculture Organization of the United Nations, Rome, Italy.
22. Avery BW (1990) Soils of the British Isles. Cambridge University Press, Cambridge.
23. Emmett BA, Frogbrook ZL, Chamberlain PM, Griffiths R, Pickup R, et al. (2008) Countryside Survey. Soils Manual. NERC/Centre for Ecology & Hydrology, (CS Technical Report No. 3/07, CEH Project Number: C03259), 180 pp.
24. Rowe EC, Emmett BA, Frogbrook ZL, Robinson DA, Hughes S (2012) Nitrogen deposition and climate effects on soil nitrogen availability: Influences of habitat type and vegetation. Sci Total Environ 434: 62–70.
25. Velioglu YS, Mazza G, Gao L, Oomah BD (1998) Antioxidant activity and total phenolics in selected fruits, vegetables and grain products. J Agric Food Chem 46: 4113–4117.
26. Smart SM, Clarke RT, van de Poll HM, Robertson EJ, Shield ER, et al. (2003) National-scale vegetation change across Britain; an analysis of sample-based surveillance data from the Countryside Surveys of 1990 and 1998. J Environ Manage 67: 239–254.
27. Lawrence GB, Fernandez IJ, Richter DD, Ross DS, Hazlett PW, et al. (2013) Measuring environmental change in forest ecosystems by repeated soil sampling: a North American perspective. J Environ Qual 42: 623–639.
28. O'Neill KP, Amacher MC, Palmer CJ (2005) Developing a national indicator of soil quality on U.S forestlands: Methods and initial results. Environ Monit Assess 107: 59–80.
29. Steinberg CEW, Hoss S, Kloas W, Lutz I, Meinelt T, et al. (2003) Hormonelike effects of humic substances on fish, amphibians, and invertebrates. Environ Toxicol 19: 409–411.
30. Zhang Q, Kuang WF, Liu LY, Li KX, Wong KH, et al. (2009) Trihalomethane, haloacetonitrile, and chloral hydrate formation potentials of organic carbon fractions from sub-tropical forest soils. J Hazard Mater 172: 880–887.
31. Matilainen A, Gjessing ET, Lahtinen T, Hed L, Bhatnagar A, et al. (2011) An overview of the methods used in the characterisation of natural organic matter (NOM) in relation to drinking water treatment. Chemosphere 83: 1431–1442.
32. Ohno T, First PR (1998) Assessment of the Folin and Ciocalteu's method for determining soil phenolic carbon. J Environ Qual 27: 776–782.
33. Sharma OK, Krishnan PS (1966). Colorimetric estimation of humic acid with phenol reagent of Folin and Ciocalteu. Anal Biochem 14: 11–17.
34. Thorn KA, Folan DW, MacCarthy P (1989) Characterization of the International Humic Substances Society standard and reference fulvic and humic acids by solution state carbon-13 (^{13}C) and hydrogen-1 (^{1}H) nuclear magnetic resonance spectrometry. U.S. Geological Survey, Water-Resources Investigations Report 89-4196, Denver, CO, USA.
35. Dougan WK, Wilson AL (1974) The absorptiometric determination of aluminium in water. A comparison of some chromogenic reagents and the development of an improved method. Analyst 99: 413–30.
36. Osborne DM, Podgorski DC, Bronk DA, Roberts Q, Sipler RE, et al. (2013) Molecular-level characterization of reactive and refractory dissolved natural organic nitrogen compounds by atmospheric pressure photoionization coupled to Fourier transform ion cyclotron resonance mass spectrometry. Rapid Comm. Mass Spectrom. 27: 851–858.
37. Bunce RGH, Barr CJ, Gillespie MK, Howard DC, Scott RA, et al. (1999) Vegetation of the British Countryside—the Countryside Vegetation System. Department of the Environment, London.

2

Effects of Land Use, Topography and Socio-Economic Factors on River Water Quality in a Mountainous Watershed with Intensive Agricultural Production in East China

Jiabo Chen[1], **Jun Lu**[1,2]*

1 Department of Natural Resources, College of Environment and Natural Resources, Zhejiang University, Hangzhou, Zhejiang Province, China, 2 China Ministry of Education Key Lab of Environment Remediation and Ecological Health, Zhejiang University, Hangzhou, Zhejiang Province, China

Abstract

Understanding the primary effects of anthropogenic activities and natural factors on river water quality is important in the study and efficient management of water resources. In this study, analysis of Variance (ANOVA), Principal component analysis (PCA), Pearson correlations, Multiple regression analysis (MRA) and Redundancy analysis (RDA) were applied as an integrated approach in a GIS environment to explore the temporal and spatial variations in river water quality and to estimate the influence of watershed land use, topography and socio-economic factors on river water quality based on 3 years of water quality monitoring data for the Cao-E River system. The statistical analysis revealed that TN, pH and temperature were generally higher in the rainy season, whereas BOD_5, DO and turbidity were higher in the dry season. Spatial variations in river water quality were related to numerous anthropogenic and natural factors. Urban land use was found to be the most important explanatory variable for BOD_5, COD_{Mn}, TN, DN, NH_4^+-N, NO_3^--N, DO, pH and TP. The animal husbandry output per capita was an important predictor of TP and turbidity, and the gross domestic product per capita largely determined spatial variations in EC. The remaining unexplained variance was related to other factors, such as topography. Our results suggested that pollution control of animal waste discharge in rural settlements, agricultural runoff in cropland, industrial production pollution and domestic pollution in urban and industrial areas were important within the Cao-E River basin. Moreover, the percentage of the total overall river water quality variance explained by an individual variable and/or all environmental variables (according to RDA) can assist in quantitatively identifying the primary factors that control pollution at the watershed scale.

Editor: Miguel Lopez-Ferber, Ecole des Mines d'Alès, France

Funding: Chinese National Key Technology R&D Program (Grant No. 2012BAC17B01). The funders had no role in study design, data collection and analysis, decision to publish, or preparation of the manuscript.

Competing Interests: The authors have declared that no competing interests exist.

* Email: jlu@zju.edu.cn

Introduction

The deterioration of river water quality has become a primary environmental concern due to unsustainable anthropogenic activities; the demand for freshwater has been rapidly increasing in many developing countries, especially in China [1–3]. River water quality is controlled by complex anthropogenic activities and natural factors at both the river and watershed scales [1], [7–9]. Understanding the temporal and spatial variations in river water quality and estimating the primary regional factors that affect water quality can assist researchers in establishing priorities for sustainable water management [2], [10–12]. The relationships between water quality parameters and land use/cover, population density and point source discharge have been frequently studied [1], [7], [9], [13–15]. Simultaneously, topography and animal waste discharge are considered important factors that affect watershed river water quality [7], [13–16]. Particularly, several countries, e.g., China, contain large mountainous areas [17] and extensive animal production that lack strict management techniques [16], [18,19].

Recently, Pearson correlations have been widely employed to determine the relationships between environmental variables and river water quality [1], [7]. This method is simple and provides quantitative information; however, Pearson correlations lack visualization. Multiple regression analysis (MRA) is a useful tool that is commonly used to determine that watershed characteristics that best explain the spatial variability of an individual river water quality variable [7], [20]. This method lacks general information regarding pollution types. Moreover, principal component analysis (PCA) has become a widely accepted method in river water quality assessment and source apportionment studies in the last decade [2]. This method is commonly used to obtain the types of specific pollution sources without explanatory variables; however, PCA can only provide preliminary information regarding pollution types. Redundancy analysis (RDA), a multivariate statistical analysis method, has been proved useful for qualitative analysis of the interactions between river water quality and watershed characteristics in highly complex systems [13], [21,22]. RDA can reveal the influences of environmental factors on overall water quality, not only on single water quality variables [20]. Several commonly used statistical methods for pollution source identification are complementary (i.e., they have their own benefits and limits). Summary of commonly used statistical methods on pollution source identification in recent years is listed in **Table S1**. However, comprehensive applications of different statistical methods to evaluate the effects of environmental variables on water quality have not been fully explored in river studies in China. Furthermore, few studies have used RDA analysis to quantitatively evaluate the effects of watershed characteristics on the overall river water quality.

The Cao-E River is located upstream of the Qiantang Estuary, which is a major riverine system in Zhejiang Province, China, and is the primary source of industrial, agricultural and domestic water supplies [10]. Furthermore, the blue algae bloom that occurred in July 2004 occurred in the Qiantang Estuary [23,24].

In this study, ANOVA, Pearson correlation analysis, MRA, PCA and RDA were applied to investigate the effects of subwatershed land use, topography and socio-economic factors (including animal waste discharge) on river water quality at the watershed scale for the Cao-E River system in a GIS environment. The objectives of this study were threefold: (1) to examine the temporal and spatial variations in water contamination within the study area, (2) to investigate the relationships between the river water quality parameters and land use, topography and socio-economic factors, and (3) to identify the primary pollution sources to estimate the possible sources that affect the water quality parameters in the Cao-E River basin. The results can be helpful for water conservation in the Cao-E River basin, and they provide a valuable tool for water quality agencies to develop assessment strategies for effective water quality management and rapid solutions for water pollution problems at the watershed scale.

Materials and Methods

1 Ethics statement

No specific permits were required for the described field studies; the sampling did not cause any disturbance to the environment or to the protected species at the sampling sites.

2 Study area

The Cao-E River (29°08′-30°15′ N, 120°30′-121°15′ E) is located in Zhejiang Province, East China. The Cao-E River is one of the main tributaries of the Qiantang Estuary, which flows into the East China Sea. The river system contains a main stream and several major tributaries (i.e., the CT River, the XC River, the CL River, the HZ River, the XS River and the YT River; **Figure 1**). The main stream is ~197 km long with an average slope of 3.0‰. The Cao-E River basin has a typical subtropical monsoon climate with an average annual rainfall of 1500 mm and an average ambient temperature of 16.2°C [25]. The watershed is within a mountainous region. Specifically, mountains and hills cover 2/3 of the area; the remaining area is covered by plains with intensive agricultural production [26].

3 Data Sources

Water quality data. The stream water samples (the number of samples used for statistical analysis is 36) were collected from July 2003 to June 2006 at 20 sampling sites throughout the watershed (Figure 1). Specific sampling stations that were largely influenced by point source pollution were excluded from further analysis. These specific sampling sites should meet the following conditions: (1) there was an obvious outlet of point pollution discharge near the sampling site, and (2) The standard deviation of the water quality parameter (BOD_5) at the specific sampling sites were triple more than the standard deviation for total sampling sites. The samples were collected once a month between 9:00 am and 16:00 pm. Moreover, the water samples were collected at approximately 0.3 m below the water surface in the central stream, placed into plastic bottles (2.5 L), transported to the laboratory and stored at 0~4°C for subsequent chemical analysis. The chemical measurements were performed in the laboratory within 24 h after collecting the water samples. Moreover, dissolved oxygen (DO), water temperature (T), pH and electrical conductivity (EC) were measured when the water samples were collected with a hand-held multi-parameter instrument (Multi 340i SETs, The Merck Co. Ltd., Germany). The turbidity was measured with a hand-held Turbiquant 1000IR (Multi 340i, The Merck Co. Ltd., Germany). The following chemical and biological water quality parameters were measured: chemical oxygen demand (COD_{Mn}), 5-day biochemical oxygen demand (BOD_5), total nitrogen (TN), total phosphorus (TP) and dissolved phosphorus (DP), which were measured using standard methods [27]. The ammonium nitrogen (NH_4^+-N) content was measured using an Astoria analyzer system (AAS, Brown Rupee CO. Ltd., Germany) after filtration through a 0.45-μm filter (Hailing Medicine, Zhejiang Province, China). The nitrate nitrogen (NO_3^--N) content was measured from the absorbance of the sample at 220 nm (A220) and 275 nm (A275) using a visible and ultraviolet spectrophotometer (Shanghai Spectrum Instruments, Shanghai, China) after filtration through a 0.45-μm filter; the results were determined by the following relationship: A220 - 2×A275 [27].

Land use/cover, topography and socio-economic data and GIS analysis. Latitude and longitude were measured using a hand-held GPS. The catchment boundaries were delineated on a 30-m spatial resolution digital elevation model (DEM); the stream network was represented using the Soil and Water Assessment Tool (SWAT) model. ArcGIS 9.3 Desktop GIS software was used to calculate the watershed characteristics for each sampling site. All datasets were converted to a common digital format (WGS 1984) and a common coordinate system (Albers Equal Area). Hydrologic units were not solely defined by the watershed above an individual point [28]; a hydrologic unit on the Cao-E River may cover 2/3 or more of the entire watershed above that sampling station (most hydrologic units on tributaries encompass the entire watershed above a sampling station).

The socioeconomic data, including land use change, the gross domestic product, animal husbandry output and human population in the watershed, were obtained from the Shengzhou (SZ), Xinchang (XC), Shangyu (SY) and Shaoxing (SX) County Statistical Yearbooks for the period 2003–2006 and were adjusted with some typical investigation in several towns within the watershed. The temporal variations in land use were considered to be minimal. The annual changes in forest areas and water areas were less than 0.05%, the annual decrease rate in cropland areas was less than 1% and the annual increase rate in urban areas was less than 2.5%., respectively, for the entire watershed and throughout the study period. Thus, Landsat Thematic Mapper imagery from 2006 was used to map the land cover in the study area [29]. The land cover was categorized into the following 4 classes: forest, water, cropland and urban land. The annual increases in human population density, gross domestic product per capita and animal husbandry output per capita were less than 1.5%, 10% and 2%, respectively, for the entire watershed and throughout the study period. Therefore, the averages of the socio-economic data for the entire study period were selected in this study. The slope map was derived from the DEM using a GIS spatial analyst tool (ESRI, 2006). The land use, slope, DEM and

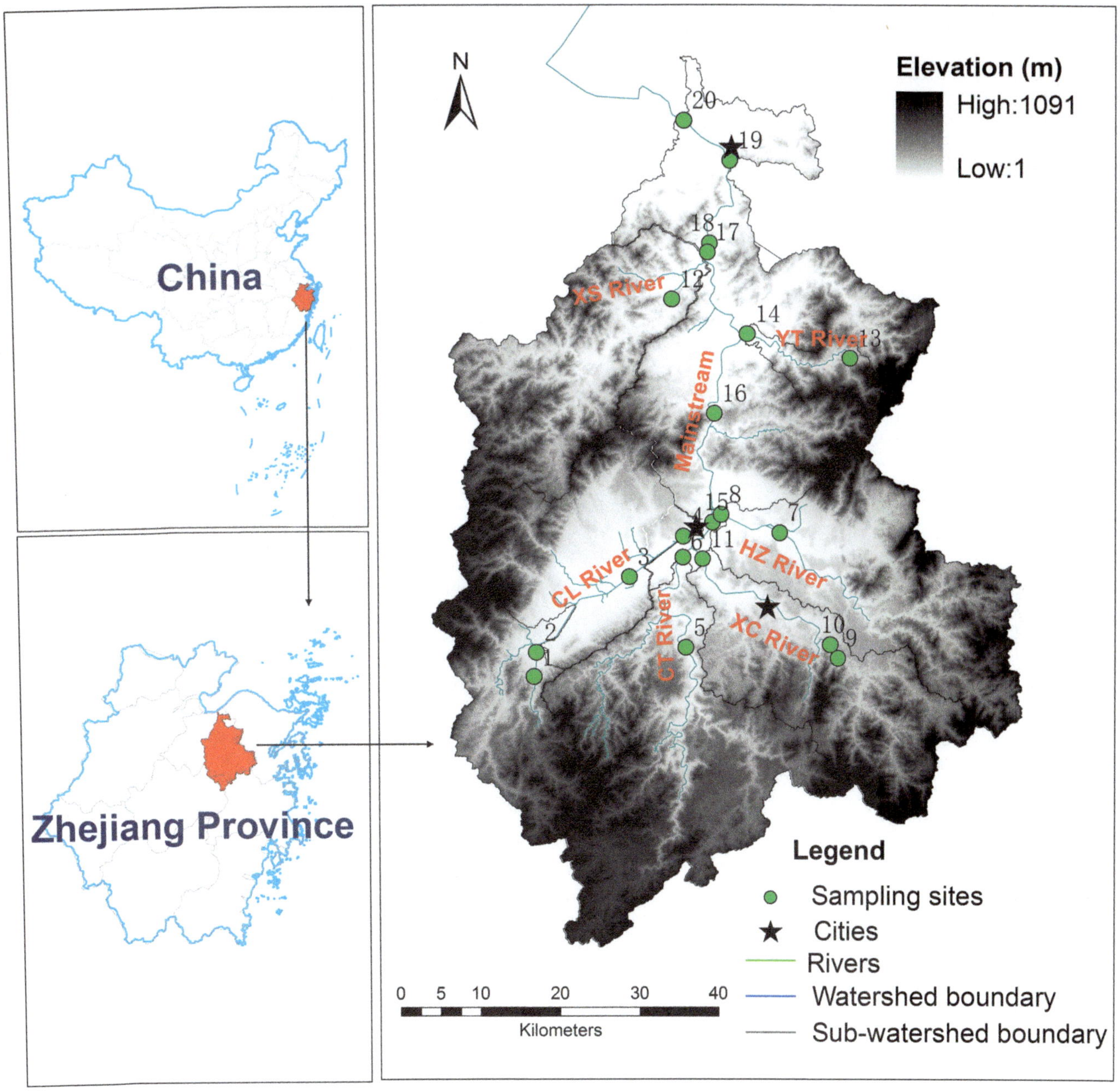

Figure 1. The Geographical Location Of The Sampling Sites In The Cao-E River System. (Sampling Stations 1-Cl1, 2- Cl2, 3-Cl3, 4-Cl4, 5- Ct 1,6-Ct2,7- Hz1,8- Hz2,9- Xc1,10- Xc2,11- Xc3,12- Xs1,13- Yt1,14- Yt2,15- M1,16- M2,17- M3,18- M4,19- M5, 20- M6).

administrative map (with attribute data, including the gross domestic product, animal husbandry output and human population for every town) were used to calculate the respective land use area (forest, water, urban land and cropland), mean elevation, mean slope, area, human population density, gross domestic product per capita and animal husbandry output per capita within each sub-watershed using a GIS spatial analyst tool. The abbreviations and statistics of the sub-watershed characteristics in the Cao-E River basin are summarized in **Table 1**.

4 Data analysis

Analysis of variance (ANOVA) was used to compare river water quality variations between different seasons (i.e., rainy and dry) via the Student-Newman-Keuls multiple-range test ($P<0.05$). Relationships among the considered variables were tested using Pearson's coefficient with statistical significance set at $P<0.05$. Major gradients and principal patterns in the water quality data between wet and dry seasons were detected using principal component analysis (PCA). PCA transforms a dataset consisting of p variables (analytical constituents) that are interrelated or correlated to various degrees into a new dataset containing p new orthogonal and uncorrelated variables; these variables are called the principal components (PCs) [30], [35]. The PCs are linear functions of the original variables in which the sum of their variances is equal to that of the original variables [30]. The PCs following a descending ordered according to their variances with PC1 corresponding to the variable with the largest variance. Algebraically, for p original variables (i.e., $x_1, x_2, \ldots, x_p$). Additional

Table 1. Abbreviations And Basic Statistical Information Of The Sub-Watershed Characteristics In The Cao-E River Basin.

	Parameters	Abbreviation	Range	Mean ± Sd
Land Use	Forest Percentage (%)	Frt	25.72–74.50	50.31±12.36
	Water Percentage (%)	Wt	0.13–3.39	1.87±0.87
	Urban Percentage (%)	Ub	0.45–14.44	4.09±3.52
	Cropland Percentage (%)	Crp	21.58–57.73	43.85±10.18
Topography	Mean Elevation (M)	Elv	125.12–556.54	306.17±116.17
	Mean Catchment Slope (°)	Slp_{mean}	10.20–17.72	13.74±2.29
	Catchment Area (Km^2)	Ar	38.39–2286.21	764.53±674.58
Socio-Economic Factors	Human Population Density (Persons/Km^2)	Hp_d	123.01–576.68	316.11±119.12
	Gross Domestic Product Per Capita (10^5 Chinese Yuan)	Gdp_{pc}	135.51–2151.16	582.38±427.88
	Animal Husbandry Output Per Capita (10^5 Chinese Yuan)	Aho_{pc}	8.07–39.40	20.56±8.61

PCs up to p can be formulated in a similar manner. The variances of the PCs are the eigenvalues; the coefficients or weights, are the eigenvectors extracted from the covariance or correlation matrix. In the example PCA, the correlation matrix was used in all cases. The goal of PCA is that the first k PCs (where $k << p$) retain most of the information in all of the p original variables, which effectively reduces the practical dimensionality of the dataset. More specifically, if the correlations are high among many of the original variables, the first few PCs contain (or explain) a large percentage of the total variance and may be used to describe multivariate patterns or water quality variations across the watershed nearly as well as using the complete set of p original variables. These patterns are often related to specific sources of contamination [30].

The stepwise regression analysis was performed to determine the environmental or socio-economic factors (TP, $P<0.10$ and the others, $P<0.05$) that best explained the spatial variability of an individual river water quality variable [20]. Variables that were strongly intercorrelated with others (variance inflation factor >10) in the initial analysis, were removed and a further analysis was carried out with the remaining environmental variables for MRA and RDA [20], [31,32]. Variance inflation factor (VIF) is a common way for detecting multicollinearity, a general rule is that the Variance Inflation Factor should not exceed 10 [49]. RDA was performed using Canoco 4.5 software to evaluate the relationships between river water quality and the environmental variables [33]. RDA was chosen because previous inspection of the data revealed a linear response rather than a unimodal response in the primary river water quality variables [34]. A Monte Carlo permutation test was used to verify the significance of the models [29]. Two series of matrices (i.e., for river water quality and environmental variables) from the RDA results were visualized as ordination diagrams using the CanoDraw software package for Windows. The river water quality parameters and the environmental/socio-economic variables were represented with arrows. Correlations between the watershed characteristics and/or the river water quality parameters were obtained by projecting the watershed characteristics onto each river water quality parameter in which higher values indicated higher correlations [34]. Application summary of 5 statistical methods in this study was listed in **Table S2**. The statistical analyses, except RDA, were completed using the SAS 9.1 software for Windows.

Results

1 Physico-chemical water quality in the Cao-E River basin

ANOVA was used to compare river water quality variations between the different seasons (i.e., rainy and dry). Temporal and spatial variations in the physico-chemical parameters are shown in **Table 2** and **Figure 2**. More specifically, BOD_5, TN, DO, pH, T and turbidity exhibited significant temporal variations (**Table 2**, $p<0.10$). Moreover, TN, pH and T were generally higher in the rainy season (April to September), whereas higher values for BOD_5, DO and turbidity occurred in the dry season (October to March). Most of the water quality parameters, except pH and T, exhibited considerable spatial variations. BOD_5, COD_{Mn} and NH_4^+-N were higher in urban areas compared with surrounding rural areas, whereas DO exhibited a reverse response to the extent of urbanization. TN, DN, NO_3^--N, TP and DP were higher in the lower part of the basin where more land has been developed and were lower in the upper-eastern part of the Cao-E River basin. Turbidity and EC were higher in the lower part of the basin (**Figure 2**) where the reach is often affected by tides and/or point source pollution [36,37].

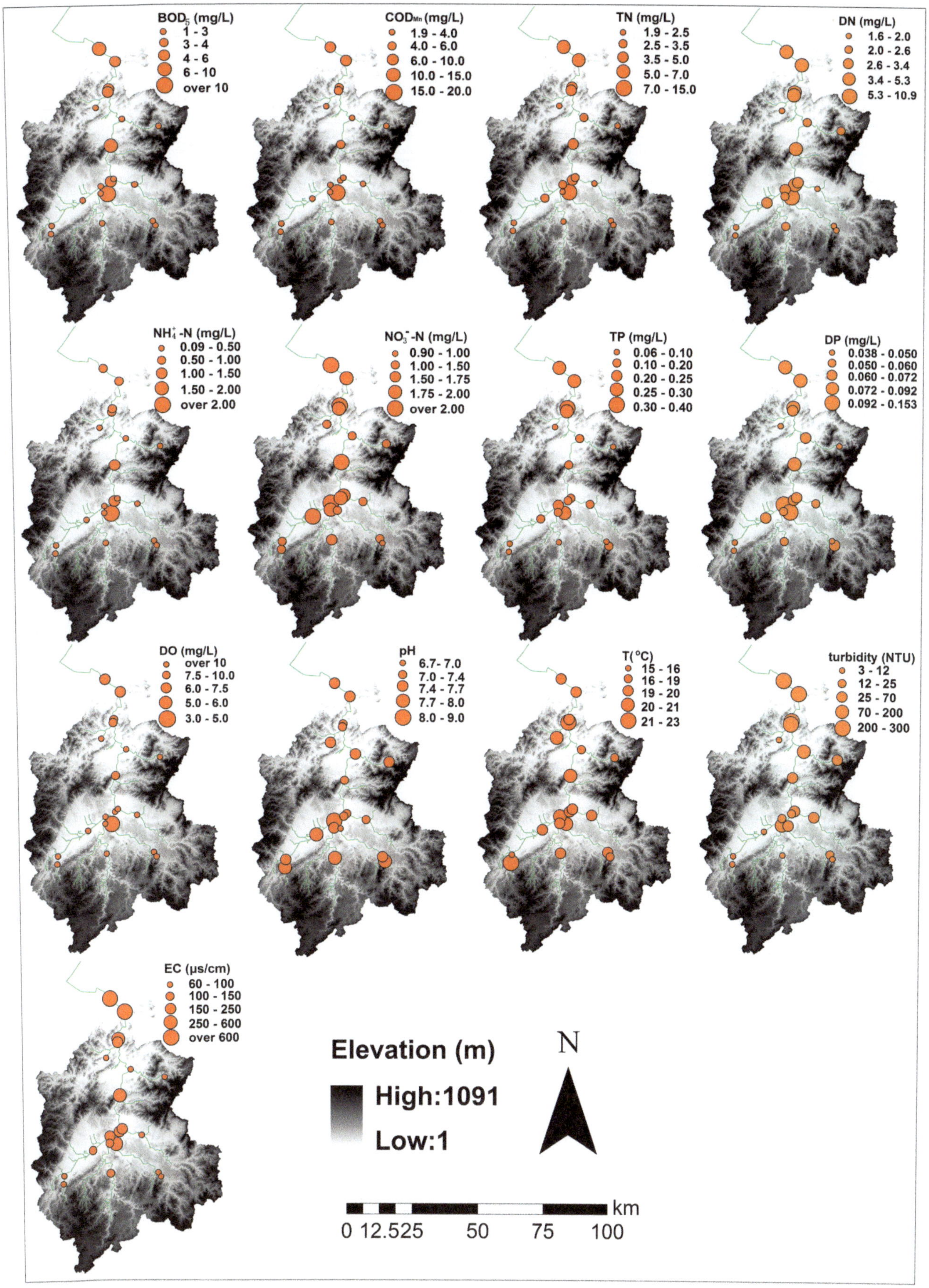

Figure 2. Spatial Variations In Bod5, Codmn, Tn, Dn, Nh_4^+-N, No_3^--N, Tp, Dp, Do, Ph, T, Turbidity And Ec In The Study Area.

Table 2. Seasonal Averages, Coefficients Of Variation [Cv (%)] And River Water Quality Parameter Ranges.

Water Quality Parameters	Rainy Season		Dry Season		
	Average [Cv (%)]	Range	Average [Cv (%)]	Range	*P-Value*
Bod_5 (Mg/L)	5.912 [183]	0.05–173.5	7.850 [181]	0–255.00	0.052
Cod_{mn} (Mg/L)	3.845 [87]	0.39–43.87	4.0361 [99]	0.49–47.75	0.557
Tn (Mg/L)	4.793 [103]	0.29–45.85	4.208 [73]	0.56–31.09	0.049
Dn (Mg/L)	3.549 [90]	0.06–22.53	3.295 [80]	0.03–30.07	0.219
Nh_4^+-N (Mg/L)	0.605 [195]	0–15.87	0.759 [2099]	0–19.58	0.156
No_3^--N (Mg/L)	1.677 [41]	0.01–4.30	1.606 [69]	0.08–15.08	0.283
Tp (Mg/L)	0.150 [108]	0–1.88	0.168 [113]	0–3.86	0.317
Dp (Mg/L)	0.061 [91]	0–0.49	0.076 [253]	0–3.49	0.163
Do (Mg/L)	7.164 [33]	0–19.55	8.045 [42]	0–17.86	<0.001
Ph	7.538 [7]	5.33–9.34	7.449 [8]	5.45–10.12	0.055
T (°C)	24.967 [24]	8.40–37.20	14.262 [64]	4.70–105.00	<0.001
Turbidity (Ntu)	70.011 [105]	0.83–586.10	92.939 [113]	0.91–1111	0.002
Ec (µs/Cm)	306.38 [347]	14–14420	348.56 [282]	14–11230	0.557

The Probabilities Associated With The Student-Newman-Keuls Multiple-Range Test Are Also Provided.

Major gradients and principal patterns in the water quality data between the wet and dry seasons were detected using PCA. The PCA based on only the dry season water quality data indicated that 4 significant factors (i.e., PCs with an eigenvalue >1) explained 87.8% of the total variance (**Table 3**). The first factor accounted for 50.2% of the total variance and had a strong positive correlation with BOD_5, COD_{Mn}, TN, DN, NH_4^+-N, and T. The PCA of the wet season water quality data showed that the eigenvalues for the 2 most significant factors were 8.442 and 1.817, accounting for 78.9% of the total variance. Factor 1 accounted for 64.9% of the total variance and was positively correlated with BOD_5, COD_{Mn}, TN, DN, TP and DP (**Table 3**).

2 Relationships between watershed characteristics and river water quality

Relationships between the river water quality variables and their corresponding sub-watershed land use, topography and socio-economic factors were explored using a Pearson correlation test, MRA and RDA. All environmental variables, including land use, topography and socio-economic factors, were utilized to

Table 3. River Water Quality Variables (13) On Rotated Pcs For Datasets During The Dry And Wet Seasons.

Water Quality Parameters	Dry Season				Wet Season	
	Pc1	Pc2	Pc3	Pc4	Pc1	Pc2
Bod_5 (Mg/L)	0.354	−0.110	0.008	0.076	0.312	−0.041
Cod_{mn} (Mg/L)	0.350	0.200	−0.187	0.128	0.310	0.263
Tn (Mg/L)	0.380	0.001	0,077	0.083	0.333	−0.054
Dn (Mg/L)	0.362	−0.044	0.147	0.127	0.311	−0.197
Nh_4^+-N (Mg/L)	0.345	−0.215	0.019	0.084	0.267	−0.346
No_3^--N (Mg/L)	0.275	−0.020	0.425	−0.079	0.283	0.174
Tp (Mg/L)	−0.221	0.409	0.077	−0.476	0.314	0.200
Dp (Mg/L)	0.097	0.329	0.609	−0.090	0.300	−0.018
Do (Mg/L)	−0.270	0.361	0.091	−0.204	0.272	0.349
Ph	−0.146	0.328	0.268	0.614	−0.184	0.546
T (°C)	0.418	−0.420	0.119	−0.212	0.010	0.246
Turbidity (Ntu)	0.222	0.340	−0.416	−0.312	0.283	0.246
Ec (µs/Cm)	0.241	0.308	−0.335	0.382	0.247	0.395
Eigenvalues	6.522	2.344	1.518	1.028	8.442	1.817
Percentage Of Variance	0.502	0.180	0.117	0.079	0.649	0.140
Cumulative Percentage Of Variance	0.502	0.682	0.799	0.878	0.649	0.789

Pcs Refer To Principal Components. The Significant Factors (I.E., Pcs With An Eigenvalue >1) Are List In The Table.

Table 4. The Pearson Correlation Coefficients For The River Water Quality Parameters Compared With The Corresponding Sub-Watershed Characteristics.

	Frt	Wt	Ub	Crp	Elv	Slp_{mean}	Ar	Hp_d	Gdp_{pc}	Aho_{pc}
Bod_5	−0.52*	0.11	0.76**	0.35	−0.31	−0.40	−0.08	0.59**	0.32	0.09
Cod_{mn}	−0.61**	0.31	0.90**	0.40	−0.58	−0.64**	0.10	0.78**	0.60**	0.42
Tn	−0.72**	0.27	0.96**	0.51*	−0.61**	−0.74**	0.37	0.87**	0.60**	0.48*
Dn	−0.70**	0.17	0.91**	0.51*	−0.53*	−0.68**	0.34	0.81**	0.48*	0.40
Nh_4^+-N	−0.60**	0.15	0.84**	0.42	−0.39	−0.51*	0.14	0.69**	0.38	0.19
No_3^--N	−0.54*	0.13	0.41	0.50*	−0.52*	−0.76**	0.70**	0.60**	0.45*	0.70**
Tp	−0.63**	0.37	0.77**	0.46*	−0.82**	−0.76**	0.46**	0.77**	0.66**	0.80**
Dp	−0.60**	0.08	0.61**	0.50*	−0.48*	−0.63**	0.30	0.64**	0.45*	0.48*
Do	0.48*	−0.43	−0.75**	−0.29	0.45*	0.59**	−0.63**	−0.71**	−0.44	−0.36
Ph	0.53*	0.01	−0.66**	−0.42	0.37	0.33	−0.44	−0.53*	−0.16	−0.15
T	−0.43	0.18	0.44	0.36	−0.44	−0.49*	0.46	0.53*	0.23	0.33
Turbidity	−0.28	0.40	0.42	0.15	−0.71**	−0.51*	0.36	0.45*	0.58**	0.81**
Ec	−0.21	0.61**	0.52*	0.02	−0.64**	−0.60**	0.20	0.57**	0.88**	0.78**

Abbreviations Are Provided In Table 1. The Figures In The Table Are Correlation Coefficients.
*Indicates Significance At The 0.05 Probability Level.
**Indicates Significance At The 0.01 Probability Level.

evaluate the response of river water quality to environmental gradients. The correlation analysis indicated that CRP was positively correlated with TN, DN, NO_3^--N, TP and DP ($r>0.45, p<0.05$; r refers to the correlation coefficient). Moreover, UB was positively correlated with BOD_5, COD_{Mn}, TN, DN, NH_4^+-N, TP, DP and EC ($r>0.50, p<0.05$) and negatively correlated with DO and pH ($r<-0.65, p<0.01$). FRT was positively correlated with DO and pH ($r>0.47, p<0.05$) and exhibited a negative correlation with BOD, COD, TN, DN, NH_4^+-N, NO_3^--N TP, and DP ($r<-0.50, p<0.05$). Furthermore, SLP_{mean} was positively correlated with DO ($r=0.59, p<0.01$) and negatively correlated with COD_{Mn}, TN, DN, NH_4^+-N, NO_3^--N, TP, DP, T, turbidity and EC ($r<-0.48, p<0.05$). The elevation was positively correlated with DO ($r=0.45, p<0.05$) and negatively correlated with COD_{Mn}, TN, DN, NO_3^--N, TP, DP, turbidity and EC ($r<-0.47, p<0.05$). Additionally, GDP_{pc} exhibited a positive correlation with COD, TN, DN, NO_3^--N, TP, DP, turbidity, and EC ($r>0.44, p<0.05$). AHO_{pc} was positively correlated with TN, NO_3^--N, TP, DP, turbidity and EC ($r>0.47, p<0.05$). Lastly, HP_d exhibited a negative correlation with DO and pH ($r<-0.52, p<0.05$) and a positive correlation to all other river water quality parameters ($r>0.44, p<0.05$) (**Table 4**).

MRA demonstrated that no individual environmental factor was able to describe the overall water quality; however, most of the physico-chemical water parameters could be sufficiently predicted using 1 to 3 environmental factors (**Table 5**). Specifically, BOD_5 could be predicted using UB, AR and GDP_{pc}; COD_{Mn} with UB and AR; TN, DN, pH and DO with UB; NH_4^+-N with UB and HP_d; DP with HP_d; turbidity with AHO_{pc}; EC with GDP_{pc}; TP with UB and AHO_{pc}; and NO_3^--N with UB and SLP_{mean}.

The RDA of the overall water quality as the dependent variable suggested that FRT, UB, CRP, ELV, SLP_{mean}, AR, HP_d, GDP_{pc} and AHO_{pc} explained 15.7–50.8% of the river water quality spatial variation, and a combination of the topography, land use and socio-economic factors explained 86.1% of the variation in overall water quality (**Table 6**). The ordination diagram of overall water quality and topography, land use and socio-economic factors indicated that the first RDA axis displayed a pollution gradient (e.g., COD_{Mn} and TN increased along the axis), which was positively correlated with UB and HP_d and negatively correlated with FRT and SLP_{mean} (**Figure 3**), accounting for 52.4% of the total variance in water quality (**Table 7**). The second axis was related to AHO_{pc} and only explained 16.5% of the total variance (**Table 7**).

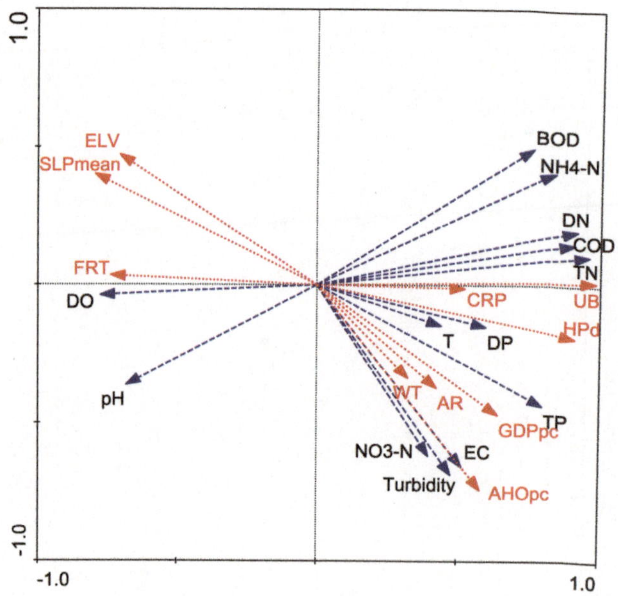

Figure 3. Biplots Of The River Water Quality Parameters And Watershed Characteristics In The Cao-E River Basin According To The Redundancy Analysis.

Table 5. Stepwise Multiple Regression Models For The River Water Quality Parameters And Watershed Characteristics In The Cao-E River Basin In Eastern China.

Parameters	Independent Variables	Regression Equation	R^2	Adjusted R^2	Significance
Bod_5	Ub, Ar, Gdp_{pc}	$1.053+566.924ub-0.012ar-0.015gdp_{pc}$	0.856	0.830	**
Cod_{mn}	Ub, Ar	1.451+88.727ub-0.002ar	0.903	0.892	**
Tn	Ub	1.668+69.168ub	0.912	0.907	**
Dn	Ub	1.231+53.629ub	0.821	0.811	**
Nh_4^+-N	Ub, Hp_d	$1.407+61.854ub-1.030hp_d$	0.805	0.782	**
No_3^--N	Slp_{mean}, Ub	$5.161012-0.233slp_{mean}-7.675ub$	0.720	0.670	**
Tp	Ub, Aho_{pc}	$0.012+1.093ub+0.050aho_{pc}$	0.794	0.770	*
Dp	Hp_d	$0.022+0.015hp_d$	0.406	0.373	**
Do	Ub	9.802-50.534ub	0.562	0.538	**
Ph	Ub	7.724012-5.814ub	0.439	0.408	**
Turbidity	Aho_{pc}	$-115.264+96.209aho_{pc}$	0.662	0.643	**
Ec	Gdp_{pc}	$-261.761+1.009gdp_{pc}$	0.778	0.766	**

Abbreviations Are Provided In Table 1. The Parameters Without Regression Models Are Not Listed.
*Indicates Significance At The 0.05 Probability Level.
**Indicates Significance At The 0.01 Probability Level.

Table 6. The Rda Results For The Percentage Of The Overall River Water Quality Variance Explained By An Individual Variable And All The Explanatory Variables.

Explanatory Variables	Variance Explained%	*P* Value
Land Use		
Frt	29.6	0.002
Wt	8.8	0.16
Ub	50.8	0.002
Crp	16.2	0.016
Topography		
Elv	29.8	0.004
Slp_{mean}	36.3	0.002
Ar	15.7	0.032
Socio-Economic Factors		
Hp_d	44.6	0.002
Gdp_{pc}	26.2	0.032
Aho_{pc}	26.9	0.002
Land Use + Topography + Social-Economic Factors	86.1	0.002

Abbreviations Are Provided In Table 1.

Discussion

1 Seasonal effects

Seasonal variations in the flow caused by the subtropical monsoon climate can partially explain the river water quality temporal variations [7], [38]. The relatively high TN concentrations in the rainy season were attributed to a flushing effect [36]. BOD_5, which corresponds to point source pollution, were lower in the rainy season due to the dilution by precipitation [2]. Relatively high T in the rainy season resulted from the specific climatic conditions during this period (T is higher in the rainy season than in the dry season; see **Table 2**). The lower values of DO and higher values of pH in the rainy season may related to some environmental factors variations, such as flow rate, water temperature, aquatic plant growth and so on [50], [39–41]. The reasons are complex and the further research is needed in the future. The turbidity was higher in the dry season, which may be primarily because the reduced flow during this period caused the river water to be easily influenced by tides. Moreover, Odokuma and Okpokwasili [40] reported that the other parameters such as phosphate showed significantly higher values in the rainy season than in the dry season in the New Calabar River, Nigeria. These different results compared with our study were mainly due to the regional climate difference. However, the absence of a significant difference in the other physico-chemical parameters between the dry and wet seasons indicated mixed and irregular influences, e.g., point sources and diffusion may have played an important role [1], [42].

2 Land cover effects

In the study area, land-use types could play important roles in affecting major river water quality parameter distributions, as reflected by their considerable variability (**Figure 2**). This result is corroborated by the strong positive correlations between urban land and BOD_5, COD_{Mn}, TN, DN, NH_4^+-N and TP (**Table 5**) and the strong negative correlations with DO and pH. Moreover, most of the nutrient variables exhibited lower concentrations in the forest-dominated region, whereas the cropland-dominated region had high nutrient concentrations. The region downstream of the urban area exhibited higher concentrations of BOD_5, COD, NH_4^+-N, TP and DP and lower pH and DO (**Figure 2**); this finding agrees with many related studies [21], [43,44]. Studies have also shown that certain river water quality parameters (e.g., TN and TP) are primarily determined by agricultural land use at

Table 7. The Rda Results Showing The Eigenvalues And Percentage Of The Overall River Water Quality Parameter Variance Explained By The First Two Rda Axes And All The Rda Axes.

	Axis 1	Axis 2	All Axes
Eigenvalues	0.524	0.165	0.861
Percentage Of Water Quality Variance Explained	52.4	16.5	86.1
Dominant Environmental Factors	Ub(0.967)	Aho_{pc}(−0.699)	
	Hp_d (0.897)		
	Slp_{mean}(−0.780)		
	Frt (−0.725)		

Abbreviations Are Provided In Table 1. The Number In Parentheses Indicates The Canonical Coefficients Of The Environmental Variables With The First Two Rda Axes.

the watershed scale in many parts of the world [1], [20]. However, this study found that agricultural land use in watersheds plays a minor role in explaining the spatial variations in river water quality (explaining 16.2% of the total overall water quality variance) in the Cao-E River basin. The different result is primarily due to the enhanced effects of urban runoff and incompletely treated industrial waste and domestic pollution on river water quality compared to agricultural runoff. In addition, vegetated buffers adjacent to the croplands can substantially mitigate agricultural runoff of nutrients and other contaminants via deposition, absorption, and denitrification [20].

Urban areas are primarily located along river networks in the Cao-E River basin. Therefore, urban effects on the river water quality were expected. Urban land use comprised a much smaller percentage of the Cao-E River basin than the cropland in this study. However, as observed from the regression model (BOD_5, COD_{Mn}, TN, DN, NH_4^+-N, NO_3^--N, TP, DO and pH) and the RDA analysis (explaining 50.8% of the total overall water quality variance) results, urban areas play a pivotal role in influencing water quality (**Tables 5 and 6**). This observation is most likely due to two factors: (1) the deficient capacity of urban sewage treatment plants leads to domestic pollution and industrial wastewater inputs into the Cao-E River system, and (2) the high percentage of impervious surfaces and over fertilization of grassy areas can increase discharge rates, sedimentation and pollutant runoff to streams [44].

3 Topography effects

Topographical factors played important roles in explaining spatial variations in river water quality within the Cao-E River basin (SLP_{mean} was 36.3% and ELV was 29.8%). Topography largely regulated the river water quality parameters. Most related studies have shown that higher SLP_{mean} and/or ELV lead to higher erosion rates, which subsequently increase the rate at which particulate matter enters a water body [13], [21], [45]. However, our results indicated an inverse relationship. Here, SLP_{mean} and ELV were both negatively correlated with TN, DN, NO_3^--N, TP and DP; SLP_{mean} exhibited a highly significant negative correlation with NO_3^--N values (**Figure 3 and Table 5**). The slope effects on water chemistry varied. Watershed physical characteristics, such as soil properties (soil texture and soil drainage), morphological variables (drainage density and elongation) and particularly surficial debris, largely affected water chemistry in river waters [37], [46]. In addition, these negative correlations existed primarily because ELV and SLP_{mean} were strongly correlated with land cover (**Table 8**). High elevations and/or steep catchments are primarily occupied by forest land cover; however, cropland is primarily located in regions in which the topography is relatively flat and has a low ELV. Therefore, the forested mountain area, which has higher SLP_{mean} and/or ELV, may export fewer nutrients than flat land (e.g., cropland). Similar results were published regarding water quality spatial variations in the Han River basin in South Korea [7] and non-point source effects on stream nutrient concentrations in the Seattle region of the USA [47].

4 The effects of socio-economic factors

The socio-economic factors commonly have different extent effects on river water quality, especially in developing countries such as China. Our study shows that socioeconomic factors have significant effects on river water quality during the study period (**Table 5**) and that HP_d, AHO_{pc}, and GDP_{pc} explain 44.6%, 26.9% and 26.2% of the total overall river water quality variance, respectively (**Table 6**). Furthermore, HP_d is a fundamental

Table 8. The Pearson Correlation Matrix For The Watershed Characteristics Of The Cao-E River Basin In Eastern China.

	Frt	Wt	Ub	Crp	Elv	Slp_{mean}	Ar	Hp_d	Gdp_{pc}	Aho_{pc}
Frt	1.00									
Wt	0.09	1.00								
Ub	−0.79**	0.31	1.00							
Crp	−0.96**	−0.31	0.59**	1.00						
Elv	0.57**	−0.48*	−0.71**	−0.41	1.00					
Slp_{mean}	0.83**	−0.33	−0.82**	−0.70**	0.74**	1.00				
Ar	−0.44	0.17	0.42	0.37	−0.44	−0.56*	1.00			
Hpd	−0.79**	0.37	0.95**	0.60**	−0.79**	−0.91**	0.55*	1.00		
Gdp_{pc}	−0.38	0.49*	0.69**	0.18	−0.65**	−0.67**	0.30	0.76**	1.00	
Aho_{pc}	−0.47*	0.20	0.55*	0.36	−0.69**	−0.72**	0.47*	0.64**	0.72**	1.00

*Indicates That The Correlation Was Significant At The 0.05 Probability Level.
**Indicates That The Correlation Was Significant At The 0.01 Probability Level.

parameter for predicting DP and AHO_{pc} for TP and turbidity (**Table 5**). This result can be attributed to high population density, the massive animal production and the lack of strict management regulations for animal waste in the study area. The population density exceeds 380 people km^{-2}, the annual animal output value exceeds 2.0 billion Yuan and the annual hog production exceeds 1.0 million heads for the entire basin [48]. Most human and animal waste with incomplete treatment is discharged into the surrounding water bodies. Moreover, EC was positively correlated with GDPpc, which was most likely related to point source pollution [37].

In addition, some researches [4–6] point out that climate condition is also an important factor of influencing river water quality. Thus, we will further study it in our future work.

Conclusions

The primary results of this study can be summarized as follows:

1) The in-stream water quality in the Cao-E River basin streams suggested that TN, pH and T were generally higher in the rainy season, whereas BOD_5, DO and turbidity were higher in the dry season.
2) Spatial variations in river water quality are typically associated with several anthropogenic and natural factors. Urban land cover was determined to be the most important explanatory variable for BOD_5, COD_{Mn}, TN, DN, NH_4^+-N, NO_3^--N, TP, DO and pH. Moreover, Animal husbandry output per capita was an important predictor for TP and turbidity, and Gross domestic product per capita largely determined the spatial variations in EC. The remaining unexplained variance resulted from other factors, such as topography.
3) 3) Pollution control for animal waste discharge in rural settlements was important in the study area. Moreover, agricultural runoff, industrial pollution and domestic pollution in urban and industrial areas were also important factors within the Cao-E River basin.
4) The percentage of total overall river water quality variance explained by individual variables and/or all environmental variables in the study area (as determined using RDA) can assist in quantitatively identifying primary pollution control factors at the watershed scale.

This study improved our understanding of the anthropogenic activities and natural factors that affect river water quality and can assist in the design of efficient strategies for controlling river water pollution at the watershed scale. Moderate-resolution remote sensing data were adopted in this study; future investigations will require high-resolution DEM maps and additional land use classes to better evaluate the effects of specific environmental variables on overall river water quality.

Supporting Information

Table S1 Summary of commonly used statistical methods on pollution source identification in recent years.

Table S2 Application summary of 5 statistical methods in this study.

Acknowledgments

We would like to acknowledge the members of the Zhejiang University soil physical laboratory for assistance in the field and laboratory.

Author Contributions

Conceived and designed the experiments: JL JC. Performed the experiments: JC. Analyzed the data: JC JL. Wrote the paper: JL JC.

References

1. Li S, Gu S, Liu W, Han H, Zhang Q (2008) Water Quality In Relation To Land Use And Land Cover In The Upper Han River Basin, China. Catena 75: 216–222.
2. Zhang Y, Guo F, Meng W, Wang X (2009) Water Quality Assessment And Source Identification Of Daliao River Basin Using Multivariate Statistical Methods. Environ Monit Assess 152: 105–121.
3. Zhou N, Westrich B, Jiang S, Wang Y (2011) A Coupling Simulation Based On A Hydrodynamics And Water Quality Model Of The Pearl River Delta, China. J Hydrol 396: 267–276.
4. Astaraie-Imani M, Kapelan Z, Fu G, Butler D (2012) Assessing The Combined Effects Of Urbanisation And Climate Change On The River Water Quality In An Integrated Urban Wastewater System In The Uk. J Environ Manage 112: 1–9.
5. Costa-Cabral M, Coats R, Reuter J, Riverson J, Sahoo G, et al. (2013) Climate Variability And Change In Mountain Environments: Some Implications For Water Resources And Water Quality In The Sierra Nevada (Usa). Clim Change 116: 1–14.
6. Aubert Ah, Gascuel-Odoux C, Merot P (2013) Annual Hysteresis Of Water Quality: A Method To Analyse The Effect Of Intra-And Inter-Annual Climatic Conditions. J Hydrol 478: 29–39.
7. Chang H (2008) Spatial Analysis Of Water Quality Trends In The Han River Basin, South Korea. Water Res 42: 3285–3304.
8. Hoorman J, Hone T, Sudman Jr T, Dirksen T, Iles J, et al. (2008) Agricultural Impacts On Lake And Stream Water Quality In Grand Lake St. Marys, Western Ohio. Water Air Soil Pollut Int 193: 309–322.
9. Ahearn Ds, Sheibley Rw, Dahlgren Ra, Anderson M, Johnson J, et al. (2005) Land Use And Land Cover Influence On Water Quality In The Last Free-Flowing River Draining The Western Sierra Nevada, California. J Hydrol 313: 234–247.
10. Huang F, Wang X, Lou L, Zhou Z, Wu J (2010) Spatial Variation And Source Apportionment Of Water Pollution In Qiantang River (China) Using Statistical Techniques. Water Res 44: 1562–1572.
11. Chang N (2005) Sustainable Water Resources Management Under Uncertainty. Stoch Environ Res Risk Assess 19: 97–98.
12. Kolovos A, Christakos G, Serre Ml, Miller Ct (2002) Computational Bayesian Maximum Entropy Solution Of A Stochastic Advection - Reaction Equation In The Light Of Site - Specific Information. Water Resour Res 38: 51–54.
13. Sliva L, Dudley Williams D (2001) Buffer Zone Versus Whole Catchment Approaches To Studying Land Use Impact On River Water Quality. Water Res 35: 3462–3472.
14. Heathcote Aj, Filstrup Ct, Downing Ja (2013) Watershed Sediment Losses To Lakes Accelerating Despite Agricultural Soil Conservation Efforts. Plos One 8: E53554.
15. Keatley Be, Bennett Em, Macdonald Gk, Taranu Ze, Gregory-Eaves I (2011) Land-Use Legacies Are Important Determinants Of Lake Eutrophication In The Anthropocene. Plos One 6: E15913.
16. Wei R, Ge F, Huang S, Chen M, Wang R (2011) Occurrence Of Veterinary Antibiotics In Animal Wastewater And Surface Water Around Farms In Jiangsu Province, China. Chemosphere 82: 1408–1414.
17. Xianghao Z (2008) Study Of Protection And Construction Of Mountain Ecological Security Barrier In China. J Mount Sci (In Chinese) 1: 4–5.
18. Jiang X, Sommer Sg, Christensen Kv (2011) A Review Of The Biogas Industry In China. Energy Policy 39: 6073–6081.
19. Tong L, Li P, Wang Y, Zhu K (2009) Analysis Of Veterinary Antibiotic Residues In Swine Wastewater And Environmental Water Samples Using Optimized Spe-Lc/Ms/Ms. Chemosphere 74: 1090–1097.
20. Liu W, Li S, Bu H, Zhang Q, Liu G (2012) Eutrophication In The Yunnan Plateau Lakes: The Influence Of Lake Morphology, Watershed Land Use, And Socioeconomic Factors. Environ Sci Pollut Res Int 19: 858–870.
21. Ye L, Cai Q, Liu R, Cao M (2009) The Influence Of Topography And Land Use On Water Quality Of Xiangxi River In Three Gorges Reservoir Region. Environ Geol 58: 937–942.
22. Tian C, Lu X, Pei H, Hu W, Xie J (2013) Seasonal Dynamics Of Phytoplankton And Its Relationship With The Environmental Factors In Dongping Lake, China. Environ Sci Pollut Res Int 185: 2627–2645.
23. Yang X, Wu X, Hao H, He Z (2008) Mechanisms And Assessment Of Water Eutrophication. J Zhejiang Univ Sci B 9: 197–209.

24. Chen D, Lu J, Shen Y, Gong D, Deng O (2011) Spatio-Temporal Variations Of Nitrogen In An Agricultural Watershed In Eastern China: Catchment Export, Stream Attenuation And Discharge. Environ Pollut 159: 2989–2995.
25. Jin S, Lu J, Chen D, Shen Y, Shi Y (2009) Relationship Between Catchment Characteristics And Nitrogen Forms In Cao-E River Basin, Eastern China. J Environ Sci (China) 21: 429–433.
26. Shen Y, L U J, Chen D, Shi Y (2011) Response Of Stream Pollution Characteristics To Catchment Land Cover In Cao-E River Basin, China. Pedosphere 21: 115–123.
27. Sepbc (2002) Water And Wastewater Analysis Method. Prepared And Published By: State Environment Protection Bureau Of China (Sepbc). Beijing: China Environmental Science Press. 784 P.
28. Dodds Wk, Oakes Rm (2004) A Technique For Establishing Reference Nutrient Concentrations Across Watersheds Affected By Humans. Limnol Oceanogr Methods 2: 333–341.
29. Chen J, Lu J (2013) Establishment Of Reference Conditions For Nutrients In An Intensive Agricultural Watershed, Eastern China. Environ Sci Pollut Res Int 4: 2496–2505.
30. Olsen Rl, Chappell Rw, Loftis Jc (2012) Water Quality Sample Collection, Data Treatment And Results Presentation For Principal Components Analysis–Literature Review And Illinois River Watershed Case Study. Water Res 46: 3110–3122.
31. Myers Rh (1990) Classical And Modern Regression With Applications: Duxbury Press Belmont, Ca. 350p.
32. Miserendino Mial, Casaux R, Archangelsky M, Di Prinzio Cy, Brand C, et al. (2011) Assessing Land-Use Effects On Water Quality, In-Stream Habitat, Riparian Ecosystems And Biodiversity In Patagonian Northwest Streams. Sci Total Environ 409: 612–624.
33. Braak Ct, Šmilauer P (2002) Canoco Reference Manual And Canodraw For Windows User's Guide: Software For Canonical Community Ordination (Version 4.5). Section On Permutation Methods. Microcomputer Power, Ithaca, New York. 300p.
34. Lepš J, Šmilauer P (2003) Multivariate Analysis Of Ecological Data Using Canoco. New York: Cambridge University Press. 283 P.
35. Johnson Ra, Wichern Dw (2002) Applied Multivariate Statistical Analysis: Prentice Hall Upper Saddle River, Nj. 755p.
36. Chen D, Lu J, Yuan S, Jin S, Shen Y (2006) Spatial And Temporal Variations Of Water Quality In Cao-E River Of Eastern China. J Environ Sci (China) 18: 680–688.
37. Li S, Xia X, Tan X, Zhang Q (2013) Effects Of Catchment And Riparian Landscape Setting On Water Chemistry And Seasonal Evolution Of Water Quality In The Upper Han River Basin, China. Plos One 8: E53163.
38. Shah Vg, Dunstan Rh, Geary Pm, Coombes P, Roberts Tk, et al. (2007) Comparisons Of Water Quality Parameters From Diverse Catchments During Dry Periods And Following Rain Events. Water Res 41: 3655–3666.
39. Karaer F, Küçükballi A (2006) Monitoring Of Water Quality And Assessment Of Organic Pollution Load In The Nilüfer Stream, Turkey. Environ Monit Assess 114: 391–417.
40. Odokuma Lo, Okpokwasili Gc (1997) Seasonal Influences Of The Organic Pollution Monitoring Of The New Calabar River, Nigeria. Environ Monit Assess 45: 43–56.
41. Kannel Pr, Lee S, Lee Y (2008) Assessment Of Spatial–Temporal Patterns Of Surface And Ground Water Qualities And Factors Influencing Management Strategy Of Groundwater System In An Urban River Corridor Of Nepal. J Environ Manage 86: 595–604.
42. Chen Js, Gao Xm, He Dw, Xia Xh (2000) Nitrogen Contamination In The Yangtze River System, China. J Hazard Mater 73: 107–113.
43. Pratt B, Chang H (2012) Effects Of Land Cover, Topography, And Built Structure On Seasonal Water Quality At Multiple Spatial Scales. J Hazard Mater 209: 48–58.
44. Mehaffey Mh, Nash Ms, Wade Tg, Ebert Dw, Jones Kb, et al. (2005) Linking Land Cover And Water Quality In New York City's Water Supply Watersheds. Environ Monit Assess 107: 29–44.
45. Xiao B, Wang Q, Wu J, Huang C, Yu D (2010) Protective Function Of Narrow Grass Hedges On Soil And Water Loss On Sloping Croplands In Northern China. Agric Ecosyst Environ 139: 653–664.
46. Lindell L, Aa Str O M M, O Berg T (2010) Land-Use Change Versus Natural Controls On Stream Water Chemistry In The Subandean Amazon, Peru. Appl Geochem 25: 485–495.
47. Brett Mt, Arhonditsis Gb, Mueller Se, Hartley Dm, Frodge Jd, et al. (2005) Non-Point-Source Impacts On Stream Nutrient Concentrations Along A Forest To Urban Gradient. Environ Manage 35: 330–342.
48. Shaoxing City Statistical Bureau (2005) Shaoxing Statistical Yearbook. Shaoxing: Shaoxing City Press. 300p.
49. Belsey Da, Kuh E, Welsch Re (1980) Regression Diagnostics: Identifying Influential Data And Sources Of Collinearity: John Wiley. 320p.
50. Zhao J, Fu G, Lei K, Li Y (2011) Multivariate Analysis Of Surface Water Quality In The Three Gorges Area Of China And Implications For Water Management. J Environ Sci (China) 23: 1460–1471.

3

Stream Macroinvertebrate Response Models for Bioassessment Metrics: Addressing the Issue of Spatial Scale

Ian R. Waite[1]*, Jonathan G. Kennen[2], Jason T. May[3], Larry R. Brown[3], Thomas F. Cuffney[4], Kimberly A. Jones[5], James L. Orlando[3]

1 U.S. Geological Survey, Portland, Oregon, United States of America, **2** U.S. Geological Survey, West Trenton, New Jersey, United States of America, **3** U.S. Geological Survey, Sacramento, California, United States of America, **4** U.S. Geological Survey, Raleigh, North Carolina, United States of America, **5** U.S. Geological Survey, Salt Lake City, Utah, United States of America

Abstract

We developed independent predictive disturbance models for a full regional data set and four individual ecoregions (Full Region vs. Individual Ecoregion models) to evaluate effects of spatial scale on the assessment of human landscape modification, on predicted response of stream biota, and the effect of other possible confounding factors, such as watershed size and elevation, on model performance. We selected macroinvertebrate sampling sites for model development (n = 591) and validation (n = 467) that met strict screening criteria from four proximal ecoregions in the northeastern U.S.: North Central Appalachians, Ridge and Valley, Northeastern Highlands, and Northern Piedmont. Models were developed using boosted regression tree (BRT) techniques for four macroinvertebrate metrics; results were compared among ecoregions and metrics. Comparing within a region but across the four macroinvertebrate metrics, the average richness of tolerant taxa (RichTOL) had the highest R^2 for BRT models. Across the four metrics, final BRT models had between four and seven explanatory variables and always included a variable related to urbanization (e.g., population density, percent urban, or percent manmade channels), and either a measure of hydrologic runoff (e.g., minimum April, average December, or maximum monthly runoff) and(or) a natural landscape factor (e.g., riparian slope, precipitation, and elevation), or a measure of riparian disturbance. Contrary to our expectations, Full Region models explained nearly as much variance in the macroinvertebrate data as Individual Ecoregion models, and taking into account watershed size or elevation did not appear to improve model performance. As a result, it may be advantageous for bioassessment programs to develop large regional models as a preliminary assessment of overall disturbance conditions as long as the range in natural landscape variability is not excessive.

Editor: James P. Meador, Northwest Fisheries Science Center, NOAA Fisheries, United States of America

Funding: This work received US Federal funding - part of the National Water Quality Assessment Program (NAWQA) of the US Geological Survey. NAWQA funded and supported all labor involved, but the authors are responsible for all design, data collection and analysis and all aspects of publication. The funders had no role in study design, data collection and analysis, decision to publish, or preparation of the manuscript.

Competing Interests: The authors have declared that no competing interests exist.

* E-mail: iwaite@usgs.gov

Introduction

Understanding the effects of human land use modification on stream biota, the processes that cause these effects, and the various spatial and temporal scales at which these effects and processes operate are fundamental goals of bioassessment in stream ecology. The use of models has increased markedly in the past decade in all areas of ecology partly to address the issues stated above. In addition, major advances have been made in conceptual models and statistical techniques [1–7], which, in turn, help practitioners derive response models that better support the needs of bioassessment programs. Models provide a useful framework for testing hypotheses, determining potential direct and indirect linkages, and directing where further research is needed. The expansion and application of multivariate models in stream ecology are helping to address these issues and hopefully will lead to a broader understanding of ecological and anthropogenic pathways and responses [3],[4],[8].

Spatial scale is important in understanding ecosystem processes and species distributions; as such, scale is an important consideration in ecological research [9–11]. At the same time, managers often need decision-making tools that can be applied to as large a region as possible. Cuffney et al. [6] showed that responses of algal and macroinvertebrate metrics to urbanization varied by taxon, geographic region, and antecedent land use. Stevenson et al. [12] modeled algal biomass and found a positive correlation with nutrients in two distinct regions, yet the amount of variation explained differed by nutrient concentration and region. Potapova and Charles [13] suggested that algal indicators and optima were improved if developed regionally rather than nationally. Ode et al. [14] found that macroinvertebrate indices developed at large regional scales, such as the western U.S., had lower precision in California than California-based indices. They found that the larger scale indices were influenced by two natural gradients that did not affect the statewide indices. Seelbach et al. [15], on the other hand, found that regions too large can sometimes give misleading results when strong natural gradients at larger scales are mismatched with ecological-scale responses. For example, their streamflow models across three states in the midwestern U.S. showed that more rainfall in the southern portion of their region created lower stream baseflows (a nonsensical relationship) and that higher northern baseflows were the result of very permeable glacial deposits that are variable across smaller scales that were not accounted for by the large scale models [15]. Overall, national or regional scale models are likely to focus on large scale natural landscape effects such as climate, typography, elevation, and geology as the primary discriminating variables, with disturbance variables such as land use, nutrients, sediments, and contaminants as secondary variables. However, current ecosystem theory indicates that models at smaller scales should allow for more insight and interpretation of disturbance related processes or mechanisms that are likely to operate at smaller watershed and site specific scales [5],[6],[16],[17].

Research documenting the effects of land-use change on stream biota indicates that as the amount of urban and(or) agricultural land use in the watershed increases, individual biological metrics and multimetric indices (e.g., Index of Biotic Integrity – IBI) that reflect compositional changes in sensitive species generally decrease [10],[11],[18],[19],[20]. However, a better understanding of the effect of spatial scale on disturbance processes and the subsequent effect on modeling performance is paramount for developing models that better support bioassessment efforts and regulatory application [21]. Waite et al. [11] developed macroinvertebrate response models for three regions in the western U.S. The best multiple linear regression (MLR) models from each individual region required only two or three explanatory variables to model macroinvertebrate metrics. In each region, their best model contained some measure of urban and(or) agricultural land use, yet often the model was improved by including a natural landscape factor such as mean annual precipitation or mean watershed slope. Brown et al. [22] were also able to develop strong models using modeling techniques such as MLR and boosted regression trees (BRT). They modeled a macroinvertebrate index of biotic integrity (BIBI) across a gradient of urbanized streams in southern California using MLR models; however, overall model prediction and performance was generally improved by using BRT.

Regression trees are one type of modeling technique within the commonly used classification and regression tree (CART) or decision tree family (e.g., [23–25]). These modeling techniques have some highly desirable properties, including the ability to handle categorical and censored data and non-normal distributions, and they can model complex interactions simply [26]. De'ath [26] and Elith et al. [27] showed that the more flexible BRT models outperform general linear models (GLM) and general additive models (GAM) in variable selection and predictive ability (higher R^2 and lower error) and can handle sharp discontinuities in data that are difficult for the other methods. Aertsen et al. [28] also showed that BRT outperformed most modeling techniques (MLR, GLM, GAM, and CART), with the exception of artificial neural networks, which the authors penalized for being complex and non-transparent.

The primary goal of this paper is to evaluate the influence of scale on predictive macroinvertebrate response models developed across four proximal ecoregions. We compare the overall performance (i.e., R^2 and cross-validation R^2) of models for predicting the responses of common macroinvertebrate metrics (e.g., EPT Richness, Tolerance, and Non-Insects) developed using BRT techniques. Potential explanatory variables in the models include common landscape-based disturbance variables such as land use, road and population density, riparian canopy, infrastructure, and hydrologic runoff as well as natural landscape factors such as climate, slope, elevation, and soil infiltration. Our objectives are to test whether (1) the small more region specific scales (individual ecoregions) will allow for better performing models compared to the larger full regional scale models and (2) the ecoregion scaled models will highlight more region-specific disturbance-related explanatory variables (e.g., hydrologic runoff and region specific landscape disturbance variables) compared to higher inclusion of more large scale natural setting variables in the full region models.

In addition, we evaluate whether elevation and watershed size classes are better predictors of common macroinvertebrate metrics than the Full Region model and hypothesize that BRT models developed for each of three a-priori watershed size classes will produce stronger predictive models than that for the full region because stream size and location along the gradient from headwaters to the mouth influences the type and distribution of aquatic fauna [29], a similar logic for the elevation classes. In this context, hydrologic variability and, subsequently, stream biota are highly dependent on their longitudinal location along the continuum; therefore, we would expect that models with greater predictive power would be developed, for example, in small low order streams (streams <27 km^2) compared to models of large watershed streams or models developed for the Full Region that blends all stream sizes.

Methods

Description of modeling regions

Macroinvertebrate data aggregated across four proximal level III ecoregions—the North Central Appalachians, Ridge and Valley, Northeastern Highlands, and Northern Piedmont, located in the northeastern U.S.—were used in our analysis. These four ecoregions have broad similarities in some features such as climate and forest land cover, yet there are stark differences among some of these regions in basic physiography, climate, and agriculture and urban land use. Mean and range of select environmental variables for each region are provided (variable list and definitions: Table 1, mean and ranges: Table 2).

The North Central Appalachians ecoregion is a partially glaciated and forested plateau of horizontally bedded shale, siltstone, sandstone, coal, and conglomerate that is punctuated by high hills and low mountains [30]. The climate can be characterized as continental, with cool summers and cold winters. The average annual precipitation ranges from 900 to 1,450 mm per year. The Ridge and Valley ecoregion is a series of alternating elongated forested ridges and narrow agricultural valleys created by tightly folded and intensely faulted bedrock composed chiefly of shale, limestone, sandstone, and dolomite under a veneer of unconsolidated materials [31], [32]. Elevations typically range from 91 to 1,220 m and precipitation ranges from 1,010 to 1,330 mm, with 20 to 30 percent of the annual precipitation falling as snow. Stream patterns are generally trellis-shaped and reflect the regular folding of the geomorphology with high-gradient streams along ridge slopes, and gentle gradient, meandering streams in the valleys. Land cover is predominantly forests (about 56 percent) with a mixture of agricultural, developed, and managed lands (e.g., state parks and wild and scenic rivers).

The Northeastern Highlands is a rugged, high elevation forested region characterized by a series of northeast-trending ridges and valleys with elevations ranging from about 43 m to over 425 m on

Table 1. Description, variable code and definition of explanatory environmental (landscape, riparian and habitat) and response (invertebrate metrics) variables used for model development.

Description	Variable Code	Definition
Explanatory Variables		
Watershed Variables		
Percent Agricultural Landuse	Ag	Percent watershed area in agricultural landuse (NLCD 2001 category 81, 82)
Percent Urban Landuse	Urban	Percent watershed area in urban landuse (NLCD 2001 categories 21, 22, 23, and 24)
Sum of Percent Ag + Urban	Ag+Urb	Sum of percent watershed area in urban (NLCD 2001 categories 21, 22, 23, and 24) and agricultural (NLCD 81, 82) landuse
Percent Forest	Forest	Percent watershed area in forest landuse (NLCD 2001 categories 41, 42, 43)
Percent Wetland	Wetland	Percent watershed area in wetland land cover (NLCD 2001 category 90 and 95)
Road Density	Rd.Density	Road density in watershed = Road length (km)/watershed area (km^2)
Mean Population Density	Pop.Density	Watershed mean population density based on 2000 census (#/km^2)
Dam Density	Dam Density	Density of dams in watershed = Number of dams/watershed area (km^2)
Percent Manmade Channel	ManMadeChan	Percentage of linear water features in stream buffer which are manmade
Riparian Variables		
Percent Agricultural Landuse	Rip.AG	Percent buffer area in agricultural landuse (NLCD 2001 category 81, 82) in riparian buffer
Percent Urban Landuse	Rip.Urban	Percent buffer area in urban landuse (NLCD 2001 categories 21, 22, 23, and 24) in riparian buffer
Sum of Percent Ag + Urban	Rip.Ag+Urb	Sum of percent buffer area in urban (NLCD 2001 categories 21, 22, 23, and 24) and agricultural (NLCD 81, 82) landuse in riparian buffer
Percent Forest	Rip.Forest	Percent buffer area in forest landuse (NLCD 2001 categories 41, 42, 43) in riparian buffer
Mean Tree Canopy Cover	Rip.Canopy	Percent canopy cover (NLCD 2001 Percent Tree Canopy dataset; 30 m pixel) in riparian buffer
Road Density	Rip.Rd.Dens	Road density in buffer = Road length (km)/riparian buffer area (km^2) in
Mean Population Density	Rip.Pop.Dens	Buffer area mean population density based on 2000 census (#/km^2)
Natural Landscape Variables		
Watershed Mean Elevation	Mn.Elev	Mean watershed elevation (m)
Watershed Mean Slope Percent	Slope	Mean percent watershed slope (%)
Mean Annual Precipitation	Mn.Ann.Precip	Mean annual precipitation (mm)
Riparian Mean Slope Percent	Rip.Slope	Mean percent riparian buffer slope (%)
Riparian Maximum Elevation	Rip.Max.Elev	Maximum riparian buffer elevation (m)
Soil Infiltration Rate B	Soil Infiltration B	Area of stream buffer having soils with moderate infiltration rates (From NRCS, STATSGO database) (km^2)
Soil Infiltration Rate C	Soil Infiltration C	Area of stream buffer having soils with slow infiltration rates (From NRCS, STATSGO database) (km^2)
Soil Infiltration Rate D	Soil Infiltration D	Area of stream buffer having soils with very slow infiltration rates (From NRCS, STATSGO database) (km^2)
Hydrologic Runoff Variables		
Average Monthly Coefficient of Variation	Ave_MonthCV	Coefficient of variation of average monthly runoff values for 2001
Maximum Monthly Runoff	Max_Monthly	Maximum monthly runoff for 2001 (mm)
Maximum Monthly Coefficient of Variation	Max_MonthCV	Coefficient of variation of maximum monthly runoff values for 2001
Maximum Runoff for January Months	Max_January	Maximum runoff for January 2001 (mm)
Average March Runoff	Ave_March	Average runoff for March 2001 (mm)
Maximum March Runoff	Max_March	Maximum runoff for March 2001 (mm)
Minimum Runoff for April	Min_April	Minimum runoff for April 2001 (mm)
Maximum Runoff for April	Max_April	Maxmum runoff for April 2001 (mm)
Average Spring Runoff	Ave_Spring	Average runoff for April and May 2001 (mm)
Maximum Runoff for May	Max_May	Maximum runoff for May2001 (mm)
Maximum Runoff for July	Ave_July	Maximum runoff for July 2001 (mm)
Maximum Runoff for November	Max_Nov	Maximum runoff for November 2001 (mm)
Average December Runoff	Ave_Dec	Average runoff for December 2001 (mm)
Response Variables: Invertebrate Metrics		

Table 1. Cont.

Description	Variable Code	Definition
Ephemeroptera, Plecoptera, and Trichoptera Richness	EPTR	Richness composed of Mayflies, Stoneflies and Caddisflies for a sample
Average Tolerance of Taxa in a Sample	RichTOL	Average USEPA tolerance values for sample based on richness
Intolerant Richness	INTOL_RICH	Number of USEPA intolerance (0 – 4 values) taxa
Noninsect Richness	NonInsectR	Total richness composed on noninsects

All variables listed were initially considered for inclusion in boosted regression tree (BRT) models. NLCD–National Land Cover Dataset, NRCS–National Resource Conservation Service, STATSGO–State Soil Geographic data base.

the highest ridges. The entire physiographic area is noteworthy as a sparsely populated corridor of forests, wetlands, and grasslands of regional importance to migratory birds and many other plant and animal communities all within close proximity of the greater New York City metropolitan area of over 20 million people [33]. The forest is dominated by upland hardwood forest types on the ridges and valley slopes, and forested wetlands in the valleys. Annual average rainfall varies from 990 to 1,610 mm, with significantly higher amounts in mountainous areas. The Northern Piedmont ecoregion trends northeast to southwest covering parts of northeastern New Jersey and southeastern Pennsylvania and consists of northwestward-dipping sedimentary rocks that form rounded hills, irregular plains, and open valleys [3]. Elevations range from about 99 m on limestone to 396 m on more resistant metamorphic rock, with some isolated, rocky and higher hills or ridges. The climate includes moderate winters and warm, humid summers (i.e., humid continental) with more than 1,000 mm of precipitation falling in an average year. Land use varies, ranging from busy urban and suburban areas, to intensely farmed and densely settled locales, to relatively quiet pastoral places, which forms a mosaic of agricultural, forest, and developed lands, but the mixture varies, depending on local conditions.

The Northern Piedmont (N_Pied) has the highest mean percent urban and agriculture (29 and 34%, respectively) among the four ecoregions, the Ridge and Valley (R&V) has the second highest (11 and 29%, respectively) (Table 2). The North Central Appalachians (NC_App) has the lowest percentage of urban and agriculture (4 and 7%, respectively), however, the proportion of urban and agriculture in the Northeastern Highlands (NE_High) is only slightly higher (10 and 12%, respectively). In general, these two ecoregions tend to be slightly less developed than the N_Pied and R&V. The N_Pied also has slightly more than five times the average population density than the next highest ecoregion (i.e., NE_High) and has the highest percent manmade channels, yet the lowest average spring runoff and mean watershed slope (Table 2). Mean annual precipitation across all ecoregions is generally similar and ranges from 1140 to 1240 mm (Table 2). Percent slope is highest for NC_App and lowest for N_Pied.

Data Aggregation and Landscape Analysis

ArcGIS, ArcMap 9.2 (Environmental Systems Research Institute, Redlands, CA), was used to create, interpret, and analyze spatial data sets representing sampling site locations, watershed area and size class, riparian zones, and potential sources of anthropogenic disturbance for all watersheds located throughout the study area (Supplemental File: Appendix S1). Site location data sets were obtained from federal and state agencies and partners, including the U.S. Geological Survey, Connecticut Department of Environmental Protection (CTDEP), New Jersey Department of Environmental Protection (NJDEP), New York Department of Environmental Conservation (NYDEC), and Pennsylvania Department of Environment Protection (PADEP). These data sets were combined and standardized to create one seamless data set with over 10,000 sample sites. The data were then screened and censored according to the following criteria: (1) sampling methodology, (2) sampling season and year (we retained only samples taken between May and October from 1996 to 2009 that best coincide with the 2001 land use data), (3) upstream watershed area of > 8.0 and <777.0 km^2, and (4) watersheds not nested (i.e., not linked by downstream flow). In addition, sites that represented multiple sampling events in the same location were removed. In total, 1058 sites met the criteria and were used either as model development (n = 591) or validation (n = 467) sites (Figure 1). The development data set included sites within the four contiguous ecoregions: North Central Appalachians (n = 167), Ridge and Valley (n = 152), Northeastern Highlands (n = 139), and Northern Piedmont (n = 133) (Figure 1). Sites in the validation data set represent a randomly selected subset of the larger data set and, for the most part, are distributed across the four ecoregions comparably to the development data set (Figure 1). To generate the best possible data set for model development, nested sites were removed but then used in the validation data set to provide the highest number and best distribution of sites for model validation.

For consistency, all watersheds were delineated for the selected sampling sites using USGS 7.5-minute quadrangle digital raster graphics (DRG) as base layers. The DRGs were overlayed with National Hydrography Data set (NHD) high resolution stream lines for each region [34]. Watershed boundaries were digitized at a scale of 1:10,000 or larger. Adjacent watershed polygons were edge matched to eliminate all overlaps and gaps resulting in a consistent geospatial data set containing all study watersheds. Riparian buffer zone polygons were created within each watershed, extending 2 km upstream from the outlet of each watershed (i.e., point where the macroinvertebrate data was collected) along the main stem and all tributaries and 90 m on either side of the stream, similar to previous buffering approaches by Brenden et al. [35]. The selected reaches and associated buffer zone polygons were then saved as a new geospatial data set. To ensure that the riparian buffer zone polygons matched the watershed boundary polygons at the outlet, the buffered riparian polygons were overlayed on the watershed polygons and any areas extending beyond the existing watershed polygons were clipped and deleted. Landscape metrics derived from the spatial data sets were created for each watershed and riparian zone buffer and included ecoregion, elevation, slope, land cover (2001), road networks, soil infiltration capacity, hydrography, dams, pollution point sources, precipitation, population density, and canopy cover (Supplemental File: Appendix S1). All abbreviations for riparian based explanatory variables begin with the letters "Rip"; otherwise, variables are watershed based (Table 1).

Table 2. Summary of select environmental variables by region, numeric variables presented as average values, range in parentheses.*

Region	Urban (%)	Agriculture (%)	Population Density (#/km^2)	Manmade Channels (%)	Ave_Spring Runoff (mm)	Mean Annual Precipitation (mm)	Mean Slope (%)
Full Region	**13**	**20**	**421**	**2.1**	**60**	**1190**	**7**
n = 591	(0 – 94)	(0 – 97)	(0 – 13300)	(0 – 40)	(31 – 116)	(900 – 1610)	(0.4 – 18.8)
North Central Appalachians	**4**	**7**	**52**	**0.7**	**63**	**1140**	**8.1**
n = 167	(0 – 37)	(0 – 50)	(0 – 390)	(0 – 17)	(52 – 87)	(900 – 1450)	(1.7 – 18.8)
Ridge and Valley	**11**	**29**	**215**	**2.0**	**55**	**1170**	**7.3**
n = 152	(0 – 88)	(0 – 75)	(0 – 2100)	(0 – 40)	(34 – 87)	(1010 – 1330)	(1.6 – 13.0)
Northeastern Highlands	**10**	**12**	**257**	**2.5**	**74**	**1240**	**8.0**
n = 139	(0 – 56)	(0 – 57)	(0 – 1800)	(0 – 34)	(44 – 116)	(990 – 1610)	(3.3 – 17.0)
Northern Piedmont	**29**	**34**	**1290**	**3.4**	**45**	**1230**	**3.5**
n = 133	(1 – 94)	(0 – 97)	(70 – 13300)	(0 – 32)	(31 – 58)	(1060 – 1350)	(0.4 – 8.3)

*Ave_Spring equal to the average runoff for the spring months (April-May).

Hydrologic Runoff Metrics

Runoff is defined here as the total flow delivered to streams and rivers expressed on a per unit area basis. Runoff has the units of length divided by time and, when multiplied by upstream basin area, has the units of volume divided by time. The estimated runoff does not differentiate various sources of streamflow; it includes all sources of flow (groundwater discharge, rapid subsurface flow, and overland flow) delivered to streams (David Wolock pers. comm.), [36]. Estimates of hydrologic-unit runoff for all flow variables (Table 1) used for developing ecological models were derived using an area-weighted flow modeling approach that combines historical daily flow data collected at U.S. Geological Survey (USGS) streamgages, the respective drainage basin boundaries of the streamgaging stations, and the boundaries of the individual eight digit hydrologic cataloging units (HUC-8s) to develop a hydrologic unit runoff grid [36]. HUC-8 runoff values were used to estimate flow metrics for ungaged streams. This geospatial grid was used as the base layer to develop streamflow runoff estimates for the study basins. To derive these estimates, all 1058 of the previously delineated study basins were overlain onto the grid and run-off values in millimeters were computed. A total of 51 hydrologic runoff metrics were computed, which included the yearly average, minimum, and maximum runoff and coefficient of variation of runoff; the average, minimum, and maximum runoff for summer (Jul–Aug), winter (Jan–Feb), and spring (Apr–May); and the monthly average, minimum, and maximum for January–December. Refer to Brakebill et al. [36] for full description of how the hydrologic-unit runoff grid was developed for the coterminous U.S.

Macroinvertebrate Data

Macroinvertebrate data from 1996 to 2009 assembled for this study were comparable in terms of sampling protocols (sampled habitat, number of composite samples, and total sampled area) and laboratory procedures, including sorting, subsample count level, and taxonomic resolution. In general, all macroinvertebrate samples were collected in wadeable riffle/run habitats using quantitative collection techniques (e.g., kick, Slack, or d-frame nets) from designated stream areas (typically five to eight separate collections combined to form a composite sample or a single running kick sample from a 5 m area) [37–42]. These data were extensively reviewed to ensure that the aggregated data included the same taxonomic groups, followed the same nomenclature, and had appropriate taxonomic resolution before data analysis was initiated. The Invertebrate Data Analysis System (IDAS) software [43] was used to resolve all taxonomic issues (taxonomic identification level and nomenclature, i.e., taxonomic harmonization), to remove ambiguous taxa [44] and to randomly subsample raw counts to an equal 100 specimen count (the highest possible subset for maintaining consistency across all ecoregions and programs). In general, data for dominant aquatic insect orders were resolved at genus level. Less common orders were often aggregated to family level. Rare organisms or those with difficult taxonomy were sometimes aggregated to order or higher. The dipteran family Chironomidae, for example, is considered an important bioindicator group, but historically a difficult group to identify to genus or species. Because of differences among the various state data sets, this group had to be aggregated to the family level. IDAS was used to generate approximately 120 macroinvertebrate richness and abundance metrics, including tolerance and functional group metrics, which were calculated using values from Barbour et al. [45]. A reduced set of metrics for analysis was obtained using a variety of exploratory analysis techniques including evaluating scatter plots for outliers, correla-

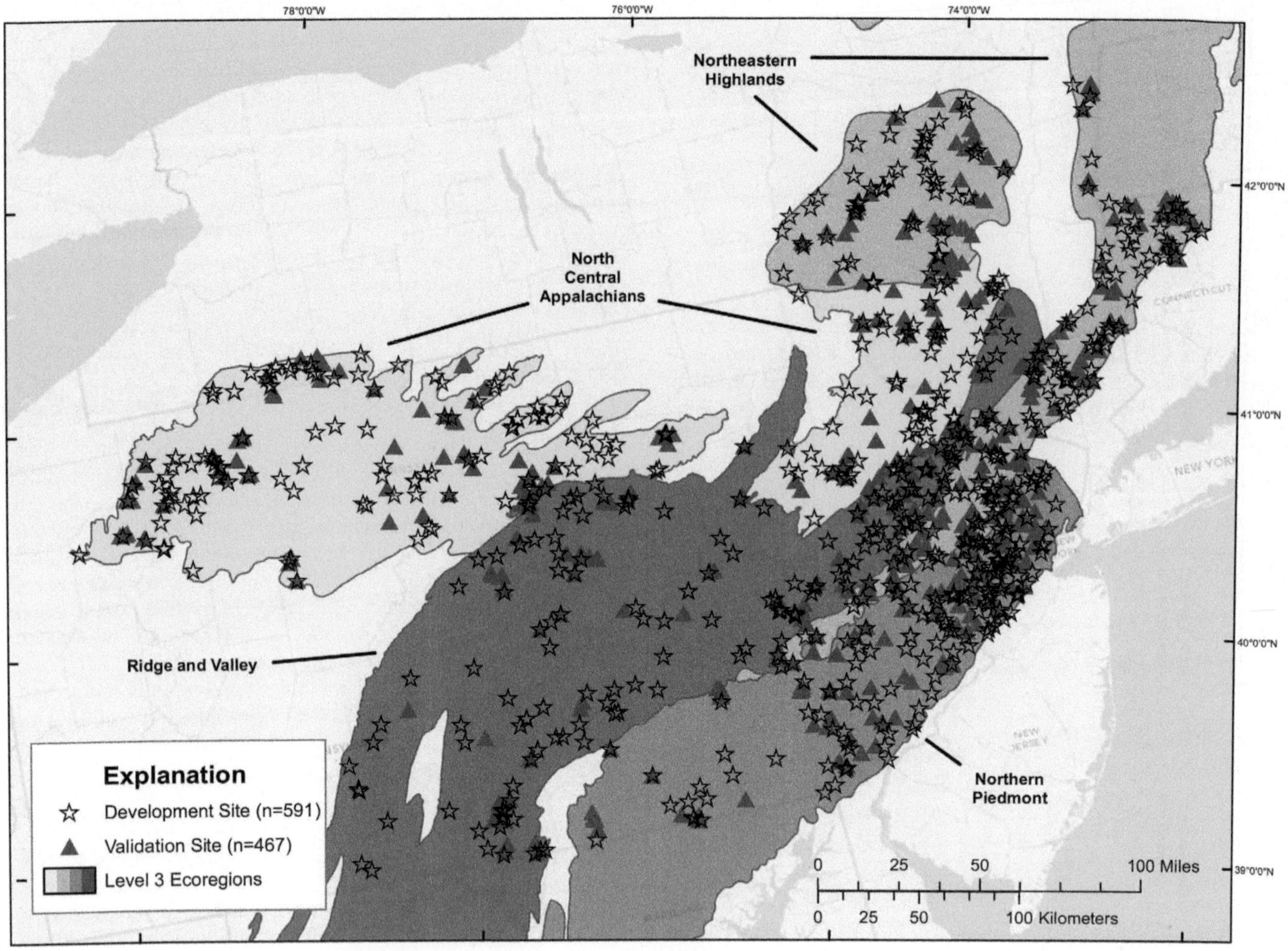

Figure 1. Map showing the stream sites used for model development (stars) and model validation (triangles). Sites used for model development (stars) (n = 591) and model validation (triangles) (n = 467) are evenly spread across the four ecoregions used in this study (N.C. Appalachian, Ridge and Valley, N.E. Highlands and N. Piedmont).

tion of metrics to core watershed disturbance variables, correlation among macroinvertebrate metrics and results of previous analyses in different regions of the U.S. [7],[11],[22],[46]. From the reduced set, we selected four macroinvertebrate metrics as response variables for model development and validation: EPT Richness (richness of the orders Ephemeroptera, Plecoptera, and Trichoptera; EPTR), average tolerance of all taxa (RichTOL) based on supplemented EPA tolerance values, Richness of Intolerant Taxa (INTOL_RICH), and Non-Insect Richness (NonInsectR) (Table 1).

Ethics

This paper uses only data already collected in previous studies; there are no ethical conflicts with animal use. All data were collected by either by the USGS or state agencies following appropriate state protocols for field sampling and permission access. No permits were required for sampling of macroinvertebrates and no protected species were sampled.

Model Development

We developed BRT models for four macroinvertebrate metrics for the full region and for each of the four ecoregions separately. Regression trees fall within the commonly used classification and regression tree (CART) or decision tree family, and their use and technical details have been described (e.g., [23],[24],[25]) and extensively applied [5],[11],[17],[22],[27],[28],[47] in the literature; therefore, we will provide only a brief description. Boosted regression trees (BRT) are among a family of techniques used to advance single classification or regression trees by averaging the results for each binary split from numerous trees or forests, thus reducing the predictive error and improving overall performance [26],[27]. In BRT, after the initial tree has been generated, successive trees are grown on reweighted versions of the data, giving more weight to cases that are incorrectly classified than those that are correctly classified within each growth sequence. Thus, as more and more trees are grown in BRT, the large number of trees increases the chance that cases that are difficult to classify initially are correctly classified, thus representing an improvement to the basic averaging algorithm used in random forest [26]. Boosted trees and random forest models retain the positive aspects of single trees seen in CART models, yet have improved predictive performance, nonlinearities and interactions are easily assessed, and they can provide an ordered list of the importance of the explanatory variables [26],[47].

Although BRT offers improved modeling performance over CART, the simple single tree obtained from CART is lost, making it more difficult to visualize the results. Partial dependency plots (PDPs) provide a way to visualize the effect of a specific explanatory variable on the response variable after accounting for the average effects of all other explanatory variables [26],[27]; PDPs for selected variables important in models appear as examples in the results. BRT models were run using the gbm library in R and specific code from Elith et al. [27]. We used R^2 and cross-validation R^2 (CV R^2) values to compare BRT models because they are well understood measures of the amount of variation explained by the models.

BRT models were developed using four macroinvertebrate response metrics (EPTR, RichTOL, INTOL_RICH, and Non-InsectR) and a set of explanatory variables (watershed, riparian, hydrologic, and natural landscape; Table 1) that were evaluated using a variety of exploratory analysis techniques, including examining scatter plots for outliers and assessing intercorrelation among variables. In addition, all explanatory variables were screened and only those that had > 50% non-zero values were retained for further analysis. The BRT model was first developed using the model development data set. We used a bag fraction of 0.75, a learning rate of 0.001 to 0.0005 for developing our models, and a tree complexity of 5; a bag fraction of 0.75 means that each tree is developed using a random selection of 75% of the data. The learning rate influences the total number of trees evaluated for the model, while tree complexity controls whether interactions are fitted, a value of 5 allows the assessment of up to 5-way interactions. Variable relative importance (VRI) was calculated using formulae developed by Friedman [48] and implemented in the R gbm library to estimate the relative influence of predictor variables. Calculations of VRI are based on the number of times a variable is selected for splitting, weighted by the squared improvement to the models as a result of each split, averaged over all trees. The relative importance of each variable is scaled so that the sum adds to 100, with higher numbers indicating stronger influence on the modeled response. We developed our BRT models using a multi-stage process: (1) BRT models were first run with all watershed, riparian, and natural landscape variables only, with the top 10 variables in the variable relative importance list retained for further analysis, (2) BRT models were run with all hydrologic runoff based variables only, with the top 10 variables in the variable importance list retained for further analysis, and (3) BRT models were run combining the top 10 variables from steps 1 and 2. Finally, explanatory variables in the final BRT models were pruned by using a combination of VRI scores, evaluation of interactions and partial dependency responses and gbm simplify scores following the approach outlined by Elith et al. [27] to minimize overfitting. In brief, we deleted all variables with relative importance values less than 7. The remaining variables were then used to develop the final BRT model. The final model was selected by sequentially deleting variables and evaluating the effects on R^2 and examining partial dependency plots; this was done independently for each invertebrate metric for each region. We then used the final BRT model to predict EPTR, RichTOL, INTOL_RICH, and NonInsectR values for the validation data set of 467 sites and calculate model performance measures. Finally, we regressed observed values against predicted values for both development and validation models as a visual measure of model precision and bias [49]. The sites in validation data set were not randomly selected from the full data set as would be ideal, but instead were extra nested sites that had been removed from the final full region data set. Nevertheless, the validation data encompassed a large number of sites equally distributed across all regions.

In order to get a large enough sample size, the data set analyzed had a large range in watershed size (8–780 km^2) and mean elevation (23–870 m); so, in addition to the tests above regarding spatial scale, we wanted to determine whether breaking the data set into watershed and elevation classes would improve model performance. Stream size, that is, location along the continuum or gradient from the headwaters to the mouth, influences the type and distribution of the aquatic fauna [29]. Therefore, three watershed size and elevation classes were developed to evaluate whether improved model performance over the Full Region model could be achieved. We divided the Full Region model data set into three watershed size (8–27.0; >27.0<66.0; and >66.0< 777.0 km^2) and mean elevation classes (23–225, 225–450, and > 450 m). This division created relatively even sample sizes among the three classes, but also followed a recognized common regional break for small watershed sizes (± 26 km^2). All BRT modeling was completed in R ([50], version 2.13.1).

Results

Comparing within a region but across metrics, the average tolerance of all taxa (RichTOL) consistently had the highest R^2 and CV R^2 for BRT models among the four metrics for all the regions studied (Table 3). Given the use of the EPT Richness (EPTR) metric as a component of many multimetric indices in the northeast, it was unexpected that it would have the lowest R^2 across all regions for BRT models when compared to the other metrics, except in the NE_High, where it had the second lowest value. The other two metrics, Richness of Intolerant Taxa (INTOL_RICH) and Non-insect Richness (NonInsectR), were, in general, intermediate in model R^2 value to RichTOL and EPTR (Table 3).

Spatial scale results showed that relatively strong explanatory models with R^2 values ranging from 0.63 to 0.77 were developed using BRT techniques for the Full Region models (Table 3). In all cases, however, the N_Pied produced stronger models and the NC_App produced weaker models than those produced in the Full Region models. In the other regions (R&V and NE_High) the results were mixed and in some cases the Individual Ecoregion models explained slightly more variation than the Full Region models. The cross-validation R^2 values, which are much more conservative than adjusted R^2, decreased significantly across all metrics and regions compared to R^2, the decrease ranging from 12 to 33 points per model (Table 3). NC_App, which had the smallest range or gradient in land use disturbance (e.g., urbanization, manmade channels, dam density, and agricultural land use) had the lowest R^2 for each metric across the regions and the largest average decrease in CV R^2.

Across the four metrics, the final Full Region BRT models had between four and six explanatory variables (Table 4): a variable related to urbanization (population density, percent urban, or percent manmade channels), a measure of hydrologic runoff (average December or maximum monthly runoff) and a natural landscape variable (slope and elevation) were in every model. Other explanatory variables in the Full Region models were percent riparian forest, riparian canopy, and percent forest in the watershed. Models developed across the four ecoregions showed the same pattern related to urbanization as the Full Region models. Every model had at least one, and in some cases, more than one, urbanization-related variable. Similar to the Full Region models, many of the Individual Ecoregion BRT models had forest or riparian canopy-related variables and over half of the models

Table 3. Comparison of model evaluation statistics for boosted regression tree models (BRT) for four macroinvertebrate metrics for development (develop) and validation (valid) data sets at two spatial scales (full region and four ecoregions), number of variables in final model in parentheses.

Metric	Model Data Set	Model Statistic	Full Region	North Central Appalachians	Ridge and Valley	Northeastern Highlands	Northern Piedmont
			n = 591	n = 167	n = 152	n = 139	n = 133
EPTR	Develop	R^2	0.63 (6)	0.54 (4)	0.65 (5)	0.70 (4)	0.76 (4)
		CV R^2	0.46	0.27	0.30	0.38	0.50
	Valid	R^2	0.63	0.68	0.82	0.57	0.78
		CV R^2	0.41	0.08	0.39	0.28	0.48
RichTOL	Develop	R^2	0.77 (5)	0.67 (5)	0.81 (4)	0.80 (4)	0.84 (4)
		CV R^2	0.64	0.42	0.51	0.56	0.65
	Valid	R^2	0.73	0.64	0.64	0.68	0.82
		CV R^2	0.59	0.23	0.34	0.44	0.62
INTOL_RICH	Develop	R^2	0.66 (5)	0.60 (7)	0.65 (5)	0.67 (4)	0.72 (6)
		CV R^2	0.49	0.27	0.33	0.43	0.42
	Valid	R^2	0.66	0.70	0.67	0.59	0.70
		CV R^2	0.48	0.26	0.42	0.30	0.39
NonInsectR	Develop	R^2	0.68 (4)	0.66 (5)	0.69 (4)	0.75 (5)	0.72 (4)
		CV R^2	0.56	0.34	0.49	0.43	0.48
	Valid	R^2	0.67	0.31	0.80	0.70	0.80
		CV R^2	0.53	0.02	0.57	0.39	0.62

Validation models run with the same variables as the final development model.*

*R^2–adjusted R-squared, CV R^2-Cross validation R^2; EPTR–Total taxa richness of Ephemeroptera (mayflies), Plecoptera (stoneflies) and Trichoptera (caddisflies); RichTOL-Average tolerance of all taxa; INTOL_RICH-Richness of intolerant taxa; NonInsectR-Noninsect taxa richness.

Table 4. Comparison of explanatory variables for boosted regression trees (BRT) models for four macroinvertebrate metrics at two spatial scales (Full Region and four Individual Ecoregions); variables are presented in descending order of variable relative importance (VRI) in each model.*

Invertebrate Metrics	Full Region		North Central Appalachians		Ridge and Valley		Northeastern Highlands		Northern Piedmont	
	n = 591		n = 167		n = 152		n = 139		n = 133	
		VRI		VRI		VRI		VRI		VRI
EPT Richness	Urban	34	Min_April	29	Rip.Forest	32	Urban	51	Urban	46
(EPTR)	Rip.Slope	19	Rip.Pop.Density	27	Ag+Urban	21	Wetland	17	Mean Elevation	26
	Ave_Dec	14	Rip.MeanCanopy	25	Pop.Density	18	ManMadeChan	16	Rip.Forest	18
	Rip.MeanCanopy	12	Road Density	20	Max_MonthCV	17	Rip.Forest	16	ManMadeChan	10
	Rip.Forest	12			Urban	12				
	ManMadeChan	9								
Average	Pop.Density	31	Rip.Forest	36	ManMadeChan	33	Ave_March	48	Mean Elevation	51
Tolerance of all	ManMadeChan	25	Rip.Pop.Density	17	Rip.Forest	29	Rip.Wetland	27	ManMadeChan	22
Taxa	Rip.Slope	16	Rip.Ag	17	Max_Nov	20	Urban	15	Urban	15
(RichTOL)	Rip.Forest	14	Pop.Density	16	Pop.Density	18	Road Density	10	Rip.Forest	12
	Ave_Dec	14	Soil Infiltration D	14						
Richness of	Rip.Forest	34	Rip.Forest	22	Rip.Forest	31	Urban	43	Urban	28
Intolerant Taxa	Pop.Density	23	Rip.Slope	18	Pop.Density	24	Ave_July	25	Rip.Max.Elev	20
(INTOL_RICH)	Forest	17	Soil Infiltration D	14	Urban	16	Rip.Forest	19	Mean Slope	18
	Rip.Max.Elev	17	Min_April	13	Rip.MeanCanopy	16	ManMadeChan	14	Soil Infiltration B	14
	Max_Monthly	10	Rip.Ag	13	Mean Slope	13			Rip.Wetland	11
			Rip.Pop.Density	12					Pop.Density	10
			Road Density	9						
Noninsect	Ave_March	41	Rip.Forest	25	Max_March	44	Max_January	24	Max_April	36
Richness	ManMadeChan	26	Pop.Density	24	Max_May	24	Ave_March	24	Rip.Max.Elev	27
(NonInsectR)	Mean Elevation	17	ManMadeChan	22	Soil Infiltration C	21	Rip.Wetland	23	Soil Infiltration C	20
	Rip.Slope	16	Dam Density	15	ManMadeChan	11	Rip.Slope	18	ManMadeChan	17
			Rip.Pop.Density	14			Pop.Density	12		

*EPTR –Total taxa richness of Ephemeroptera, Plecoptera and Trichoptera; B–Soils with a moderate infiltration rate, C–Soils with a slow infiltration rate, D–Soils with a very slow infiltration rate, Rip –Riparian. See Table 1 for variable definitions.

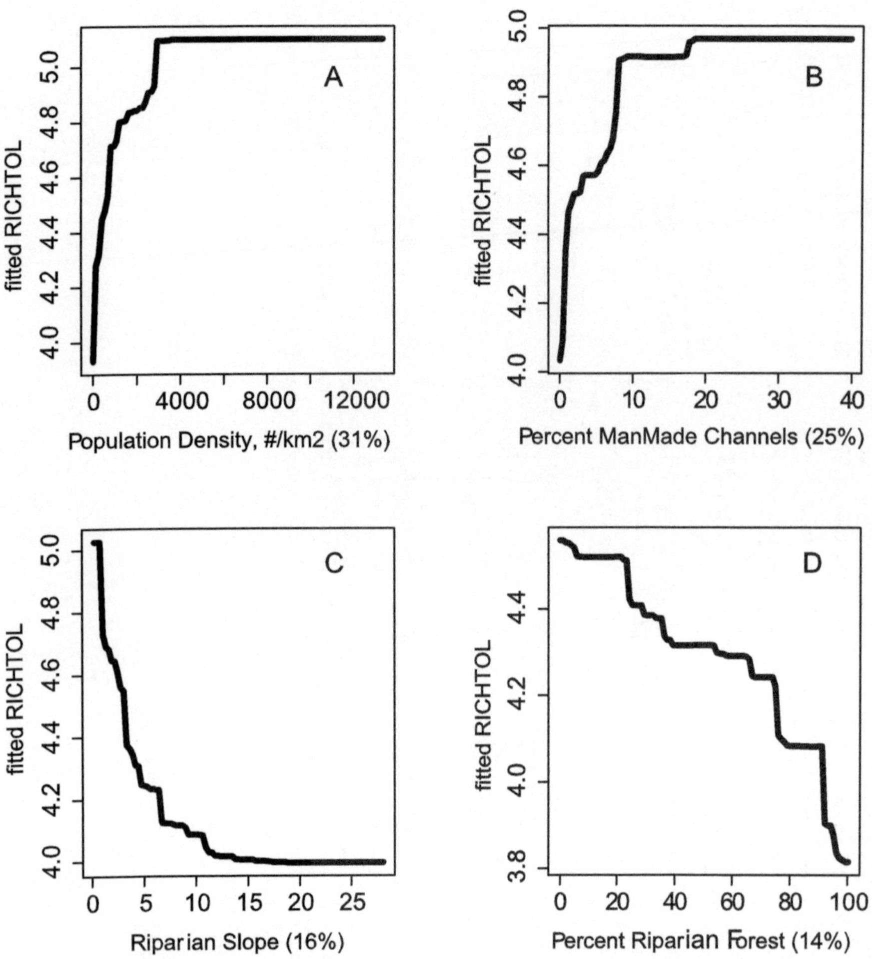

Figure 2. Partial dependency plots for variables in BRT model for RichTOL for the Full Region. Boosted regression tree partial dependency plots show the response form of average taxa tolerance (y-axis = fitted function of RichTOL) based on the effect of individual explanatory variables with the response of all other variables removed (development data set). Shown in order of model importance: (A) population density (numbers/km^2), (B) percent manmade channels, (C) riparian slope and (D) percent riparian forest, model $R^2 = 0.77$. The relative contribution of each explanatory variable is reported in parentheses. Refer to Table 1 for variable definitions. The top two variables for the Full Region model showed potential threshold type responses for urbanization variables and the third variable, a natural factor, is likely acting as an urban surrogate (riparian slope). The final variable, a measure of riparian disturbance follows a more linear response and along with urbanization variables was a common explanatory variable in most models among the different regions.

had some form of natural landscape factors. In addition, 9 of the 16 Individual Ecoregion models (56.2%) had a hydrologic runoff variable, and in some cases 2 hydrologic variables (e.g., the NonInsectR models for R&V and NE_High). In six out of the nine (66%) cases where a hydrologic variable entered the BRT model for the Individual Ecoregions, it was ranked either first or second in importance. Urbanization-related, riparian forest, and hydrologic runoff variables were consistently selected as the top variable with almost equal frequency across all 20 BRT models; however, only in one (the RichTOL model for N_Pied) was a natural landscape factor selected first (Table 4).

There were some common response patterns in the urbanization-related explanatory variables related to RichTOL. For example, the partial dependency plots for population density, road density, percent manmade channels, and percent urban showed a relatively abrupt threshold-type response (Figs. 2A,B; 3B,D; 4A,D; 5C,D; and 6B,C) (Note: to reduce complexity, only the top four variables are shown in each partial dependency plot). However, response patterns for percent riparian forest were mixed and showed a more linear response for the Full Region (Figure 2D) as compared to the stepped response and strong threshold seen for the NC_App and R&V and N_Pied regions, respectively (Figs. 3A, 4B, and 6D). The responses for hydrologic runoff variables also tended to show a threshold or stepped type response (Figure 4C and 5A, respectively). There were also some clear interactions among important explanatory variables, including natural landscape factors and disturbance variables. For example, at low elevations in the N_Pied region there are a range of values of manmade channels, with high values (threshold at about 15% manmade channels) resulting in high values of tolerant taxa (RichTOL), but at high elevations there are only low values of manmade channels, resulting in low values of RichTOL (Figure 7). As another example, the combination of high values of average March runoff and high values of riparian wetlands interact in the result of the highest values of RichTOL. At low values of March runoff there are relatively low values of RichTOL even across a range of values for riparian wetlands, except at the highest wetland values, yet these values are not as high as when there is high March runoff in combination with high wetland values (Figure 8). For brevity, the response form as shown in the partial dependency

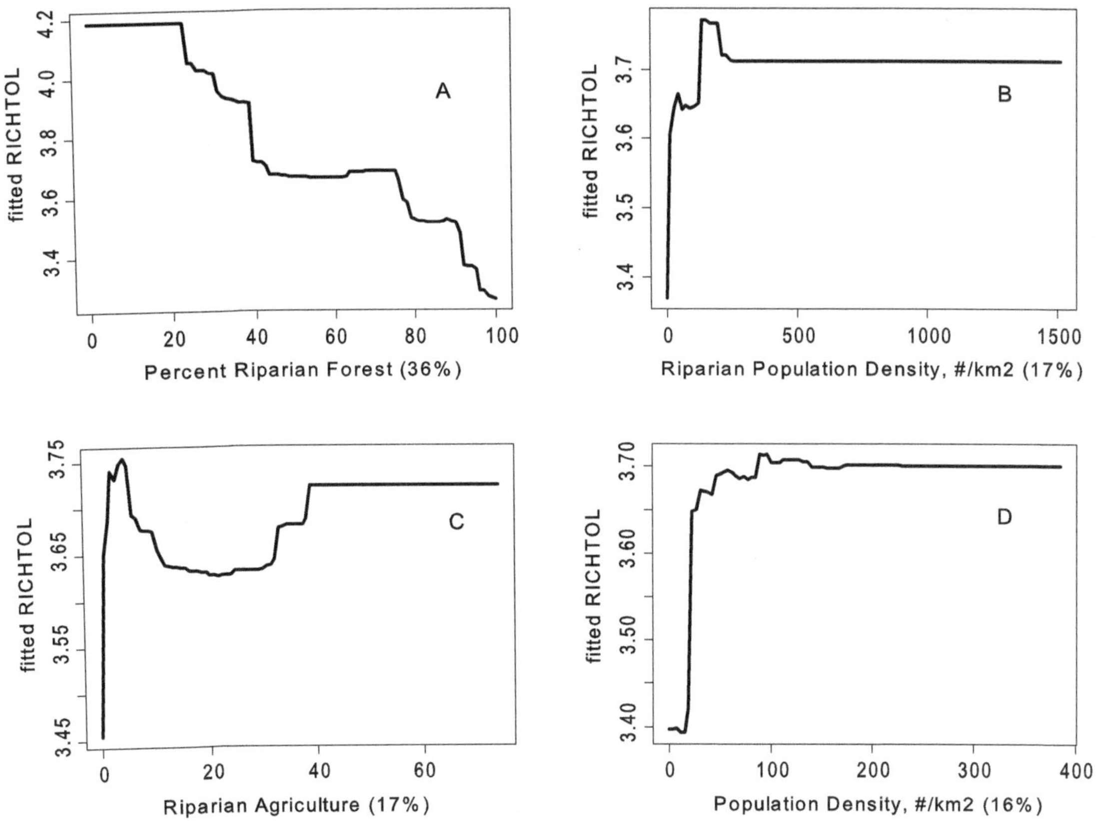

Figure 3. Partial dependency plots for variables in BRT model for RichTOL for North Central Appalachian Region. Boosted regression tree partial dependency plots show the response form of average taxa tolerance (y-axis = fitted function of RichTOL) based on the effect of individual explanatory variables with the response of all other variables removed (development data set). Shown in order of model importance: (A) percent riparian forest, (B) riparian population density (#/km^2), (C) percent riparian agriculture and (D) population density (#/km^2). The relative contribution of each explanatory variable is reported in parentheses. Refer to Table 1 for variable definitions. Three of the four variables can be interpreted as disturbance variables, two directly assessing urban land use (population density) and the third, riparian forest, which measures the amount of disturbance in the riparian zone was the top variable modeled. However, this region had the shortest disturbance gradient and the lowest modeled R^2 (0.67), though still relatively strong.

plots for the other invertebrate metrics for each region are not shown here but provided as supplemental material online (Figure S1).

Validation model results varied across metrics and regions, 10 out of the 20 total models had R^2 values for the validation models within five points of the development models. Of the remaining 10 models, five had validation R^2 values lower than the development models (range −8 to −35) and five had higher values (range +8 to + 17). Models with higher performance for the development data set tended to also have higher R^2 values in the validation data, the exception were those models where the validation values were higher. Observed vs. predicted plots (Figure 9) are shown only for RichTOL, primarily because it consistently represented stronger BRT models with greater predictive power (highest R^2 values) than the other metrics assessed in this study and to avoid adding unnecessary length to the manuscript. Observed vs. predicted plots for all other metrics and regions are provided as supplemental material online (Figure S2). In general, the observed-predicted plots reveal that for all metrics and regions, the models over predict at low values and under predict at high values for each invertebrate metric, the extent of the bias varied with the R^2 of the models. RichTOL for the Full Region models represented the largest range in the disturbance gradient and reflected only a slight under prediction bias in the upper end of the values for the development model (n = 591; Figure 9 left panel) as compared to wider and more even distribution of points across the full range of values for the validation model (n = 467; Figure 9 right panel). As would be expected, the Individual Ecoregion models, which have an overall smaller sample size, reflected a shorter disturbance gradient that varied across regions. The NC_App ecoregion encapsulated the shortest gradient (range in RichTOL values; Figure 9) and reflected a distinct bias for the development and validation models, under and over predicting tolerance values near the high and low end of its range, respectively (n = 167). R&V and NE_High covered a slightly longer portion of the disturbance gradient than NC_App with values more tightly clustered along the 1:1 line for the development models, yet also less tightly clustered around the 1:1 line for the validation models (Figure 9). The N_Pied region which had the strongest development model encapsulated a relatively long portion of the disturbance gradient and reflected only slight bias at low and high tolerance values. However, the validation model for N_Pied showed slightly more scatter around the 1:1 line compared to the development model, although still less bias than all the other regional validation models (Figure 9).

BRT models developed for watershed size classes for the same metrics tended to show similar patterns in response as the ecoregion models. For example, RichTOL had the highest overall

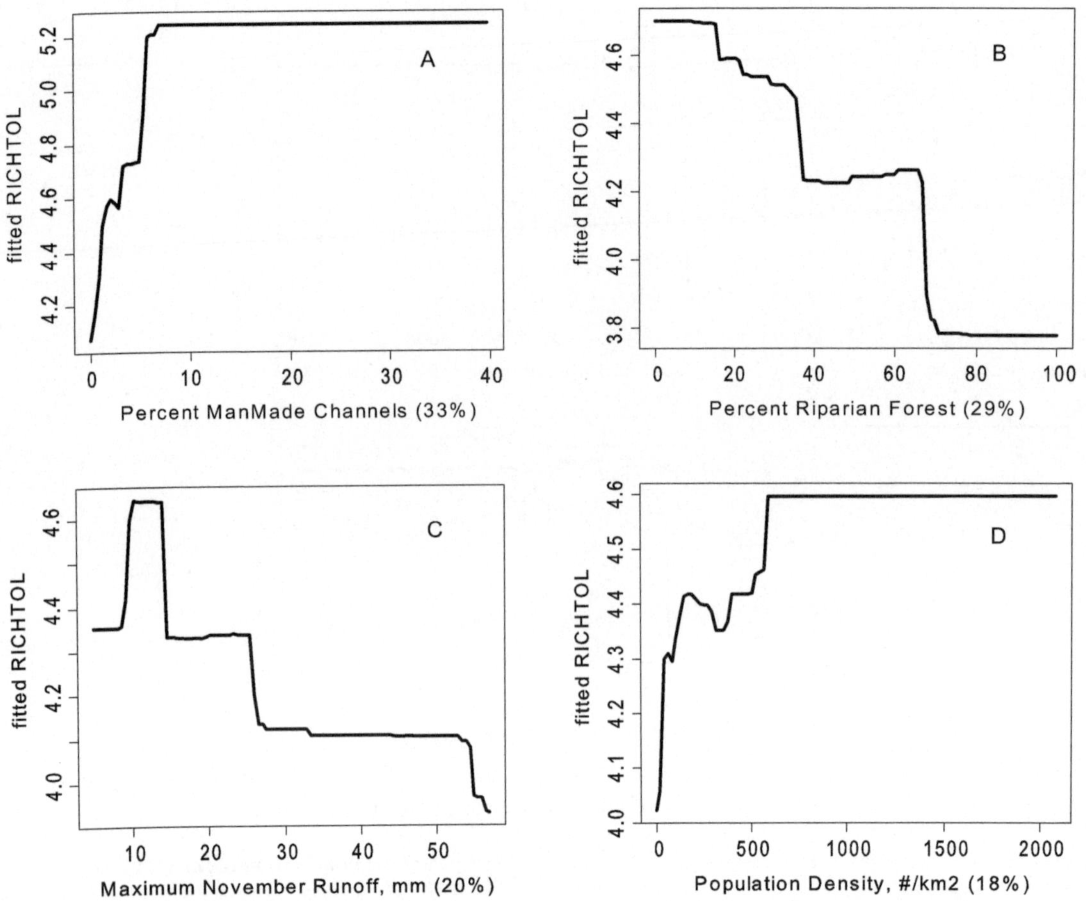

Figure 4. Partial dependency plots for variables in BRT model for RichTOL for Ridge and Valley Region. Boosted regression tree partial dependency plots show the response form of average taxa tolerance (y-axis = fitted function of RichTOL) based on the effect of individual explanatory variables with the response of all other variables removed (development data set). Shown in order of model importance: (A) percent manmade channels, (B) percent riparian forests, (C) maximum November runoff (mm) and (D) population density (#/km^2), model $R^2 = 0.81$. The relative contribution of each explanatory variable is reported in parentheses. Refer to Table 1 for variable definitions. Three of the four variables measure the effects of disturbance, two measure the response to urban land use and the other disturbance in the riparian zone due to either agriculture or urbanization. The fourth variable shows the response due to maximum November runoff.

model performance of the four metrics across all watershed size classes and EPTR had the lowest performance (Table 5). There was little or no overall improvement in model performance for the majority of the watershed size classes compared to the Full Region models. The largest improvements in R^2 occurred for RichTOL (watershed size class 2), NonInsectR (watershed size class 1), and EPTR (watershed size class 2) which slightly exceeded the Full Region model results by 0.06, 0.09, and 0.06, respectively (Table 5). We also tested model performance for three classes of elevation (mean elevation: 23–225, 225–450, and > 450 m) and similar to the result for models for watershed size classes, found little or no improvement compared to the Full Region models (for brevity, data not shown).

Discussion

The primary objective of this study was to evaluate the influence of scale (Full Region models versus four Individual Ecoregion models) on model performance by developing predictive benthic-macroinvertebrate boosted regression tree (BRT) response models using readily delineated and commonly applied watershed variables. Our working hypothesis was that small scale Individual Ecoregion models would outperform larger, Full Region models. Previous research has shown that large scale assessments, in general, do not perform as well as smaller local or regional scale models largely due to inherent differences in biogeography and disturbance type or extent [5],[6],[12],[14],[15],[16]. Waite [51] and Riseng et al. [52] also found that the strength of correlations of macroinvertebrate and algal metrics to agricultural disturbance variables (agricultural intensity and nutrient concentrations), varied with spatial scale and geographic region. Based on this weight of evidence, we hypothesized that the larger scale Full Region models would be more generic, have lower predictive power, and would likely include more natural landscape variables (e.g., slope and elevation, etc.) as compared to the smaller scale Individual Ecoregion models. Our modeling results, however, did not support this hypothesis. For example, the Full Region BRT models were, in some cases, as strong or stronger than the Individual Ecoregion models (e.g., they were consistently stronger than all the North Central Appalachian models; Table 3), and land use, hydrologic runoff, and other disturbance related variables were frequently important explanatory variables in these models (Table 4). The Full Region models, with the exception of NonInsectR, only had one natural landscape variable out of the four to six variables in each of the final models. Some Individual Ecoregion BRT models did show significant improvement in

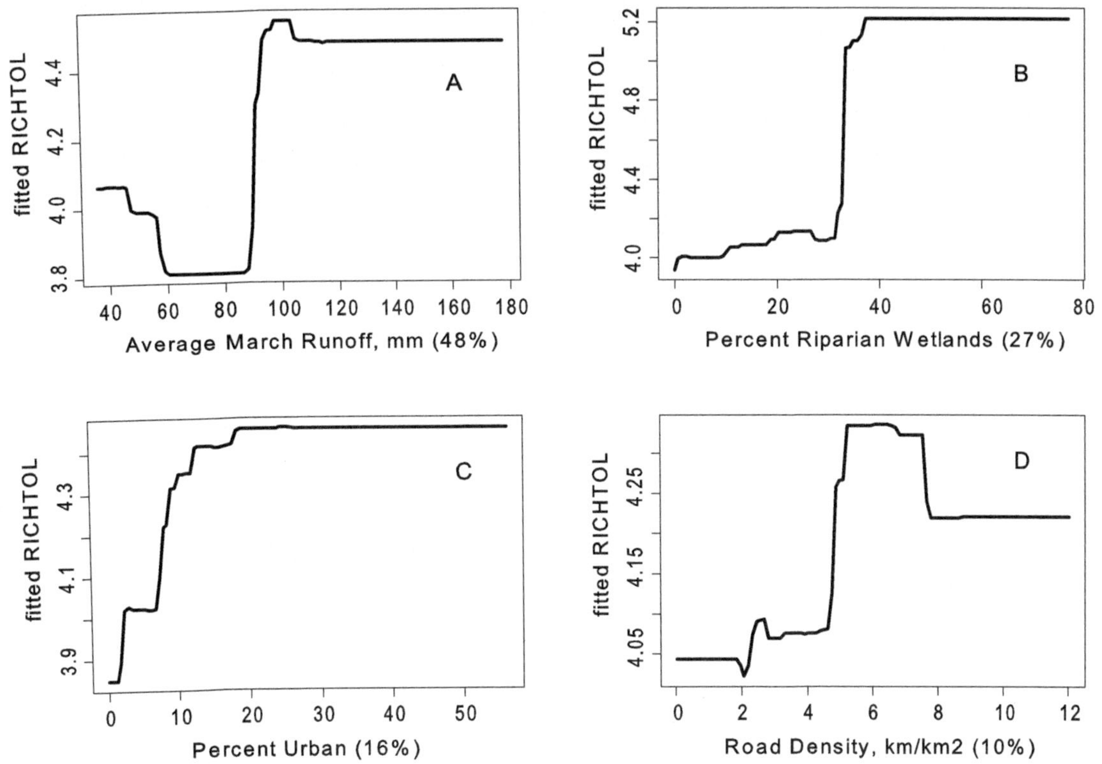

Figure 5. Partial dependency plots for variables in BRT model for RichTOL for NE Highlands Region. Boosted regression tree partial dependency plots show the response form of average taxa tolerance (y-axis = fitted function of RichTOL) based on the effect of individual explanatory variables with the response of all other variables removed (development data set). Shown in order of model importance: (A) average March runoff (mm), (B) percent riparian wetlands, (C) percent urban and (D) road density (km/km^2), model $R^2 = 0.80$. The relative contribution of each explanatory variable is reported in parentheses. Refer to Table 1 for variable definitions. All four explanatory variables modeled can be interpreted as an urbanization land use effect. Average March is expected to increase due to higher imperviousness with higher urbanization and percent riparian wetlands we believe is acting as a surrogate for urbanization; higher wetlands commonly occur in lower elevation valleys where there is commonly more urban development. The last two variables measure urban land use directly.

model performance (e.g., gains ranged from a minimum in R^2 of 0.01 for the Northeastern Highlands to a maximum of 0.13 for Northern Piedmont); however, natural landscape variables were equally as common in the Individual Ecoregion models as they were in the Full Region models. One possible explanation for why these findings tended not to coincide with our original hypothesis may be that the Individual Ecoregions used in this study did not have as strong a biogeographic gradient when compared to regions modeled in other recent studies (e.g., Western U.S. in [14],[17]) or perhaps the larger sample size used to develop the Full Region models resulted in longer and more defined disturbance gradients for model development. In either case, our results may indicate that models of broader spatial extent, such as the Full Region models evaluated in this study, could represent an important starting point for prediction of changes in macroinvertebrate response at unsampled stream sites. These results tend to support the need to develop models at the larger regional scales, but also indicate that some improvement in model performance may be provided by smaller scale efforts using Individual Ecoregions or other common surrogates (e.g., physiographic provinces). The extent of improvement, however, may directly depend upon the extent of the gradient encapsulated by the data set relative to the scale of the investigation and in the synchronicity in scales between the biota and the landscape (i.e., are the biota responding at one scale while landscape data and related variability occurs at another scale).

Even though the validation data set was not ideal since it was created from "left over" nested sites and not sites randomly subset from a larger pool of sites, it still provides a measure of how well the models are likely able to predict at unsampled sites. Fifteen out of the 20 validation models had R^2 values that were no lower than five points less than the development models or had values that were actually higher. The Full Region and N_Pied region still had some of the higher R^2 values for the validation data than the other regions but there was more variability among the metrics within a region and among regions as to which had the higher performing validation models. Looking at the Observed vs Predicted plots for the development and validation models, it becomes clear that even the strongest models have bias at the low and high ends of the metric ranges and that the validation data had a slightly different range in observed values than the development data (Figure 9 and supplemental material). For example, the models frequently would over predict richness values at low values (three metrics were based on richness), rarely predicting zero or one's even though these were observed in the two data sets (seen as values above the 1:1 line in plots). The opposite occurred at higher values, the models often under predicted at the higher ranges (seen as values below the 1:1 line in plots). As would be expected, this bias was directly related to measures of model performance, models with higher R^2 had less low and high bias than those models with lower R^2 values. However, a lot could be learned by investigating why some sites were predicted to have poor or higher invertebrate values than

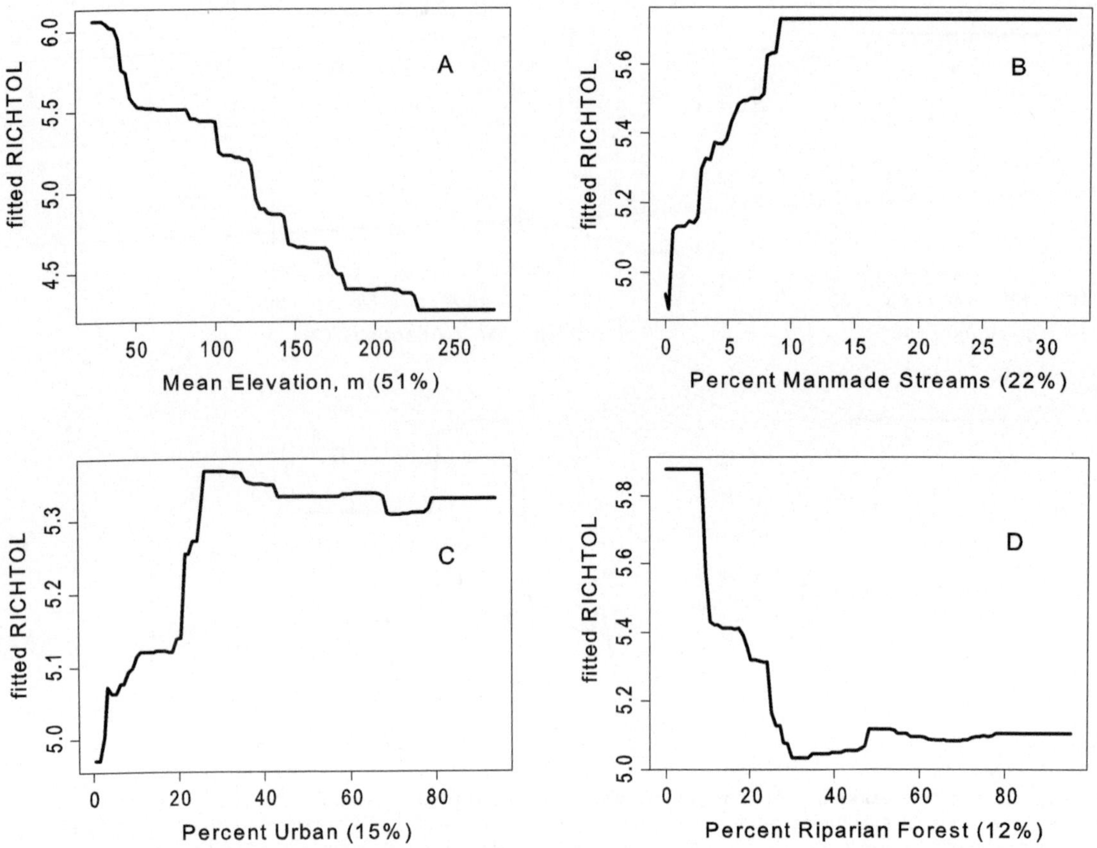

Figure 6. Partial dependency plots for variables in BRT model for RichTOL for Northern Piedmont Region. Boosted regression tree partial dependency plots show the response form of average taxa tolerance (y-axis = fitted function of RichTOL) based on the effect of individual explanatory variables with the response of all other variables removed (development data set). Shown in order of model importance: (A) mean elevation (m), (B) percent manmade channels, (C) percent urban and (D) percent riparian forest, model $R^2=0.84$. The relative contribution of each explanatory variable is reported in parentheses. Refer to Table 1 for variable definitions. This was the only model that had a natural factor as the top explanatory variable, however, we believe elevation is acting as a strong surrogate in this region for urbanization, though it is likely more complex than a strictly one for one surrogate. The other three variables all measure directly or indirectly urban land use disturbance and show a relatively strong potential threshold type response.

were observed. Since the models primarily use land use at the watershed and riparian scales, perhaps these sites were better or worse than expected in water quality or habitat condition than most sites with similar land use condition. Across all models, the more conservative CV R^2 showed significant decrease in values compared to the standard adjusted R^2 model performance measure, suggesting that there may be some level of over-fitting in the models.

Across all the models, two clear patterns emerged: (1) some measure of urbanization, the amount of riparian disturbance (e.g., percent riparian forest, riparian population density), and hydrologic runoff were important explanatory variables, and (2) many of the explanatory variables showed a distinct threshold-type response (shown in the PDPs as transition points in ecological condition [53–55]). For example, a threshold response was commonly seen for the metric RichTOL as predicted by population density, percent urbanization, and percent manmade channels (Figs. 2A, 5C, and 6B, respectively). Note that these are only potential thresholds; more analysis and evaluation would have to be conducted to determine if there are actual thresholds in this data [55]. Land use in parts of the Northeast tends to be dominated by urbanization rather than agriculture, so it is not unexpected that variables representing urbanization or urban-related disturbance processes (e.g., population density and road density) were dominant factors in many of the predictive models developed in this study. The effects of urbanization have been well documented by other researchers ([16],[18],[19],[56–61], and many more); therefore, that topic will not be addressed here. The amount of stream channel alteration characterized in this study by the percentage of manmade channels, however, is seldom discussed in the urbanization literature, yet it was one of the variables commonly modeled in this study and represents an important anthropogenic pathway affecting stream communities that needs to be more broadly considered in future ecological modeling studies. In addition, more research is emerging on the importance of riparian integrity related to stream condition (e.g., [17],[51],[52],[62]) thus revealing riparian's critical influence on stream integrity. For example, Clapcott et al. [5] recently found that the amount of native riparian vegetation was the dominant variable in BRT models describing macroinvertebrate metrics in New Zealand and that riparian and urbanization variables showed threshold responses, similar to the responses seen for urbanization-related and riparian variables in this study. However, variables like population and road density and percent manmade channels, which are frequently highly skewed, are notoriously difficult to analyze via linear parametric techniques like MLR and this difference may be another reason why BRT models generally outperform MLR models.

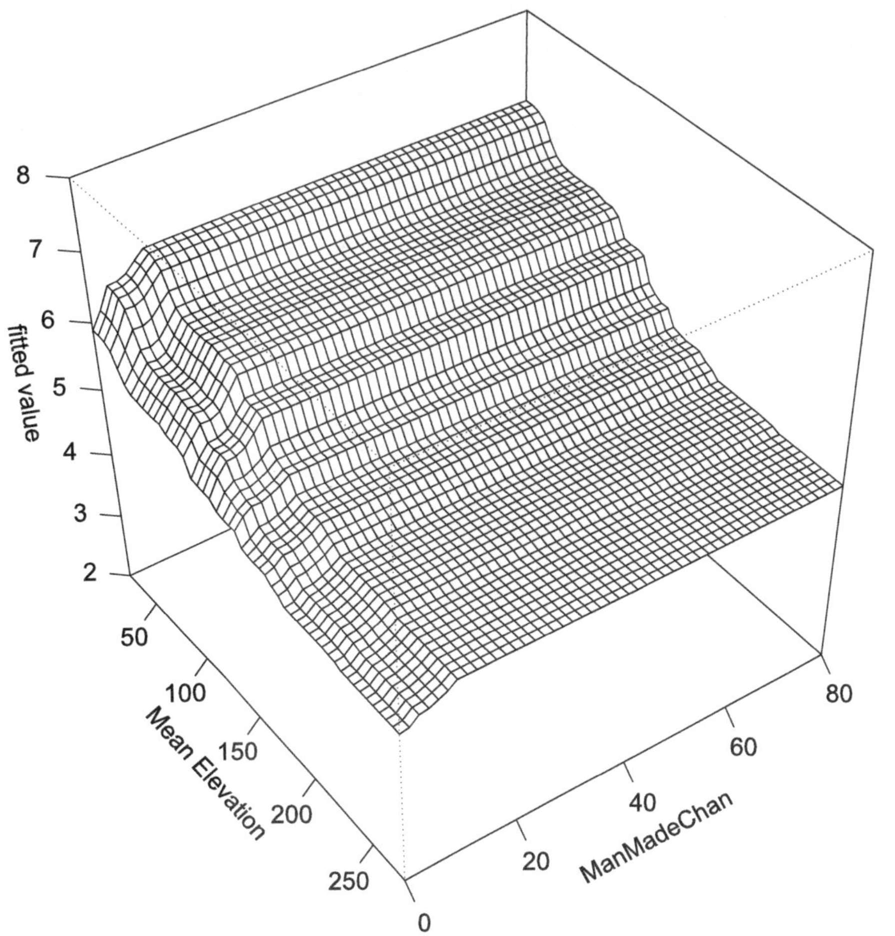

Figure 7. Interaction of manmade streams and mean elevation on RichTOL for Northern Piedmont BRT model. Boosted regression tree partial dependency plot shows the response form of average taxa tolerance (y-axis = fitted function of RichTOL) based on the effect of the interaction of two individual explanatory variables along the response variable (all other variable responses removed). There is a relatively strong interaction acting on RichTOL at low values of mean elevation and high values of percent manmade streams that cause high values of tolerant taxa to occur. This is a common pattern, higher urbanization occurring in the lower elevation valleys.

The average tolerance of all the taxa in a sample (RichTOL) generally had the best model performance (highest R^2) across all regions when compared to the other metrics. The richness of Ephemeroptera, Plecoptera, and Trichoptera taxa (EPTR), a common component of multimetric indices in the northeastern U.S., on the other hand, generally had the worst performance across all regions when compared to the other metrics. This is in contrast to what we found in development of models in the western U.S., where some measure of EPT richness and RichTOL were the two best response variables of all the macroinvertebrate metrics tested [11]. Most water resource agencies in the U.S. use a multimetric approach to evaluate the effect of anthropogenic stressors on aquatic systems [45], primarily because multimetric indices (MMIs) incorporate information from a number of metrics to provide a meaningful measure of overall biological condition. In addition, most state agencies that develop MMIs use some form of EPT as a component metric (e.g., percent EPT or EPT richness). Our finding that RichTOL was the most predictive metric is interesting (and to some degree confounding) because it may indicate that RichTOL could potentially be a highly useful component of state MMIs or BIBIs, especially given its strong predictive capability across all northeastern ecoregions studied. RichTOL, although sometimes considered in MMI development, is seldom used as a component of MMIs, more often a measure of total taxa richness or Hilsenhoff's Biotic Index [63] is used instead (e.g., New Jersey Impairment Score [40] and Pennsylvania Index of Biotic Integrity [64]). Bioassessment programs are continuously trying to improve criteria in support of MMI development and BRT models that explicitly show the utility of metrics such as RichTOL for predicting ecological changes along a disturbance gradient further support such efforts.

Because many of the variables used in the analysis did not follow a linear response form (see Figures 2–6), they are not likely to be identified as significant in multiple linear regression models, even when transformed. Therefore, it is possible that since MLR models assume linearity that they may sometimes underestimate the explanatory power of nonlinear relations in some variables. BRT is more robust and allows the inclusion of more variables in the model building phase than MLR, which permits easier testing for interaction effects and produces a list of variables explaining the importance of variation in the response variable. In addition, partial dependency plots from BRT can offer valuable insight into the pattern or form of the response based on select explanatory variables, thus improving model interpretation. Although BRT models have only recently begun to be applied in ecology, a number of researchers have shown the strength and promise of this

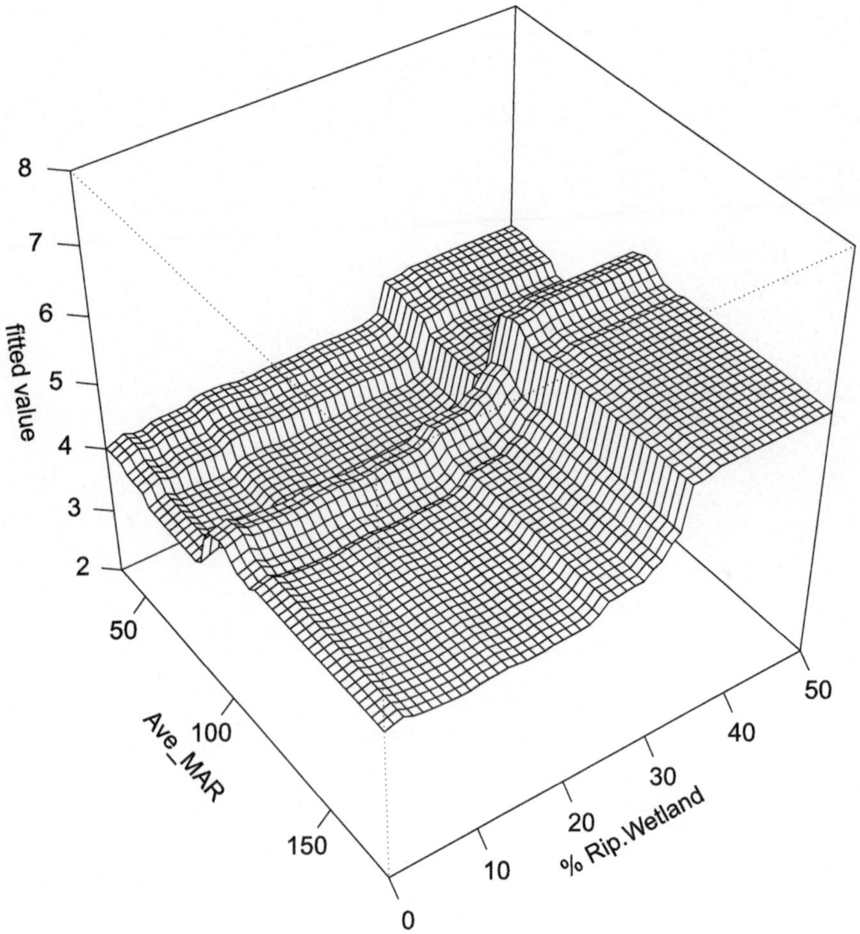

Figure 8. Interaction of riparian wetlands and average March runoff (mm) on RichTOL for N.E Highlands BRT model. Boosted regression tree partial dependency plot shows the response form of average taxa tolerance (y-axis = fitted function of RichTOL) based on the effect of the interaction of two individual explanatory variables along the response variable (all other variable responses removed). There is a relatively large interaction at high values of average March runoff when there are also high values of percent riparian wetland thus resulting in higher values of tolerant taxa (RichTOL) than would be expected. We believe that high values of riparian wetland are acting as a surrogate for high values of percent urban land use.

modeling technique compared to other methods [5],[17],[26]–[28],[46]. It is evident from the results of this and other studies that BRT represents a more powerful modeling technique than MLR (and many other methods) for evaluating the effects of human alteration of the environment on aquatic communities.

For this study, we were also interested in determining whether predictive models could be improved by breaking the sites up into three distinct watershed size classes. Recent studies have shown that human disturbance variables (i.e., land use and water quality) in combination with natural landscape factors such as climate, stream size, and elevation may help account for the greatest amount of variation in biotic indicators [5],[11],[46],[65]. Thus, to broaden our understanding of scale effects on macroinvertebrate response model development and prediction, we felt it important to take factors such as stream size into account as it is well established that processes along the river continuum, including variation in landscape and riparian disturbance, water quality, habitat, and streamflow, affect the distribution and abundance of macroinvertebrates ([29] and numerous other authors). However, BRT models developed for the individual watershed size classes did not greatly improve overall model performance when compared to the Full Region model results (Table 5). Two plausible explanations for this are (1) the range in watershed size classes (8.0–777.0 km^2) were not large enough to elicit a significant response and (2) changes along the stream size gradient tended to follow changes in landscape disturbance and elevation. Although stream size range was not extreme, it was large enough that we anticipated we would see an improved response; however, human landscape disturbance often follows an elevation gradient that may be acting as a surrogate for stream size. That is, smaller headwater streams tend to be in higher gradient areas (foothills and mountains) and larger streams are often in low elevation valleys [10],[11],[66]. However, we also tested model performance for three classes of elevation (mean elevation: 23–225, 225–450, and > 450 m) and similar to the result for models for watershed size classes, found little or no improvement compared to the Full Region models (data not shown). On its own, the range in elevation in this study did not seem to affect model performance; nevertheless, we believe there was interaction between elevation and land use variables that was important. This was exemplified in the PDPs, where there was interaction between elevation (and by extension, stream size) and percent man-made channels, a surrogate for urbanization. This interaction between elevation or other natural landscape variables and land use variables, which

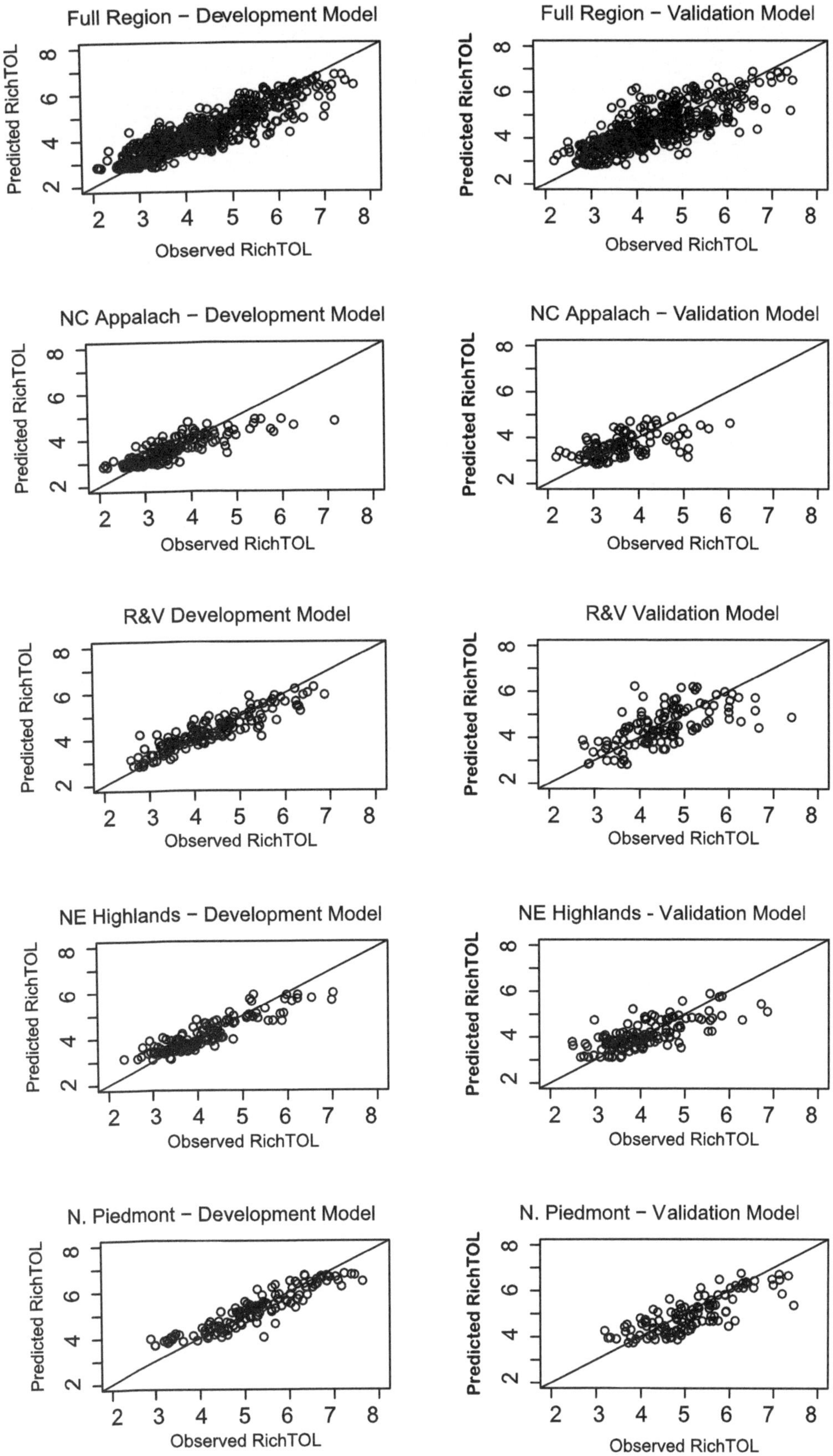

Figure 9. Observed versus predicted plots for BRT models for development (left) and validation (right) data sets. The observed versus predicted plots are based on the boosted regression models developed for average taxa tolerance (RichTOL) for five models: Full Region and four individual ecoregions (NC Appalachian, Ridge and Valley, NE Highlands, and N. Piedmont). The Full Region and N. Piedmont region plot relatively

tight to the 1:1 line for both the development and validation models indicating a good predictive fit with only slight bias at high and low values of RichTOL. The other regions in general showed more scatter and the N.C. Appalachian region which had the lowest modeled R^2, had had the shortest disturbance gradient (narrow range of RichTOL values) compared to the other regions.

was common in this study, has also been shown in many other recent studies [6],[10],[11],[19],[22]. In essence, the greater predictability anticipated by developing models based on watershed size classes, was largely obscured by the interaction between watershed elevation and human disturbance. In the Northern Piedmont, elevation and urbanization closely follow each other, so even though there is not a large range in elevation in this region, elevation was the most important variable because it also expresses the effects of urbanization.

Stream hydrology, as represented by the proportion of hydrologic runoff, was an integral component of many of the predictive BRT models developed for this study (Table 4). The relation between aquatic community response and changes in hydrologic processes is well established. Poff et al. [67] emphasized the fundamental importance of inter- and intra-annual hydrologic variation as the primary controlling factor for sustaining the ecological integrity of streams, and nearly two decades of hydroecological research (see review by [68]) have substantiated this connection. Previous predictive modeling studies in the western U.S. using BRT (e.g., [11],[17],[22]) did not include hydrologic variables in their models predicting macroinvertebrate metrics or BIBIs, respectively, due to a dearth of hydrologic data availability across the region of study. Although speculative, it may have been possible to improve model performance in these studies by the inclusion of hydrologic attributes, especially hydrologic attributes that are commonly altered by anthropogenic processes. Many of the hydrologic runoff variables identified in this study were an integral part of the models developed and were strong predictors of macroinvertebrate-assemblage composition. For example, 75.0% of the Full Region and 56.3% Individual Ecoregion models developed included a hydrologic runoff variable with a high degree of predictive power and high relative variable importance (Table 4). These findings are highly consistent with the results of other studies that point to changes in annual streamflow processes as being a significant driver of changes in assemblage structure and function (e.g., [7],[69–75]. In some cases, hydrologic runoff was the most important variable (e.g., Average March and Maximum April for the NE_High and N_Pied, respectively) and notably, many of these hydrologic runoff variables represent periods of annual streamflow variability that are important to the timing of macroinvertebrate emergence and reproduction (spring high and low flows; maximum and minimum April, average March, and maximum May runoff; Table 4). Changes in assemblage structure due to modified annual flow patterns have been commonly associated with a disruption or alteration of life-history or behavioral cues [76–78]. For example, emergence periods for more sensitive taxa (taxa with less plastic life histories) like Ephemeroptera, Plecoptera, and some Trichoptera may be affected by alterations in mean annual spring runoff. Therefore, it is no surprise that in three of five EPTR and four of five NoninsectR predictive models (Table 4), hydrologic runoff was of high relative importance and in a number of cases was either the first or second most important variable. This finding further emphasizes the importance of developing models that include streamflow attributes, which encapsulate variability across the full hydrologic regime, especially for parts of the hydroperiod that are critical to the health and survival of aquatic communities.

Implications

Even though spatial scale of the Full Region modeling area in this study was not extensive (i.e., it did not encapsulate the entire northeastern U.S.), it did include parts of four eastern U.S. states and contained four proximal ecoregions. In contrast to our initial hypothesis, however, we found that the Full Region models did almost as well in predicting macroinvertebrate structure as the more area specific Individual Ecoregion models. It may, therefore, be valuable for large regional models to be developed as a good preliminary assessment of major factors affecting the biological condition of streams in a geographic region as long as the range in natural landscape variability (e.g., climate, slope, elevation) can be minimized. The regional models can then be tested and compared against smaller subregional or watershed models to determine whether there is a large enough improvement in model performance or variable specificity to justify the general application of the regional models or more specificity is needed from local models.

Table 5. Comparison of model evaluation statistics for four macroinvertebrate metrics for three watershed size classes, number of variables in final model in parentheses.*

Macroinvertebrates	Model Type	Model Statistic	Full Region	WS Size Class 1 (8 – 27 km^2)	WS Size Class 2 (>27<66 km^2)	WS Size Class 3 (> 66<777 km^2)
			n = 591	n = 282	n = 188	n = 121
EPT Richness	BRT	R^2	0.63 (6)	0.64 (5)	0.69 (4)	0.65 (4)
(EPTR)		RMSE	2.34	2.68	2.30	2.18
Average Tolerance of	BRT	R^2	0.77 (5)	0.78 (4)	0.83 (4)	0.68 (4)
all Taxa (RichTOL)		RMSE	0.53	0.58	0.46	0.39
Richness of Intolerant	BRT	R^2	0.66 (5)	0.68 (4)	0.67 (4)	0.63 (4)
Taxa (INTOL_RICH)		RMSE	2.36	2.55	2.12	2.09
Noninsect Richness	BRT	R^2	0.68 (4)	0.77 (4)	0.73 (4)	0.72 (4)
(NonInsectR)		RMSE	1.39	1.28	1.25	2.09

*BRT – Boosted Regression Trees. R^2–adjusted R-squared, CV R^2—cross-validation R^2, EPTR–Total taxa richness of Ephemeroptera, Plecoptera and Trichoptera.

There is an obvious trade-off in development time and model specificity between the Full Region and Individual Ecoregion models. In some instances, the Full Region model may be sufficient for understanding broader implications of human-derived disturbance; however; smaller subregional or local models that provide more detailed information on cause-effect processes (e.g., exemplify the utility of specific metrics such as RichTOL) are generally more advantageous for bioassessment programs and may ultimately be more useful from a management or regulatory perspective. We cannot emphasize enough the importance and critical nature of riparian vegetation related to stream integrity revealed in this and other studies. Although our models were relatively successful in predicting macroinvertebrate metrics based on land use and hydrologic runoff, there may be additional explanatory variables that we did not use that could improve model prediction (other landscape data, instream habitat, water chemistry, etc.). These additional variables would not typically be available for all unsampled streams, where one of the benefits of the models developed in this study is that they can be used to develop maps of macroinvertebrate metric response at unsampled stream reaches while including some measure of error. This last point is where future modeling assessments could benefit both the science and management communities: development and testing of models predicting biological condition at unsampled streams under various scenarios. If possible, models should be tested on independent validation data sets to assess potential over-fitting and model bias.

Supporting Information

Figure S1 Partial dependency plots for variables in BRT model for the other three invertebrate metrics for all the Regions. Boosted regression tree partial dependency plots show the response form for invertebrate metrics (y-axis = fitted function of invertebrate metric) based on the effect of individual explanatory variables with the response of all other variables removed (development data set); variables shown in order of model importance. The relative contribution of each explanatory variable is reported in parentheses. Refer to Table 1 for variable definitions.

Figure S2 Observed versus predicted plots for BRT models for development (left) and validation (right) data sets for the other three invertebrate metrics. The observed versus predicted plots are based on individual boosted regression models developed separately for three invertebrate metrics for all five models: Full Region and four individual ecoregions (NC Appalachian, Ridge and Valley, NE Highlands, and N. Piedmont).

Acknowledgments

We thank the individuals that provided macroinvertebrate data and associated advice: Brian Chalfant, Pennsylvania Department of Environmental Protection; Chris Bellucci, Connecticut Department of Environmental Protection; AJ Smith, New York State Department of Environmental Conservation; and Dean Bryson; New Jersey Department of Environmental Protection. We also thank the many dedicated U.S. Geological Survey colleagues who acquired the biological samples used in this analysis. We gratefully acknowledge Dr. Yangdong Pan and Chris Parker (Portland State University) for advice on R programming and statistics. We also thank David Wolock (USGS) for developing the hydrologic runoff metrics and his advice in their interpretation. Finally, we thank one U.S. Geological Survey reviewer for suggestions that improved the manuscript prior to journal submission. Any use of trade, product, or firm names is for descriptive purposes only and does not imply endorsement by the U.S. Government.

Author Contributions

Conceived and designed the experiments: IW JK JM LB TC. Analyzed the data: IW. Contributed reagents/materials/analysis tools: IW. Wrote the paper: IW. Helped write the manuscript: JK. Helped analyze the data: JK JM TC KJ JO.

References

1. Leathwick JR, Rowe D, Richardson J, Elith J, Hastie T (2005) Using multivariate adaptive regression splines to predict the distributions of New Zealand's freshwater diadromous fish. Freshw Biol 50: 2034–2052.
2. Austin M (2007) Species distribution models and ecological theory: a critical assessment and some possible new approaches. Ecol Modell 200: 1–19.
3. Cabecinhaa E, Silva-Santosa P, Cortesb R, Cabral J (2007) Applying a stochastic-dynamic methodology (StDM) to facilitate ecological monitoring of running waters, using selected trophic and taxonomic metrics as state variables. Ecol Modell 207: 109–127.
4. Turak E, Ferrier S, Barrett T, Mesley E, Drielsma M, et al. (2010) Planning for the persistence of river biodiversity: exploring alternative futures using process-based models. Freshw Biol 56: 39–56.
5. Clapcott J, Young R, Goodwin E, Leathwick J, Kelly D (2011) Relationships between multiple land-use pressures and individual and combined indicators of stream ecological integrity. NZ Department of Conservation, DOC Research and Development Series 365. Wellington, New Zealand.
6. Cuffney TF, Kashuba R, Qian SS, Alameddine I, Cha Y, et al. (2011) Multilevel regression models describing regional patterns of invertebrate and algal responses to urbanization across the USA. J North Am Benthol Soc 30: 797–819.
7. Kennen JG, Riva-Murray K, Beaulieu KM (2010) Determining hydrologic factors that influence stream macroinvertebrate assemblages in the northeastern US. Ecohydrology 3: 88–106.
8. Oberdorff T, Pont D, Hugueny B, Chessel D (2001) A probabilistic model characterizing fish assemblages of French rivers: a framework for environmental assessment. Freshw Biol 46: 399–415.
9. Levin SA (1992) The Problem of Pattern and Scale in Ecology: The Robert H. MacArthur Award Lecture. Ecology 73: 1943–1967.
10. Allan JD (2004) Landscapes and riverscapes: The influence of land use on stream ecosystems. Annu Rev Ecol Evol Syst 35: 257–284.
11. Waite IR, Brown LR, Kennen JG, May JT, Cuffney TF, et al. (2010) Comparison of watershed disturbance predictive models for stream benthic macroinvertebrates for three distinct ecoregions in western US. Ecol Indic 10: 1125–1136.
12. Stevenson RJ, Rier ST, Riseng CM, Schultz RE, Wiley MJ (2006) Comparing effects of nutrients on algal biomass in streams in 2 regions with different disturbance regimes and with applications for developing nutrient criteria. Hydrobiologia 561: 149–165.
13. Potapova M, Charles DF (2007) Diatom metrics for monitoring eutrophication in rivers of the United States. Ecol Indic 7: 48–70.
14. Ode PR, Hawkins CP, Mazor RD (2008) Comparability of biological assessments derived from predictive models and multimetric indices of increasing geographic scope. J North Am Benthol Soc 27: 967–985.
15. Seelbach PW, Hinz LC, Wiley MJ, Cooper AR (2011) Use of multiple linear regression to estimate flow regimes for all rivers across Illinois, Michigan, and Wisconsin. Michigan Department of Natural Resources, Fish Res Report 2095, Ann Arbor.
16. Kennen JG, Chang M, Tracy BH (2005) Effects of landscape change on fish assemblage structure in a rapidly growing metropolitan area in North Carolina, USA. In: Brown LR, Gray RH, Hughes RM, Meador MR, editors. Effects of urbanization on stream ecosystems. Am Fish Soc Symp 47: 39–52.
17. Waite IR, Kennen JG, May JT, Brown LR, Cuffney TF, et al. (2012) Comparison of stream invertebrate response models for bioassessment metrics. J Am Water Resour Assoc 48: 570–583.
18. Paul MJ, Meyer JL (2001) Streams in the urban landscape. Annu Rev Ecol Syst 32: 333–365.
19. Cuffney TF, Zappia H, Giddings EMP, Coles JF (2005) Urbanization effects on benthic macroinvertebrate assemblages in contrasting environmental settings: Boston, Birmingham, and Salt Lake City. In: Brown LR, Gray RH, Hughes RM, Meador MR, editors. Effects of urbanization on stream ecosystems. Am Fish Soc Symp 47: 361–408.

20. Van Sickle J, Baker J, Herlihy A, Bayley P, Gregory S, et al. (2004) Projecting the biological condition of streams under alternative scenarios of human land use. Ecol Appl 14: 368–380.
21. Cao Y, Hawkins CP (2011) The comparability of bioassessments: a review of conceptual and methodological issues. J North Am Benthol Soc 30: 680–701.
22. Brown LR, May JT, Rehn AC, Ode PR, Waite IR, et al. (2012) Predicting biological condition in southern California streams. Landsc Urban Plan 108: 17–27.
23. Breiman L, Friedman JH, Olshen RA, Stone CJ (1984) Classification and regression trees. Wadsworth and Brooks/Cole, Monterey, California, USA.
24. De'ath G, Fabricius KE (2000) Classification and regression trees: a powerful yet simple technique for ecological data analysis. Ecology 81: 3178–3192.
25. Prasad AM, Iverson LR, Liaw A (2006) Newer classification and regression tree techniques: bagging and random forests for ecological prediction. Ecosystems 9: 181–199.
26. De'ath G (2007) Boosted trees for ecological modeling and prediction. Ecology 88: 243–251.
27. Elith J, Leathwick JR, Hastie T (2008) A working guide to boosted regression trees. J Anim Ecol 77: 802–813.
28. Aertsen W, Kint V, Van Orshoven J, Özkan K, Muys B (2010) Comparison and ranking of different modelling techniques for prediction of site index in Mediterranean mountain forests. Ecol Modell 221: 1119–1130.
29. Vannote RL, Minshall GW, Cummins KW, Sedell JR, Cushing CE (1980) The river continuum concept. Can J Fish Aquat Sci 37: 130–137.
30. Woods AJ, Omernik JM, Brown DD, Kiilsgaard CW (1996) Level III and IV ecoregions of Pennsylvania and the Blue Ridge Mountains, the Ridge and Valley, and Central Appalachians of Virginia, West Virginia, and Maryland: EPA/600/R-96/077, U.S. Environmental Protection Agency, National Health and Environmental Effects Research Laboratory, Corvallis, OR, 50 p.
31. Omernik JM (1987) Ecoregions of the conterminous United States. Map (scale 1:7,500,000). Annals of the Association of American Geographers 77: 118–125.
32. U.S. Environmental Protection Agency (1997) Descriptions of level III ecological regions for the CEC report on ecological regions of North America. Available: http://www.epa.gov/wed/pages/ecoregions/na_eco.htm#Downloads. Accessed 2013 Feb 4.
33. Phelps MG, Hoppe MC (Compilers) (2002) New York - New Jersey Highlands Regional Study: 2002 Update. United States Department of Agriculture, Forest Service, Northeastern Area State and Private Forestry, Newtown Square, PA 19073, NA-FR-02-03, 209 p. Available: http://na.fs.fed.us/pubs/stewardship/ny_nj_highlands02_lr.pdf. Accessed 2010 Oct 19.
34. U.S. Geological Survey (2007) National hydrography dataset (high resolution), digital data. Available: http://nhd.usgs.gov/data.html. Accessed 2010 Nov.
35. Brenden TO, Clark Jr RD, Cooper A, Seelbach PW, Wang L, et al. (2006) A GIS framework for collecting, managing, and analyzing multi-scale landscape variables across large regions for river conservation and management. In Hughes R, Wang L, Seelbach PW, editors. Landscape influences on stream habitats and biological communities. Am Fish Soc Symp 48: 49–74.
36. Brakebill JW, Wolock DM, Terziotti SE (2011) Digital hydrologic networks supporting applications related to spatially referenced regression modeling. J Am Water Resour Assoc 47: 916–932.
37. Moulton SR II, Kennen JG, Goldstein RM, Hambrook JA (2002) Revised protocols for sampling algal, invertebrate and fish communities as part of the National Water-Quality Assessment Program: U.S. Geological Survey Open-File Report 02-150, 75 p. Available: http://pubs.usgs.gov/of/2002/ofr-02-150/index.html. Accessed 2010 Nov 10.
38. Connecticut Department of Environmental Protection (CTDEP) (1999) Ambient monitoring strategy for rivers and streams: Rotating basin approach. Connecticut Department of Environmental Protection, Bureau of Water Management, Planning and Standards Division, Hartford, CT. Available: http://dep.state.ct.us/wtr/wq/rotbasinplan.pdf). Accessed 2010 Oct 20.
39. Pennsylvania Department of Environmental Protection (PADEP) (2012) Pennsylvania's Surface Water Quality Monitoring Network (WQN). Pennsylvania Department of Environmental Protection, Bureau of Point and Non-Point Source Management, Harrisburg, PA. 106 p. Available: http://www.elibrary.dep.state.pa.us/dsweb/Get/Document-91443/3800-BK-DEP0636.pdf. Accessed 2010 Oct 21.
40. New Jersey Department of Environmental Protection (NJDEP) (2007) Standard Operating Procedures Ambient Biological Monitoring Using Benthic Macroinvertebrates Field, Lab, Assessment Methods. Department of Environmental Protection. Available: http://www.state.nj.us/dep/wms/bfbm/download/AMNET_SOP.pdf. Accessed 2010 Oct 15.
41. New York State Department of Environmental Conservation (NYDEC) (2012) Standard Operating Procedure: Biological Monitoring of Surface Waters in New York State. New York State Department of Environmental Conservation, Division of Water; Albany, NY. 164 p. Available: http://www.dec.ny.gov/docs/water_pdf/sbusop12.pdf. Accessed 2010 Oct 17.
42. Bellucci CJ, Becker M, Beauchene M (2011) Characteristics of macroinvertebrate and fish communities from 30 least disturbed small streams in Connecticut. Northeastern Naturalist 18: 411–444.
43. Cuffney TF (2003) User's manual for the National Water-Quality Assessment Program invertebrate data analysis system (IDAS) software: Version 3. U.S. Geological Survey Open-File Report 03–172, Raleigh, North Carolina. Available: ftp://ftpext.usgs.gov/pub/er/nc/raleigh/tfc/IDAS/. Accessed 2011 Feb 14.
44. Cuffney TF, Bilger MD, Haigler AM (2007) Ambiguous taxa: effects on the characterization and interpretation of invertebrate assemblages. J North Am Benthol Soc 26: 286–307.
45. Barbour MT, Gerritsen J, Snyder BD, Stribling JB (1999) Rapid bioassessment protocols for use in streams and wadeable rivers: periphyton, benthic macroinvertebrates, and fish. 2nd edition. U.S. Environmental Protection Agency, Office of Water, EPA Report 841-B-99-002, Washington, D.C.
46. Cuffney TF, Brightbill RA, May JT, Waite IR (2010) Responses of benthic macroinvertebrates to environmental changes associated with urbanization in nine metropolitan areas. Ecol Appl 20: 1384–1401.
47. Leclere J, Oberdorff T, Belliard J, Leprieur F (2011) A comparison of modeling techniques to predict juvenile 0+ fish species occurrences in a large river system. Ecol Inform 6: 276–285.
48. Friedman JH (2001) Greedy function approximation: A gradient boosting machine. Ann Stat 29: 1189–1232.
49. Pineiro G, Perelman S, Guerschman JP, Paruelo JM (2008) How to evaluate models: Observed vs. predicted or predicted vs. observed? Ecol Modell 216: 316–322.
50. R Development Core Team (2010) R: a language and environment for statistical computing. R Foundation for Statistical Computing, Vienna, Austria. Available: http://www.R-project.org.
51. Waite IR (2013) Development and application of an agricultural intensity index to invertebrate and algal metrics from streams at two scales. J Am Water Resour Assoc 49: 431–448.
52. Riseng CM, Wiley MJ, Black RW, Munn MD (2011) Impacts of agricultural land use on biological integrity: a causal analysis. Ecol Appl 21: 3128–3146.
53. Dodds WK, Clements WH, Gido K, Hilderbrand RH, King RS (2010) Thresholds, breakpoints, and nonlinearity in freshwaters as related to management. J North Am Benthol Soc 29: 988–997.
54. Hilderbrand RH, Utz RM, Stranko SA, Raesly RL (2010) Applying thresholds to forecast potential biodiversity loss from human development. J North Am Benthol Soc 29: 1009–1016.
55. Qian SS, Cuffney TF (2012) To threshold or not to threshold? That's the question. Ecol Indic 15: 1–9.
56. Wang LJ, Lyons J, Kanehl P, Bannerman R (2001) Impacts of urbanization on stream habitat and fish across multiple scales. Environ Manage 28: 255–266.
57. Roy AH, Rosemond AD, Paul MJ, Leigh DS, Wallace JB (2003) Stream macroinvertebrate response to catchment urbanization (Georgia, U.S.A.). Freshw Biol 48: 1–18.
58. Morgan RP, Cushman SF (2005) Urbanization effects on stream fish assemblages in Maryland, USA. J North Am Benthol Soc 24: 643–655.
59. Walsh CJ, Fletcher TD, Ladson AR (2005) Stream restoration in urban catchments through redesignin stormwater systems: looking to the catchment to save the stream. J North Am Benthol Soc 24: 690–705.
60. Wang L, Brendon T, Seelbach P, Cooper A, Allan D, et al. (2008) Landscape based identification of human disturbance gradients and reference conditions for Michigan streams. Environ Monit Assess 141: 1–17.
61. Brown LR, Cuffney TF, Coles JF, Fitzpatrick F, McMahon G, et al. (2009) Urban streams across the USA: lessons learned from studies in 9 metropolitan areas. J North Am Benthol Soc 28: 1051–1069.
62. Carlisle DM, Falcone J, Meador MR (2008) Predicting the biological condition of streams: use of geospatial indicators of natural and anthropogenic characteristics of watersheds. Environ Monit Assess 151: 143–160.
63. Hilsenhoff WL (1987) An improved biotic index of organic stream pollution. Great Lakes Entomol 20: 31–39.
64. Pennsylvania Department of Environmental Protection (PADEP) (2002) A Benthic Index of Biotic Integrity for Wadeable Freestone Riffle-Run Streams in Pennsylvania. Pennsylvania Department of Environmental Protections, Division of Water Quality Assessment and Standards, Harrisburg, PA, 52 p. plus appendices. Available: http://files.dep.state.pa.us/Water/Drinking%20Water%20and%20Facility%20Regulation/WaterQualityPortalFiles/ibi_rifflerun2009.pdf. Accessed 2010 Oct 18.
65. Maloney KO, Weller DE, Russell MJ, Hothorn T (2009) Classifying the biological condition of small streams: an example using benthic macroinvertebrates. J North Am Benthol Soc 28: 869–884.
66. Munn MD, Waite IR, Larsen DP, Herlihy AT (2009) The relative influence of geographic location and reach-scale habitat on benthic invertebrate assemblages in six ecoregions. Environ Monit Assess 154: 1–14.
67. Poff NL, Allan JD, Bain MB, Karr JR, Prestegaard KL, et al. (1997) The natural flow regime: A paradigm for river conservation and restoration. BioScience 47: 769–784.
68. Poff NL, Zimmerman JKH (2010) Ecological responses to altered flow regimes: a literature review to inform environmental flows science and management. Freshw Biol 55: 194–20.
69. Poff NL, Allan JD (1995) Functional organization of stream fish assemblages in relation to hydrologic variability. Ecology 76: 606–627.
70. Clausen B, Biggs BJF (1997) Relationships between benthic biota and hydrological indices in New Zealand streams. Freshw Biol 38: 327–342.
71. Pusey BJ, Kennard MJ, Arthington AH (2000) Discharge variability and the development of predictive models relating stream fish assemblage structure to habitat in northeastern Australia. Ecol Freshw Fish 9: 30–50.
72. Bunn SE, Arthington AH (2002) Basic principles and ecological consequences of altered flow regimes for aquatic biodiversity. Environ Manage 30: 492–507.

73. Roy AH, Freeman MC, Freeman BJ, Wenger SJ, Ensign WE, Meyer JL (2005) Investigating hydrologic alteration as a mechanism of fish assemblage shifts in urbanizing streams. J North Am Benthol Soc 24: 656–678.
74. Chessman BC, Jones HA, Searle NK, Growns IO, Pearson MR (2010) Assessing effects of flow alteration on macroinvertebrate assemblages in Australian dryland rivers. Freshw Biol 55: 1780–1800.
75. Freeman MC, Buell GR, Hay LE, Hughes WB, Jacobson RB, et al. (2012) Linking River Management to Species Conservation using Dynamic Landscape-Scale Models. River Res Appl, Published online April 20, 2012. DOI: 10.1002/rra.2575.
76. Lytle DA (2002) Flash floods and aquatic insect life history evolution: evaluation of multiple models. Ecology 83: 370–385.
77. Boulton AJ (2003) Parallels and contrasts in the effects of drought on stream macroinvertebrate assemblages. Freshw Biol 48: 1173–1185.
78. Lytle DA, Poff NL (2004) Adaptation to natural flow regimes. Trends in Ecol Evol 19: 94–100.

4

Low-Intensity Agricultural Landscapes in Transylvania Support High Butterfly Diversity: Implications for Conservation

Jacqueline Loos[1]*, Ine Dorresteijn[1], Jan Hanspach[1], Pascal Fust[2], László Rakosy[3], Joern Fischer[1]

1 Institute of Ecology, Leuphana University, Lueneburg, Germany, **2** Organic Agricultural Science Group, University Kassel, Witzenhausen, Germany, **3** Department Taxonomy and Ecology, Babes-Bolay University, Cluj-Napoca, Romania

Abstract

European farmland biodiversity is declining due to land use changes towards agricultural intensification or abandonment. Some Eastern European farming systems have sustained traditional forms of use, resulting in high levels of biodiversity. However, global markets and international policies now imply rapid and major changes to these systems. To effectively protect farmland biodiversity, understanding landscape features which underpin species diversity is crucial. Focusing on butterflies, we addressed this question for a cultural-historic landscape in Southern Transylvania, Romania. Following a natural experiment, we randomly selected 120 survey sites in farmland, 60 each in grassland and arable land. We surveyed butterfly species richness and abundance by walking transects with four repeats in summer 2012. We analysed species composition using Detrended Correspondence Analysis. We modelled species richness, richness of functional groups, and abundance of selected species in response to topography, woody vegetation cover and heterogeneity at three spatial scales, using generalised linear mixed effects models. Species composition widely overlapped in grassland and arable land. Composition changed along gradients of heterogeneity at local and context scales, and of woody vegetation cover at context and landscape scales. The effect of local heterogeneity on species richness was positive in arable land, but negative in grassland. Plant species richness, and structural and topographic conditions at multiple scales explained species richness, richness of functional groups and species abundances. Our study revealed high conservation value of both grassland and arable land in low-intensity Eastern European farmland. Besides grassland, also heterogeneous arable land provides important habitat for butterflies. While butterfly diversity in arable land benefits from heterogeneity by small-scale structures, grasslands should be protected from fragmentation to provide sufficiently large areas for butterflies. These findings have important implications for EU agricultural and conservation policy. Most importantly, conservation management needs to consider entire landscapes, and implement appropriate measures at multiple spatial scales.

Editor: Robert B. Srygley, USDA-Agricultural Research Service, United States of America

Funding: The study was funded through a Sofja Kovalevskaja Award by the Alexander von Humboldt Foundation to Joern Fischer, financed by the German Ministry for Research and Education. The funders had no role in study design, data collection and analysis, decision to publish, or preparation of the manuscript.

Competing Interests: The authors have declared that no competing interests exist.

* Email: loos@leuphana.de

Introduction

Almost half of Europe's terrestrial surface consists of farmland, and many species, including rare and endangered ones, depend on farmland as habitat [1,2]. The loss of cultural-historic landscapes through intensification or abandonment of farming practices is causing declines of farmland biodiversity [1,3,4,5,6]. To effectively design conservation strategies, knowledge is needed about which variables influence species richness and distribution at different spatial scales [7,8,9].

In Western Europe, species loss in farmland has been associated with an increase of agricultural productivity [9,10,11], most likely caused by the use of agrochemicals [12] and the loss and fragmentation of semi-natural patches, especially grasslands [8,13]. In Eastern Europe, socio-economic conditions and land use have been rapidly changing since the breakdown of communism and accession of new member states to the European Union (EU) [14,15,16]. Current changes involve a dual threat to biodiversity, with a trend towards structural simplification on the one hand and abandonment of low-intensity practices on the other hand [17,18]. The current situation in Eastern Europe thus differs in important ways from Western European countries [1,19,20], and a better understanding is needed of how organisms respond to landscape features within low-intensity farming areas of Eastern Europe.

Heterogeneous landscapes typically harbour greater species richness than homogenous landscapes [3,21,22], most likely because of their greater niche diversity, as well as spillover effects and habitat complementation [23]. Agricultural simplification and land abandonment typically lead to a loss of landscape connectivity, which may not only dissect the habitats for species, but also causes flow-on effects on the composition and configuration of the landscape as a whole [24,25].

A particularly interesting cultural-historic region in Eastern Europe is Transylvania, which supports extraordinarily high levels of farmland biodiversity [26,27]. Especially in its South, Transylvania is characterised by a small-scale mosaic of different low-intensity land-uses that provide many different, well-connected structures such as field margins and roadside vegetation. The historic management of the area has created heterogeneity at multiple spatial scales: within tens of metres (hereafter termed the local scale), in the immediate surroundings around any given location (the context scale), as well as over thousands of metres (the landscape scale) [28,29].

Here, we focus on butterflies as a taxonomic group that rapidly responds to environmental changes [30] and is known to be sensitive to land use change worldwide [4]. In Europe, many butterflies use anthropogenic landscape elements [31], but species with different traits are expected to respond differently to land use change [8,32]. For example, Öckinger & Smith [33] found that the effects of landscape composition differed between species of different mobility classes, and Börschig et al. [34] found that intensively used agricultural landscapes mostly support generalists. Yet, evidence on the responses of butterflies to gradients of spatial heterogeneity is sparse, and more thorough studies at multiple scales are needed [22,35].

We sought to understand the responses of butterfly diversity to key landscape gradients in Southern Transylvania, using a snapshot natural experiment [36,37] that spanned the full range of environmental conditions with respect to heterogeneity and woody vegetation cover across multiple scales. Our overarching aim was to understand drivers of species richness and composition. Specifically, we asked (i) how landscape structures affect the composition of butterfly communities; (ii) which landscape structures explain butterfly species richness at various spatial scales; and (iii) which landscape structures affect abundance patterns of selected species. We discuss our findings in the context of possible landscape changes that may take place in Transylvania.

Materials and Methods

Ethics Statement

We obtained the necessary permit for surveying butterflies within the EU *Natura 2000* network from Progresul Silvic, the organization officially entrusted with the custody of the protected area by the Romanian government. The survey procedure was approved beforehand by the ethics committee of Leuphana University Lueneburg.

Data Availability Statement

All data underlying the findings reported in this study are available from the Dryad Digital Repository.

Study area and experimental design

The study area covered approximately 7,000 km^2 in the lowlands of Southern Transylvania, Romania (Figure 1). We followed the notion of a natural experiment [36], with randomised site selection in pre-defined strata at two levels: study villages and survey sites within villages.

To select study villages, we first allocated each raster pixel of the study area to different "village catchments". These were calculated using a cost-distance algorithm in ARCGIS with the village centre as the reference point and the slope and the distance to the next village as the cost variables. Information about village locations was extracted from CORINE land cover data 2006 (http://www.eea.europa.eu/data-and-maps/data#c12=corine+land+cover+version+13&b_start=0&c17=CLC2006), and slope was calculated from the digital elevation model ASTER (Advanced Spaceborne Thermal Emission and Reflection Radiometer). Topographically based village catchments were used instead of administrative boundaries because administrative boundaries were only available at the commune level (typically 3–5 villages). However, we found that the resulting polygons accurently reflected historical land use responsibilities. Second, we stratified village catchments along a gradient of terrain ruggedness and according to their protection status under the EU Birds and Habitats Directives. Terrain ruggedness was calculated as the standard deviation of the altitude of the catchment, and we used quantiles to classify ruggedness as either low, medium or high. Protection status of the catchments was either unprotected, SCI (Site of Community Importance) or SPA (Special Protection Area). Third, we randomly chose 30 villages, covering all combination of ruggedness and protection status (Table S1).

To select survey sites, we stratified the agricultural area within these 30 villages according to CORINE land cover as grassland or arable land and excluded other land cover classes. Within these strata, we spanned two gradients that we assumed sensitive to change in the future as a result of structural simplification, namely woody vegetation cover and heterogeneity. We estimated woody vegetation cover in a circular one hectare area based on classified 10 m SPOT data (CNES, ISIS programme). To assess heterogeneity, we used the standard deviation of 2.5 m panchromatic SPOT data within a one hectare circle. We assigned each hectare of the agricultural landscape to a combination of three classes of woody vegetation cover by three classes of heterogeneity. We distinguished low (0–5%), medium (>5–15%) and high (>15%) woody vegetation cover and used the lower, middle and upper third of percentiles to classify heterogeneity. Within these combinations, we randomly selected replicates for each cross-combination (except for the combination of high heterogeneity and low woody vegetation cover, which did not exist (Table S2)). In total, we selected 120 circular 1 ha survey sites, with 60 in grassland and 60 in arable land, and an average of four survey sites per village catchment. Notably, sites in arable land in this context were consciously placed not to represent only arable fields specifically, but rather to capture the whole range of conditions within the mosaic of arable land [38], including field margins and fallow land.

Data collection

Butterfly surveys (response data). We assessed species richness and abundance of butterflies (Rhopalocera) and diurnal burnet moths (Zygaenidae) by walking four transects of 50 m length per survey site [39]. We included burnet moths because they are comparable to butterflies in their ecology [33,40]. These transect pointed north, east, south and west, and started 6 m from the centre of a given site. In a given transect walk, each butterfly observed within 2.5 m of each side of the transect and 5 m in front of the observer was identified and counted. Species that we could not identify in the field were treated as compound species: *L. sinapis/juvernica*, *C. alfacariensis/hyale* and *Zygaena minos/purpuralis*. *Adscita*, *Jordanita* and *Carcharodus* occurred within the study region, and are represented by two, two and three species, respectively [41]. However, these species are difficult to distinguish and therefore were only identified to the genus level. Surveys were repeated on four occasions between May and August 2012 by four different, trained observers. Surveys were conducted under suitable weather conditions (no rain, <90% cloud cover, > 17°C, no strong wind), between 9 am and 5 pm.

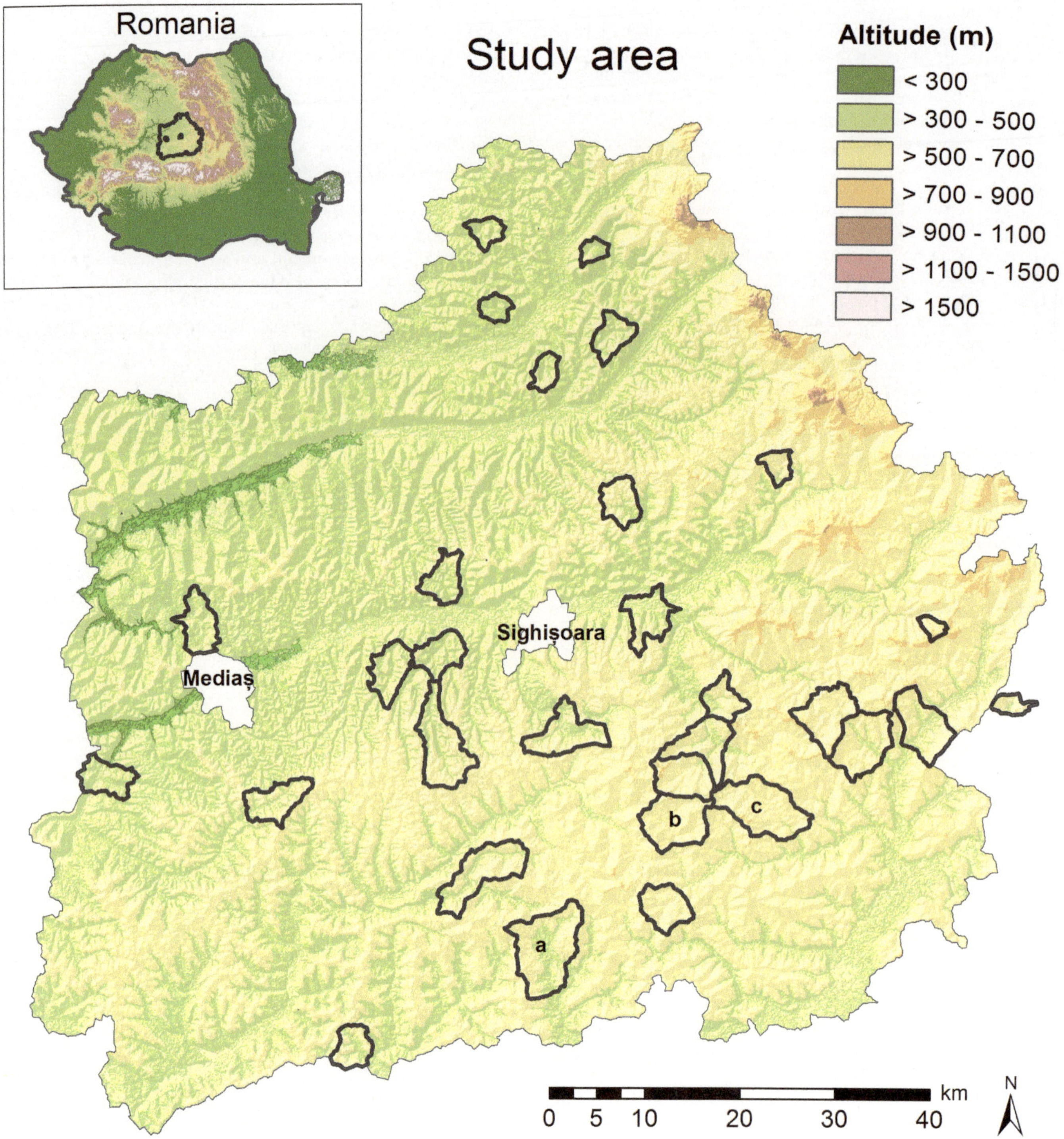

Figure 1. Location of the study area with investigated village catchments in Transylvania, Romania. The small letters indicate the village catchments illustrated for predictions in Figure 4 (a = Cincu, b = Granari, c = Viscri).

Environmental data (explanatory variables). We followed a multi-scale approach and included explanatory variables that could potentially explain butterfly distribution at the local (1 ha), context (50 ha) and landscape scale (i.e. village catchments, ranging from 430 to 4963 ha). An overview of all variables included in the analysis is presented in Table 1.

At the local scale, we collected data on vascular plant species richness in eight randomized quadrants (1×1 m). We used cumulative plant species richness per site as an explanatory variable. We also calculated indices for heatload (after [42]) and terrain wetness as a measure of potential soil wetness, and included heterogeneity assessed by the spectral variance of SPOT data (see Table 1 for details). We calculated percent woody vegetation cover at local and context scales, and used CORINE land cover to calculate percent forest at the landscape scale. For the context and landscape scales, we calculated the terrain ruggedness as the

Table 1. Definition of environmental variables used in the study at three different scales and method of obtaining those. Abbreviations are used in Figure 2 and Table 2.

Scale	Variable (abbreviation)	Definition and method
local (1 ha)	Number of plants species (NoPlant)	Vascular plant species richness assessed by eight randomly distributed quadrants of one by one meter
	Heterogeneity (het_1 ha)	Heterogeneity measured as the standard deviation of 2.5 m panchromatic SPOT picture (CNES, ISIS programme)
	Woody vegetation cover (woody_1 ha)	Proportion of woody vegetation cover based on classified 10 m SPOT satellite image (CNES, ISIS programme)
	Heat index (heatload)	Potential for ground heating calculated after Parker [42]: Heat index = cos (slope aspect − 225) * tan (slope angle)
	Terrain Wetness Index (TWI)	Measure of potential soil wetness, estimated as the position in the landscape and the slope from ASTER digital elevation model with 30 m resolution.
	Land Cover (LU_type)	Land use classification as arable land, grassland or forest based on CORINE land cover
context (50 ha)	Ruggedness (rugg_50 ha)	Terrain ruggedness, calculated as standard deviation of altitude
	Woody vegetation cover (woody_50 ha)	Proportion of woody vegetation cover based on classified 10 m SPOT satellite image
	Configurational heterogeneity (ED_50 ha)	Configuration of different land covers, calculated as the edge density with FRAGSTATS v4.2 based on CORINE land cover
landscape (village catchment)	Amount of pasture (past_catch)	Proportion of pasture, based on CORINE land cover
	Woody vegetation cover (woody_catch)	Proportion of forest cover based on CORINE land cover
	Ruggedness (catch_rugg)	Terrain ruggedness, calculated as the standard deviation of the altitude
	Compositional heterogeneity (SIDI)	Composition of different land covers, calculated as Simpson index of diversity with FRAGSTATS v4.2 based on CORINE land cover
	Configurational heterogeneity (ED)	Configuration of different land covers, calculated as edge density with FRAGSTATS v4.2 based on CORINE land cover
Random effects	Village catchment	Classification of the landscape into social-ecological units according to a cost distance algorithm of proximity to the nearest village as reference point and the slope of the terrain as cost factor
	Level	Observation level random effect

standard deviation of altitude. We also quantified compositional or configurational heterogeneity of the different land covers grassland, arable land and forest as provided by CORINE land cover data. At the context scale, our chosen heterogeneity measures (Simpson index of land cover diversity, edge density) were correlated ($r = 0.76$). Hence, we included only edge density as an explanatory variable (following [7]). At the landscape scale, we used both edge density and the Simpson index of diversity and added the amount of pasture and forest per village catchment, based on CORINE land cover data. Variables on compositional and configurational heterogeneity were calculated using FRAGSTATS v4.2 [43] and all other variables using ARCGIS 10.1 (ESRI Inc., Redland, CA).

Analysis

We pooled all observed butterfly species and individuals from the four survey rounds for each survey site. First, we tested for differences in species richness and abundance between different levels of official protection by using Analysis of Variance (ANOVA). Second, we conducted a detrended correspondence analysis (DCA) to describe species composition and its relation to environmental variables. We used a permutation test to fit and test the correlation of environmental variables with the ordination.

Third, we used generalized linear mixed effects models (GLMMs) to assess effects of environmental variables on butterfly species richness. Beforehand, we tested the explanatory variables for collinearity (all $r<0.7$; Table S4; [44]). We log-transformed woody vegetation cover at local and context scales and heterogeneity at the local scale because these variables were highly skewed. All numerical explanatory variables were scaled to mean zero and unit variance. We included the variables listed in Table 1 to model species richness of butterflies. To test for a unimodal relationship in response to woody vegetation cover, we included a quadratic term of local woody vegetation cover. We furthermore expected that the effect of heterogeneity may differ between grassland and arable land and therefore included an interaction term between land cover type and heterogeneity. Grasslands are also interesting to look at separately because they are among the most species rich biotopes for butterflies in Europe [45]. We assessed the variance inflation factor (VIF) of the generalized linear model (GLM) and tested for spatial auto-correlation in the residuals. We included the village catchment as a random effect and corrected for over-dispersion by adding an observation level random effect. We simplified the model by stepwise backwards selection retaining all variables with $p<0.1$. For GLMMs, significance levels are only approximations, hence many statisticians suggest using a significance level of $p<0.1$ [46].

Likewise, we modelled species richness of functional groups. To this end, we distinguished between species of low mobility (Bink's mobility classes 1 and 2) and high mobility (Bink's mobility classes 7, 8 and 9; [47]). Highly mobile species were *Colias crocea*, *Pieris brassicae*, *Vanessa atalanta* and *Vanessa cardui*. Low-mobility species were *Brenthis daphne*, *Brenthis ino*, *Coenonympha glycerion*, *Cupido minimus*, *Euphydryas aurinia*, *Hamaeris lucina*, *Heteropterus morpheus*, *Lopinga achine*, *Melitaea britomartis*, *Melitaea diamina* and *Satyrium acaciae*. As a third group we modelled the richness of grassland specialists, namely *Euphydryas aurinia*, *Polyommatus coridon*, *Cyaniris semiargus*, *Lysandra bellargus*, *Phengaris arion*, *Cupido minimus* and *Erynnis tages* [48].

We also modelled the abundance of individual species considered to be declining in Western and Northern Europe, but that are widespread or even increasing in Eastern Europe [48,49,50,51]. We only used species that were common enough in the study area to obtain reliable models, namely *Maniola jurtina*, *Coenonympha pamphilus*, *Polyommatus Icarus*, *Lycaena dispar* and *Glaucopsyche alexis*. We performed all statistical analyses in R [52], using the packages MASS, ade4, vegan, gdata and lme4.

Results

In total, we counted 19,878 individuals of 112 species of butterflies (Table S3). Site-level species richness varied between three and 45, and the number of individuals between seven and 452. Eighty-five percent of all individuals belonged to 12 species: *Colias alfacariensis/hyale, Minois dryas, Aphantopus hyperantus, Pieris rapae, Everes argiades, Coenonympha glycerion, Leptidea sinapis/juvernica, Melanargia galathea, Coenonympha pamphilus, Maniola jurtina, Polyommatus icarus,* and *Plebeius argus*. SCI, SPA and unprotected sites did not differ in species richness ($F = 0.54$, $p = 0.58$) but SCI sites appeared to have a slightly lower abundance of individuals than unprotected sites ($F = 2.37$, $p = 0.09$). Arable land and grassland did not differ in species richness ($F = 1.32$, $p = 0.25$) nor abundance of individuals ($F = 1.51$, $p = 0.22$).

Multivariate analysis showed substantial overlap in species composition between arable land and grassland (Figure 2), with less than one complete species turnover (length of first axis = 2.9). The first axis (Eigenvalue = 0.21) described a gradient from sites with a low terrain wetness index in homogenous landscapes to sites with a high terrain wetness index within highly heterogeneous landscapes. The second DCA axis (Eigenvalue = 0.18) described a gradient from low to high richness of vascular plants, ruggedness, woody vegetation cover and context-level heterogeneity and landscape-level woody vegetation cover.

Butterfly species richness was positively related to local plant species richness and local woody vegetation cover, but negatively to local heatload (Table 2). It increased in response to local heterogeneity in arable sites, but not in grasslands (Figure 3). Species richness furthermore increased with configurational heterogeneity and ruggedness at the context scale, but decreased with landscape woody vegetation cover. The models show suitable areas for species of conservation interest exist throughout village catchments, especially in large grassland areas and boundary areas of arable land (Figure 4).

Species richness of mobile butterflies was highest in arable land, and responded positively at the landscape scale to both compositional heterogeneity and ruggedness. By contrast, richness of low-mobility species was negatively related to landscape configurational heterogeneity, but responded positively to local-scale plant species richness and context heterogeneity (for additional details, see Table 2). Richness of specialist species was higher in grassland, in landscapes with high terrain ruggedness and at sites with high plant species richness.

For individual species, both *L. dispar* and *G. alexis* were more abundant in arable land, and were positively related to local plant species richness. *L. dispar* also responded positively to local woody vegetation cover, but negatively to local heatload, whereas *G. alexis* showed a positive response to context ruggedness and the amount of grassland in the landscape. The abundances of *P. icarus, M. jurtina* and *C. pamphilus* increased with heterogeneity in arable land, but not in grassland, and decreased with increasing heatload. Abundance of *M. jurtina* and *C. pamphilus* were positively related to local plant species richness, and negatively to landscape woody vegetation cover. *P. icarus* responded positively to the amount of grassland in the landscape. Abundance of *C. pamphilus* was unimodally related to local woody vegetation cover.

Discussion

We found a high diversity of butterflies in the cultural-historic landscape of Southern Transylvania. This is especially the case considering that we did not seek out sites expected a priori to harbour great diversity, but rather surveyed randomly selected sites within the agricultural matrix. An even greater diversity of butterflies, including rare and endangered species, would be expected to occur in dry grassland patches and traditionally managed hay-meadows, which occur within our study area but which we did not specifically target. Our findings suggest that some types of land use change could pose serious threats to butterfly diversity in Transylvania. Our findings can be summarised within four themes, which we discuss in the following: (i) both grassland and arable land have conservation value; (ii) low-intensity landscapes provide important resources for butterflies; (iii) heterogeneity has a different effect in arable land than in grassland; and (iv) it is important to consider multiple scales for effective butterfly conservation.

Both grassland and arable land have conservation value

Our findings revealed a high conservation value for butterflies of the small-scale farming system in the lowlands of Transylvania. Interestingly, butterfly species richness and abundance were similar in arable land and grassland. This is a surprising result and suggests a need to broaden the emphasis of conservation activities from grassland protection towards the maintenance of heterogeneous mosaic farmland, including cropland [38]. This is particularly important in the context of criticisms that the recent reform of the European Union's Common Agricultural Policy, for example, falls far short of what is needed in terms of biodiversity conservation [53]. Throughout Europe, grasslands are considered most important for butterfly conservation (e.g. [8,45]). Arable land, on the other hand, has received far less attention. In Western Europe, arable land has been found to support lower species richness and more homogenous butterfly communities than grassland [9,54]. Our results indicate that this situation may be different in Eastern Europe, and that certain types of arable land can in fact support similar levels of butterfly diversity as grasslands. A possible explanation for the similar species richness in arable land and grassland in Transylvania may be spillover effects [23], which may be more likely in small-scale mosaics of land covers. The mosaic character of the landscape also could explain the strong overlap in butterfly communities between arable land and grassland.

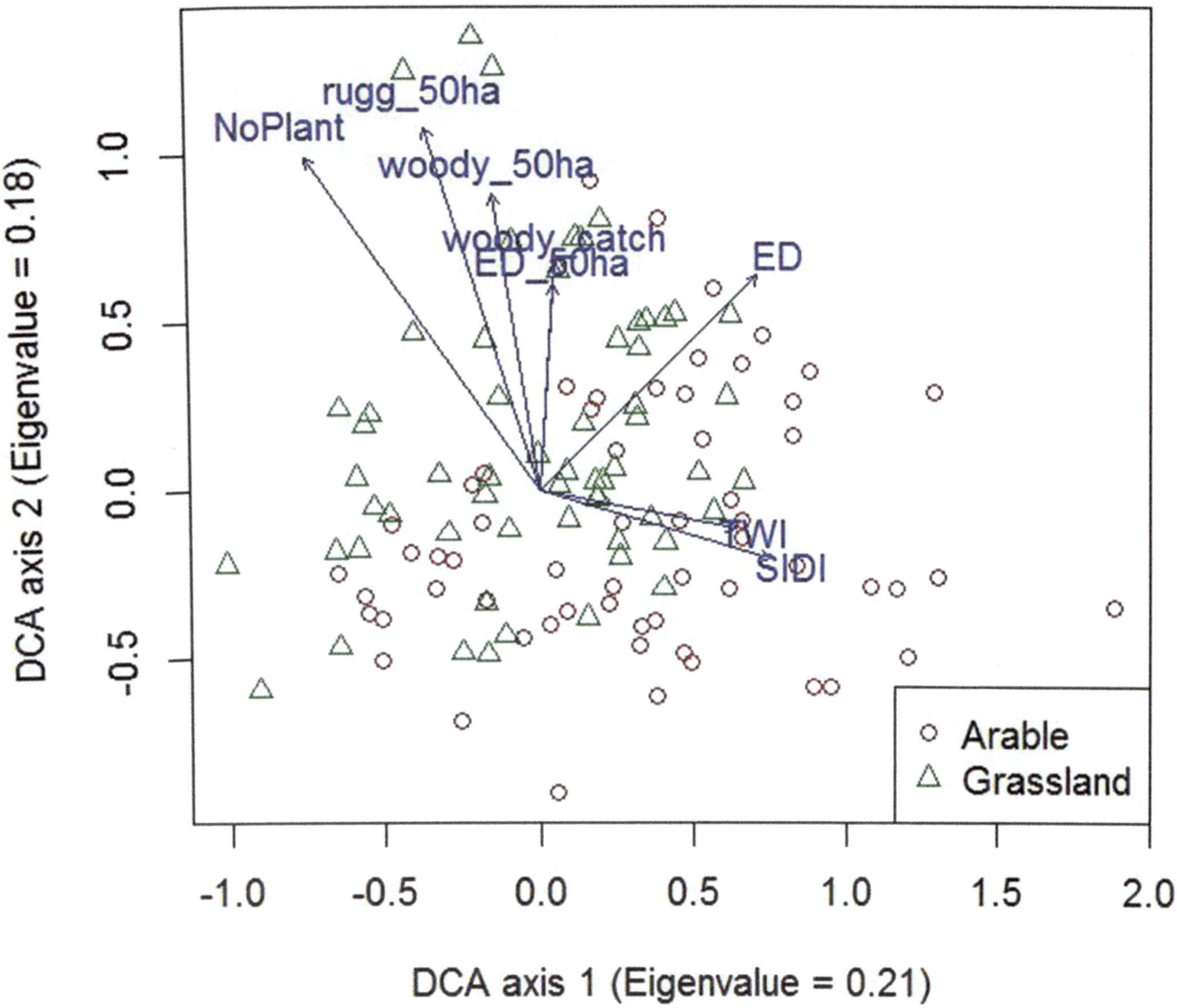

Figure 2. DCA ordination plot of butterfly species, with significant environmental variables superimposed ($p<0.05$) (Abbreviations: NoPlant = Local plant species richness; TWI = Local terrain wetness index; rugg_50 ha = context terrain ruggedness; woody_50 - ha = context woody vegetation cover; ED_50 ha = context edge density; woody_catch = landscape woody vegetation cover; SIDI = landscape compositional heterogeneity; Table 1).

Low-intensity landscapes provide important resources for butterflies

The fine-grained mosaic nature of arable land and the low-intensity nature of grassland in Southern Transylvania emphasize that low-intensity land use practices have major benefits for butterfly conservation. Semi-natural elements occur throughout the landscape, and are a likely reason why species richness is high throughout different land covers [22]. Furthermore, species richness of vascular plants can be high in field margins, which in turn may indicate high quality habitat for butterflies [55]. Consistent with the findings of Kumar, Simonson & Stohlgren [7], we found plant species richness strongly related to butterfly species richness. Currently, Transylvania contains some of the world's most species rich areas for plants [56], which is partly linked to the low use of fertilizers [57]. Agricultural intensification, by contrast, would likely lead to increased use of fertilizers and hence reduced plant species richness [58,59,60]. Furthermore, intensification is typically associated with the use of fewer, high yielding crop varieties. Interestingly, many butterflies in Transylvania use the common crop *Medicago sativa ssp. sativa* (Alfalfa), a leguminous species that provides nectar and that we also observed to serve as a host plant for several butterfly species (e.g. *Glaucopsyche alexis*). Alfalfa is grown in small parcels, is primarily used as winter fodder for livestock, and may easily be lost as a result of intensification. However, high amounts of floral resources are critically important to maintain butterfly diversity. Similarly, woody vegetation offers important resources for butterflies, including shelter and space for thermoregulation [61]. At present, Transylvania contains many scattered trees and hedgerows, and we found that butterfly species richness responded positively to these structures at the local scale. By contrast, a large amount of

Table 2. Parameter estimates of the species distribution models with significance levels indicated by: †P<0.1; *P<0.005; **P<0.01; ***P<0.001.

	Species Richness	High mobile species	Low mobile species	Specialists	*L. dispar*	*G. alexis*	*P. icarus*	*M. jurtina*	*C. pamphilus*
Intercept	3.026	−0.739	0.681	−0.436	0.581	−1.172	2.914	2.637	1.959
NoPlant	0.261***		0.600***	0.685***	0.581*	0.941**		0.597***	0.313***
LU_type	−0.243**	−1.250***		0.052†	−2.059***	−1.554*	−0.040	0.046	0.210
het_1 ha	0.109*						0.278**	0.235*	0.224*
LU_type*het_1 ha	−0.140*						−0.326*	−0.415**	−0.387*
TWI									
woody_1 ha	0.072*		−0.054		0.443†				0.102
woody_1 ha^2			0.232†						0.177*
Heatload	−0.057*				−0.622†		−0.167*	−0.321***	−0.207**
rugg_50 ha	0.064†					0.511*			
woody_50 ha									
ED_50 ha	0.077*		0.261*						
woody_catch	−0.079*							−0.412***	−0.232†
past_catch						0.721**	0.256.		
rugg_catch		−0.423*	0.249*	0.051**					
ED			−0.374**						
SIDI		0.448*							

Arable land was used as the baseline land cover in all models. See Table 1 for abbreviations.

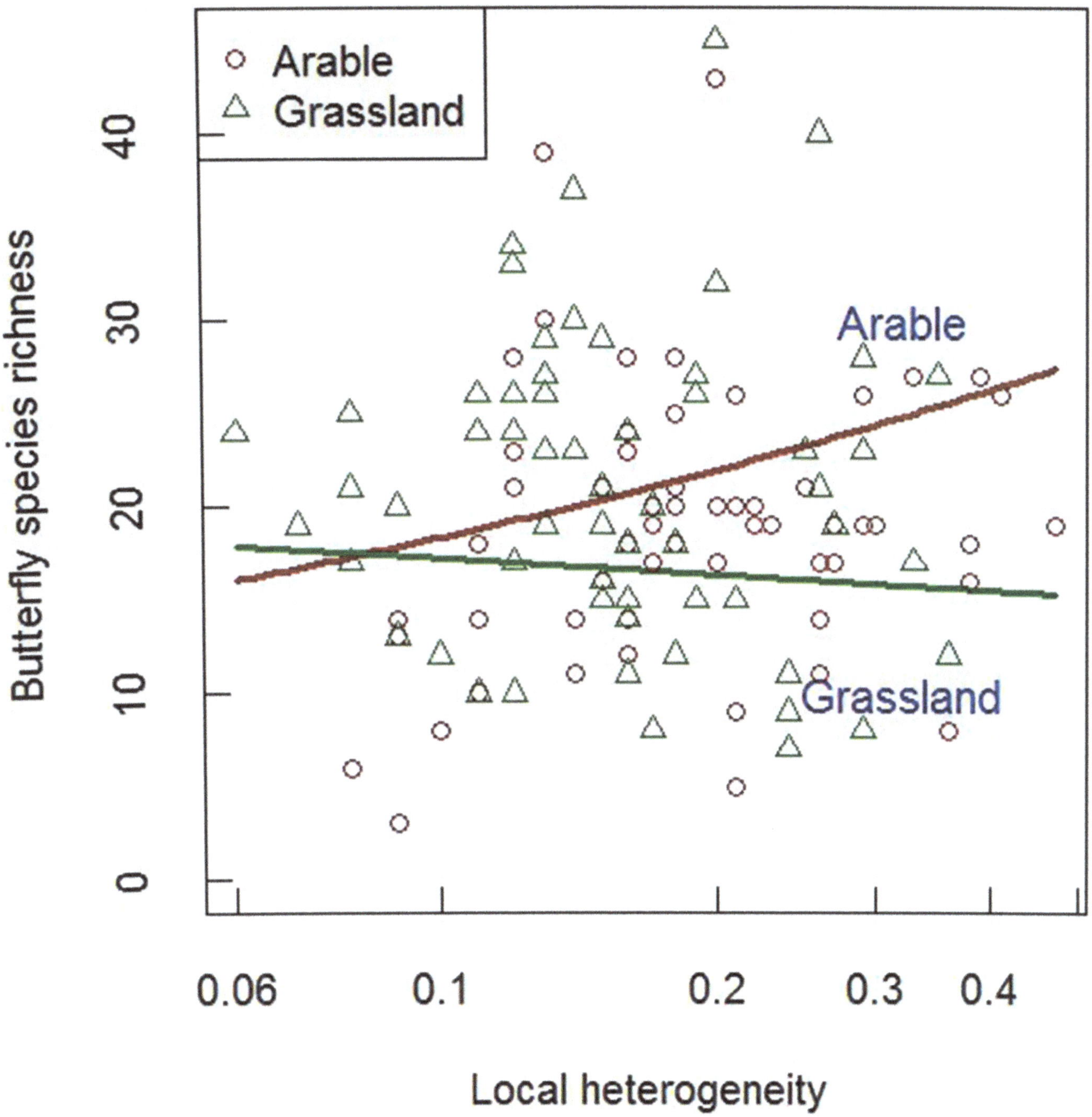

Figure 3. Predicted effect of local heterogeneity on species richness in arable land versus grassland, based on the simplified generalized linear mixed model (Table 2).

woody vegetation at the landscape scale may lead to decreased species richness, probably due to a lack of open habitat.

Heterogeneity has a different effect in arable land than in grassland

We considered heterogeneity and woody vegetation cover at the local scale as two potentially important gradients describing the structure of the landscape. Interestingly, our results showed that the effect of local heterogeneity on species richness depended on land cover. In arable land, species richness increased with heterogeneity, supporting our hypothesis that small-scale farming benefits biodiversity by providing a range of different resources for butterflies. Notably, our land use class of "arable land" reflected the highly heterogeneous nature of traditional farmland, and included cropped areas as well as fallows and uncultivated field margins. These non-cropped areas are likely to be particularly important to maintain butterfly diversity in arable land. By contrast, in grassland, high heterogeneity was associated with reduced butterfly diversity. A possible explanation for this pattern is that heterogeneity of grassland may correspond to a higher

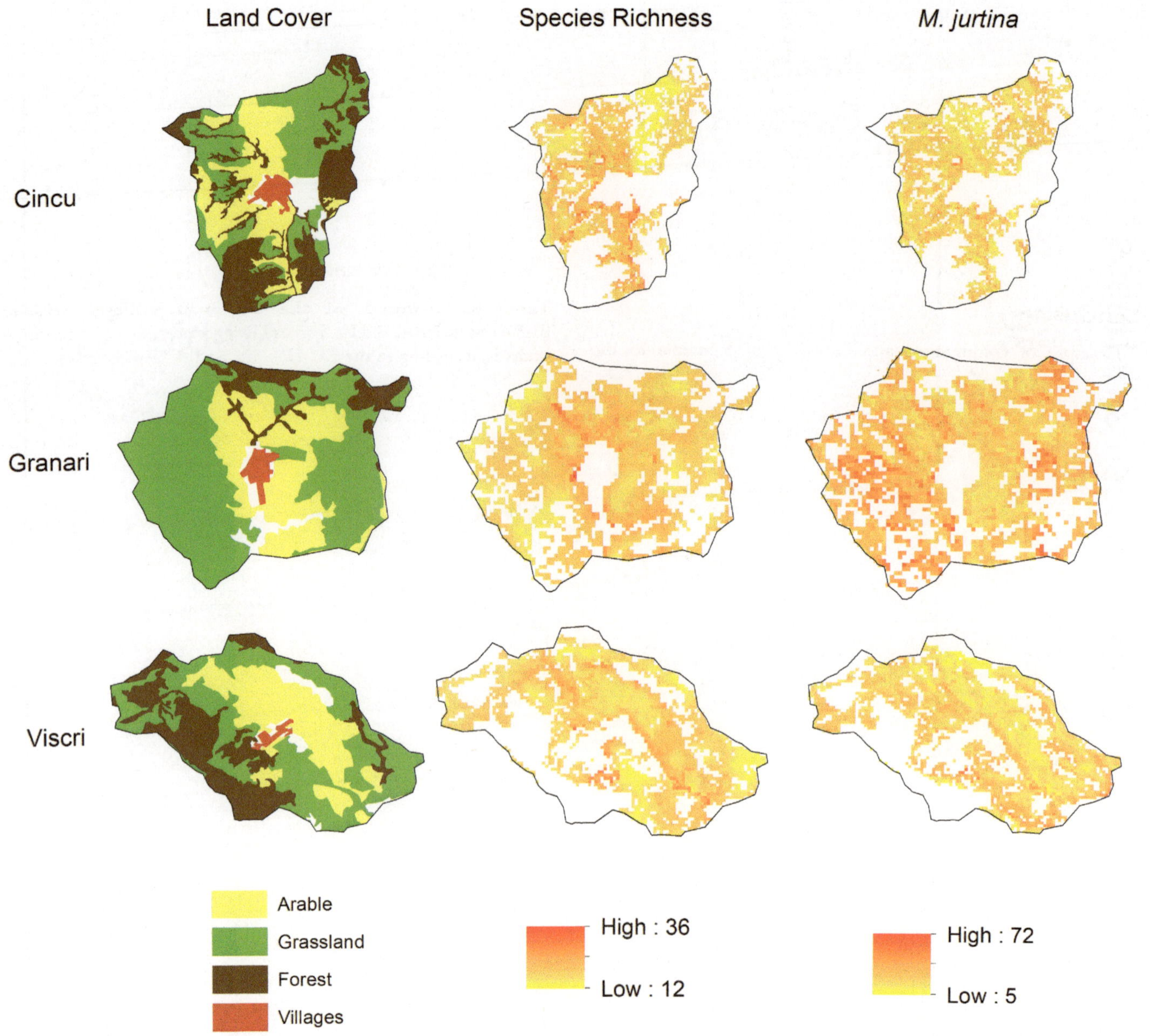

Figure 4. Maps of predicted butterfly distributions in three example villages. Left: Land cover map according to CORINE 2006; middle: predicted species richness for arable and grassland areas within each village catchment; right: predicted abundance of the Meadow Brown (*Maniola jurtina*).

degree of fragmentation of butterfly habitat, with likely negative consequences for species diversity [62]. Our study thus confirms that heterogeneity *per se* is not universally beneficial for species richness (see also [63]), although most work to date has focused on its positive effects (e.g. [64]).

The importance of considering multiple scales

To date, results from studies investigating multiple scales have been disparate and difficult to generalize [65]. We included three spatial scales in our study which we considered relevant for butterfly diversity and distribution. Our study revealed that all investigated scales affected butterfly community composition. Previous studies found local factors affecting butterfly species composition, with local heterogeneity in land cover being a good predictor for species composition in Canada [54,64]. Butterfly species composition in Transylvania also showed a significant correlation with local factors, but was explained by heterogeneity and woody vegetation cover only at the two larger scales. Butterfly species richness also responded to variables at all different spatial scales, especially at the local scale, but also at the two larger scales (see also [9]). This suggests that local habitat conditions are particularly important, yet these cannot be considered in isolation from the surrounding landscape [33,66].

Our models also showed that the different functional groups of butterflies were affected by variables from different spatial scales. For example, landscape heterogeneity appeared to benefit mobile species but not low-mobility species. Furthermore, we found that woody vegetation cover was related to species richness. Land abandonment induces natural succession, whereas intensification leads to loss of scattered woody vegetation, and both have negative effects on butterfly richness in the long term [67]. Both processes decrease structural heterogeneity, which is important for viable butterfly populations in agricultural landscapes. In our study, only *Coenonympha pamphilus* showed a unimodal relationship to local woody vegetation cover. For such low-mobile species, presence of woody vegetation is crucial for wind shield and thermoregulation. *C. pamphilus* is abundant in Transylvania, however its population state in other European countries is declining [68]. Habitat heterogeneity from different spatial scales, including the presence of woody vegetation, should be further investigated as possible key elements in landscapes to halt biodiversity loss in farmland.

Conclusion

Collapse of communism and accession of Romania to the European Union have accelerated land use change in the rural areas of Transylvania, in particular towards land abandonment and agricultural intensification. The two key gradients considered in this study, namely woody vegetation cover and heterogeneity, would fundamentally change as a result of these two land use change processes. Along the gradients of woody vegetation cover and heterogeneity, we were able to show that butterfly abundance and distribution were affected by a range of different variables operating at multiple spatial scales. Not only local conditions, but the composition and configuration of the landscape as a whole need to be considered for effective conservation management of butterflies in low-intensively managed farming landscapes such as in Transylvania.

Our results showed that, unlike in Western Europe, species richness of butterflies was not only high in grassland, but also in arable land. This suggests that more emphasis needs to be placed on low-intensity farming practices and management of the landscape mosaic, and that arable land needs to be actively considered in butterfly conservation strategies. In our study area, butterfly richness would likely benefit from (1) the continuation of small-scale farming; (2) the production of a variety of crops, including legume species; and (3) the maintenance of broad field margins and uncultivated ruderal areas. New payment schemes under the Common Agricultural Policy have recently been criticised as grossly inadequate [53]. Our findings suggest that even measures considered adequate in Western Europe may not be directly transferable to Transylvania – in low-intensity landscapes, it will be particularly important to consider the high nature value that entire agro-ecosystems provide, both inside and outside of protected areas (see also [69]). Ultimately, the continued existence of historic-cultural landscapes such as those in Transylvania hinges on the successful transfer of its appreciation and historic management to future generations of farmers. Substantial efforts are therefore needed in environmental education and in developing alternative ways for local people to make a living, for example through the development of cultural and ecological tourism.

Supporting Information

Table S1 Number of the 30 focal villages within different strata. TRI = Terrain Ruggedness index, Protection status is according to the EU Habitats and Birds Directives.

Table S2 Number of survey sites along the two gradients local heterogeneity and local woody vegetation cover.

Table S3 Species list of butterfly species observed in the transects.

Table S4 Correlation matrix of the variables used in the study.

Acknowledgments

We warmly thank Rémi Bigonneau, Paul Kirkland, Joerg Steiner, Kimberley Pope and Elek Telek for help in the field. Sincere thanks to the farmers, landowners, mayors and the Mihai Eminescu Trust for their support. We are grateful for analytical support by Henrik von Wehrden and Dave Abson. Comments by two anonymous reviewers greatly helped to improve an earlier version of this manuscript.

Author Contributions

Conceived and designed the experiments: JL JF JH ID LR. Performed the experiments: JL. Analyzed the data: JL JH ID. Contributed reagents/materials/analysis tools: PF. Wrote the paper: JL.

References

1. Stoate C, Baldi A, Beja P, Boatman ND, Herzon I, et al. (2009) Ecological impacts of early 21st century agricultural change in Europe - A review. Journal of Environmental Management 91: 22–46.
2. Kleijn D, Rundlöf M, Scheper J, Smith HG, Tscharntke T (2011) Does conservation on farmland contribute to halting the biodiversity decline? Trends in Ecology & Evolution 26: 474–481.
3. Benton TG, Vickery JA, Wilson JD (2003) Farmland biodiversity: is habitat heterogeneity the key? Trends in Ecology & Evolution 18: 182–188.
4. Thomas JA, Telfer MG, Roy DB, Preston CD, Greenwood JJD, et al. (2004) Comparative losses of British butterflies, birds, and plants and the global extinction crisis. Science 303: 1879–1881.
5. Foley JA, DeFries R, Asner GP, Barford C, Bonan G, et al. (2005) Global consequences of land use. Science 309: 570–574.
6. Cremene C, Groza G, Rakosy L, Schileyko AA, Baur A, et al. (2005) Alterations of steppe-like grasslands in Eastern Europe: a threat to regional biodiversity hotspots. Conservation Biology 19: 1606–1618.
7. Kumar S, Simonson SE, Stohlgren TJ (2009) Effects of spatial heterogeneity on butterfly species richness in Rocky Mountain National Park, CO, USA. Biodiversity and Conservation 18: 739–763.
8. Brückmann SV, Krauss J, Steffan-Dewenter I (2010) Butterfly and plant specialists suffer from reduced connectivity in fragmented landscapes. Journal of Applied Ecology 47: 799–809.
9. Weibull AC, Ostman O, Granqvist A (2003) Species richness in agroecosystems: the effect of landscape, habitat and farm management. Biodiversity and Conservation 12: 1335–1355.
10. Van Dyck H, Van Strien AJ, Maes D, Van Swaay CAM (2009) Declines in Common, Widespread Butterflies in a Landscape under Intense Human Use. Conservation Biology 23: 957–965.
11. Maes D, Van Dyck H (2001) Butterfly diversity loss in Flanders (north Belgium): Europe's worst case scenario? Biological Conservation 99: 263–276.
12. McLaughlin A, Mineau P (1995) The impact of agricultural practices on biodiversity. Agriculture, Ecosystems & Environment 55: 201–212.
13. Bergman KO, Askling J, Ekberg O, Ignell H, Wahlman H, et al. (2004) Landscape effects on butterfly assemblages in an agricultural region. Ecography 27: 619–628.
14. Kluvánková-Oravská T, Chobotová V, Banaszak I, Slavikova L, Trifunovova S (2009) From government to governance for biodiversity: the perspective of central and Eastern European transition countries. Environmental Policy & Governance 19: 186–196.
15. Kuemmerle T, Muller D, Griffiths P, Rusu M (2008) Land use change in Southern Romania after the collapse of socialism. Regional Environmental Change 9: 1–12.
16. Mikulcak F, Newig J, Milcu AI, Hartel T, Fischer J (2013) Integrating rural development and biodiversity conservation in Central Romania. Environmental Conservation 40: 129–137.
17. Schmitt T, Rákosy L (2007) Changes of traditional agrarian landscapes and their conservation implications: a case study of butterflies in Romania. Diversity and Distributions 13: 855–862.

18. Young J, Richards C, Fischer A, Halada L, Kull T, et al. (2007) Conflicts between Biodiversity conservation and human activities in the central and eastern European countries. Ambio 36: 545–550.
19. Pullin AS, Baldi A, Can OE, Dieterich M, Kati V, et al. (2009) Conservation Focus on Europe: Major Conservation Policy Issues That Need to Be Informed by Conservation Science. Conservation Biology 23: 818–824.
20. Tryjanowski P, Hartel T, Baldi A, Szymanski P, Tobolka M, et al. (2011) Conservation of farmland birds faces different challenges in Western and Central-Eastern Europe. Acta Ornithologica 46: 1–12.
21. Tscharntke T, Klein AM, Kruess A, Steffan-Dewenter I, Thies C (2005) Landscape perspectives on agricultural intensification and biodiversity - ecosystem service management. Ecology Letters 8: 857–874.
22. Ekroos J, Rundlöf M, Smith HG (2013) Trait-dependent responses of flower-visiting insects to distance to semi-natural grasslands and landscape heterogeneity. Landscape Ecology 28: 1283–1292.
23. Dunning JB, Danielson BJ, Pulliam HR (1992) Ecological Processes That Affect Populations in Complex Landscapes. Oikos 65: 169–175.
24. Fahrig L, Baudry J, Brotons L, Burel FG, Crist TO, et al. (2011) Functional landscape heterogeneity and animal biodiversity in agricultural landscapes. Ecology Letters 14: 101–112.
25. Persson AS, Olsson O, Rundlöf M, Smith HG (2010) Land use intensity and landscape complexity-Analysis of landscape characteristics in an agricultural region in Southern Sweden. Agriculture Ecosystems & Environment 136: 169–176.
26. Fischer J, Hartel T, Kuemmerle T (2012) Conservation policy in traditional farming landscapes. Conservation Letters 5: 167–175.
27. Page N, Bălan A, Popa SHR, Rákosy L, Sutcliffe L (2012) România/Romania. In: Oppermann R, Beaufoy GJ, G., editors. High Nature Value Farming in Europe. Ubstadt-Weiher: Verlag Regionalkultur. pp. 346–358.
28. Akeroyd JR, Page JN (2006) The Saxon villages of southern Transylvania: Conserving biodiversity in a historic landscape. In: Gafta D, Akeroyd JR, editors. Nature Conservation: Concepts and Practice. Heidelberg, Germany: Springer Verlag. pp. 199–210.
29. Hartel T, Moga CI, Öllerer K, Sas I, Demeter L, et al. (2008) A proposal towards the incorporation of spatial heterogeneity into animal distribution studies in Romanian landscapes. North-Western Journal of Zoology 4: 67–74.
30. Erhardt A (1985) Diurnal Lepidoptera - Sensitive Indicators of Cultivated and Abandoned Grassland. Journal of Applied Ecology 22: 849–861.
31. van Swaay C, Warren M, Lois G (2006) Biotope use and trends of European butterflies (vol 10, pg 189, 2006). Journal of Insect Conservation 10: 305–306.
32. Krauss J, Steffan-Dewenter I, Tscharntke T (2003) How does landscape context contribute to effects of habitat fragmentation on diversity and population density of butterflies? Journal of Biogeography 30: 889–900.
33. Öckinger E, Smith HG (2006) Landscape composition and habitat area affects butterfly species richness in semi-natural grasslands. Oecologia 149: 526–534.
34. Börschig C, Klein A-M, von Wehrden H, Krauss J (2013) Traits of butterfly communities change from specialist to generalist characteristics with increasing land-use intensity. Basic and Applied Ecology.
35. Öckinger E, Franzen M, Rundlof M, Smith HG (2009) Mobility-dependent effects on species richness in fragmented landscapes. Basic and Applied Ecology 10: 573–578.
36. Diamond JM (1986) Overview: laboratory experiments, field experiments, and natural experiments. In: Diamond JM, Case TJ, editors. Community Ecology. New York: Harper & Row. pp. 3–22.
37. Lindenmayer DB, Cunningham RB, MacGregor C, Crane M, Michael D, et al. (2008) Temporal changes in vertebrates during landscape transformation: a large-scale "natural experiment". Ecological Monographs 78: 567–590.
38. Bennett AF, Radford JQ, Haslem A (2006) Properties of land mosaics: Implications for nature conservation in agricultural environments. Biological Conservation 133: 250–264.
39. Pollard E, Yates TJ (1993) Monitoring butterflies for ecology and conservation: the British butterfly monitoring scheme; Institute of Terrestrial Ecology JNCC, editor. London: Chapman & Hall. 274 p.
40. Naumann CM, Tarmann GM, Tremewan WG (1999) The Western Palaearctic Zygaenidae (Lepidoptera). Stenstrup: Apollo Books. 304 p.
41. Rakosy L, Goia M, Kovács Z (2003) Catalogul Lepidopterelor Romaniei/ Verzeichnis der Schmetterlinge Rumäniens. In: Romana SL, editor. Cluj-Napoca: Romsver.
42. Parker KC (1991) Topography, Substrate, and Vegetation Patterns in the Northern Sonoran Desert. Journal of Biogeography 18: 151–163.
43. McGarigal K, Cushman S, Ene E (2012) FRAGSTATS v4: Spatial Pattern Analysis Program for Categorical and Continuous Maps. Computer software program produced by the authors at the University of Massachusetts, Amherst. Available: http://www.umass.edu/landeco/research/fragstats/fragstats.html.
44. Dormann CF, Elith J, Bacher S, Buchmann C, Carl G, et al. (2013) Collinearity: a review of methods to deal with it and a simulation study evaluating their performance. Ecography 36: 27–46.
45. van Swaay C, Maes D, Collins S, Munguira ML, Sasic M, et al. (2011) Applying IUCN criteria to invertebrates: How red is the Red List of European butterflies? Biological Conservation 144: 470–478.
46. Bolker BM, Brooks ME, Clark CJ, Geange SW, Poulsen JR, et al. (2009) Generalized linear mixed models: a practical guide for ecology and evolution. Trends in Ecology & Evolution 24: 127–135.
47. Bink FA (1992) Ecologische Atlas van de Dagvlinders van Noordwest-Europa. Haarlem: Schuyt & Co.
48. van Swaay C, Strien Av, Harpke A, Fontaine B, Stefanescu C, et al. (2013) The European Grassland Butterfly Indicator: 1990–2011. Copenhagen, Denmark: European Environment Agency. No 11/2013.
49. Franzén M, Ranius T (2004) Occurrence patterns of butterflies (Rhopalocera) in semi-natural pastures in southeastern Sweden. Journal for Nature Conservation 12: 121–135.
50. van Swaay C, Warren MS (1999) Red data book of European butterflies (Rhopalocera): Council of Europe.
51. Konvicka M, Maradova M, Benes J, Fric Z, Kepka P (2003) Uphill shifts in distribution of butterflies in the Czech Republic: effects of changing climate detected on a regional scale. Global Ecology and Biogeography 12: 403–410.
52. R Core Team (2013) R: A Language and Environment for Statistical Computing. Vienna, Austria: R Foundation for Statistical Computing.
53. Pe'er G, Dicks LV, Visconti P, Arlettaz R, Báldi A, et al. (2014) EU agricultural reform fails on biodiversity. Science 344: 1090–1092.
54. Weibull AC, Ostman O (2003) Species composition in agroecosystems: The effect of landscape, habitat, and farm management. Basic and Applied Ecology 4: 349–361.
55. Steffan-Dewenter I, Tscharntke T (2000) Butterfly community structure in fragmented habitats. Ecology Letters 3: 449–456.
56. Wilson JB, Peet RK, Dengler J, Partel M (2012) Plant species richness: the world records. Journal of Vegetation Science 23: 796–802.
57. Jones A (2009) Wildflower species indicators for lowland grassland habitat conservation in Transylvania (Romania). Contributii Botanice 44: 57–66.
58. Zechmeister HG, Schmitzberger I, Steurer B, Peterseil J, Wrbka T (2003) The influence of land-use practices and economics on plant species richness in meadows. Biological Conservation 114: 165–177.
59. Kleijn D, Kohler F, Baldi A, Batary P, Concepcion ED, et al. (2009) On the relationship between farmland biodiversity and land-use intensity in Europe. Proceedings of the Royal Society B-Biological Sciences 276: 903–909.
60. Van Landuyt W, Vanhecke L, Hoste I, Hendrickx F, Bauwens D (2008) Changes in the distribution area of vascular plants in Flanders (northern Belgium): eutrophication as a major driving force. Biodiversity and Conservation 17: 3045–3060.
61. Dover J, Sparks T, Greatorex-Davies J (1997) The importance of shelter for butterflies in open landscapes. Journal of Insect Conservation 1: 89–97.
62. Krauss J, Klein AM, Steffan-Dewenter I, Tscharntke T (2004) Effects of habitat area, isolation, and landscape diversity on plant species richness of calcareous grasslands. Biodiversity and Conservation 13: 1427–1439.
63. Ekroos J, Piha M, Tiainen J (2008) Role of organic and conventional field boundaries on boreal bumblebees and butterflies. Agriculture, Ecosystems & Environment 124: 155–159.
64. Kerr JT, Southwood TRE, Cihlar J (2001) Remotely sensed habitat diversity predicts butterfly species richness and community similarity in Canada. Proceedings of the National Academy of Sciences of the United States of America 98: 11365–11370.
65. Flick T, Feagan S, Fahrig L (2012) Effects of landscape structure on butterfly species richness and abundance in agricultural landscapes in eastern Ontario, Canada. Agriculture Ecosystems & Environment 156: 123–133.
66. Steffan-Dewenter I, Munzenberg U, Burger C, Thies C, Tscharntke T (2002) Scale-dependent effects of landscape context on three pollinator guilds. Ecology 83: 1421–1432.
67. Baur B, Cremene C, Groza G, Rakosy L, Schileyko AA, et al. (2006) Effects of abandonment of subalpine hay meadows on plant and invertebrate diversity in Transylvania, Romania. Biological Conservation 132: 261–273.
68. Conrad KF, Fox R, Woiwod IP (2007) Monitoring Biodiversity: Measuring Long-term changes in Insect Abundance. In: Stewart AJA, New TR, Lewis OT, editors. Insect Conservation Biology: Proceedings of the Royal Entomological Society's 23rd symposium. Oxford: The Royal Entomological Society. pp. 203–225.
69. González-Estébanez FJ, García-Tejero S, Mateo-Tomás P, Olea PP (2011) Effects of irrigation and landscape heterogeneity on butterfly diversity in Mediterranean farmlands. Agriculture, Ecosystems & Environment 144: 262–270.

5

Political Systems Affect Mobile and Sessile Species Diversity – A Legacy from the Post-WWII Period

Sara A. O. Cousins[1]*, Mitja Kaligarič[2,3], Branko Bakan[2], Regina Lindborg

1 Landscape Ecology, Department of Physical Geography and Quaternary Geology, Stockholm University, Stockholm, Sweden, 2 University of Maribor, Biology Department, Faculty of Natural Sciences and Mathematics, Maribor, Slovenia, 3 Faculty of Agriculture and Life Sciences, University of Maribor, Pivola 10, Hoče, Slovenia

Abstract

Political ideologies, policies and economy affect land use which in turn may affect biodiversity patterns and future conservation targets. However, few studies have investigated biodiversity in landscapes with similar physical properties but governed by different political systems. Here we investigate land use and biodiversity patterns, and number and composition of birds and plants, in the borderland of Austria, Slovenia and Hungary. It is a physically uniform landscape but managed differently during the last 70 years as a consequence of the political "map" of Europe after World War I and II. We used a historical map from 1910 and satellite data to delineate land use within three 10-kilometre transects starting from the point where the three countries meet. There was a clear difference between countries detectable in current biodiversity patterns, which relates to land use history. Mobile species richness was associated with current land use whereas diversity of sessile species was more associated with past land use. Heterogeneous landscapes were positively and forest cover was negatively correlated to bird species richness. Our results provide insights into why landscape history is important to understand present and future biodiversity patterns, which is crucial for designing policies and conservation strategies across the world.

Editor: David L. Roberts, University of Kent, United Kingdom

Funding: Funding was provided by the program group "Biodiversity" (P1-0078) and granted by Slovenian Research Agency to MK and by FORMAS to SC and RL. MK was supported by program group P1-0164, funded by Slovenian Research Agency. The funders had no role in study design, data collection and analysis, decision to publish, or preparation of the manuscript.

* Email: sara.cousins@natgeo.su.se

Introduction

It is increasingly recognized that conservation biology should have a "landscape perspective" [1–4]. This is generally understood in a spatial context when considering targets for conservation, but a temporal dimension of the landscape is also necessary to understand effects of delayed species responses. This is however rarely considered. Land use change, either by intensification or abandonment, is one of the main drivers causing deterioration of species-richness across the world [5]. Land use and vegetation structure and composition, commonly used as explanatory factor for biodiversity patterns, is to a large extent outcomes of political and socio-economic decisions or constraints. However, the same driving forces may lead to different effects depending on the physical landscape [6,7], e.g. differences in soil fertility, topography or water availability. Land use effects on biodiversity are highly debated topics, especially in conservation research [8,9], because a lack of spatially explicit historical biodiversity data. However, there is a consensus that the decline in traditional agriculture often has negative effects on biodiversity, as the low intensive utilisation of grasslands and forests in the past has been a prerequisite for much of the high small-scale species richness found in the rural landscape of Europe today [10–12]. Species' richness, abundance and composition may respond directly to land use changes but a delayed response has been detected in several studies [13,14]. Such responses often differ depending on organism group, where many mobile organisms respond more quickly to landscape change compared to long-lived sessile organism [15,16].

Despite the increased awareness of social-ecological linkages [17] in conservation, few studies have used large scale *in situ* experimental designs to analyze direct or indirect effects of non-ecological drivers on biodiversity patterns. One reason is the difficulty to find suitable study systems as the divisions into countries or regions often are a result of underlying physical landscape differences [18]. Furthermore, magnitude and timing of intensifications or abandonment is also constrained by physical properties at local or regional scales. For example, Cousins [19] found that areas with a larger proportion of clayey soils changed towards intensive crop-production earlier than areas with smaller proportions of clayey soil. Landscapes on more marginal soils or locations have shown a tendency to be abandoned and afforested [20,21]. However, because of geopolitical reasons, during the last 100 years, there are regions all over the world that have been divided without considering physical landscape divisions or uniformities. Beside former colonies in Africa and Asia some recent examples are Korea and New Guinea. Considerable differences between Eastern and Western Europe, regarding bird and plant diversity, and forestry, have previously been highlighted by several authors [22–24], but are rarely addressed in the design and interpretation of research or policy, but see [25,26]. In Europe, the division between socialist and non-socialist states since WWII, and the recent agenda to privatize or re-privatize the former socialist economies, has especially affected agricultural systems. During the socialist era, there was a widespread collectivization resulting in few but large farming units, contrasted by many small co-existing non-industrial farms outside the industrial food production [27]. Also in Western Europe, driven by market economy, the majority of traditional small scale farms have disappeared due to intensified and specialized agriculture [28,29] or abandonment and afforestation [19].

In this study, we compare current biodiversity patterns in a physically uniform area in the borderland where three countries meet, Austria, Hungary and Slovenia. The historical management in this part of Europe was dominated by small-scale traditional farming 100 years ago when it was part of the same state; the Austrian-Hungarian Empire. Although the area was divided into three countries after 1921, the economic system (market oriented traditional agriculture) remained more or less the same until 1945. After 1945 the difference between countries became prominent, due to changed political systems. In Austria almost all land were privately owned, driven by a free market Western economic policy, whereupon fields became more intensively used and larger in general. Slovenia (part of Yugoslavia) became a pronounced socialist market economy. Large farms on the lowlands were confiscated and collectivised, but in the hilly remote region near the borders to Hungary and Austria the majority of the land remained private and fairly small-scaled. In contrast, the command economy in Hungary eroded the traditional farming system through collectivisation of farms, where former landowners became employed as workers in big state-owned farms. Thus, both free-market economy and collectivization resulted in larger and more intensively used farms where the physical landscape made it possible. Along the Iron Curtain (i.e. here the border to Austria and Slovenia) there was a policy of depopulation. Since the fall of former Yugoslavia and the Eastern Block in the 1990s, Hungary and Slovenia has moved to a free market economy and are both part of the European Union since 2004, whereas Austria became a member state in 1995. Today all nations are part of EU and the "borderless" Schengen region. It is important to note that after the socialist era, no quick, abrupt or radical changes in land use or land ownership occurred in the studied region.

Our primary focus is on landscape matrix effects (as a consequence of land use change) on biodiversity. The rationale behind this study is that the different political systems during the last 70 years will be reflected in the land use and hence also in biodiversity patterns [13], here analysed by using mobile (birds) and sessile (plants) species. We hypothesize that traditionally managed landscape has highest biodiversity and intensified agricultural landscape has the lowest biodiversity [29–31] but see [32]. Furthermore, mobile organisms will be more associated with current land use structure than less mobile organisms [15], exposing a legacy from the land use prior to the post-war era.

Methods

Land cover

Because mobile organisms are expected to respond more quickly to landscape change compared to long-lived sessile organism we chose to investigate both birds and plants. Methods were chosen to best fit the mobility of the organisms i.e. transect mapping for birds and plot inventory for plants. Birds and plants were inventoried in 2-kilometers long cross-transects along three 10 kilometre long main transects radiating out from the point (46,869° N; 16,114° E) where Austria, Slovenia and Hungary meet (Fig. 1). Each main transect ends in Austria at 46,556 N; 16,116 E, in Hungary at 46,532 N; 16,143 E and in Slovenia at 46,465 N; 16,637 E. The climate is moderate continental or sub-Pannonic, with relatively dry winters and with an average annual rainfall of 900 mm. Mean temperature in January is −2°C and in July 19°C. Geologic substrates are mainly tertiary sediments, which forms a soft hilly landscape of sandy-acid soils with networks of running fresh water. The landscape is a mixture of forest and open areas with small farms scattered along hilltops. The investigated landscapes belong to the Trilateral Park: Raab (Austria), Goričko (Slovenia) and Őrség (Hungary). Raab is a nature park (established 1997) aiming to preserve traditional landscapes, Goričko is a Natura 2000 area (established 2002) with the goal to keep traditional and extensive small-scale farming, and Őrség National Park and Natura 2000 area was established 2004 to promote wildlife and tourism and preserve the unique Oak-Pine forests.

Hereafter these different landscapes will be referred to as Raab (Austria), Goričko (Slovenia) and Őrség (Hungary) although it should be noted that the study does not encompass the whole of each park. We used a historical map from 1910 (Fig. 1) from the Austrian-Hungarian Empire (3rd Military Mapping Survey of Austria-Hungary, sheet "Szombathely"), to estimate the relationship between open and forest land in the past. Unfortunately it was not possible to carry out any detailed analyses on land cover composition because of low thematic resolution. To link current biodiversity to land use, we calculated current land use in the study region using CORINE land-cover data from 2000 with a resolution of 50×50 m. We used each 10-km transect (Fig. 1) with sample cross-section width, i.e. 10×2 km, to calculate percentages of different land-cover classes from CORINE in a geographical information system (GIS). Maps of topography and soils were cross-checked to detect dissimilarities between the different transects.

Field survey

A field inventory of birds and plants was designed to capture the differences in land use as well as biodiversity patterns within the countries (File S1). Bird diversity was investigated by slowly walking along the cross-section transect and noting all birds seen or heard following the transect mapping method [33]. All cross-transects were visited twice during the breeding season (spring and late summer), which resulted in 5 samples for each country. Bird classifications follows the nomenclature by Geister [34] and Svensson & Mullarney [35].

The plant inventory was conducted during the field-season 2010. Along the cross-section transects 21 sampling points were placed evenly every 50 meters, in total 105 sampling points in each country. First, all vascular plants found within a 2×2 m square were noted, and then all additional plants found in a circle with radius of 10 m around the plot were added. In addition, we noted the main land-cover type for each 10 m radius plot: grassland (grazed or mown), forest, field (for crop-production) and ruderal or urban surfaces land (for example house, road, playground), which is hereafter referred to as ruderal. For each sampling point in forest the age of trees were categorized as: >30 years old, between 15–30 years or <15 years. Plant nomenclature followed Martinčič et al. [36]. No specific permissions were required for any of the field studies. Only observational studies were performed without interference of plants or birds.

Statistical analysis

Number of bird and plant species in relation to nation and land use type was analysed in separate ANOVAs. To examine how the number of bird species was affected by land use type we used the proportion of present day forest-cover in an ANCOVA, using forest cover as explanatory variable and nationality as covariate.

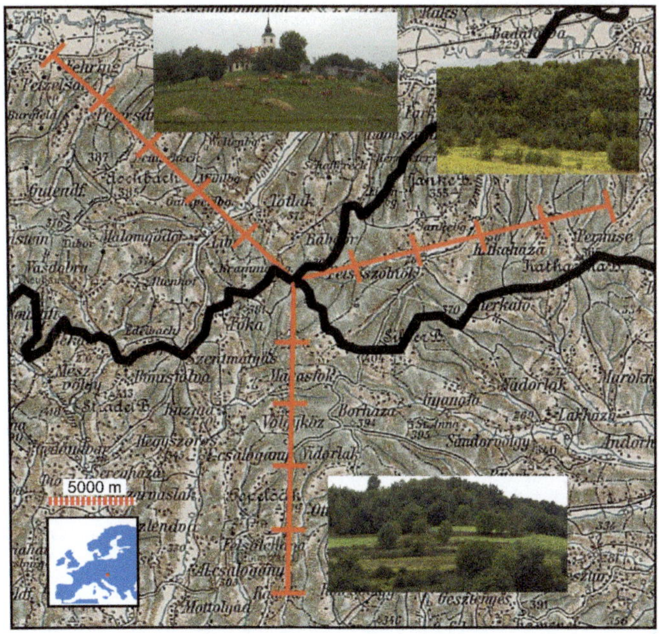

Figure 1. Sampling design for investigating bird and plant diversity in three bordering landscapes in Austria, Slovenia and Hungary. Birds and plants were inventoried in 2-kilometers long cross-transects along three 10 kilometer long main transects radiating out from the point where Austria, Slovenia and Hungary meet today. The investigated landscapes belong to the Trilateral Park: Raab (Austria), Goričko (Slovenia) and Őrség (Hungary). The historical land cover map is from 1910, under the Austrian-Hungarian Empire, showing the study area with the present-day borders (black) superimposed. Green areas are forested land and pink areas are arable land.

"Nationality" (Austria, Hungary, and Slovenia) is arbitrarily used here to reflect past land use history i.e. past political system. To investigate differences in plant composition, nationality and each cross transect distance from the border were predictor variables. We used the model Generalized Estimating Equations (GEE) [37] which allows for terms specifying autocorrelation and are well suited for evaluating landscape processes. First we did a Principal Coordinates Analysis (PCoA), based on dissimilarity (Bray-Curtis) of plant species at the 10 m scale, using the first three axes. We would expect similarity to decrease with distance from where the transects meet, both due to distance in itself and because of differences in land use. Models including both predictor variables transect and nationality, as well as a model with nationality only, were tested against the null model (only including random effects). The significance of each model was tested with likelihood ratio test and post hoc test to separate each variable within the predictors. Data deviating from a normal distribution was log10 transformation and all numbers were increased by one before analysis. The statistical software R 2.13.0 was used for the analyses using the *geepack* package for GEE modeling [37].

Results

Land use change

In 1910, the landscape along each transect was dominated by open agricultural land with a forest cover between 30–39%; a landscape composition that today is inverted with forests covering between 57–74% of the landscapes (Fig 2). Based on the plot surveys, Raab (Austria) has the highest current percentage (23%) of forest older than 30 years, whereas 65% of the forests in Goričko (Slovenia) and 67% in Őrség (Hungary) are between 15 and 30 years old. Young forests (<30 years) are primarily on former arable fields or grasslands. In the investigated landscapes grasslands are few and arable fields even more rare: 11% and 4% Raab (Austria), 19% and 1% Őrség (Hungary) and 15% and 9% Goričko (Slovenia) for grassland and fields, respectively.

Bird diversity patterns

We found 53 different birds species in total (Table 1; File S1). The bird species composition was 40% in total overlap between the three countries, and 55% to 58% when comparing countries pair wise. There was a significant difference between countries ($F_{2,12} = 6.747$, $p = 0.0108$) (Fig. 2), where Goričko (Slovenia) had significantly higher diversity of birds than Raab (Austria) (Tukey HSD, $p<0.0087$), but there were no significant difference between Őrség (Hungary) and Raab (Austria) or Goričko (Slovenia) (Tukey HSD, $p = 0.37$ and $p = 0.1$ respectively). When associating different bird species to habitat most birds were classified as forest species; Goričko (Slovenia) had 28% (29 species), Őrség (Hungary) 25% (27 sp.) and Raab (Austria) 24% (28 sp.). Only a few percent of the bird species were associated to open grassland habitats: Raab (5%), Goričko (15%) and Őrség (7%). Number of birds was clearly related to forest cover in each transect, with a significant difference between frequencies of birds found in transects depending on forest cover and land use history (i.e. nationality) as bird diversity declined with an increase in forest cover (Fig. 2). Although the trend is similar for all countries it is only significant for Slovenia ($F_3 = 4.7$, $p = 0.02$, adjusted $R2 = 0.45$, ANCOVA).

Plant diversity patterns

In total, we found 407 vascular plant species with relatively few endangered plant species, according to (separately considered) national Red Lists: Raab (Austria) 7% (18 sp.), Goričko (Slovenia) 1% (3 sp.) and Őrség (Hungary) 5% (14 sp.) (File S1). Only 180 out of 407 species occurred in all the three countries. There was a significant difference in plant species richness among countries at the larger (314 m^2) scale. Őrség (Hungary) had more plant species

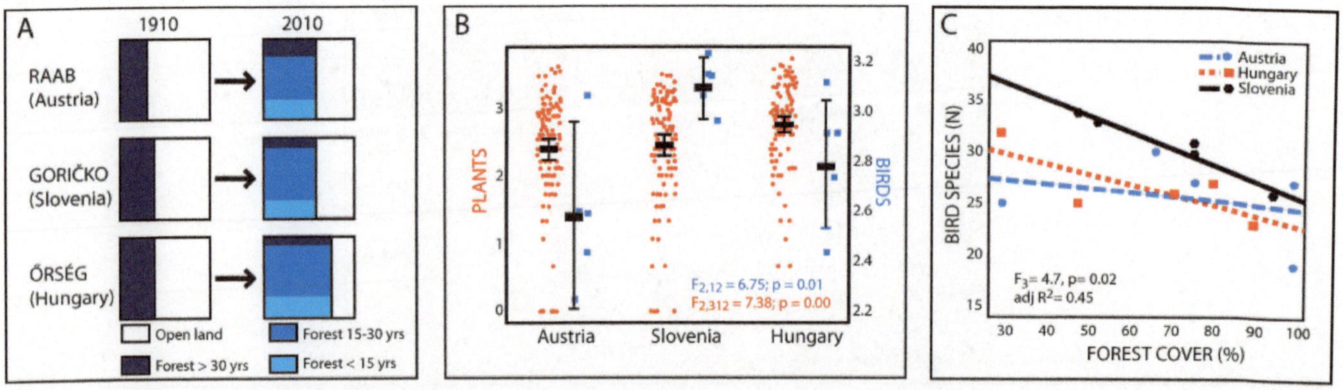

Figure 2. 100 year of land cover change and present-day plant and bird diversity in the border landscapes in Austria, Slovenia and Hungary. (A) Open and forest land cover within the transects between 1910 to 2010. The map from 1910 is thematically coarser so it is not possible to separate the difference in forest age. (B) Mean numbers of plants (red circles) and birds (blue squares) from plots along transects in each country. (C) The percentage of forest cover along each cross-transect and the relationship to number of bird species. There were 5 cross-transects investigated in each country.

than Goričko (Slovenia) and Raab (Austria) ($F_{2,312} = 7.38$; $p = 0.0007$) (Fig. 2). At the smaller scale (4 m^2) no significant difference could be detected for plant richness between countries ($p > 0.11$; Table 1). Plots in grassland habitats had significantly higher plant richness compared to all other habitats, irrespective of country ($F_{3,310} = 14.9$; $p < 0.0001$).

In the GEE analyses, using distance of cross-transects and country as predictive variables on plant composition, we found that distance from the border clearly affected plant community structure were there also was a clear effect of nationality (likelihood ratio test: $\chi^2 = 19.1$, $DF = 4$, $p = 0.0008$). The post hoc test showed that the species composition in first cross-transect (closest to the border, Fig. 3b) is separated from the second which are both separate from the third to fifth. Furthermore the species composition in Raab (Austria) is different compared to both Őrség (Hungary) and Goričko (Slovenia).

Discussion

To understand the processes behind, and to predict, biodiversity patterns we need to analyse landscape history [38]. Unfortunately there are hardly any records of historical species diversity patterns, making it impossible to analyse direct effects of landscape changes on species richness and composition but see [21,39]. However, historical maps together with landscape change trajectories have been used to indirectly analyse how biodiversity patterns are affected [10,40,41]. Here we show, by using 100 year old landscape data, that the difference in historical political systems during the last 70 years can be detected on present-day species diversity patterns. Hundred years ago the landscape along the investigated transects was dominated by open agricultural land with a forest cover between 30–40%, whereas it is today inverted with forests covering between 60–70%. Many arable fields and grasslands have become afforested, particularly in Hungary. The current landscape composition is fairly similar in the three landscapes, considering the percentage of forest cover to open land (Fig. 2a) although land use changes and different conservation strategies have resulted in clear differences in species composition (Fig. 2b). In the Hungarian Őrség, rewilding [42,43] has led to increasing forest area, whereas Goričko (Slovenia) is more similar to the traditional landscape before the changes after 1945. Despite Austria and Slovenia having similarities in conservation goals, higher biodiversity of both plants and birds was noted in Slovenia compared to Austria. These biodiversity patterns could be indirectly linked to differences in political system and economic drivers, where the market oriented agriculture in Austria, compared to the more subsistence agriculture in the area in Slovenia during the whole period, has resulted in an intensification of agriculture through the EU Common Agricultural Policy (CAP) funding structures.

The distribution pattern of mobile and sessile organisms varied depending on country (i.e. nationality). The more traditional

Table 1. Species richness of plants and birds based on surveys along three 10 km long transects in Hungary, Slovenia and Austria.

	Austria		Slovenia		Hungary	
	plants	birds	plants	birds	plants	birds
Total (n)	256	43	281	47	299	42
2×2 plot (SD)	7.2 (5.3)	-	8.5 (5.8)	-	9.3 (6.6)	-
10 m radius (SD)	14.8 (9.7)	-	15.6 (9.4)	-	19.7 (10.4)	-
Cross-transect (SD)	-	13.4 (4.6)	-	21.8 (2.3)	-	16.0 (3.1)

Total (n) is the total number of different species found in each country. Richness of plants is recorded at two different spatial scales, 2×2 m and 10×10 m (105 plots in each country), and bird richness is recorded per cross-transect (5 transects in each country). Grassland/forest species is the number of specialists species for grassland and forest habitats found. (-) indicate not applicable.

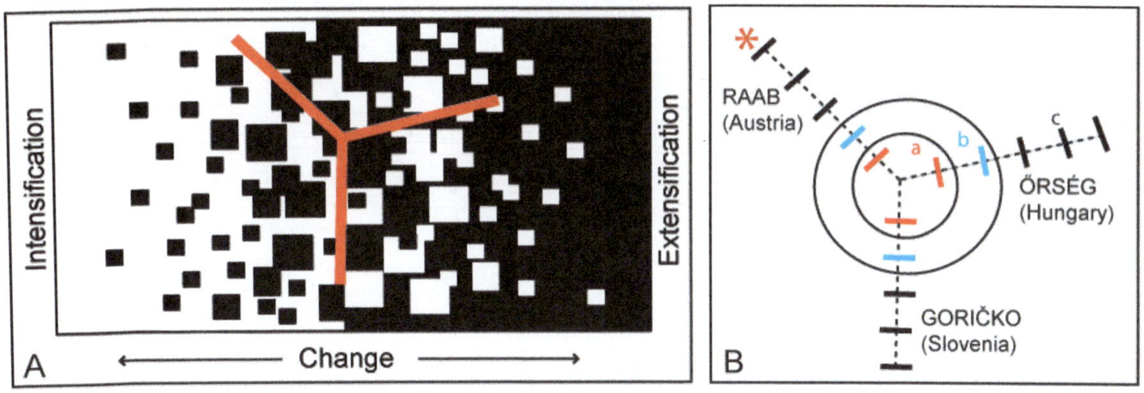

Figure 3. A conceptual model (A) shows how the three areas have either changed because of intensification (Austria) or extensification (Hungary) or remained more or less status quo (Slovenia). Similarity in plant species composition along cross-transects (B) spanning out from the point where the three countries meet. Significant difference in composition is marked with different letters a, b or c. An * marks the significant difference in species composition between countries.

landscape in Goričko (Slovenia) had significantly higher diversity of birds compared to the more intensively managed Raab (Austria), but there were no difference between the Hungarian landscape and the other two. The bird composition similarity among areas was low compared to other studies [44,45], especially considering the high mobility of birds that can move across national borders for nesting and feeding. The frequencies of birds found in transects depended on forest cover, where bird diversity declined with increasing forest cover. Small-scale heterogeneity, i.e. structurally more complex landscapes, in contemporary landscapes favour bird species richness, whilst a denser forest-cover has a negative effect on bird diversity. Several models suggest that more wildlife-friendly farming and heterogeneous landscapes, with many small natural or semi-natural habitats, help to support a relative high diversity compared to large scale farming and commercial forestry which is negative for biodiversity [46,47].

The highest plant species richness was found in Őrség (Hungary), with many typical grassland plant species, despite being primarily conserved and managed as forest. Prior to the Eastern Bloc policy to depopulate and reforest the area it was managed as a traditional agricultural landscape with many orchards and grasslands, and today remnant grassland communities intermingle with colonizing forest species. Particularly long-lived organisms, like plants, may survive as remnant populations for a long time after management has ceased [48], creating a so-called extinction debt [14,15,40,49]. Cousins [50] estimated a threshold for extinction debt in plant communities in grasslands to be settled after around 70 years, in Northern Europe. Many rural ecosystems have a long history of co-evolution with human management and today the survival of many species depends on the maintenance of low intensity farmland practices [51–53]. A historical dimension is hence a necessary complement to the spatial conservation perspective, particularly in landscapes where biodiversity is associated to traditional management. Here, we expect that many plant species associated to the remnant grassland habitats will disappear in Őrség (Hungary). However, other organisms might benefit and those associated to forest habitats should increase with succession of young forests [54]. We envisage based on the past and current trajectories that in the future the rewilding in Hungary will lead to that the legacy from past grassland composition will disappear in favour of forest biodiversity. The traditional landscape in Austria will probably remain fairly stable, but for the traditional landscape in Slovenia to remain, subsidies are needed and a functioning infrastructure to increase retailing of farm products. Thus, the long-term legacies from the pre-war landscape will disappear slowly and differences in plant diversity patterns become even more pronounced in the future also close to the border at a local scale, despite that there formally are no borders any more.

Although we cannot explain the direct causes for the differences in biodiversity patterns, due to possible co-variation of unknown environmental variables, the study area is relatively small and the abiotic conditions similar, i.e. bedrock and soil types, topography, climate; which strongly infers that the effects are a consequence of land use change, linked to past political systems. Both intensification and abandonment are clear results of political systems during the last 70 years. Similar effects have been observed also within the countries that used to lie within the Eastern Bloc where the different political systems established after the collapse of Soviet Union. For example, a comparison between e.g. Poland, as a EU country, to Russia and Belarus showed different trajectories of landscape transformations caused by agricultural abandonment [55]. Studies from areas, not confounded by underlying abiotic landscape differences, but driven by different political and economical policies can further disentangle how biodiversity patterns may change in the future [25,26]. There are several other comparable political divisions outside Europe that potentially can be used as experimental sites for investigating landscape history's effect on biodiversity and conservation.

In this study we give an *in situ* example of how national political priorities for social structure and economy may drive regional changes in land use that affects species diversity and composition. As expected, traditional agricultural landscape in Slovenia had the highest diversity but only for birds. Heterogeneity (here traditional agricultural landscape) at a landscape scale is expected to favour also plants, but we found that the heterogeneity in time (new land uses superimposed on former land use) created higher plant richness (Fig. 3a and 3b). Thus the hypothesis that mobile species richness is more associated to current land use and many sessile species are more associated to past land use is confirmed, as shown for plant diversity and composition in the Hungarian transect. We stress that awareness of how political and economical decisions directly or indirectly affect land use and biodiversity is crucial information, not only for the managers of these particular conservation areas, but also for designing sustainable policies and conservation strategies across the world.

Supporting Information

File S1 Species data. Plants and birds species found in Austria (AUS), Slovenia (SLO) and Hungary (HUN). + indicates occurrence during the field survey in respectively country. If a plant specie occurs on the National Red List (Anonymous [56] for Slovenia, Gergely [57] for Hungary and Niklfeld [58] for Austria) it is indicated with a 1. The typical habitats for plants were classified as forest (F), ruderal (R) and grassland (G) on the basis of the local flora monograph [59]. The birds were classified; [43] into species typical for forest (F), open and grassland habitats (O), settlements (S) or mixed habitats (M) including both open and forest landscape habitats. Bird abundance was classified as very common (vC), common (C) or rare (R) and breeding status as resident (rB), migratory (mB) and possible breeding species (rB?). Total numbers of plant and bird taxa found were 407 and 53 respectively.

Acknowledgments

This research is part of the EkoKlim-project at Stockholm University. J. Plue and R. Schmucki gave valuable advice on statistical analyses and M. Hjernqvist on bird inventories. We also thank M. Stenseke, and P. Batáry for valuable comments.

Author Contributions

Conceived and designed the experiments: SAOC RL MK. Performed the experiments: BB. Analyzed the data: SAOC RL. Contributed reagents/materials/analysis tools: SAOC BB MK RL. Wrote the paper: SAOC BB MK RL. Provided expert knowledge on the flora and bird fauna of the region: BB.

References

1. Tscharntke T, Tylianakis JM, Rand TA, Didham RK, Fahrig L, et al. (2012) Landscape moderation of biodiversity patterns and processes - eight hypotheses. Biological Reviews 87: 661–685.
2. Ibbe M, Milberg P, Tuner A, Bergman KO (2011) History matters: Impact of historical land use on butterfly diversity in clear-cuts in a boreal landscape. Forest Ecology and Management 261: 1885–1891.
3. Eriksson O, Cousins SAO (2014) Historical landscape perspectives on grasslands in Sweden and the Baltic Region. Land 3: 300–321.
4. Lindborg R, Bengtsson J, Berg Å, Cousins SAO, Eriksson O, et al. (2008) A landscape perspective on conservation of semi-natural grasslands. Agriculture Ecosystems & Environment 125: 213–222.
5. Sala OE, Chapin FS, Armesto JJ, Berlow E, Bloomfield J, et al. (2000) Global biodiversity scenarios for the year 2100. Science 287: 1770–1774.
6. Lambin EF, Turner BL, Geist HJ, Agbola SB, Angelsen A, et al. (2001) The causes of land-use and land-cover change: moving beyond the myths. Global Environmental Change-Human and Policy Dimensions 11: 261–269.
7. Bürgi M, Hersperger AM, Schneeberger N (2004) Driving forces of landscape change - current and new directions. Landscape Ecology 19: 857–868.
8. Balmford A, Green RE, Scharlemann JPW (2005) Sparing land for nature: exploring the potential impact of changes in agricultural yield on the area needed for crop production. Global Change Biology 11: 1594–1605.
9. Vandermeer J, Perfecto I (2007) The agricultural matrix and a future paradigm for conservation. Conservation Biology 21: 274–277.
10. Kull K, Zobel M (1991) High species richness in an Estonian wooded meadow. Journal of Vegetation Science 2: 715–718.
11. Cousins SAO, Eriksson O (2008) After the hotspots are gone: Land use history and grassland plant species diversity in a strongly transformed agricultural landscape. Applied Vegetation Science 11: 365–374.
12. Wilson JB, Peet RK, Dengler J, Pärtel M (2012) Plant species richness: the world records. Journal of Vegetation Science 23: 796–802.
13. Kuussaari M, Bommarco R, Heikkinen RK, Helm A, Krauss J, et al. (2009) Extinction debt: a challenge for biodiversity conservation. Trends in Ecology & Evolution 24: 564–571.
14. Tilman D, May RM, Lehman CL, Nowak MA (1994) Habitat destruction and the extinction debt. Nature 371: 65–66.
15. Krauss J, Bommarco R, Guardiola M, Heikkinen RK, Helm A, et al. (2010) Habitat fragmentation causes immediate and time-delayed biodiversity loss at different trophic levels. Ecology Letters 13: 597–605.
16. Bommarco R, Lindborg R, Marini L, Öckinger E (2014) Extinction debt for plants and flower-visiting insects in landscapes with contrasting land use history. Diversity and Distributions 20: 591–599.
17. Folke C (2006) Resilience: The emergence of a perspective for social-ecological systems analyses. Global Environmental Change-Human and Policy Dimensions 16: 253–267.
18. Serra P, Pons X, Sauri D (2008) Land-cover and land-use change in a Mediterranean landscape: A spatial analysis of driving forces integrating biophysical and human factors. Applied Geography 28: 189–209.
19. Cousins SAO (2009) Landscape history and soil properties affect grassland decline and plant species richness in rural landscapes. Biological Conservation 142: 2752–2758.
20. Bender O, Boehmer HJ, Jens D, Schumacher KP (2005) Analysis of land-use change in a sector of Upper Franconia (Bavaria, Germany) since 1850 using land register records. Landscape Ecology 20: 149–163.
21. Hooftman DAP, Bullock JM (2012) Mapping to inform conservation: A case study of changes in semi-natural habitats and their connectivity over 70 years. Biological Conservation 145: 30–38.
22. Baldi A, Batary P (2011) Spatial heterogeneity and farmland birds: different perspectives in Western and Eastern Europe. Ibis 153: 875–876.
23. Liira J, Schmidt T, Aavik T, Arens P, Augenstein I, et al. (2008) Plant functional group composition and large-scale species richness in European agricultural landscapes. Journal of Vegetation Science 19: 3–14.
24. Mikusinski G, Angelstam P (1998) Economic geography, forest distribution, and woodpecker diversity in central Europe. Conservation Biology 12: 200–208.
25. Kuemmerle T, Hostert P, Radeloff VC, van der Linden S, Perzanowski K, et al. (2008) Cross-border comparison of post-socialist farmland abandonment in the Carpathians. Ecosystems 11: 614–628.
26. Kuemmerle T, Muller D, Griffiths P, Rusu M (2009) Land use change in Southern Romania after the collapse of socialism. Regional Environmental Change 9: 1–12.
27. Primdahl J, Swaffield SR (2010) Globalisation and agricultural landscapes: change patterns and policy trends in developed countries. Cambridge, UK, New York: Cambridge University Press. xv, 275 p.
28. Sutherland WJ (2002) Restoring a sustainable countryside. Trends in Ecology & Evolution 17: 148–150.
29. Stoate C, Baldi A, Beja P, Boatman ND, Herzon I, et al. (2009) Ecological impacts of early 21st century agricultural change in Europe - A review. Journal of Environmental Management 91: 22–46.
30. Benton TG, Vickery JA, Wilson JD (2003) Farmland biodiversity: is habitat heterogeneity the key? Trends in Ecology & Evolution 18: 182–188.
31. Fahrig L, Baudry J, Brotons L, Burel FG, Crist TO, et al. (2011) Functional landscape heterogeneity and animal biodiversity in agricultural landscapes. Ecology Letters 14: 101–112.
32. Batáry P, Fischer J, Baldi A, Crist TO, Tscharntke T (2011) Does habitat heterogeneity increase farmland biodiversity? Frontiers in Ecology and the Environment 9: 152–153.
33. Robbins CS (1970) An international standard for a mapping method in bird census work. International Bird Census Committee. Audubon Field Notes 24: 722–726.
34. Geister I (1995) Ornitološki atlas Slovenije. Ljubljana: DZS.
35. Svensson L, Mullarney K (2009) Birds of Europe. Princeton, NJ [u.a.]: Princeton Univ. Press. 448 S. p.
36. Martinčič A, Wraber T, Jogan N, Podobnik A, Turk B, et al. (2007) Mala flora Slovenije. Ključ za določanje praprotnic in semenk. Četrta, dopolnjena in spremenjena izdaja. Ljubljana: Tehniška založba Slovenije. 968 p.
37. Højsgaard S, Halekoh U, Yan J (2006) The R Package geepack for Generalized Esimating Equations. Journal of Statistical Software 15: 1–11.
38. Ewers RM, Didham RK, Pearse WD, Lefebvre V, Rosa IMD, et al. (2013) Using landscape history to predict biodiversity patterns in fragmented landscapes. Ecology Letters 16: 1221–1233.
39. Aggemyr E, Cousins SAO (2012) Landscape structure and land use history influence changes in island plant composition after 100 years. Journal of Biogeography 39: 1645–1656.
40. Lindborg R, Eriksson O (2004) Historical landscape connectivity affects present plant species diversity. Ecology 85: 1840–1845.
41. Cousins SAO, Ohlson H, Eriksson O (2007) Effects of historical and present fragmentation on plant species diversity in semi-natural grasslands in Swedish rural landscapes. Landscape Ecology 22: 723–730.
42. Donlan CJ, Berger J, Bock CE, Bock JH, Burney DA, et al. (2006) Pleistocene rewilding: An optimistic agenda for twenty-first century conservation. American Naturalist 168: 660–681.
43. Navarro LM, Pereira HM (2012) Rewilding Abandoned Landscapes in Europe. Ecosystems 15: 900–912.
44. Franklin JF, Lindenmayer DB (2009) Importance of matrix habitats in maintaining biological diversity. Proceedings of the National Academy of Sciences of the United States of America 106: 349–350.

45. Zurita GA, Bellocq MI (2010) Spatial patterns of bird community similarity: bird responses to landscape composition and configuration in the Atlantic forest. Landscape Ecology 25: 147–158.
46. Tscharntke T, Klein AM, Kruess A, Steffan-Dewenter I, Thies C (2005) Landscape perspectives on agricultural intensification and biodiversity - ecosystem service management. Ecology Letters 8: 857–874.
47. Fischer J, Lindenmayer DB, Montague-Drake R (2008) The role of landscape texture in conservation biogeography: a case study on birds in south-eastern Australia. Diversity and Distributions 14: 38–46.
48. Eriksson O (1996) Regional dynamics of plants: a review of evidence for remnant, source-sink and metapopulations. Oikos 77: 248–258.
49. Vellend M, Verheyen K, Jacquemyn H, Kolb A, Van Calster H, et al. (2006) Extinction debt of forest plants persists for more than a century following habitat fragmentation. Ecology 87: 542–548.
50. Cousins SAO (2009) Extinction debt in fragmented grasslands: paid or not? Journal of Vegetation Science 20: 3–7.
51. Bignal EM, McCracken DI (1996) Low-intensity farming systems in the conservation of the countryside. Journal of Applied Ecology 33: 413–424.
52. Plieninger T, Hochtl F, Spek T (2006) Traditional land-use and nature conservation in European rural landscapes. Environmental Science & Policy 9: 317–321.
53. Katoh K, Sakai S, Takahashi T (2009) Factors maintaining species diversity in satoyama, a traditional agricultural landscape of Japan. Biological Conservation 142: 1930–1936.
54. Öckinger E, Lindborg R, Sjödin NE, Bommarco R (2012) Landscape matrix modifies richness of plants and insects in grassland fragments. Ecography 35: 259–267.
55. Gobster PH, Nassauer JI, Daniel TC, Fry G (2007) The shared landscape: what does aesthetics have to do with ecology? Landscape Ecology 22: 959–972.
56. Anonymous (2002) Pravilnik o uvrstitvi rastlinskih in živalskih vrst v rdeči seznam: Official Gazette of the Republic of Slovenia.
57. Gergely K (2007) Vörös Lista. A magyarorszagi edenyes flora veszelyeztett fajai. Red list of the vascular flora of Hungary. Sopron.
58. Niklfeld H (1986) Rote Listen gefährdeter Pflanzen Oesterreichs: 1. Fassung. Wien: Bundesministerium für Gesundheit und Umweltschutz. 202 S. p.
59. Bakan B (2006) Slikovni pregled višjih rastlin Prekmurja. Prispevek k poznavanju flore Prekmurja. Lendava: RC.

6

Conservation of Sandy Calcareous Grassland: What Can Be Learned from the Land Use History?

Anja Madelen Ödman*, Pål Axel Olsson

Biodiversity, Department of Biology, Lund University, Lund, Sweden

Abstract

Understanding the land use history has proven crucial for the conservation of biodiversity in the agricultural landscape. In southern Sweden, very small and fragmented areas of the disturbance-dependent habitat xeric sand calcareous grassland support a large number of threatened and rare plants and animals. In order to find out if historical land use could explain variation in present-day habitat quality, the land use on eight such sites was traced back to the 18th century and compared with key factors such as the amount of bare sand, lime content and P availability. There was no support for the common explanation of the decline in xeric sand calcareous grassland being caused by abandonment of agricultural fields during the last century. Instead, fertilization history was the main explanation for the difference in depletion depth of $CaCO_3$ seen between the sites. The decline in xeric sand calcareous grassland since the 18th century is most probably the result of the drastic changes in land use during the 19th century, which put an end to the extensive sand drift. Since cultivation was shown to have played an important role in the historical land use of xeric sand calcareous grassland, grazing alone may not be the optimal management option for these grasslands. Instead more drastic measures are needed to restore the high calcium content and maintain proper disturbance levels.

Editor: Peter Shaw, Roehampton university, United Kingdom

Funding: Financial support from The County Administrative Board, The Swedish Research Council, and the Gyllenstiernska Krapperup Foundation made this study possible. The funders had no role in study design, data collection and analysis, decision to publish, or preparation of the manuscript.

Competing Interests: The authors have declared that no competing interests exist.

* E-mail: Anja.Odman@biol.lu.se

Introduction

There is a growing awareness among conservation biologists regarding the importance of land use history and local management practices when implementing conservation and restoration measures for biodiversity and threatened species [1,2,3]. It is also recognised that land use changes long ago may have caused the vegetation patterns observed today [4,5]. For example, Eberhardt et al. [2] found that large areas of the sand forests at Cape Cod had been agricultural fields in the middle of the 19th century. They concluded that it was important to "mimic past agricultural practices in order to maintain and restore important sand-plain habitatsÓ. In particular for sandy areas, erosion and vegetation burial have been an important part of habitats history, and plants are adapted to this [6,7].

Throughout Europe, sandy grassland habitats support a large number of threatened species. Perhaps the best example of such sandy habitats is the threatened xeric sand calcareous grassland (Natura 2000 code 6120, 2002/83/EC Habitat Directive), which is home to several endangered species of vascular plants [8,9,10], bryophytes [11], fungi [12] and invertebrates [13]. These habitats are dry, open grasslands on calcareous, more or less humus-free, nutrient-poor and well-drained sandy soils with a discontinuous vegetation cover. The xeric sand calcareous grassland specialists are adapted to high pH [14] and low nutrient content [8,15], and many of the species depend on the open sand for regeneration [8,9]. In eastern Skåne (southern Sweden), these grasslands developed on lime-rich glaciofluvial sand [8,16], which was deposited by glacial meltwater when the ice retreated around 14,000 years ago, and later re-deposited by wind and waves [16]. Sites with natural erosion such as river banks, steep slopes, sand dunes and dune slacks have been proposed as sites where the type of species found in dry grasslands could have occurred before deforestation [17].

Before 3000 BC, eastern Skåne (Fig. 1) was predominantly covered by forest [18]. After 3000 BC, coppice agriculture led to an opening of the landscape, and there was a concentration of settlements to the flat coastal areas, and around 800 BC large areas of the coastal plains were deforested [18]. From 800 BC, permanent settlements replaced the mobile settlements and fields previously practiced [18]. At this time, intensified land use also opened up the landscape on other places in the Baltic region [19]. An infield/outland system with fertilisation was introduced in the area and there was an expansion of cereal cultivation, which also increased erosion. The traditional method of farming was extensive and with long (up to 30 years) periods of fallow when the arable fields were used for grazing [20,21,22]. During the 17th and 18th centuries the human population in southern Sweden grew, which caused further deforestation and intensified agricultural practices [20,22], and in the 18th century, Linné [23] noted an extensive sand drift and the occurrence of sand fields all over eastern Skåne. The problem with sand drift was for the most part solved at the end of the 18th century and during the 19th century by the introduction of ley plants and more efficient use of fertilisers, which increased the production per surface area and thereby stopped over-exploitation. At the same time, the afforestation of Skåne started [22] and pine was planted to

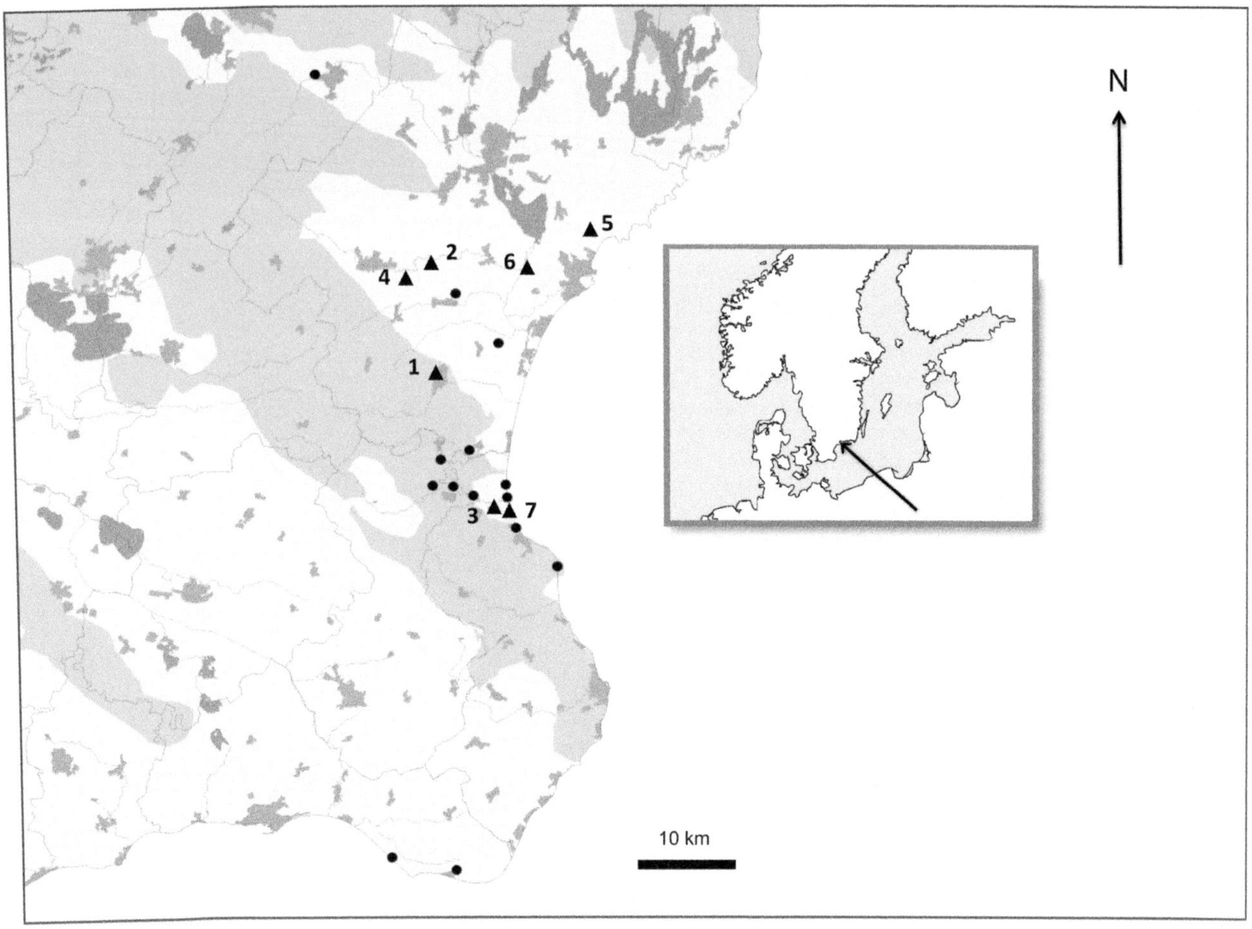

Figure 1. Map showing the existing xeric sand calcareous grassland sites in Skåne. Sites included in the study are represented by triangles and circles represent sites not included. Grey shading indicates forested land and unshaded represents open land. Scale bar represent 10 km. Enclosed is an overview map for southern Scandinavia, where an arrow points out the study area. The sites chosen for the study were: (1) Degeberga, (2) Everöd, (3) Klammersbäck, (4) Lyngsjö, (5) Rinkaby, (6) Ripa and (7) Vitemölla.

prevent wind erosion. Similar trends with deforestation until the middle of the 19th century, and then afforestation again has also been found in other places in Sweden [24] as well as in New England, USA [25].

Today, the old agricultural practices in sandy grasslands have been abandoned, and the arable fields have been turned into pine plantations or pastures [26,27], leaving the vegetation cover to close and thereby ending sand drift [28]. A natural accumulation of nutrients and organic matter follows [10], which can be further accelerated by atmospheric nitrogen deposition [29]. In addition, there is a threat of acidification of the topsoil [8]. The remaining areas of xeric sand calcareous grassland are small and fragmented, which poses a further threat to the rarest species (Fig 1).

Linné [23] described a number of xeric sand calcareous grassland species growing in eastern Skåne. In particular *Dianthus arenarius* ssp *arenarius*, which is a key species in these grasslands and protected by the Natura 2000 network (EU Habitat Directive, Annex II), was mentioned several times and was said to grow on all the sand fields, almost like a weed. In fact, all xeric sand calcareous grassland specialists but two, *Silene conica* and *Alyssum alyssoides*, which came to Skåne in the 19th century, were present in Skåne in 1749 [23,30,31]. Observations by Linné [23] and Campbell [20] suggest that xeric sand calcareous grassland was widespread in eastern Skåne during the 18th century, but there are no records of the actual area before the 1970s. Since the 1970s, reported areas range between 30 and 50 ha, with no consistent trend between the different studies [26, 27 32]. However, there is a strong indication of decline in xeric sand calcareous grassland species since the 18th century, which is reflected in the high density of nationally red-listed (and thereby declining) plant species [8].

Changes in land use during the last century, i.e. that the agricultural practices on these grasslands were abandoned, has been the common explanation for the decline of xeric sand calcareous grassland in Skåne [26 27, 33]. However, this hypothesis had never been tested prior to this study. The aim of this study was to improve the conservation of dry calcareous grassland by increasing the understanding of how xeric sand calcareous grasslands came to be and why they have disappeared. Land use change and decalcification were compared at eight sites in eastern Skåne in areas with a range of different quantities and qualities of xeric sand calcareous grassland. The amount of bare sand and the $CaCO_3$ content of the soil were used as proxies for favourable conditions and, in the case of open sand, as a proxy for erosion. Historical maps were used to determine the former land use and aerial photographs were used to map the extent of bare sand during the last 70 years. By analysing soil profiles, possible links between land use history and decalcification in the studied grasslands were investigated at specific sites within each studied landscape. Extractable P concentrations were also measured as a proxy for fertilisation history, since P remains in the soil for a longer period than N and causes problems during restoration [34,35].

Materials and Methods

Site descriptions

The sites studied were situated in the coastal area of eastern Skåne in southernmost Sweden. The area is 80 km long and 35 km wide, situated between 6172130 and 6204470 N, and 441330 and 458270 E (SWEREF99 TM). Sites were selected on the basis that they contained at least some remnants of xeric sand calcareous grassland vegetation. They should also be, or be part of, a larger open area. Since the focus of the study was the effects of agriculture, sites known to have been created by sand pits or other digging activities were excluded. Still, the majority of landscapes with xeric sand calcareous grassland in the area were included. The sites chosen for the study were: 1. Degeberga (6188081, 442627), 2. Everöd (6198995, 443126), 3. Klammersbäck (61749171, 448414), 4. Lyngsjö (6198615, 442108), 5. Rinkaby (6202616, 455831), 6. Ripa (6198003, 452133) and 7. Vitemölla (6173465, 449873) (Fig 1). The sites are listed in Table 1, where information about their topography, present day management and the amount and quality of the remaining xeric sand calcareous grassland vegetation can be found. These sites are all under nature conservation management, and the conducted study was made in collaboration with the county administrative board in Skåne and followed site-specific regulations. The Everöd, Klammersbäck, Lyngsjö, Rinkaby and Ripa sites are all more or less flat areas. The vegetation at these sites predominantly consists of Fennoscandian lowland species-rich, dry to mesic grassland (N6270, EU Habitat Directive) as well as small areas of heath-like vegetation resembling continental dunes with open Corynephorus and Agrostis grasslands (N2330, EU Habitat Directive) [8,28]. All of the flat sites contain patches of xeric sand calcareous grassland of varying sizes, depending on the level of degradation. Degeberga and Vitemölla are hilly sites with large presence of xeric sand calcareous grassland vegetation on the south facing slopes, and with Fennoscandian lowland species-rich, dry to mesic grassland on flatter areas.

Soil sampling and analyses

Soil was sampled at the following locations: Everöd (6199099, 443104), Klammersbäck (6174279, 448413), Lyngsjö (6198665, 442138), Rinkaby (6202600, 455808), Ripa pasture (6198191, 452028), and Ripa airfield (6197740, 452277) (SWEREF99 TM). At each of these sites a smaller area was selected that was assumed to consist of degraded xeric sand calcareous grassland. The vegetation at most sampling sites was either Fennoscandian dry to mesic grassland or dry Corynephorus grassland. The soil was sampled on 8-18th June 2009 and soil cores of 20 cm length (Ø 15 cm) were taken down to 1.6 m using a soil auger. Three profiles were sampled at each site, located approximately 10 m from each other in a triangle. At the Ripa airfield site, the soil could only be sampled down to 1.2 m for two of the three profiles because bedrock or boulders were encountered at this depth.

For pH measurements, 10 g of soil was mixed with 50 ml distilled water, and the pH was measured electrometrically in supernatants obtained by 2-h extraction in a rotator. The total amounts of Ca in the samples were measured by digesting the soil in concentrated HNO_3, followed by ICP-AES analysis. One g of soil at field moisture and 50 ml HNO_3 were placed in a 300 ml flask, covered with a watch glass and heated on a hot plate for

Table 1. A summary of the characteristics of the eight sites analyzed in this study, including the total area analysed and the area of xeric sand calcareous grassland (Natura 2000 habitat 6120) according to Olsson [32].

Site	Topography	Present day management/Disturbance	Area (ha)		Vegetation (vascular plants)				
			Total	Habitat type 6120	Total cover (%)	Specialists cover (%)	Red-listed cover (%)	Red-listed (Nr/m^2)	N
1. Degeberga	Hilly	Ungulate grazing	13.5	2.6	47±20	23±13	17±14	2.7±0.6	3
2. Everöd	Flat	Ungulate grazing	15	4.9	57±23	14±16	9.2±15	1.0±0.7	10
3. Klammersbäck	Flat	Ungulate grazing	38	0.09	34±16	4±1.4	2.5±3.5	0.5±0.7	2
4. Lyngsjö	Flat	Horse grazing	3	0.1	58±23	12±16	10±16	0.6±0.7	17
5. Rinkaby	Flat	Ungulate grazing/military training	155	4.5	64±20	21±15	14±12	1.4±0.8	25
6a. Ripa airfield	Flat	Fallow/wild grazers (rabbits)	14	0.2	71±23	20±26	17±29	0.3±0.6	3
6b. Ripa pasture	Flat	Ungulate grazing	22	3.0	52±6.4	7.0±5.7	5.5±6.4	1.0±0.0	2
7. Vitemölla	Hilly	Fallow/recreation, wild grazers (rabbits)	38	4.8	59±22	23±21	16±18	1.7±1.3	12

Vegetation data comes from a survey in 2004-2005 [9] and includes the total vegetation cover, as well as the cover of specialist species and nationally red-listed species as defined in Olsson and Ödman [55]. In addition, the number of red-listed species per square meter (Nr/m^2) is given. Means are displayed (± SD), and N denotes the number of analysed squares (1 m^2) at each site. The dominant red-listed and specialist species were *Koeleria glauca*, *Dianthus arenarius* ssp *arenarius* and *Alyssum alyssoides*.

72 h. The solution was evaporated to 2 ml and diluted to 25 ml with distilled water. Plant-available (extractable) ortho-phosphate (P_{ext}) in the top 20 cm was determined using flow injection analysis, with stannous chloride and ammonium molybdate as reactant (measured at 720 nm, method ISO/FDIS 15681-1). Ten g fresh soil was extracted for 1 min with Bray-1 solution [36].

Soil in Degeberga and Vitemölla was not sampled during this study and samples collected for earlier studies by Olsson et al. [8] and Bahr et al. [37] respectively were used instead. At Degeberga, two samples were used: one from Degeberga south slope (xeric sand calcareous grassland) and one from Degeberga north slope (Fennoscandian dry to mesic grassland), which had been sampled down to 1.5 m. At Vitemölla, four samples were used from a mosaic of xeric sand calcareous grassland and dry Corynephorus grassland. The samples were taken down to 0.6 m at three sites and 0.5 m for one of the sites. Data for soil pH and Ca content comes from data collected for earlier studies by Olsson et al. [8] and Bahr et al. [37]. At Vitemölla, Ca concentration values were only available for the top 0.1 m. Additional measurements of P_{ext} were performed on the soil samples for this study, using the methods described above. The samples from the top 0.1 m for all samples at Vitemölla had unfortunately been used up in the previous study, and the 0.1–0.2 m layer was used for the P_{ext} analysis.

Analysis of land use history

Information on land use in the past was obtained through visual analysis of historical maps and aerial photographs. The area at each site was delimited so that it enclosed the soil sampling plots, had clearly defined borders (arable field margins, roads, water etc.) and only included open land without trees in the 2007 aerial photograph. The sizes of the delimited areas are presented in Table 1. Historical maps were obtained from Lantmäteriet (the Swedish land surveying agency), thru a thorough search in their archives [38]. The historical maps used dated from the middle of the 18th century (oldest maps available from Lantmäteriet) until 1974 and consisted of economic and property redistribution maps (Table 2). The analyses used arable field margins and other objects to locate the sites, and explanatory text or legends were used to determine land use. The land use was either printed on the map or found in a supplementary document accompanying the map. The land uses recorded were: arable field, pasture, meadow, forest and other land uses of interest such as military training or recreation. Areas named "utmark" (outland), "fålad" (old word for common pasture in southern Sweden) and "allmänning" (common land) were classified as pastures, and areas named "wång" (old word for common arable fields in southern Sweden) were classified as arable fields. In the economic map from 1974, only arable fields were mapped and all other land uses were treated as one class (other). The aerial photographs used dated from the 1940s until 2007 (Table 2) and were obtained from Lantmäteriet [39]. The resolution varied between 0.2 m^2 and 1 m^2. The analysis, which was done visually, was based on the following criteria: forest (>50% tree cover), arable field (open, with furrows) or pasture (open, without furrows). Furrows were interpreted as signs of ploughing. The map and aerial photograph analyses were performed at each site, both for the whole area (landscape level) and at the exact sites (plot level) where the soil was sampled. The results from the analysis were used to calculate the number of years since the area was last used for cultivation. The number of years was counted from the last year the area was recorded as having been cultivated. In the case of Degeberga and Vitemölla, which had never been cultivated, the number of years from the first available record of land use was used.

Analysis of the amount of bare sand and forest

The aerial photographs described above were used to analyse how the amount of bare sand changed between the 1940s and 2007. The aerial photographs were rectified and classified using ArcGIS Desktop 9.3 (ESRI 2008). The delimited areas were the same as those described above for the historical land use analysis. The amount of bare sand was estimated by dividing the surface features shown on the aerial photographs into two classes: bare sand and other (anything which was not bare sand). The classes were identified separately for each photograph using supervised classification where areas of bare sand were identified based on colour (and previous knowledge of the areas for the 2007 photographs), and then used as training areas for the software. The delimited areas were then classified using maximum likelihood classification. The classification for 2007 was evaluated in the field by comparing classified and real patches, and the older ones by comparing the classified patches of bare sand with the visual ones in the aerial photographs. The analysis identified patches of bare sand, but did not consider the overall density of the vegetation. Consequently, only patches with close to 100% bare sand could be identified, and areas with partly bare sand of varying amounts were classified as "other" since they could not be distinguished.

The amount of forest (>50% tree cover) in the surrounding landscape was also analysed from the aerial photographs. The size of the aerial photographs varied considerably between sites and years, and four different areas around the sites, 1×1, 1.5×1.5, 2×2 and 3×3 km, were used. The Everöd and Lyngsjö sites are situated within the same area and were analysed together. Analysis showed that, in those areas where squares larger than 1×1 km could be fitted, the amount of forest for all years increased slightly with increasing size of areas (although not significantly so). However, differences between the years were constant and so only results from the analyses using the 1×1 km squares are presented. Polygons were drawn manually around all forest and the percentage of forest in the total area was calculated using ArcGIS.

Statistical analyses

Correlations between land use history, amount of forest, amount of bare sand, depletion depth for $CaCO_3$, maximum $CaCO_3$, pH and P_{ext} content were performed using Pearson correlation or Spearman's correlation when parameters were not normally distributed. For the amount of forest and amount of bare sand, correlations were also performed with year, using the same method as described above. Differences between sites, land use history and depths in terms of the amount of $CaCO_3$ and P were tested using a general linear model followed by Tukey's post hoc test. All statistical analyses were performed using IBM SPSS Statistics (IBM Corporation, New York, US). An Arcsine square root transformation of the parameter amount of bare sand was performed to obtain normal distribution. For the correlation between pH and $CaCO_3$, the $CaCO_3$ values were log transformed. Results presented are mean values and standard deviations.

Results

Historical land use changes

The results from the analysis of land use change at the soil sampling plots showed that the last record of cultivation at the six flat sites was between 1940 and 1957, except for the Ripa pasture, where indications of cultivation could be seen also in aerial photographs from 1985 (Table 2). At the sampling plots at the two

Table 2. Results from visual analysis of land use history through historical maps (economic and property redistribution maps, with original names in Swedish) and aerial photographs for all eight soil-sampling sites.

	1. Degeberga	2. Everöd	3. Klammersbäck	4. Lyngsjö	5. Rinkaby	6a. Ripa airfield	6b. Ripa pasture	7. Vitemölla
Aerial photographs								
2007	Pasture	Pasture	Pasture	Pasture	Military training/Pasture	Pasture	Pasture	Pasture/recreation
2001					Military training/Pasture			
1999	Pasture	Pasture	Pasture	Pasture		Pasture	Pasture	Pasture/recreation
1985	Pasture	Pasture	Pasture	Pasture	Military training/Pasture	Airfield/Pasture	Arable field?	Pasture/recreation
1970					Military training/Pasture	Airfield/Pasture	Arable field?	
1969	Pasture	Pasture	Pasture	Arable field				Pasture/recreation
1957	Pasture	Arable field		Arable field		Pasture?	Arable field	
1956			Pasture		Arable field			Pasture/recreation
1940	Pasture	Arable field	Arable field	Arable field	Arable field	Arable field	Arable field	Pasture/recreation
Historical maps								
1974, Ekonomiska kartan	Other	Other	Other	Other	Other	Other	Other	Other
1926-34, Härads ekonomiska	Pasture	Arable field	Arable field	Arable field	Arable field	Arable field	Arable field	Pasture
1845, Laga skifte					Plantation			Meadow, drift sand
1826, Enskifte		Arable field		Arable field				
1822, Enskifte			Arable field					
1818, Storskifte						Arable field	Arable field	
1811, Enskifte	Pasture							
1803, Avmätning		Arable field		Arable field				
1751, Ägodelning					Arable field			Pasture
Years since cultivation	196	50	67	38	51	67	22	256

"Other" means an area not classified as arable field or real estate in the economic maps (Ekonomiska kartan). Maps and aerial photographs were obtained from Lantmäteriet [38,39].

hilly sites, none of the maps or aerial photographs showed signs of agricultural cultivation (Table 2).

At the landscape level, the six flat areas, Everöd, Lyngsjö, Klammersbäck, Ripa pasture, Ripa airfield and Rinkaby, have all been cultivated but, according to the maps, cultivation ceased between 1940 and the 1990s. They have gone through a historical land use change from a mixture of cultivation and grazing and mowing, to a mixture of forest and grazing. During the 19th century arable fields were common, but at Everöd, Rinkaby and Klammersbäck, pastures occurred as well, and these pastures were managed as commons. At the two coastal sites Rinkaby and Klammersbäck, the common pastures occurred closest to the sea, while fields occurred inland. During the 20th century, the agricultural fields were gradually abandoned and turned into pastures. During this time, forests started to emerge at the Everöd, Rinkaby and Ripa sites.

The two hilly sites, Vitemölla and Degeberga, have a rather similar history. In 1756, the northern part of the site at Vitemölla was classified as very poor arable fields spoiled by sand drift and therefore unusable. The southern part was classified as useless and left as a common in 1778. In 1828 and 1845, the southern part of the area was described as sandy soil and drift sand, and the lower parts close to the sea as useless. According to the maps from 1926-34 and onward, the area was used as pasture and for recreation. The Degeberga site has been used as pasture continually from the first map in 1811 until today. Consequently, according to the maps used in this study, the two hilly sites were not cultivated during the analysed period. However, the note from 1756 about parts of Vitemölla being arable field spoiled by sand drift indicates that some parts had previously been cultivated.

Forest cover and amount of bare sand from 1940 to 2007 at landscape level

There was no significant temporal trend in the mean forest cover between 1940 and 2007, when analysed in 1×1 km squares within each landscape. In three areas (Ripa, Everöd/Lyngsjö and Rinkaby) there was an increase in forest cover, while the forest cover decreased in Klammersbäck, and almost no changes could be seen in the other two areas (Fig 2a). The area with the highest mean forest cover over the years ($n = 6$) was Klammersbäck (50±6%), followed by Rinkaby (44±4%), Degeberga (28±1%), Ripa (20±21%), Everöd/Lyngsjö (18±6%) and Vitemölla (11±2%).

The proportion of bare sand was analysed for eight sites, as Ripa airfield and Ripa pasture were treated separately, ranging in size between 3 and 155 ha. At Vitemölla, Rinkaby and Degeberga, the analysis showed a decrease in the amount of bare sand between 1940 and 2007 (Fig 2b). However, the general trend for all sites was that changes since the 1940s were not drastic, and that the mean amount of bare sand at the studied sites was never more than about 3% of the total study area. There was no significant change in the mean amount of bare sand between 1940 and 2007. The largest change was found at Vitemölla, where the amount of bare sand decreased from 11.2% to 4.4%. There was no

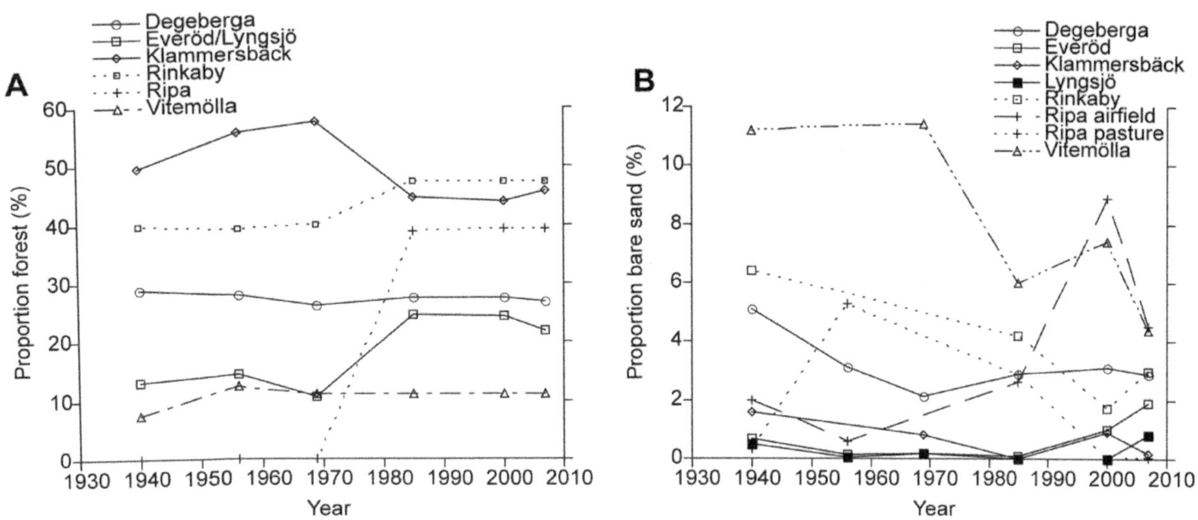

Figure 2. The proportion of land area occupied by forest (a) and bare sand (b) between 1940 and 2009 as analysed from aerial photographs (for simplification we did not differ between 1956/1957 and 1969/1970 in the graph). The proportion of forest was analysed for a 1×1 km square around the sites (Everöd/Lyngsjö and Ripa airfield/Ripa pasture were each contained within the same square), and the proportion of bare sand for the whole area of each of the eight sites.

correlation between the amount of bare sand and the forest cover or the number of years since cultivation.

Historical land use and soil properties

Extractable P (P_{ext}) showed large variation between sites. Lyngsjö had the highest mean (n = 3) concentration of P_{ext} (79±18 μg g^{-1} soil), followed by Ripa pasture (51±4.4 μg g^{-1} soil), Ripa airfield (48±4.4 g^{-1} soil) and Everöd (47±6.8 μg g^{-1} soil). Degeberga north (26 μg g^{-1} soil) and the four sampling points at Vitemölla (13±8.2 μg g^{-1} soil) had intermediate concentrations, while the sites at Rinkaby (6.5±4.4 μg g^{-1} soil), Degeberga south (6.3 μg g^{-1} soil) and Klammersbäck (5.1±1.6 μg g^{-1} soil) showed considerably lower P_{ext} values than the other sites. Lyngsjö had significantly higher P_{ext} values than all the other sites ($p<0.05$) and the Everöd, Ripa pasture and Ripa airfield sites had significantly higher P_{ext} values than Degeberga, Klammersbäck, Rinkaby and Vitemölla ($p<0.05$). No significant differences were found between the other sites. There was a positive correlation between the P_{ext} and the depletion depth in the sites that had been cultivated (N = 18, $R^2 = 0.53$, $p<0.001$, Fig. 3). There was no significant correlation between the P_{ext} and the number of years since cultivation or the amount of bare sand.

None of the sites had an enrichment of $CaCO_3$ at any particular depth. Most of the topsoils were depleted but, when the lime horizon was reached, the $CaCO_3$ content remained rather constant further down (Fig. 4). The lowest $CaCO_3$ values were found at Lyngsjö (0.2±0.03%), which was also the only site where the $CaCO_3$ content never exceeded 10%, even in the deepest layer. The only sites with a $CaCO_3$ content in the topsoil above 2% were Rinkaby (13.1±3.4%) and Vitemölla (5.4% in sampling point 2), and at 0.2–0.4 m only Everöd, Klammersbäck and Rinkaby had $CaCO_3$ concentrations above 2%. The pH values were strongly correlated with the $CaCO_3$ values (N = 156, $R^2 = 0.93$, $p<0.001$).

The maximum $CaCO_3$ concentration for each profile was identified as well as the depletion depth (the first depth where the $CaCO_3$ concentrations exceeded 2%). The maximum $CaCO_3$ value was significantly higher at Rinkaby and Degeberga than at Everöd, Klammersbäck, Lyngsjö and Ripa pasture ($p<0.05$). The maximum $CaCO_3$ value at Ripa airfield was significantly higher than that of Lyngsjö ($p<0.001$), but did not differ from that of Rinkaby and Degeberga. Lyngsjö had a significantly lower maximum $CaCO_3$ value than all other sites except Klammersbäck ($p<0.001$).

There was a positive correlation between the maximum $CaCO_3$ and the mean amount of open sand (N = 7, $R^2 = 0.73$, $p<0.01$). The maximum $CaCO_3$ value and the depletion depth were not significantly correlated but there was a weak negative correlation between maximum $CaCO_3$ and the depth at which the maximum $CaCO_3$ was found (N = 20, $R^2 = 0.30$, $p<0.01$). The maximum $CaCO_3$ and the pH in the top 20 cm were also positively correlated (N = 20, $R^2 = 0.28$, $p<0.01$). There was no correlation between the number of years since cultivation and the depletion depth, or between the mean amount of bare sand and the depletion depth.

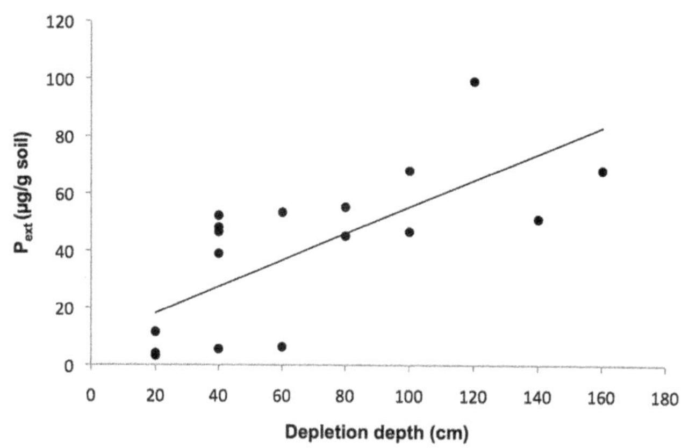

Figure 3. The correlation between P_{ext} (μg g-1 soil) and depletion depth (cm) at the previously cultivated sites (N = 18, $R^2 = 0.53$, $p<0.001$). The number of years since cultivation was counted from the last year the area was recorded as having been cultivated. In the case of Degeberga and Vitemölla the number of years from the first available record of land use was used.

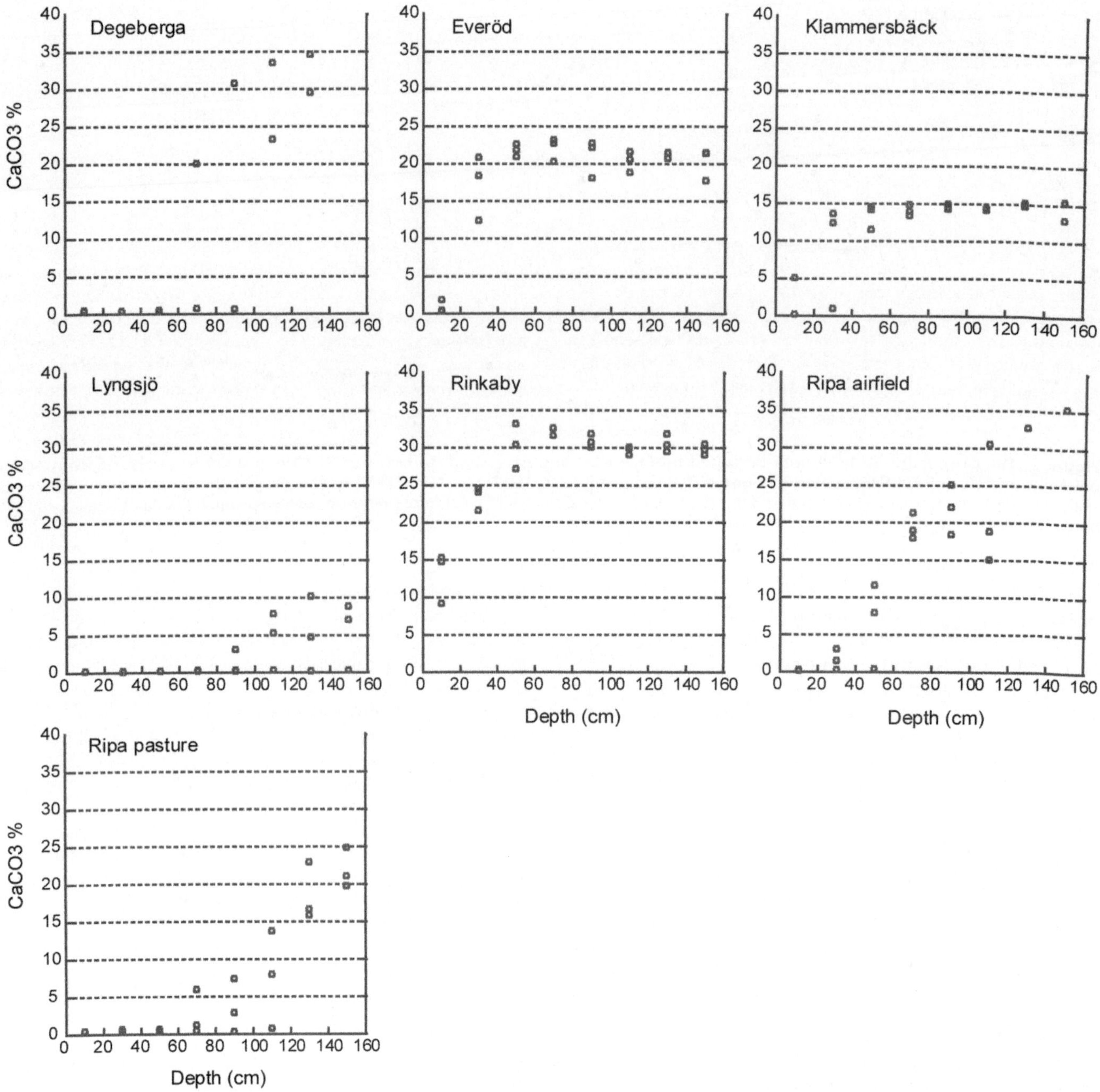

Figure 4. The $CaCO_3$ content as percentage of the soil weight for 20 cm at a time down to 160 cm. Vitemölla was not included since samples were only available for the top 20 cm.

Discussion

Most of the remaining xeric sand calcareous grasslands in Skåne are found on eroded sites [27], where the underlying lime-rich soil is exposed. This gives an indication about the great importance of erosion and other soil disturbances in the restoration and preservation of these grasslands. The importance of sand pits and military training areas for the conservation of biodiversity and threatened species has also been demonstrated in other parts of Europe [40,41,42,43]. An important finding of this study was that all the flat sites with xeric sand calcareous grassland included in the study had been cultivated in the past, which suggests that former agricultural practices were important in creating the habitat where natural erosion does not occur. Similarly, the chalk grasslands of Salisbury Plain in the UK, which were previously only thought to have a history as sheepwalks, were found to have a history of cultivation [44]. In other types of semi-natural grasslands, grazing continuity is the most important factor for diversity of grazing-dependent species [45,46]. In the case of sandy grasslands it may instead be that continuous traditional low-intensity cultivation is a very important factor in creating the proper conditions [47], particularly on flat areas where natural erosion is low when covered with vegetation and trampling is not enough to create bare soil. Therefore, the result that disturbance-dependent species were found in areas with long-term continuous soil disturbance created by cultivation should not be surprising, nor should the decline of dry sandy grassland species as soil disturbance decreased.

The cultivation of the soil is thought to have prevented the vegetation cover in the xeric sand calcareous grassland from closing and the $CaCO_3$ from being depleted by mixing the soil [26]. The highest average amount of bare sand during the period 1940 to 2007 was just over 3%, and there was no drastic change in the amount of bare sand during this period. The descriptions of Linné [23] and Campbell [20] indicate that there was much more bare sand in the area in the 18th century, and the most important decline probably occurred during the 19th century. Although there was a difference between the sites in amount of bare sand, there was no negative correlation between the amount of bare sand and the number of years since cultivation, which would be expected if lack of cultivation is an important factor in explaining recent declines in xeric sand calcareous grassland species. In fact, one of the sites not cultivated during the last 260 years had the highest amount of bare sand even in the 1940s. It must be acknowledged that there is a problem when comparing flat and hilly sites, as sloping ground will naturally be more prone to soil erosion. However, even when the two hilly sites were removed from the analysis, no correlation between bare sand and number of years since cultivation could be seen. There was a possible constraint to the method used in this study to detect the amount of bare sand. Assuming that cultivation was the important factor and it only occurred every 8–20 years, the analysis made using aerial photographs from just one date every fifteen years in this study could have failed to spot the bare sand produced. Completely bare sand might have been visible only in the first years, although the vegetation cover could have been sparse for many more. This might still have been enough of disturbance to sustain xeric sand calcareous grassland, and the failure to detect semi-bare sand was a weakness in the method used.

Another explanation for the decline in xeric sand calcareous grassland species could be nutrient enrichment. In the present study, some of the more recently cultivated sites had much higher P_{ext} value than sites that were cultivated a long time ago or not at all. However, Rinkaby and Klammersbäck stood out with their low P_{ext} values, suggesting that they have not been fertilized despite their recent cultivation history. The higher P_{ext} values found at some of the more recently cultivated plots could be a reason for the lack of correlation between the amount of bare sand and the number of years since cultivation, since high P values could speed up succession [48], thereby counteracting the effects of soil disturbance. In addition, the fertilisers would also have contained N, which is known to accelerate the decalcification process [49,50,51]. This was supported by the positive correlation between P_{ext} and depletion depth of $CaCO_3$, which suggests that fertilization could be the main explanation for the difference in depletion depth seen between the sites.

The depletion depth of $CaCO_3$ was not correlated with the time since cultivation, which would be expected if the important factor for retention of $CaCO_3$ in the topsoil was indeed cultivation. No correlation could be found between the mean amount of bare sand and the depletion depth, indicating a lack of connection between depletion and the amount of soil disturbance during the last century. The positive correlation between P and depletion depth of $CaCO_3$ would in fact suggest the opposite, i.e. that cultivation in the last 70 years, if including fertilization, might have had a negative effect. The maximum $CaCO_3$ value and the depth at which it was found were negatively correlated, although the correlation was weak. Still, this suggests that the depletion depth of $CaCO_3$ partly depended on the soil type. Different glaciofluvial deposits have different $CaCO_3$ content depending on the lithology of the local bedrock and the previously deposited soil types [16], which would explain the observed variation in $CaCO_3$ content between sites. None of the above-mentioned explanations seems entirely satisfactory in explaining differences in depletion depth between sites and they could not explain the presence of $CaCO_3$ in the topsoil of some sites. It is important to keep in mind that depletion has been going on throughout the 14,000 years since the soil was deposited by glacial melt-water. Consequently, there must have been a process counteracting the depletion.

The results discussed above indicate that ploughing was not directly preventing the $CaCO_3$ from being depleted. However, it is probable that the wind erosion resulting from the cultivation could have counteracted depletion by exposing the $CaCO_3$ rich sand. It seems to be a likely scenario, since wind erosion has already been found to remove fertile topsoil and cause organic matter content to decline [52,53]. Some sand drift still occurs in southern Sweden [22], but it is limited compared to the extensive sand drift of the 18th century described by Linné [23] and Campbell [20]. The most important factor determining the amount of sand drift is how much bare sand is available for erosion. During the time period from the 1940s until today, the amount of bare sand has been very low, but in the 18th century extensive areas of bare sand were observed due to the intensified cultivation [20,22,23]. One other factor that would affect wind erosion is the forest cover in the surroundings. Although the amount of forest at some sites increased during the last 70 years, no correlation was found with the amount of bare sand or the $CaCO_3$ content. However, a great change in forest cover took place in the 19th century, when forest was planted to bind the sand [22].

The results strongly suggest that the causes of present-day decline in threatened species may have origins far back in time and that conservation should take this into account. Human activity, such as expansion of arable fields and intensive livestock grazing, also caused sand drift in Germany [9,10], the UK [54] and the Netherlands [52]. In Germany and the Netherlands, dry sandy grasslands, including xeric sand calcareous grassland and Corynephorus grassland, are believed to have developed in areas where arable farming or livestock grazing created bare sand, which was then exposed to wind erosion and resulted in drifting sand [9,52]. Today, these sandy grasslands, in common with similar areas in Sweden, are suffering from changes in land use including afforestation, intensification and abandonment [9,52].

Conclusions

There was no support for the common explanation of the decline in xeric sand calcareous grassland being caused by abandonment of agricultural fields during the last century, i.e. that the old agricultural practices in these grasslands were abandoned. Instead, fertilization history (as indicated by soil P levels) was the main explanation for the difference in depletion depth of $CaCO_3$ seen between the sites. The decline in xeric sand calcareous grassland since the 18th century is most probably the result of the drastic changes in land use during the 19th century, which put an end to the extensive sand drift. The importance of sand movement is shown by the high density of red-listed vascular plant species at the hilly sites (Table 1). Since cultivation played a very important role in the creation of xeric sand calcareous grassland, grazing alone is not a good conservation option for these grasslands [28,55]. Instead more drastic measures are needed to restore the high calcium content, remove the accumulated nutrients and create bare sand for regeneration

[56]. An example of such a restoration measure is topsoil removal, where the decalcified and nutrient rich topsoil is removed much like it would be during sand drift [57]. Not surprisingly, topsoil removal has been shown to be a successful method for restoring xeric sand calcareous grasslands [55].

Acknowledgments

The authors would like to thank Merit Kindström and Micael Runnström for input on the methods used in the map and aerial photograph analysis, and for helpful comments on the manuscript.

Author Contributions

Conceived and designed the experiments: AMÖ PAO. Performed the experiments: AMÖ. Analyzed the data: AMÖ PAO. Contributed reagents/materials/analysis tools: AMÖ PAO. Wrote the paper: AMÖ PAO.

References

1. Poschlod P, WallisDeVries MF (2002) The historical and socioeconomic perspective of calcareous grasslands – lessons from the distant and recent past. Biol Conserv 104: 361–376.
2. Eberhardt RW, Foster DR, Motzkin G, Hall B (2003) Conservation of changing landscapes: vegetation and land-use history of Cape Cod National Seashore. Ecol Appl 13: 68–84.
3. Antrop M (2005) Why landscapes of the past are important for the future. Landscape Urban Plan 70: 21–34.
4. Motzkin G, Foster D, Allen A, Harrod J, Boone R (1996) Controlling site to evaluate history: vegetation patterns of New England sand plain. Ecol Monogr 66: 345–365.
5. Lunt ID, Spooner PG (2005) Using historical ecology to understand patterns of biodiversity in fragmented agricultural landscapes. J Biogeogr 32: 1859–1873.
6. Kent M, Owen NW, Dale P, Newnham RM, Giles TM (2001) Studies of vegetation burial: a focus for biogeography and biogeomorphology? Progress in Physical Geography 25: 455–482.
7. Owen NW, Kent M, Dale MP (2004) Plant species and community responses to sand burial on the machair of the Outer Hebrides, Scotland. J Veg Sci 15: 669–678.
8. Olsson PA, Mårtensson L-M, Bruun HH (2009) Acidification of sandy grasslands – consequences for plant diversity. Appl Veg Sci 12: 350–361.
9. Poschlod P, Baumann A, Biedermann H, Bugla B, Neugebauer K (2009) Dry sandy grasslands in Southern Germany – a case study of how to re-develop remnants to achieve the former status of a high nature value landscape. In Veen P, Jefferson R, de Smidt J, van der Straaten J, editors. Grasslands in Europe of High Nature Value. KNNV Publishing, the Netherlands. pp. 113–121.
10. Eichberg C, Storm C, Stroh M, Schwabe A (2010) Is the combination of topsoil replacement and inoculation with plant material an effective tool for the restoration of threatened sandy grassland? Appl Veg Sci 13: 425–438.
11. Tyler T (2005) The Bryophyte flora of Scanian sand-steppe vegetation and its relation to soil pH and phosphate availability. Lindbergia 30: 11–20.
12. Olsson PA, Schnoor TK, Hanson S-Å (2010) pH preferences of red-listed gasteromycetes in calcareous sandy grasslands: Implications for conservation and restoration. Fungal Ecol 3: 357–365.
13. Ljungberg H (1999) Skalbaggar och andra insekter på sandstäppslokaler i östra Skåne (Beetles and other insects on xeric sand calcareous grassland sites in eastern Scania). Länsstyrelsen i Skåne län, Malmö, 1–56.
14. Ellenberg H (1991) Indicator values of vascular plants in Europe 1. Indicator values of plants not including *Rubus*. Scripta Geobot 18: 9–166
15. Mårtensson L-M, Olsson PA (2010) Soil chemistry of local vegetation gradients in sandy calcareous grasslands. Plant Ecol 206: 127–138.
16. Persson KM (2000) Description to the quaternary map 2D Tomelilla NO. The Swedish Geological Survey, Uppsala, Sweden.
17. Pigott CD, Walters SM (1954) On the interpretation of the discontinuous distributions shown by certain British species of open habitats. J Ecol 42: 95–116.
18. Berglund BE, Larsson L, Lewan N, Olsson EGA, Skansjö S (1991) Ecological and social factors behind the landscape changes, in: Berglund BE (Ed.), The cultural landscape during 6000 years in southern Sweden. Ecol Bull 41: 425–444.
19. Poska A, Saarse L (1999) Holocene vegetation and land-use history in the environs of Lake Kahala, northern Estonia. Veget Hist Archaeobot 8: 185–197.
20. Campbell Å (1929) Skånska bygder under förra hälften av 1700-talet. Etnografisk studie över den skånska allmogens äldre odlingar, hägnader och byggnader (The Scanian countryside during the first half of the 18th century). Thesis, Uppsala University, Sweden.
21. Dahl S (1989) Studier i äldre skånska odlingssystem (Studies of the old Scanian agricultural systems). Meddelande serie B 69, Kulturgeografiska institutionen, Stockholms Universitet. Stockholm, Sweden.
22. Emanuelsson U (2002) Det skånska kulturlandskapet (The Scanian Cultural Landscape). Naturskyddsföreningen i Skåne, Lund.
23. von Linné C (1751) Carl Linnaei skånska resa på höga öfwerhetens befallning förrättad år 1749 (Carl Linnaeus's Journey to Scania as ordered by the high authorities in 1749). Stockholm, Sweden.
24. Björkman L, Sjögren P (2003) Long-term history of land-use and vegetation at Ire, an agriculturally marginal area in Blekinge, south Sweden. Veget Hist Archaeobot 12: 61–74.
25. Foster DR (1992) Land-use history (1730-1990) and vegetation dynamics in central New England, USA. J Ecol 753–772.
26. Mattiasson G (1974) Sandstäpp - Vegetation, dynamik och skötsel (Xeric sand calcareous grassland – vegetation, dynamics and management). Meddelanden från Avdelningen för Ekologisk Botanik, Lunds universitet 2: 1–40.
27. Tyler T (2003) The state of the sand-steppe vegetation of eastern Scania (Skåne) in early spring 2003. Bot Notiser 136: 1–22.
28. Schnoor TK, Olsson PA (2010) Effects of soil disturbance on plant diversity of calcareous grasslands. Agr Ecosyst Environ 139: 714–719.
29. Roem WJ, Berendse F (2000) Soil acidity and nutrient supply ratio as possible factors determining changes in plant species diversity in grassland and heathland communities. Biol Conserv 92: 151–161.
30. Snogerup S, Jörgensen M (2000) Från skånska resa till Skånes flora (From the journey to Scania to the Scanian flora). Lunds Botaniska Förening, BTJ Tryck AB, Lund, Sweden.
31. Tyler T, Olsson K-A, Johansson H, Sonesson M (2007) *Floran i Skåne: arterna och deras utbredning (The Scanian flora: the species and their distribution)*. Lunds Botaniska Förening, Grahns Tryckeri AB, Lund, Sweden.
32. Olsson K-A (1994) Sandstäpp i Skåne – ett upprop (Xeric sand calcareous grassland in Scania – a proclamation). Lunds Botaniska Förenings Medlemsblad 1994: 4–7.
33. Ljungberg KO, Löfroth M, Nitare J (1994) Åtgärdsprogram för Sandstäpp 1994-1996 (Action Plan for Xeric Sand Calcareous Grassland 1994-1996). Report no. 390-2540-94. Swedish Environmental Protection Agency, Stockholm, Sweden.
34. Wassen MJ, Venterink HO, Lapshina ED, Tanneberger F (2005) Endangered plants persist under phosphorus limitation. Nature 437: 547–550
35. Smits NAC, Willems JH, Bobbink R (2008) Long-term after-effects of fertilization on the restoration of calcareous grasslands. Appl Veg Sci 11: 279–286.
36. Bray RH, Kurtz LT (1945) Determination of total, organic, and available forms of phosphorus in soils. Soil Sci 59: 39–45.
37. Bahr A, Ellström M, Schnoor TK, Påhlsson L, Olsson PA (2012) Long-term changes in vegetation and soil chemistry in a calcareous and sandy semi-natural grassland. Flora 207: 379–387.
38. Lantmäteriet (2014a) (Swedish land surveying agency), http://historiskakartor.lantmateriet.se/arken/s/search.html?locale = en_US. Accessed 2014 Feb 15.
39. Lantmäteriet (2014b) (Swedish land surveying agency), http://www.lantmateriet.se/Kartor-och-geografisk-information/Flyg--och-satellitbilder/Flygbilder/ Accessed 2014 Feb 15.
40. Gazenbeek A (2005) LIFE, Natura 2000 and the Military. LIFE Focus. European Commission, Luxembourg.
41. Warren SD, Holbrook SW, Dale DA, Whelan NL, Elyn M, et al. (2007) Biodiversity and the heterogeneous disturbance regime on military training lands. Restor Ecol 15: 606–612.
42. Řehounková K, Prach K (2008) Spontaneous vegetation succession in gravel-sand pits: a potential for restoration. Restor Ecol 16: 305–312.
43. Jentsch A, Friedrich S, Steinlein T, Beyschlag W, Nezadal W (2009) Assessing conservation action for substitution of missing dynamics on former military training areas in Central Europe. Restor Ecol 17: 107–116.
44. Wells TCE, Sheail J, Ball DF, Ward LK (1976) Ecological studies on the Porton Ranges: relationships between vegetation, soils and land-use history. J Ecol 64: 589–626.
45. Dahlström A, Cousins SAO, Eriksson O (2006) The history (1620–2003) of land use, people and livestock, and the relationship to present plant species diversity in a rural landscape in Sweden. Environment and History 12: 191–212.
46. Johansson LJ, Hall K, Prentice H, Ihse M, Reitalu T, et al. (2008) Semi-natural grassland continuity, long-term land-use changes and plant species richness in an agricultural landscape on Öland, Sweden. Landscape Urban Plan 84: 200–211.
47. Kent M, Dargie T, Reid C (2003) The management and conservation of machair vegetation. Bot J Scot 55: 161–176.
48. Carroll JA, Caporn SJM, Johnson D, Morecroft MD, Lee JA (2003) The interactions between plant growth, vegetation structure and soil processes in semi-natural acidic and calcareous grasslands receiving long-term inputs of simulated pollutant nitrogen deposition. Environ Pollut 121: 363–376.
49. Oenema O (1990) Calculated rates of soil acidification of intensively used grassland in the Netherlands. Fert Res 26: 217–228.

50. Bolan NS, Adriano DC, Curtin D (2003) Soil acidification and liming interactions with nutrient and heavy metal transformation and bioavailability. Adv Agron 78: 216–272.
51. Stroia C, Morel C, Jouany C (2010) Nitrogen fertilization effects on grassland soil acidification: consequences on diffusive phosphorus ions. Soil Sci Soc Am J 75: 112–120.
52. Riksen M, Ketner-Oostra R, van Turnhout C, Nijssen M, Goossens D, et al. (2006) Will we lose the last inland drift sands of Western Europe? The origin and development of the inland drift-sand ecotype in the Netherlands. Landscape Ecol 21: 431–447.
53. Poortinga A, Visser SM, Riksen MJPM, Stroosnijder L (2011) Beneficial effects of wind erosion: Concepts, measurements and modelling. Aeolian Research 3: 81–86.
54. Bateman MD, Godby SP (2004) Late-Holocene inland dune activity in the UK: a case study from Breckland, East Anglia. The Holocene 14: 579–588.
55. Olsson PA, Ödman AM (2014) Natural establishment of specialist plant species after topsoil removal and soil perturbation in degraded calcareous sandy grassland. Restor Ecol 22: 49-56.
56. Ödman AM, Schnoor TK, Ripa J, Olsson PA (2012) Soil disturbance as a restoration measure in dry sandy grasslands. Biodiv Cons 21: 1921–1935.
57. Riksen M, Spaan W, Stroosnijder L (2008) How to use wind erosion to restore and maintain the inland drift-sand ecotype in the Netherlands? J Nat Conserv 16: 26–43.

7

Whither the Rangeland?: Protection and Conversion in California's Rangeland Ecosystems

D. Richard Cameron[1]*, Jaymee Marty[1¤], Robert F. Holland[2]

1 The Nature Conservancy, San Francisco, California, United States of America, **2** Geobotanical Phenomenology, Auburn, California, United States of America

Abstract

Land use change in rangeland ecosystems is pervasive throughout the western United States with widespread ecological, social and economic implications. In California, rangeland habitats have high biodiversity value, provide significant habitat connectivity and form the foundation for a number of ecosystem services. To comprehensively assess the conservation status of these habitats, we analyzed the extent and drivers of habitat loss and the degree of protection against future loss across a 13.5 M ha study area in California. We analyzed rangeland conversion between 1984 and 2008 using time series GIS data and classified resulting land uses with aerial imagery. In total, over 195,000 hectares of rangeland habitats were converted during this period. The majority of conversions were to residential and associated commercial development (49% of the area converted), but agricultural intensification was surprisingly extensive and diverse (40% across six categories). Voluntary enrollment in an agricultural tax incentive program provided widespread protection from residential and commercial conversions across 37% of the remaining rangeland habitat extent (7.5 M ha), though this program did not protect rangeland from conversion to more intensive agricultural uses. Additionally, 24% of the remaining rangeland was protected by private conservation organizations or public agencies through land or easement ownership while 38% had no protection status at all. By developing a spatial method to analyze the drivers of loss and patterns of protection, this study demonstrates a novel approach to prioritize conservation strategies and implementation locations to avert habitat conversion. We propose that this approach can be used in other ecosystem types, and can serve as a regional conservation baseline assessment to focus strategies to effect widespread, cost-effective conservation solutions.

Editor: Adina Maya Merenlender, University of California, Berkeley, United States of America

Funding: DC and JM were Nature Conservancy employees during this study and RF was a contractor to The Nature Conservancy. The funders had no role in study design, data collection and analysis, decision to publish, or preparation of the manuscript.

Competing Interests: JM is currently an employee of Marty Ecological Consulting. RH is an employee of Geobotanical Phenomenology. There are no patents, products in development or marketed products to declare.

* Email: dcameron@tnc.org

¤ Current address: Marty Ecological Consulting, Sacramento, California, United States of America

Introduction

The conversion of habitat from a natural state to developed human uses has fundamentally altered both the spatial configuration and function of ecosystems worldwide [1]. Compared to other ecosystems, the effects of habitat conversion on rangeland ecosystems have only received recent attention in the scientific literature [2–4]. This is despite the global vastness of rangelands, covering nearly one quarter to one third of the world's habitable land area [5], their high levels of biodiversity, and the large economic and social benefit provided by rangeland ecosystems [6,7].

Rangeland habitats (grasslands, shrublands and woodlands) have also received less attention from conservation efforts than other major habitat types. The primary global biomes that support extensive rangelands– 1) Temperate grasslands, savannas and shrublands; and 2) Mediterranean forests, woodlands, and scrub– have experienced widespread conversion to anthropogenic land uses (45.8%, 41.4% converted respectively) and have the lowest proportion of legal conservation protection among all biomes (4.6%, 5.0% respectively) [8]. This habitat conversion is driven by population growth and associated residential and commercial development, as well as agricultural production. High levels of habitat loss are especially detrimental for biodiversity in Mediterranean regions of the world that have some of the highest population densities of all global biodiversity hotspots [9].

The economic and social benefits that grazed ecosystems provide vary by societal context. Many livestock-based economies operate in a subsistence mode, including 120 million pastoralists worldwide dependent on livestock as a primary food source [10]. In the coterminous western United States, rangelands are one of the most extensive land types covering over 163 million hectares, with approximately half of the land in non-federal ownership [11]. In addition to the subsistence and market-based economic values provided by rangelands, numerous ecosystem services are provided by grassland and woodland ecosystems as well, such as habitat for native pollinators [12], carbon sequestration [13–15] and water supply [16].

Rangelands contribute to global biodiversity by providing habitat for numerous species as well as critical habitat connectivity. Many iconic, wide-ranging species rely on the habitat connectivity provided by unfragmented rangelands to complete annual migrations in diverse regions such as eastern Africa, western North America and the Mongolian steppe [17]. In many rangeland ecosystems, native or livestock grazers can play a critical role in maintaining habitat diversity [18] or habitat for

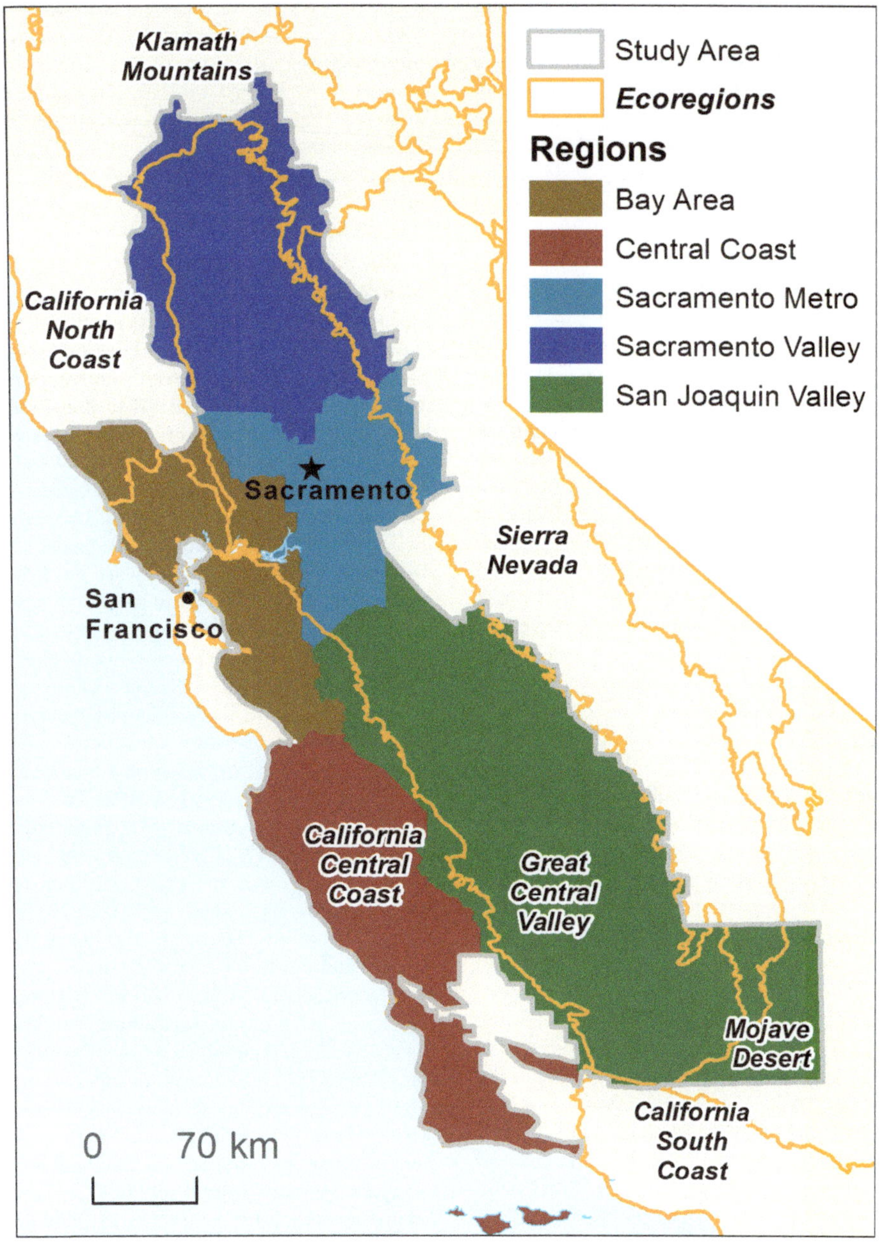

Figure 1. Study area. Extent of study area, ecoregions and reporting regions used for conversion analysis.

restricted endemic species [19,20], yet impacts to ecosystems due to incompatible grazing are well-known, especially for riparian habitats in arid regions [21].

Despite being widespread with a rich historical legacy in the Western U.S., ranching is an economically marginal activity at small to medium scales of operation. For example, Wetzel et al. found that more than 70% of surveyed California ranchers make less than $10,000 profit annually [22]. In addition to the economic challenges of running smaller ranching operations and the demographic reality that the average age of ranchers is high, other factors have contributed to a declining population of ranchers, including impediments to intergenerational land-based wealth transfer, rise in rural residential amenity-based land purchases, and loss of critical industry infrastructure (such as processing facilities) [23–25]. These demographic and economic pressures increasingly lead to the dissolution of large land holdings into smaller parcels often resulting in the conversion of rangeland habitats into uses not compatible with ecosystem functioning [4,26,27].

The resulting fragmentation of rangeland habitat has many negative biological and social effects including loss of remote habitat, increase in human commensal species, severing of wildlife connectivity between core habitats, and an increasingly difficult social environment for ranching (e.g. new neighbors are affronted by certain aspects of a livestock operation) [2,23,28]. While the decision to break-up and sell a large ranch is made at the landowner or family level, the consequences become evident ecologically at the landscape scale with large areas converting to low density "exurbs" [3]. These diffuse development patterns represented the fastest growing land use in the country in recent decades [29].

To counteract these trends, various conservation strategies have been used, including voluntary, publicly-funded conservation and restoration incentive programs (such as the Grassland Reserve Program and the Wildlife Habitat Incentive Program), and acquisition of development rights through purchase of a conservation easement [30,31]. These strategies can provide a range of conservation benefits including legal protection of land from conversion and improvement of habitat conditions for wildlife [31], but lack of transparency of terms and reporting makes a broad evaluation difficult [32]. Acquisition of fee title or certain rights to the land through a conservation easement requires the most upfront capital, but provides long-term assurance that the land will remain open habitat, if not actively grazed rangeland.

Conservation strategies such as reduction in property taxes for agricultural land uses through voluntary incentive programs can represent a significant boost to the economic viability of ranching [25]. One notable example of such a program was the California Land Conservation Act, commonly known as the Williamson Act (WA), that provided property tax reductions across six million hectares of the state as of 2009 [33]. Counties had been compensated by the State for lost property tax revenues ($35.1 M in 2009) but payments from the state to participating counties to compensate enrolled landowners stopped in 2009. Since then, the program has been implemented at the discretion of individual counties. Before 2009, to remove enrolled land from Williamson Act contracts, the landowner had to initiate a process of non-renewal which gradually restored the full tax burden over a nine year period. A primary reason that private landowners began the process of non-renewal was to plan for future subdivision and development, and as such it was a useful proxy for the threat of conversion to residential or commercial development. Other conservation tools such as habitat mitigation banks provide credits to offset development impacts to habitat for protected species and can provide conservation-friendly revenue streams for landowners. Future revenue from monetized ecosystem services including water banking and carbon sequestration is an area of widespread interest and active science and policy development throughout the western United States, yet substantial constraints remain [34,35].

To determine the scope and drivers of rangeland habitat conversion, this study uses geographic data for development and agricultural land uses across a 13.5 M hectare portion of the California Floristic Province. California is an ideal place to look at rangeland habitat loss and conservation because of the globally significant biodiversity in the Mediterranean climate-influenced part of the state, high rate of population growth and progressive conservation policies and funding programs. While many aspects of California rangelands are unique relative to other western United States ranching landscapes, many of the social and economic drivers of land use change are similar to other regions, including demographic trends of ranchers and displacement of grazing for more profitable land uses. This study quantifies the amount of rangeland conversion that has occurred over 21 years in California and the level of protection afforded to the remaining rangelands. This information is critical to understanding the vulnerability of this ecosystem and associated social and economic benefits provided by ranching. This study also provides needed data for conservation planning efforts in California.

Using data from 1984 to 2008 on the extent of rangelands, we mapped areas that were converted from rangelands to other uses and, through aerial imagery interpretation, determined the resulting land use at the end of the time period. To assess the conservation status of remaining rangelands, we quantified the overlap between different short-term and permanent protective measures and rangeland ecosystem types. We present this approach as an attempt to convey the full spectrum of conservation status options, moving beyond traditional gap analyses that quantify the percentage of habitats that are in various categories of permanent legal protection [36,37]. Such an adaptation of traditional analyses is critical for rangelands because of the importance of capturing the scope of voluntary programs that provide conservation benefits to landowners and society.

Study area

The study area encompasses the majority of the rangeland habitat within the California Floristic Province, including all of the Great Central Valley and California Central Coast ecoregions where data were available [38]. The study area also includes the full extent of Kern County which extends east into the southern part of the Sierra Nevada Mountains and western Mojave Desert because this portion of these ecoregions has similar habitats and land use to the southern Great Central Valley (Figure 1). The majority of the North and South Coast ecoregions were excluded from the analysis because they are not primarily in rangeland land cover. We defined our study area to align with the focus area of the California Rangeland Conservation Coalition (CRCC), a group of federal and state agencies, conservation organizations and ranchers working to develop collaborative solutions to make ranching economically viable and ecologically beneficial. As such, we focused on private rangelands and excluded the desert rangelands in the Northeast and Southeast part of California that are extensive, but have low productivity due to low rainfall and shorter growing seasons. All potential rangeland habitats in California cover 23.1 M hectares, of which approximately 60% was actually grazed in 2005 ([39].Four counties within the study area did not have data and were excluded from the analysis, Tuolumne and Calaveras in the Sierra Nevada foothills and San Mateo and Santa Cruz in the Central Coast ecoregion.

The rangeland habitats within the study area are dominated by Mediterranean annual grasslands and mixed hardwood woodlands, and are characterized by a strong latitudinal precipitation gradient. The highest average annual precipitation in the study area is near Redding, CA with 113 cm annually with the lowest in Kings County, near Taft with 14 cm annually. Coupled with the length of the growing season and differences in soil productivity, precipitation is a strong driver of forage productivity and quality, a major factor for the economic viability in ranching [40]. Forage productivity varies from 560 kg/ha to over 3900 kg/ha in areas closer to the coast [41].

Methods

Land use data

The California Farmlands Mapping and Monitoring Program (FMMP) was established in 1982 to provide consistent, objective information on agricultural land uses throughout California [42]. Eight land use categories are used to map general land use and evaluate the suitability of the land for irrigated agriculture (Prime, Statewide Importance, Unique, Locally Important) and to show the extent of grazing land, developed areas, water and other land uses. The grazing land category was used in this study to assess conversion patterns and is defined as "land on which the existing vegetation is suited to the grazing of livestock" (http://www.conservation.ca.gov/dlrp/fmmp/mccu/Pages/map_categories.aspx). The number of counties included in FMMP has increased over the years, now including over 95% of the privately held lands in California. The program uses soil survey GIS data, aerial imagery and field

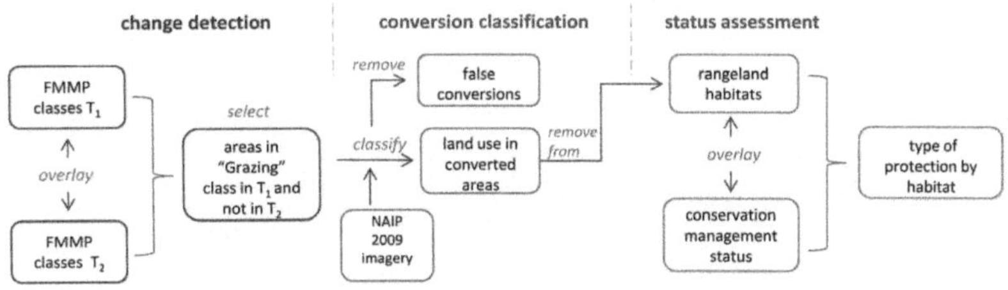

Figure 2. Sequence of core spatial analyses used in study. The three primary elements of the analysis are shown in sequence. The "change detection" step provided the full set of areas to classify in the "conversion classification." Some lands where a change was detected but were not actually converted were removed from the analysis as false conversions. Converted lands were removed from the habitat data for the "status assessment", which quantified the proportion of habitats under various types of protection. The year of the data used in FMMP overlays between time point 1 and time point 2 varied by survey area. Nineteen of the thirty-six survey areas spanned the period from 1984–2008, five spanned the period 1984–2006; and the remaining twelve covered various years with an average of seventeen years across them. The most protective conservation management status was assigned to land if multiple statuses existed, in order of precedence: fee ownership, easement, Williamson Act enrollment.

reconnaissance to map the land use categories described above every two years. Additional classes were added in 2002 for select San Joaquin Valley counties as part of the Rural Land Mapping project to further distinguish the Other Land category. We used the FMMP dataset because it represents a consistent land use change data source over a long time period at a relatively high spatial resolution (4 ha minimum mapping unit). Our study period matches the development of Geographic Information Systems (GIS) during which time the mapping protocol for the FMMP changed accordingly. Originally the mapping process used transparent aerial photos overlaid onto 100,000 scale paper topographic sheets. The resulting land use classification was then digitized for use in a GIS. Such changes introduce some error into any analysis of long-term data.

We assembled FMMP data for 33 counties throughout California for two points in time. The time period covered for each survey area (three counties were split into two survey areas) varied based on when FMMP mapping was initiated in a survey area and data availability at the time of the analysis. The beginning year was as early as 1984 and as late as 2000 and the end year was either 2006 or 2008, for a maximum range of 24 years. The range of years by survey area is shown in Table S1, and has a mean of 21 years across all survey areas. To map areas that were converted during this time, we overlaid the datasets from the beginning and end years and selected areas that were classified as Grazing Land in the first time period and another mapping category in the later period. We applied processing techniques to the composite data to reduce any errors resulting from small misalignments of the boundaries from the two years that may have resulted from the migration to a GIS-based workflow. Specifically, we forced any point along a line within 100 meters of a point from the other time period to be snapped together in a GIS. This resulted in simpler (less sinuous) polygon shapes and the removal of smaller and more linear features from the dataset. While this processing step was essential to be able to compare the data from the two time periods, it does limit analysis of the fine-scale fragmentation patterns resulting from conversion. Polygons smaller than four hectares were removed from the combined data set, as that was the minimum mapping unit for the FMMP land use data. A flow chart illustrating our analysis steps for the study is shown in Figure 2.

To assign a land use to the converted polygons, we used one-meter resolution, true color 2009 aerial imagery from the National Agriculture Imagery Program (NAIP) [43]. To distinguish the relative density of residential development, two categories were used: urban (housing density greater than 1 unit per .65 ha) and rural residential covering lower densities. During photo interpretation, a number of polygons were determined to be still in grazing land and not actual conversions and were a result of a change in classification. These were excluded from the analysis. Because the original FMMP classification does not identify crop types, we developed a new classification for this study that captures more variation in agricultural and developed land classes. We provide an assessment of the degree of overlap between these two classifications in Table S2.

To investigate the potential influence of the global financial crisis on land conversion rates, we assessed the rate of annual change from 2008 to 2010 for five counties that have different dominant land use types: Santa Barbara, Kern, Sacramento, Placer and Merced. We compared the rate of change for the years covered in this study and the rate for 2008–2010 to see if rates of conversion declined with economic activity.

Private and public conservation management data

To assess the degree to which rangeland ecosystems are protected from conversion through legal mechanisms, we assembled land status data representing land held in fee or under conservation easement with a private conservation organization or public agency [44]. To estimate the conservation benefit of a voluntary, non-permanent conservation program in rangelands, we compiled spatial data on lands enrolled in the California Land Conservation (Williamson) Act [45] using the latest data available for each county, as recent as 2009. Yuba County was the only county in our study area that did not participate in the program. Williamson Act contracts were between a landowner and the county in which the enrolled land exists, and lasted a period of at least ten years after which point they needed to be renewed. We also included land covered by the "Super Williamson Act" contracts that lasted 20 years.

Rangeland habitat data

We used the Multi-Source Land Cover data assembled by California Department of Forestry and Fire Protection [46] to represent the rangeland habitats of interest in the study area, classified using the California Wildlife Habitat Relationships (CWHR) standard [47]. Input data sources for this composite layer range from 1997–2002. We used a classification of rangelands used by the UC Davis Rangeland Watershed Lab as the primary list of habitats for this study (http://californiarangeland.ucdavis.edu/). We added the montane hardwood habitat type to the analysis as it is extensive within our study

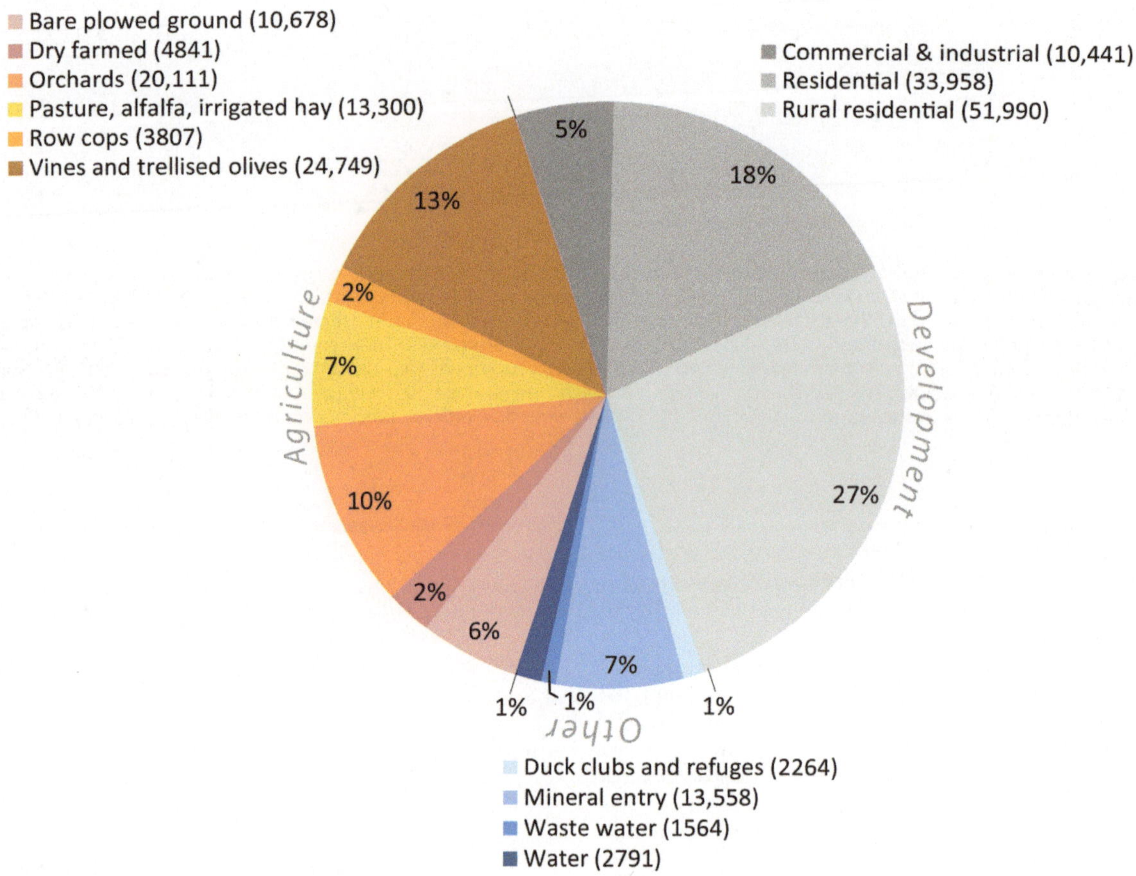

Figure 3. Area (Ha) of land cover types on land converted from rangelands during the study period. Converted areas were mapped using the California Department of Conservation's Farmland Mapping and Monitoring Program data (1984–2008) and classified using the National Agriculture Imagery Program imagery from 2009. The development categories (shown in gray) account for 49% of total area converted, agricultural conversions account for 40% and various other conversions account for about 10%.

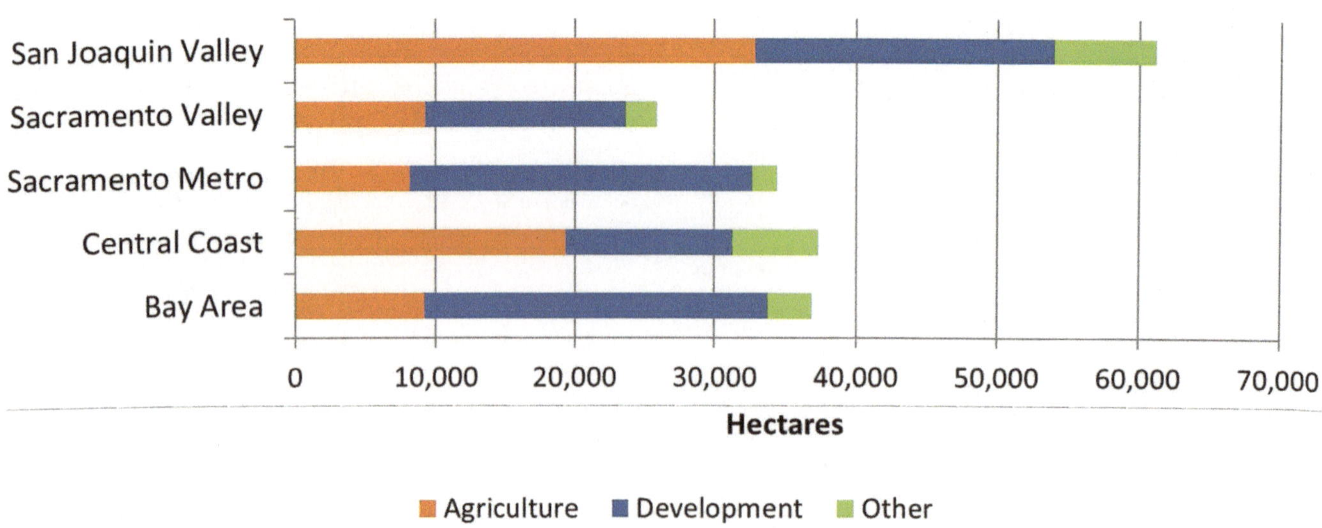

Figure 4. Regional conversion patterns by major type. The primary type of conversion for each region was variable, but can be explained partly by the existing land use in the region. For example, the more developed Sacramento Metro and Bay Area regions saw the majority of the converted land end up as residential and commercial development, while the majority of conversions in the San Joaquin Valley and Central Coast regions were to intensive agriculture classes. Conversion data are from an analysis of Farmland Mapping and Monitoring Program data, and land cover assigned using the National Agriculture Imagery Program imagery for 2009.

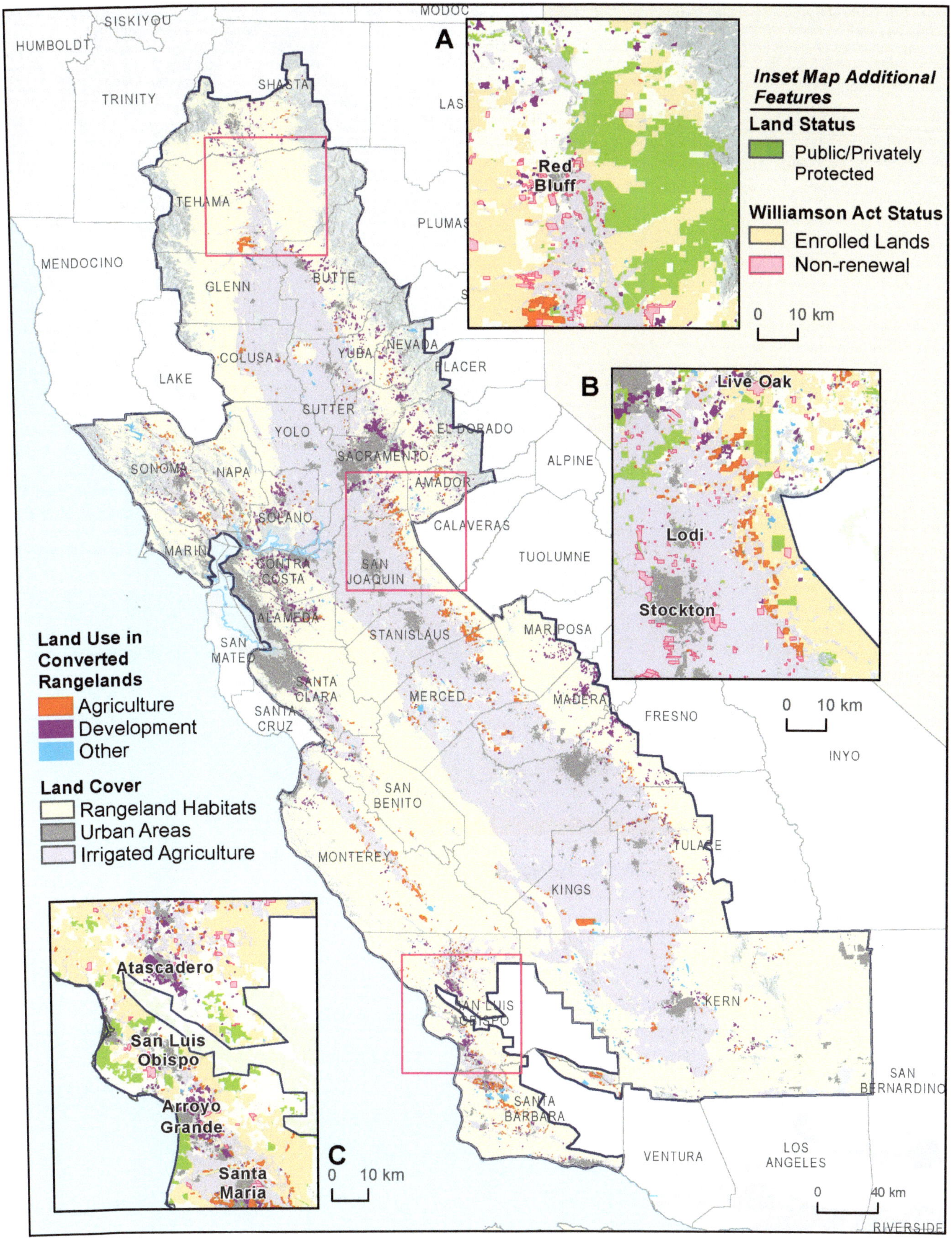
A
B
C
Inset Map Additional Features
Land Status
Public/Privately Protected
Williamson Act Status
Enrolled Lands
Non-renewal
0 10 km
Red Bluff
Live Oak
Lodi
Stockton
Atascadero
San Luis Obispo
Arroyo Grande
Santa Maria
Land Use in Converted Rangelands
Agriculture
Development
Other
Land Cover
Rangeland Habitats
Urban Areas
Irrigated Agriculture
0 40 km
SISKIYOU
HUMBOLDT
MODOC
SHASTA
TRINITY
LASSEN
TEHAMA
PLUMAS
MENDOCINO
GLENN
BUTTE
COLUSA
YUBA
NEVADA
PLACER
LAKE
SUTTER
YOLO
EL DORADO
SACRAMENTO
ALPINE
SONOMA
NAPA
AMADOR
SOLANO
CALAVERAS
MARIN
TUOLUMNE
CONTRA COSTA
SAN JOAQUIN
ALAMEDA
STANISLAUS
MARIPOSA
SAN MATEO
SANTA CLARA
MERCED
MADERA
SANTA CRUZ
FRESNO
INYO
SAN BENITO
MONTEREY
TULARE
KINGS
SAN LUIS OBISPO
KERN
SAN BERNARDINO
SANTA BARBARA
VENTURA
LOS ANGELES
RIVERSIDE

Figure 5. Conversion patterns by major type. The converted rangelands spanned a large geographic area. In some areas large protected area complexes were created where conversion was less widespread, possibly enabled by lower land prices (A). Agricultural conversions were more common adjacent to existing cropland (B). Conversions to developed land uses are dominated by new housing at lower densities at higher elevations in the Sierra Nevada foothills (Madera, Fresno, Butte counties) and at higher densities in the San Francisco Bay Area and near Sacramento. The areas in the Central Coast around Atascadero and Arroyo Grande experienced a large amount of conversion from a variety of sources (C).

area and does experience some grazing depending on the location in the state, especially where California black oak (*Quercus kelloggii*) is dominant. We chose to use a habitat-based definition of rangeland extent and not the Grazing Land category from the FMMP data for the conservation status assessment because we wanted to quantify the variability of protective status across ecosystems that support domestic livestock. The diversity of ecological and social contexts for these ecosystems is high and differences in patterns of protection are useful to interpret the results across such a large study area.

We combined the data of rangeland habitats with the data for public land and privately protected areas, Williamson Act lands and converted rangeland areas within the GIS. To assess the conservation status of remaining rangelands we summarized the habitat distribution by five categories of land holdings: 1. private land, 2. private land enrolled in the Williamson Act, 3. private land not renewing Williamson Act contracts, 4. private land with a conservation easement, and 5. land held in fee title by a public agency or a non-governmental conservation organization. To facilitate reporting and interpretation, we divided the counties in the study area into distinct regions: the Bay Area, Sacramento Metro, Sacramento Valley, San Joaquin Valley, and Central Coast (Figure 1). Because geographic data on rangeland habitats do not exist before 1984, we were not able to determine which habitats were converted.

Results

Rangeland conversion

Across the study area, 195,594 ha of rangelands were converted to other uses during the time period covered in the assessment. The most common conversions were to developed land classes (96,389 ha, 49% of the total, Figure 3) with the most extensive to rural residential followed by residential land. Loss of rangelands to more intensive agricultural land uses covered 78,793 ha (40% of total) with four land uses making up 87% of all conversions, including, in descending order, vines and trellised olives, orchards, pasture/alfalfa, and the transitional category of bare plowed ground. Other conversion types covered just over 20,000 ha and were dominated by mineral extraction (gravel, clay, oil/gas, tailings ponds) accounting for 66% of all area within this category.

The San Joaquin Valley region experienced the largest amount of conversion (61,139 ha) with 54% of the converted area going to agricultural land uses (Figure 4). Conversion in the Bay Area and Sacramento Metro region was dominated by development, with the ratio of development to agricultural conversions at 2.7 and 3, respectively. In the Bay Area, the counties to the east of San Francisco Bay have the largest converted area collectively, but Sonoma County had the largest area converted of any of the Bay Area counties analyzed (Table S3).

The vast majority of the development in the Sacramento Metro region occurred in the grasslands and woodlands leading to the Sierra Nevada foothills east of Sacramento, with large conversions directly adjacent to the existing urbanized area (Figure 5). The Central Coast experienced slightly more conversion than the Bay Area or the Sacramento Region, from a diverse mix of conversion types. Agricultural conversions dominated Monterey County (primarily to vineyards) (Figure 5). In Santa Barbara County, large areas were converted to oil and gas extraction. San Luis Obispo County experienced dramatic residential growth around Arroyo Grande, Paso Robles and Atascadero totaling over 7269 ha in residential and rural residential classes (Figure 5B, Table S3).

The counties with more residential development had a slowdown in conversion rates from 2008–2010 with Sacramento and Placer declining 88% and 43%, respectively (Table 1). Santa Barbara County also experienced a drop of 40%, while Merced County's rate was 80% higher than historical suggesting that agricultural conversions did not slow in that county after the onset of the recession. Kern County also experienced a slight increase in the rate of conversions (18%) relative to the long-term average.

Conservation status

Of the remaining rangelands, 1.8M ha (24%) were protected against further conversion in fee title ownership or conservation easement held by a public agency or private conservation organization (Figure 6). In contrast, 2.8M ha (38%) of the rangeland area had no conservation status and was potentially subject to conversion to alternative land uses. Roughly the same amount of land (37%) was temporarily protected from conversion to development through enrollment in the Williamson Act.

Table 1. Rate of conversion of rangeland for years up to 2008 and for 2008–2010 based on FMMP data.

County	Number of Years Included in Study (2008 and earlier)	Rate of conversion pre-2008 (Ha/Yr)	Rate of conversion 2008–2010 (Ha/Yr)
Santa Barbara	24	576	346
Kern	20	572	674
Merced	20	548	994
Sacramento	20	485	56
Placer	24	363	208

Counties with more agricultural land (Kern, Merced) experienced an increase in the rate of conversions compared to the long-term average, while Sacramento County experienced a decline in conversions likely due to the drop in demand for new housing.

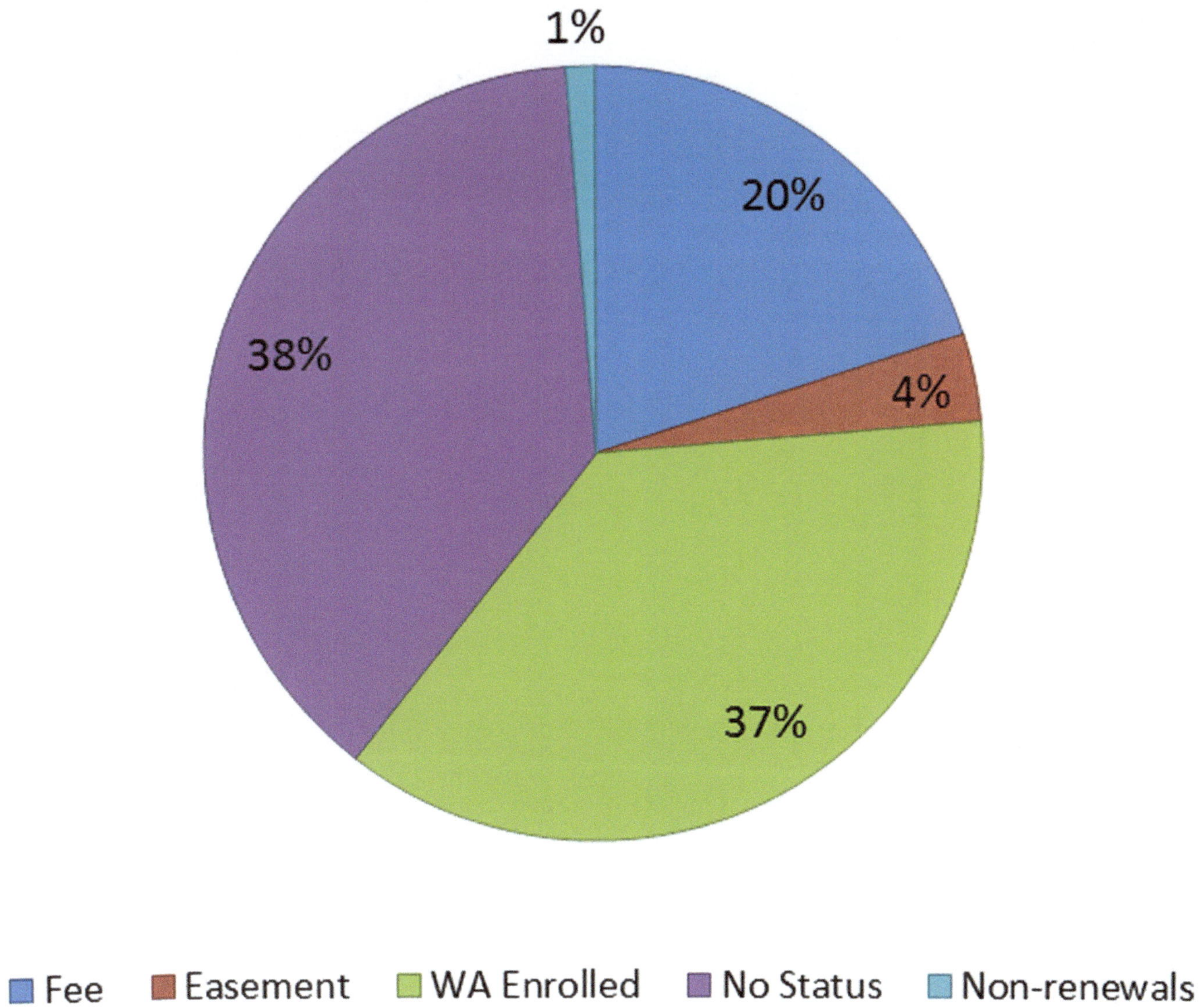

Figure 6. Conservation status of rangeland habitats. Over 60% of the total area in rangeland habitat have some protective status against various types of conversion with the majority provided by enrollment in the Williamson Act which prevented conversion to non-agricultural uses, but was defunded starting in 2009. This chart is for the portion of the study with rangeland habitats. WA = Williamson Act.

The five most common habitat types made up 80% of the rangeland extent in the study area (Table 2), with a highly variable mix of permanent conservation status and WA enrollment. Lower elevation habitat in the Central Valley and Central Coast (annual grasslands and blue oak woodlands *(Quercus douglasii)*) were primarily conserved through enrollment in Williamson Act (49% and 46%, respectively), while higher elevation types in these regions (montane hardwood and mixed chaparral) were more commonly owned by a public entity, such as the U.S. Forest Service (23 and 44% fee, respectively). Desert scrub is most common in the southern San Joaquin Valley and western Mojave and had very little enrollment in the Williamson Act (1%) but substantial area in fee conservation status (42%). Other habitats that had a substantial percentage of area (>30%) enrolled in Williamson Act included coastal oak woodland, coastal scrub, blue oak-foothill pine and valley oak woodland.

Discussion

While rangeland conversion occurs at the ranch level, the decision to convert is driven by a complex and interacting network of economic and social factors including the incentives and disincentives provided by land use policy. The ecological and economic threats to rangelands are numerous (invasive species, climate change, lack of profitability), but the conversion to development or intensive agriculture is often the most permanent. Conversion to more intensive crop types, such as orchards and vineyards, often eliminates critical endangered species habitat (e.g. vernal pools) and fundamentally alters the structure and function of these lands for wildlife use [48]. In a similar study, the extent of vernal pools in California rangelands decreased by 13% from 1976 to 2005 [49]. Agricultural conversion requiring the irreversible mechanical ripping of impervious soil layers characteristic of vernal pools accounted for over 80 percent of this habitat loss.

While conversion of rangelands to plowed, irrigated crops represents a loss of terrestrial habitat, it also impacts aquatic

Table 2. Area (Ha) by conservation status for California rangeland habitat types.

Habitat type	Total area	Fee Title	% Fee	Cons. Easement	% Easement	WA Enrolled	% WA	No status	% No status	WA Non-renewals	% Non-Renewal
Annual Grassland	3,186,845	355,034	11%	134,645	4%	1,586,176	50%	1,057,150	33%	53,840	1.68%
Blue Oak Woodland	1,073,109	120,142	11%	42,506	4%	503,270	47%	399,267	37%	7,923	0.74%
Montane Hardwood	676,434	157,316	23%	26,101	4%	144,555	21%	344,642	51%	3,819	0.56%
Mixed Chaparral	620,111	276,183	45%	17,283	3%	117,348	19%	207,128	33%	2,170	0.35%
Desert Scrub	445,750	190,521	43%	25	0%	5,936	1%	248,690	56%	578	0.13%
Coastal Oak Woodland	363,381	92,433	25%	19,971	5%	127,224	35%	114,981	32%	8,772	2.40%
Coastal Scrub	236,450	46,940	20%	9,762	4%	93,466	40%	81,460	34%	4,821	2.03%
Blue Oak-Foothill Pine	233,051	38,564	17%	7,262	3%	89,556	38%	95,968	41%	1,702	0.73%
Chamise-Redshank Chaparral	195,295	75,223	39%	6,749	3%	56,655	29%	55,751	29%	917	0.47%
Alkali Desert Scrub	173,106	19,532	11%	2,533	1%	10,114	6%	139,150	80%	1,777	1.03%
Pinyon-Juniper	71,857	51,727	72%		0%	5,948	8%	14,175	20%	7	0.01%
Valley Oak Woodland	42,150	3,933	9%	1,683	4%	14,300	34%	21,204	50%	1,030	2.43%
Montane Chaparral	40,077	18,615	46%	133	0%	4,196	10%	17,069	43%	64	0.16%
Sagebrush	29,510	19,742	67%	40	0%	1,846	6%	7,866	27%	15	0.05%
Valley Foothill Riparian	29,119	8,779	30%	2,378	8%	4,979	17%	12,711	44%	271	0.93%
Montane Riparian	26,431	4,427	17%	799	3%	7,188	27%	13,731	52%	285	1.08%
Juniper	15,505	4,850	31%		0%	3,067	20%	7,579	49%	9	0.06%
Bitterbrush	7,768	3,013	39%	38	0%	721	9%	3,996	51%		0.00%
Wet Meadow	5,818	1,966	34%	186	3%	1,192	20%	2,427	42%	47	0.81%
Desert Succulent Shrub	2,804	180	6%		0%	364	13%	2,259	81%		0.00%
Joshua Tree	1,704	848	50%		0%	45	3%	811	48%		0.00%
Desert Wash	1,606	1,111	69%	14	1%	23	1%	458	29%		0.00%
Perennial Grassland	1,409	387	27%	0	0%	213	15%	808	57%		0.00%
Total Study Area	7,479,287	1,491,466		272,106		2,778,384		2,849,281		88,050	

Data sources for fee and easement protected areas from GreenInfo Network. Williamson Act enrollment data from Department of Conservation, and rangeland habitat data from Calfire. WA = Williamson Act.

ecosystems through surface or groundwater use [50]. Conversion of oak savanna to vineyards has been shown to represent a net increase in groundwater stress, especially in the dry summer months [51]. The conversion of rangeland to irrigated crops puts excessive demands on already depleted groundwater resources, particularly during drought conditions where managed surface water resources are restricted and groundwater is the only source of agricultural water [52]. These secondary effects of conversion need to be considered when conducting land use planning for rural parts of the state, particularly in the context of projected changes due to climate change. Compared to annual crops, nuts and vines constitute a "hardening" of water use in that the crops will fail without water whereas fields of annual crops can be fallowed to conserve water resources.

Changes in consumption patterns in wine, almonds, olives and other orchard crops may be driving the expansion of these crop types in rangeland ecosystems. Wine grape acreage increased in the state during the time period covered in this study by 56% [53] with the contribution of the northern Central Valley outpacing the increase in more traditional wine regions such as Napa and Sonoma counties, due to the region's role in producing varietals whose popularity has grown in recent decades, such as merlot, chardonnay and cabernet sauvignon [54]. Increases in international consumption of almonds and olives may be contributing to the expansion of these crops in the region. Almonds are primarily grown for international export (67% of production exported in 2010) with the European Union, China and India being the largest importers. The harvested area for almonds in California increased by 78% (over 120,000 ha) from 1984 to 2008 [55].

While this analysis classified the land use in 2009 in converted areas, we cannot necessarily assume that the 2009 land use was the original driver for conversion, as the original conversion may have occurred 20 or more years earlier. Other studies have shown that land use change in California is very dynamic with many conversions between grasslands/shrublands ecosystems and agriculture due to fallowing of agricultural land and the impact of longer drought cycles [27,56,57]. Additionally, development conversions are often preceded by agricultural conversions [58]. Given that, we assume that some of the areas that were classified as residential development in 2009 were initially converted from rangeland to agricultural types, though likely to annual crops rather than perennial crops (e.g. orchards, vineyards). Interactions between conversion types have been observed elsewhere [59] and have been proposed to explain some rangeland conversion in California [27]. The theory holds that when prime farmland is converted to residential and commercial development at the edge of cities, it causes a displacement of agricultural land uses to areas with more marginal soils, such as rangelands [59]. It is likely that as more marginal soils are exploited, the fertilizer inputs would increase to maintain productivity and that the sandier, well-drained soils would require more irrigation than a similar crop in better soils [50]. Thus, the cumulative effects of development of prime farmland need to be a factor in future land use planning in the Central Valley, and other rapidly growing regions with extensive farmland.

Post-recession rates of change revealed variability in the response to the economic downturn in 2008. Yet, further investigation of the response of different land uses to catastrophic events, both economic and natural, is needed to help interpret this limited comparison. Higher temporal resolution studies of economic and climatic drivers of land use change can yield helpful information for planners, and can provide a complementary perspective to this long-term assessment of change [57].

Our results show that over a third of the rangeland area has no protective legal status, suggesting that some of these areas may be at risk to future habitat loss. Yet the drivers on land use change are spatially variable, often corresponding to proximity to existing development and infrastructure, and unprotected land is not necessarily at risk of development within the scale of a few decades. There may be local protective ordinances or zoning that benefit ranching landowners in these areas, but such an assessment is beyond the scope of this study. Permanent habitat protection is relatively low across all rangeland types (24%) with a strong bias toward more montane and desert habitat types. Some of these areas may be protected from development but not grazed, such as State Parks, because of other resource priorities (e.g. recreation or water quality) and as such represent an area of trade-off between these values and livestock production, and any ecological benefits associated with grazing.

Conservation easements are currently an underutilized strategy for rangeland protection, with only small proportions of annual grasslands (4%) and blue oak woodlands (3%) covered by these agreements. While a lack of standardized data collection on easement locations likely make this an underestimate, easements still cover a very small percentage of foothill habitats. As it is a voluntary solution that can provide conservation benefits to landowners and society while providing an infusion of money into ranching operations, conservation easements should be prioritized in public and private funding for rangeland conservation [60].The foothills of Mt. Lassen in the northern Central Valley is a region where extensive rangeland easements have helped limit the expansion of development from the city of Red Bluff while creating a connected network of protected lands stretching from valley floor to subalpine habitats (Figure 5A).

The results of this study can be used to prioritize general areas to implement conservation strategies such as purchasing easements or focusing ecological restoration efforts. The data from this study can be integrated into habitat connectivity assessments for example, to quantify the level of fragmentation from conversion in potential linkages between existing core habitat. While other factors need to be considered for site-specific project implementation, broad priority areas can be identified by analyzing patterns of conversion and types of land protection such as presented here. Similar analyses can be used to identify areas where land use policies that maintain extensive agriculture such as livestock grazing may be needed to prevent loss of rangelands, if that is an objective of the local jurisdiction.

The Williamson Act had benefited landowners in California across a large area covered by rangelands for 44 years, especially for annual grasslands and blue oak woodlands, the two most extensive habitat types (49% and 46% respectively). Other studies have highlighted the importance of these property tax benefits for maintaining ranching operations [22,25], but it is not a panacea for rangeland ecosystem protection. An analysis of the conservation status and conversion datasets used in this study showed that 53% of the areas converted to vineyards were enrolled in Williamson Act in 2009. As such, it is clear that the Act did not prevent rangeland loss to more intensive agricultural land uses. Refinements to a restored Williamson Act, or subsequent legislation, that protect the broader array of public benefits including wildlife habitat, may be needed to prevent the agricultural intensification of rangelands. Yet, given the extensive area of enrolled lands, the benefits of the law for habitat protection were immense and restoring the Williamson Act should be a statewide land use policy priority. It is highly likely that increased development would result if subvention payments from the State or counties to landowners disappear altogether [22]. In fact, we see some evidence of that occurring in the most recent report on Williamson Act enrollment [61]. There was a decline of 14.5% in annual enrollment in 2010–2011 and a sharp increase in non-renewals in 2011 of 69,000 hectares compared to the annual average of non-renewals of 30,000 for the previous decade.

While we have taken one step toward a full spectrum conservation status analysis by including enrollment in the Williamson Act and conservation easements, the status assessment highlights the need to integrate multiple types of data to develop a comprehensive picture of conversion and conservation. Because conservation in private lands relies on a diverse array of voluntary, incentive-based programs that provide direct payments, tax relief or regulatory streamlining in exchange for beneficial management, often at the scale of individual ranches, we need better systems to track and account for the cumulative benefit of these programs. Integrating such programmatic data with change detection analyses such as this one can help develop a regional conservation accounting system that can be used to generate an ecological "P&L" statement, one of *protection* and *loss*. Inventorying and mapping the broad range of beneficial restoration actions that are publicly funded to promote agricultural best management practices or habitat restoration will provide a more nuanced and complete picture of what conservation benefits are realized by what interests [62].

Ultimately, the public pays to replace the goods and services lost when natural habitats are converted, such as pollination of crops and development of water storage and delivery systems [12,52]. In addition, when conversions take place near existing protected land, the ecological connectivity value of that land is eroded due to habitat fragmentation. If the protected land was acquired with public funds, then society pays twice for that investment, once for the initial purchase and again to replace the ecosystem service benefits that are lost by adjacent habitat loss. Developing alternative markets to compensate landowners for the additive ecosystem benefit of beneficial management decisions (e.g. increased soil carbon storage as an emissions offset) may ultimately provide critical revenue to improve the economic viability of ranching operations [15,63]

Conclusion

Creating and implementing solutions to the problem of rangeland habitat loss and associated impacts will require an understanding of the economic and social drivers that lead to the breakup of larger land holdings, particularly those drivers related to global agricultural commodities. Solving such challenges requires cross-disciplinary collaboration to create strategies that address the various social contexts for ranching. The numerous ecological and social benefits provided by rangeland ecosystems in the Western United States can only be sustained if economic incentives are promoted to maintain ecologically sustainable grazing operations across large land ownerships. Accounting for the full economic, ecological and social costs of rangeland conversion would provide decision-makers with a greater ability to weigh trade-offs associated with land use change.

Supporting Information

Table S1 Time range by county or survey area for FMMP data. The average range of all survey area is 21 years.

Table S2 Classification agreement between the Farmland Mapping and Monitoring Program (FMMP) categories and the classification used in this study. B- Bare Plowed Ground; CI- Commercial and Industrial; DB- Dirt Bikes; DF- Dry farmed; DCR- Duck clubs and refuges; ME- Mineral entry: gravel, clay, oil fields, tailings; N- Nurseries; O- Orchards; P- Pasture, alfalfa, irrigated hay; R- Residential; Ri- Rice; RC- Row crops; RR- Rural residential; U- Unknown; V- Vines and trellised olives; WW- Wastewater; W- Water.

Table S3 Conversion types by counties. Area in hectares.

Acknowledgments

We would like to thank Molly Penberth at the California Department of Conservation for assistance with the Farmland Mapping and Monitoring Program data and GreenInfo Network for assistance with easement data. This paper was greatly improved from comments and suggestions from Adina Merenlender and two anonymous reviewers. We greatly benefited from discussions and reviews of early drafts by Sasha Gennet, Scott Butterfield, Scott Morrison, and Stacey Solie.

Author Contributions

Conceived and designed the experiments: DC JM. Performed the experiments: DC RH. Analyzed the data: DC RH. Wrote the paper: DC JM RH.

References

1. Vitousek PM (1994) Beyond global warming: ecology and global change. Ecology 75: 1861–1876.
2. Maestas JD, Knight RL, Gilgert WC (2003) Biodiversity across a Rural Land-Use Gradient. Conservation Biology 17: 1425–1434.
3. Theobald DM (2005) Landscape Patterns of Exurban Growth in the USA from 1980 to 2020. Ecology and Society 10.
4. Hansen AJ, Knight RL, Marzluff JM, Powell S, Brown K, et al. (2005) Effects of Exurban Development on Biodiversity: Patterns, Mechanisms and Research Needs Ecological Applications 15: 1893–1905.
5. Blench R, Sommer F (1999) Understanding Rangeland Biodiversity. London: Overseas Development Institute.
6. Knight RL (2007) Ranchers as a Keystone Species in a West That Works. Rangelands 29: 4–9.
7. FAO (2011) World Livestock 2011: Livestock in Food Security. Rome.
8. Hoekstra JM, Boucher TM, Ricketts TH, Roberts C (2005) Confronting a biome crisis: global disparities of habitat loss and protection. Ecology Letters 8: 23–29.
9. Cincotta RP, Wisnewski J, Engelman R (2000) Human population in the biodiversity hotspots. Nature 404: 990–992.
10. Rass N (2006) Policies and Strategies to Address the Vulnerability of Pastoralists in Sub-Saharan Africa. Rome: FAO.
11. USFS (1989) An Analysis of the Land Base Situation in the United States 1989–2040. U.S. Forest Service. 77 p.
12. Chaplin-Kramer R, Tuxen-Bettman K, Kremen C (2011) Value of Wildland Habitat for Supplying Pollination Services to Californian Agriculture. Rangelands 33: 33–41.
13. Christensen GA, Campbell SJ, Fried JS (2008) California Forest Resources 2001–2005. USDA Forest Service Pacific Northwest Research Station. 183 p.
14. Silver WL, Ryals R, Eviner V (2010) Soil Carbon Pools in California's Annual Grassland Ecosystems. Rangeland Ecology & Management 63: 128–136.
15. Derner JD, Schuman GE (2007) Carbon sequestration and rangelands: A synthesis of land management and precipitation effects. Journal of Soil and Water Conservation 62: 77–85.
16. Havstad KM, Peters DPC, Skaggs R, Brown J, Bestelmeyer B, et al. (2007) Ecological services to and from rangelands of the United States. Ecological Economics 64: 261–268.
17. Harris G, Thirgood S, Hopcraft JGC, Cromsigt J, Berger J (2009) Global decline in aggregated migrations of large terrestrial mammals. Endangered Species Research 7: 55–76.
18. Olff H, Ritchie ME (1998) Effects of herbivores on grassland plant diversity. Trends in Ecology & Evolution 13: 261–265.
19. Murphy DD, Weiss SB (1988) Ecological studies and the conservation of the bay checkerspot butterfly, *Euphydryas editha bayensis*. Biological Conservation 46: 183–200.
20. Marty JT (2005) Effects of cattle grazing on diversity in ephemeral wetlands. Conservation Biology 19: 1626–1632.

21. Belsky AJ, Matzke A, Uselman S (1999) Survey of livestock influences on stream and riparian ecosystems in the western United States. Journal of Soil and Water Conservation 54: 419–431.
22. Wetzel WC, Lacher IL, Swezey DS, Moffitt SE, Manning DT (2012) Analysis reveals potential rangeland impacts if Williamson Act eliminated. California Agriculture 66.
23. Liffmann RH, Huntsinger L, Forero LC (2000) To ranch or not to ranch: home on the urban range? Journal of Range Management 53: 362–370.
24. Gosnell H, Haggerty J, Travis W (2006) Ranchland Ownership Change in the Greater Yellowstone Ecosystem, 1990–2001: Implications for Conservation. Society and Natural Resources 19: 743–758.
25. Brunson MW, Huntsinger L (2008) Ranching As A Conservation Strategy: Can Old Ranchers Save The New West? Rangeland Ecology & Management 61: 137–147.
26. Mitchell JE, Knight RL, Camp RJ (2002) Landscape Attributes Of Subdivided Ranches. Rangelands 24: 3–9.
27. Sleeter B, Wilson T, Soulard C, Liu J (2011) Estimation of late twentieth century land-cover change in California. Environmental Monitoring and Assessment 173: 251–266.
28. Odell EA, Knight RL (2001) Songbird and Medium-Sized Mammal Communities Associated with Exurban Development in Pitkin County, Colorado. Conservation Biology 15: 1143–1150.
29. Brown DG, Johnson KM, Loveland TR, Theobald DM (2005) Rural Land-Use Trends in the Conterminous United States, 1950–2000. Ecological Applications 15: 1851–1863.
30. Sokolow AD, Lemp C (2002) Agricultural easement programs. Saving agriculture or saving the environment? California Agriculture 56: 9–14.
31. Rissman AR, Lozier L, Comendant T, Kareiva P, Kiesecker JM, et al. (2007) Conservation Easements: Biodiversity Protection and Private Use. Conservation Biology 21: 709–718.
32. Merenlender AM, Huntsinger L, Guthey G, Fairfax SK (2004) Land Trusts and Conservation Easements: Who Is Conserving What for Whom? Conservation Biology 18: 65–76.
33. CDOC (2010) California Land Conservation (Williamson) Act Status Report. Sacramento. CA: CA Department of Conservation
34. Diaz D, Rashford B, De Gryze S, Zakreski S, Dell R, et al. (2012) Evaluation of Avoided Grassland Conversion and Cropland Conversion to Grassland as Potential Carbon Offset Project Types. Portland, Oregon: The Climate Trust.
35. Booker K, Huntsinger L, Bartolome JW, Sayre NF, Stewart W (2013) What can ecological science tell us about opportunities for carbon sequestration on arid rangelands in the United States? Global Environmental Change 23: 240–251.
36. Davis FW, Stoms DM, Hollander AD, Thomas KA, Stine PA, et al. (1998) The California Gap Analysis Project–Final Report. Santa Barbara UC Santa Barbara.
37. Scott JM, Davis F, Csuti B, Noss R, Butterfield B, et al. (1993) Gap Analysis: A geographic approach to protection of biodiversity. Wildlife Monographs 123: 1–41.
38. TNC (2006) California Ecoregions. San Francisco, CA: The Nature Conservancy.
39. Calfire (2010) California's Forests and Rangelands: 2010 Assessment. Sacramento, CA: CA Department of Forestry and Fire Protection
40. George M, Bartolome J, McDougald N, Connor M, Vaughn C, et al. (2001) Annual Range Forage Production. University of California, Division of Agriculture and Natural Resources. 9 p.
41. NRCS (2012) Soil Survey Washington, D.C. : USDA, Natural Resources Conservation Service.
42. CDOC (2008) Farmlands Mapping and Monitoring Program. Sacramento, CA: CA Dept. of Conservation.
43. FSA (2009) National Agriculture Imagery Program. Washington, D.C.: USDA Farm Services Agency.
44. GIN (2011) California Protected Areas Database. San Francisco, CA: GreenInfo Network.
45. CDOC (2010) Williamson Act land status. Sacramento, CA: CA Dept. of Conservation.
46. Calfire (2006) Multisource Land Cover dataset, fveg06. Calfire, Sacramento, CA.
47. Mayer KE, Laudenslayer WF (1988) A Guide to the Wildlife Habitats of California. Sacramento, CA.
48. Kucera TE, Barrett RH (1995) Displaced by agriculture, urban growth: California wildlife faces uncertain future. California Agriculture 49: 23–27.
49. AECOM (2009) Loss of Central Valley Vernal Pools: Land Conversion, Mitigation Requirements, and Preserve Effectiveness. Sacramento, CA. 15 p.
50. Charbonneau R, Kondolf GM (1993) Land use change in California, USA: Nonpoint source water quality impacts. Environmental Management 17: 453–460.
51. Grismer ME, Asato C (2012) Converting oak woodland or savanna to vineyards may stress groundwater supply in summer. California Agriculture 66: 144–152.
52. Famiglietti JS, Lo M, Ho SL, Bethune J, Anderson KJ, et al. (2011) Satellites measure recent rates of groundwater depletion in California's Central Valley. Geophysical Research Letters 38: L03403.
53. USDA (2011) California Wine Grapes 1920–2010. Washington, D.C.
54. Volpe RJ, Green R, Heien D, Howitt R (2010) Wine-grape production trends reflect evolving consumer demand over 30 years. California Agriculture 64: 42–46.
55. NASS (2011) California Historic Commodity Data: Almonds 1909–2010. U.S. Department of Agriculture.
56. Sleeter BM (2008) Late 20th century land change in the Central California Valley Ecoregion. The California Geographer 48: 27–60.
57. Soulard CE, Wilson TS (2013) Recent land-use/land-cover change in the Central California Valley. Journal of Land Use Science: 1–22.
58. Huston M (2005) The Three Phases of Land-use Change: Implications for Biodiversity. Ecological Applications 15: 1864–1878.
59. Greene RP, Stager J (2001) Rangeland to cropland conversions as replacement land for prime farmland lost to urban development. The Social Science Journal 38: 543–555.
60. Rissman AR, Reiner R, Merenlender AM (2007) Monitoring Natural Resources on Rangeland Conservation Easements. Rangelands 29: 21–26.
61. CDOC (2013) The California Land Conservation Act 2012 Status Report. Sacramento. CA: CA Department of Conservation
62. Huntsinger L, Johnson M, Stafford M, Fried J (2010) Hardwood Rangeland landowners in California from 1985 to 2004: Production, Ecosystem Services, and Permanence. Rangeland Ecology & Management 63: 324–334.
63. Gosnell H, Robinson-Maness N, Charnley S (2011) Engaging Ranchers in Market-Based Approaches to Climate Change Mitigation: Opportunities, Challenges, and Policy Implications. Rangelands 33: 20–24.

8

Temporal Variation of Streamflow, Sediment Load and Their Relationship in the Yellow River Basin, China

Guangju Zhao[1,2], Xingmin Mu[1,2]*, Alex Strehmel[3], Peng Tian[4]

1 Institute of Soil and Water Conservation, Northwest A&F University, Yangling, Shaanxi Province, China, **2** Institute of Soil and Water Conservation, Chinese Academy of Sciences & Ministry of Water Resources, Yangling, Shaanxi Province, China, **3** University of Kiel, Institute of Natural Resource Conservation, Department of Hydrology and Water Resources Management, Kiel, Germany, **4** College of Resource and Environment, Northwest A&F University, Yangling, Shaanxi Province, China

Abstract

Variation of streamflow and sediment load in the Yellow River basin has received considerable attention due to its drastic reduction during the past several decades. This paper presents a detailed investigation on the changes of streamflow and sediment load from 1952 to 2011 using monthly observations at four gauging stations along the Yellow River. The results show significant decreasing trends for both streamflow and sediment load at all four gauging stations over the past 60 years. The wavelet transform demonstrated discontinuous periodicities from 1969 to 1973 and after 1986 due to the construction of large reservoirs and implementation of numerous soil and water conservations practices. The sediment rating curves with the power-law function was applied to investigate the relationship between discharge and sediment load. The results indicate distinct variations of the relationship between streamflow and sediment and implied significant hydro-morphological changes within different periods. The reducing sediment supply from the source region and the increased erosive power of the river are detected at Lanzhou station, while the decrease of the transport capacity at Toudaoguai is caused by severe siltation. Significant changes in the relationship between streamflow and sediment load are found at Huayuankou and Gaocun stations, which are largely induced by evident sediment income and trapping effects of large reservoirs. It is estimated that numerous reservoirs have strongly altered the regime and magnitude of streamflow and trapped large amount of sediment, leading to severe siltation and evident reduction of their total volumes. A decrease in precipitation, incoming water from the upper reaches, soil and water conservation measures as well as water consumption contribute most to the significant reduction of streamflow. The decrease of sediment load mainly resulted from various soil and water conservation measures and trapping in reservoirs from 1986 to 2011.

Editor: João Miguel Dias, University of Aveiro, Portugal

Funding: This work was supported by the National Natural Sciences Foundation of China (Grant Nos.: 41201266, 41271295), the Key Research Program of the Chinese Academy of Sciences (No. KZZD-EW-04), the Key Project of the National Science and Technology Ministry (No. 2012BAB02B05) and the Program for Excellent Youth Scholars in the Institute of Soil and Water Conservation (A315021394). The funders had no role in study design, data collection and analysis, decision to publish, or preparation of the manuscript.

Competing Interests: The authors have declared that no competing interests exist.

* E-mail: xmmu@ms.iswc.ac.cn

Introduction

Rivers are considered as the major link between continents and oceans and play a critical role in geological, biological and chemical processes on the land surface [1,2]. Water discharge and sediment transport in the rivers greatly affect the geomorphology of the river channels, alluvial plains and deltas, and deliver numerous terrestrial materials to oceans, which sustain the coastal and marine ecosystems [3]. During the past century, river systems encountered significant changes at a global scale due to climate change and intensive human activities (land use change, river regulations, water abstraction, damming, sand mining etc.) [2,4,5]. As a result, a decrease or increase in streamflow and sediment load in many of the rivers throughout the world can be observed. Thus, it is of great scientific and practical importance to understand the changes of river systems and evaluate their possible controlling factors. Quantitative analysis on the spatial and temporal variation of streamflow and sediment load can provide a good reference for flood mitigation, river channel training and river basin management in the future.

In recent decades, variation in streamflow and sediment load has attracted considerable attention since human activities have greatly influenced the river systems worldwide [6–9]. These studies have reported that many large rivers throughout the world (i.e. Nile River, Colorado River, Amazon River, Yangtze River and Yellow River) show a significant decline in sediment load following the construction of reservoirs and dams and land use/cover changes [1,10–12]. Walling and Fang (2003) addressed that statistically significant increasing and decreasing trends were both detected in approximately half of the 145 major rivers' long-term sediment records that they examined [2]. Syvitski et al. (2005) computed the flux of terrestrial sediment to the oceans under contemporary and pre-human conditions, and predicted a global suspended sediment yield of 14.0×10^9 t/a under pristine level without human activities, reducing to 12.6×10^9 t/a for the anthropogenic phase [6]. Furthermore, the estimation indicated that on a global scale 26% of the sediment that would be transported to the coast and to deltas has been intercepted by retention in reservoirs.

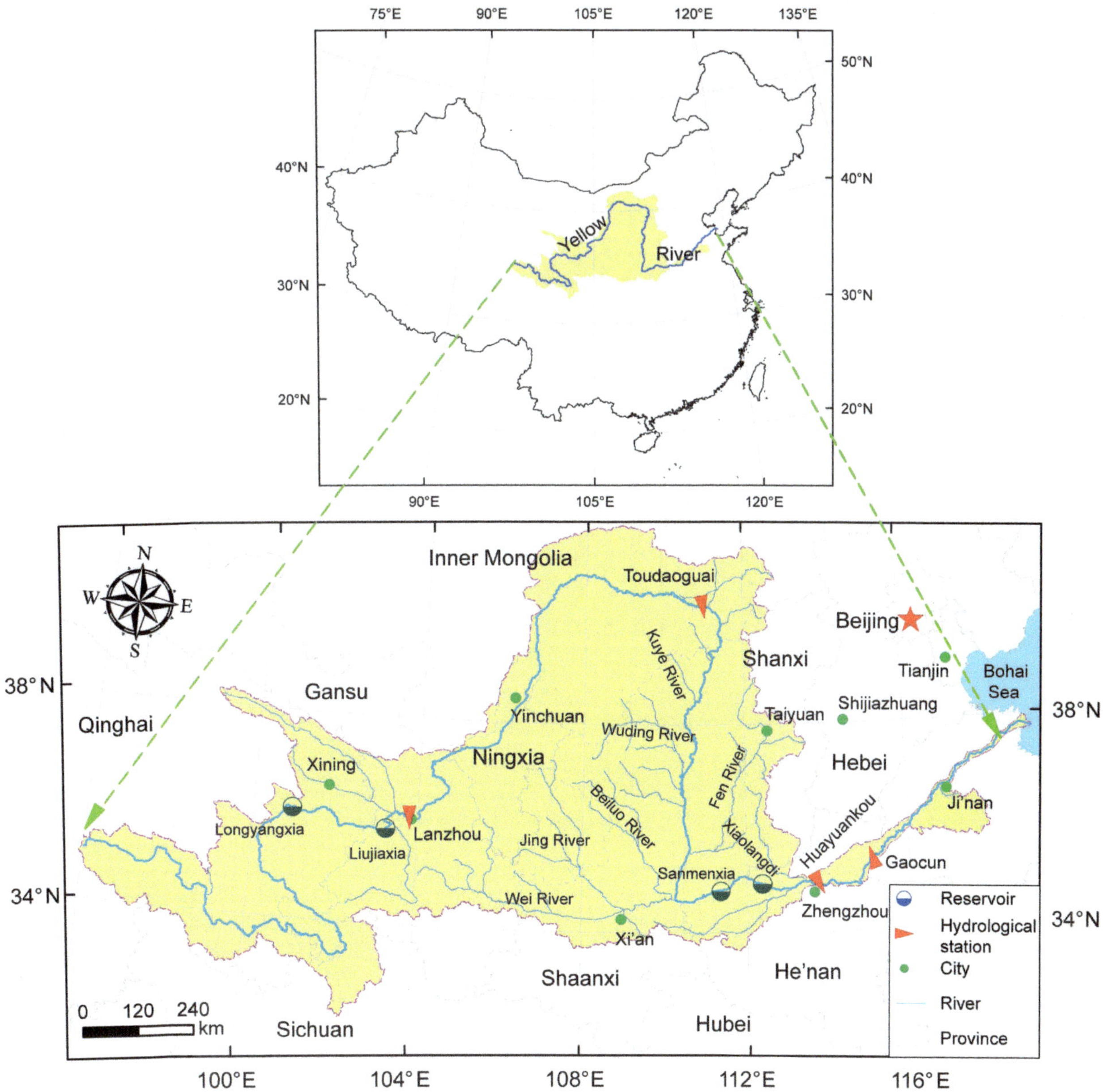

Figure 1. Location of the Yellow River basin, its major tributaries, key gauging stations and reservoirs.

The Yellow River is the second largest river in China, and is regarded as the "cradle of the Chinese civilization". The Yellow River plays a critical role in the development of regional economy as the major source of freshwater for about 1.07×10^8 people in the river basin. The total surface water consumption within the whole basin has increased from 12.9 km^3 during 1950s to 18.2 km^3 between 2000 and 2010 [7,10]. Since the 1950s, the water discharge in the Yellow River has decreased due to impacts of climate changes and anthropogenic influences coupled with the harsh natural conditions and fragile ecosystem [13–17]. In the recent ten years (2001–2010), the average annual streamflow decreased to 242.6 $\times 10^8$ m^3/a, account for only 53% of the discharge observed in the 1950s. Similarly, sediment load showed a synchronous decline with the dramatic changes in streamflow. The mean observed annual sediment load from 2001 to 2010 at Huayuankou station was 1.42×10^8 t/a, only 9.7% of in the load in the 1950s.

Because of its important geographical, cultural and environmental functions and great decline of water resources and sediment flux to the sea, numerous studies have been conducted to quantitatively assess the changes in streamflow and sediment load as well as their potential causes [7,10,18]. Zheng et al. (2009) reported that land uses changes were predominant factors for the reduction in mean annual streamflow in the headwater catchment of the Yellow River basin [19]. Gao et al. (2011) investigated the changes of streamflow and sediment load in the middle reaches of the Yellow River, and found that the decreasing precipitation accounted for 28% of the reduction in streamflow from 1986 to 2008, and the remaining 72% is resulted from intensive human activities [20]. Peng et al. (2010) conducted a detailed assessment on the variation of sediment load in the upper, middle and lower

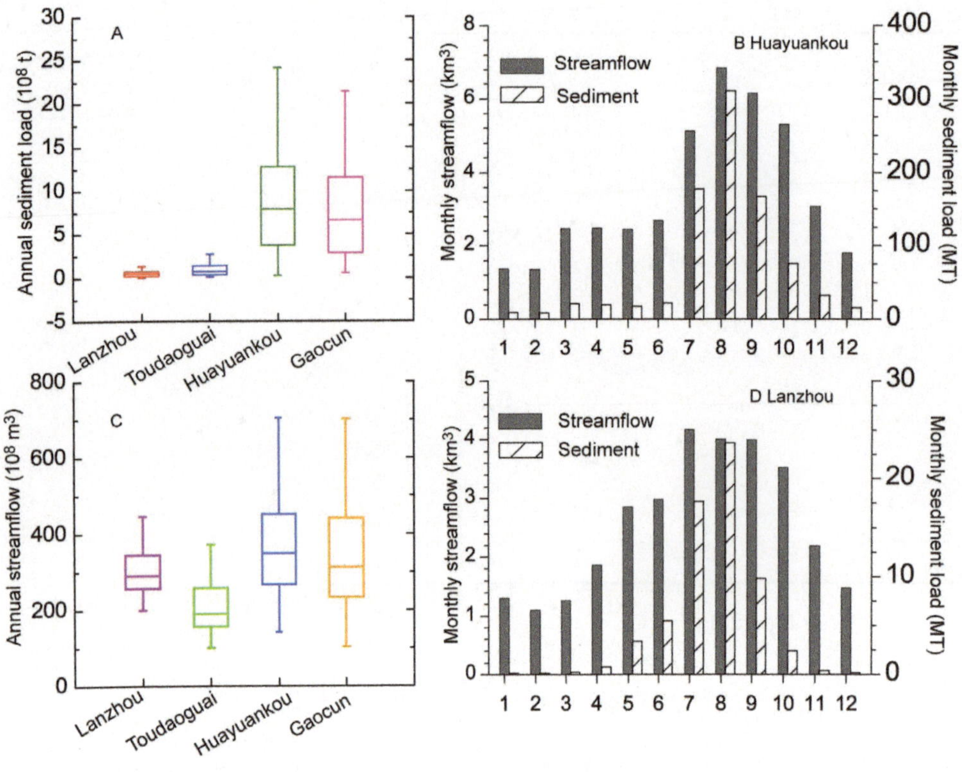

Figure 2. Streamflow and sediment load at four gauging stations along the Yellow River.

reaches of the Yellow River during the past 60 years [21]. The results indicated that the construction of reservoirs and soil-water conservation measures were the dominant factors, contributing approximately 80% of sediment reduction, and the remainder was attributed to decreasing precipitation.

Many studies have been conducted on the changes of water discharge and sediment load at different spatial scales in the Yellow River basin. However, the studies were mostly conducted by making use of average annual records [15,20]. Few studies were undertaken at a monthly scale with continuously updated data. Furthermore, very few studies have been conducted to address the relationship between streamflow and sediment load and the hydro-morphological responses to the changes of water discharge and sediment load in the Yellow River basin. In addition, the "Grain for Green" Program (GFGP) and the "Water-Sediment Regulation Project" (WSRP) were launched by the Chinese government in 1999 and 2002, respectively. The GFGP was intended to conserve the slope lands and effectively control soil erosion from the sources, while the WSRP aimed to scour large amounts of sediment, which are deposited in the reservoirs and channels in the middle and lower reaches. Until now, the changing properties of streamflow and sediment load have not been extensively addressed at the monthly scale, particularly after the operation of the GFGP and the WSRP. The objectives of this study are to investigate the changes in monthly streamflow and sediment load regime and their relationship using long-term observations along the Yellow River. In connection to that, potential influencing factors such as dams construction, soil and water conservation measures and other human activities are analyzed.

Study Area and Data

The Yellow River originates from the Qinghai-Tibetan Plateau in western China, and flows eastward through the Loess Plateau and the North China Plain with a total length of 5,464 km (Fig. 1). The whole river basin covers approximately 75.2×10^4 km^2 and drains into the Bohai Sea. The river can be divided into three segments. The upper reaches extend over a length of 3,471 km from the river source to the Toudaoguai gauging station covering an area of 38.6×10^4 km^2, flowing through a mountainous region

Table 1. General characteristics of the four gauging stations along the Yellow River.

Stations	Length of segment (km)	Controlled area (10^4 km^2)	Period	Mean annual discharge (10^8 m^3/a)	Mean annual sediment (10^8 t/a)
Lanzhou	2119	22.26	1952–2011	307.01	0.64
Toudaoguai	1352	38.6	1952–2011	212.11	1.01
Huayuankou	1206	73.00	1952–2011	373.09	8.79
Gaocun	185	73.41	1952–2011	347.73	7.92

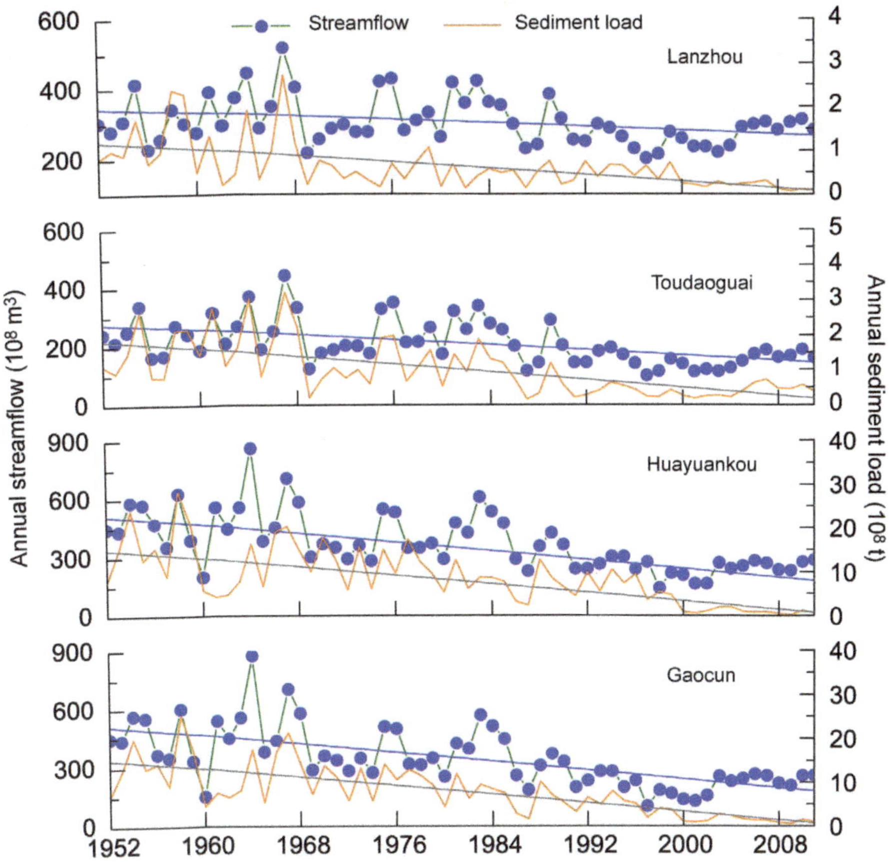

Figure 3. Trend analysis for annual runoff and sediment load in the Yellow River basin.

to the fluvial Hetao plain. The middle reaches are located between Toudaoguai and Huayuankou, and flows over 1,206 km through the highly erodible Loess Plateau with an area of 34.4×10^4 km^2. The lower reaches stretch from Huayuankou to the river mouth through a relatively flat alluvial plain with a drainage area of 2.3×10^4 km^2 (Fig. 1).

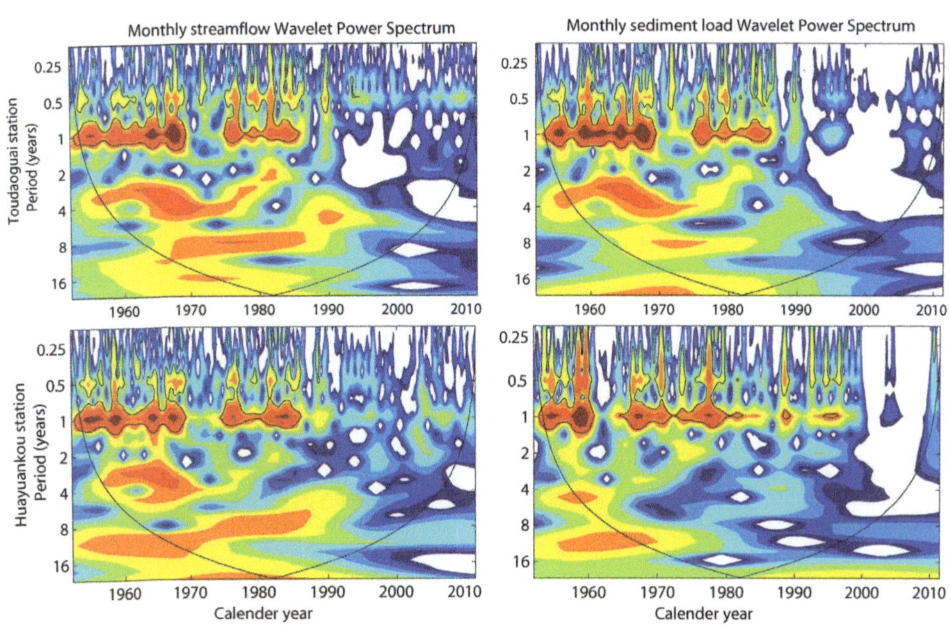

Figure 4. Wavelet analysis of monthly runoff and sediment load at Huayuankou stations.

Table 2. Mean water discharge (Q) and sediment concentration (Cs) at four gauging stations along the Yellow River within different periods.

Stations	Lanzhou		Toudaoguai		Huayuankou		Gaocun	
	Q (m^3/s)	Cs (g/l)	Q (m^3/s)	Cs (g/l)	Q (m^3/s)	Cs (g/l))	Q (m^3/s)	Cs (g/l))
1952–1968	1084.64	3.84	837.98	5.64	1614.25	43.98	1562.07	41.55
1969–1974	857.43	1.54	570.22	2.36	1038.76	36.55	1002.96	30.32
1975–1986	1125.14	1.61	849.66	4.06	1390.41	31.20	1286.34	29.05
1987–1999	842.01	1.61	516.63	1.41	880.97	22.55	753.55	16.66
2000–2011	865.06	0.63	481.37	1.31	764.31	3.20	696.06	4.47

The average annual streamflow from the upper and middle reaches recorded at Lanzhou and Toudaoguai stations are 30.7 km^3/a and 21.2 km^3/a, respectively, and relatively lower than those at Huayuankou and Gaocun station in the downstream reach. While average annual sediment load amounts to nearly 0.64 $\times 10^8$ t/a at Toudaoguai station, and is much lower than that at Huayuankou station. More than 30 tributaries with a catchment area larger than 1,000 km^2 (including Huangfu River, Kuye River, Tuwei River, Wuding River, Yan River, Wei River and Fen River and others) contribute approximately 80% of the sediment to the Yellow River [22]. The lower reach has become a narrow corridor and "suspended" river segment since the dominant deposition process of sediment led to the raising of the channel bottom above the ground level, sometimes by more than 10 m [23]. As a result, the mean annual sediment load at Huayuankou station averages to 8.79×10^8 t/a, and is higher than the measurements at the downstream stations (i.e. Gaocun and Lijin stations). Accordingly, the sediment load at all four gauging stations along the mainstream indicates that the sediment is predominantly derived from the middle reaches of the Yellow River (Fig. 2a), where severe soil erosion occurs in the gullies and inter-gullies region [24]. However, the water from the middle reaches accounts for less than 50% of the discharge to the Bohai Sea (Table 1, Fig. 2c).

The majority of the Yellow River basin is influenced by the temperate continental climate. The upper and lower reaches are influenced by the southwest Indian monsoon and southeast Asian summer monsoon, which control the spatial and temporal variability of precipitation in the river basin. The summer monsoon normally starts to influence the river basin in June and retreats in October, accounting for more than 60% of annual total rainfall. Streamflow is relatively higher during the rainy season, particularly from July to October. In contrast, high sediment loads are observed in the period between July and September due to frequently occurring rainstorms. The seasonal distributions of sediment load at the four gauging stations show a strong association with the patterns of water discharge, as characterized by the co-occurrence of high sediment loads and high water discharge during the rainy season (Fig. 2b, d).

The consecutive monthly data of streamflow and sediment yields at Lanzhou, Toudaoguai, Huayuankou and Gaocun stations from 1952 to 2011 were provided by the Yellow River Water Resources Commission (Table 1), and were partly collected from the Yellow River Sediment Bulletin [25]. The daily discharge was computed from the water level by using previously calibrated discharge-water level curves. Water was sampled at fixed intervals, and suspended sediment concentration was obtained by measuring water samples in the laboratory. All the measurements of water level, discharge, SSC followed national standards issued by the Ministry of Water Conservancy, and were printed in the Hydrological Year-book of the Yellow River. The monthly and annual streamflow and sediment load at the gauging stations were derived from the daily measured data. The accuracy and consistency of all the data used in this study have been checked out by the corresponding agencies before their release.

Methodology

In this study, we applied a simple linear regression to detect the temporal trends of annual streamflow and sediment load, and significances were detected by F-test at $p<0.05$. The discrete wavelet transform and flow duration cure method were respectively used to identify the periodical characteristics and changes of hydrological regime for monthly time series. Additionally, flow-sediment relationship at four main gauges was examined by using sediment rating curve.

Wavelet Analysis for Changes in the Streamflow and Sediment Regime

The wavelet transform is a spectral method of decomposing a time series into time and frequency, and is used to identify and analyze the dominant localized variation of power. The wavelet analysis has been widely applied to characterize the frequency, intensity, time position and duration of the variations for hydro-climatic time series [9,26]. Given the time series of Xn, with equal time interval Δt (one month in our study), a corresponding Morlet wavelet function $\psi_0(\eta)$, with a mean value of zero and localized in time and frequency space is defined as:

$$\psi_0(\eta)=\pi^{-1/4}e^{i\omega_0\eta}e^{-\eta^2/2} \quad (1)$$

Where ω_0 denotes the dimensionless frequency, and η is the non-dimensional time frequency [27,28]. The continuous wavelet transform of a discrete sequence X_n is defined as the convolution of X_n with a scaled and translated version of $\psi_0(\eta)$:

$$W_n(s)=\sum_{n'}^{N-1} X_n\psi^*\left[\frac{(n'-n)\delta t}{s}\right] \quad (2)$$

Where the * represents the complex conjugate, s is the wavelet scale, and n is the translated time index.

The statistical significance of wavelet spectra can be assessed under the null hypothesis that the signal is generated by a stationary process given the background power spectrum (P_k). It is assumed that the time series has a mean power spectrum:

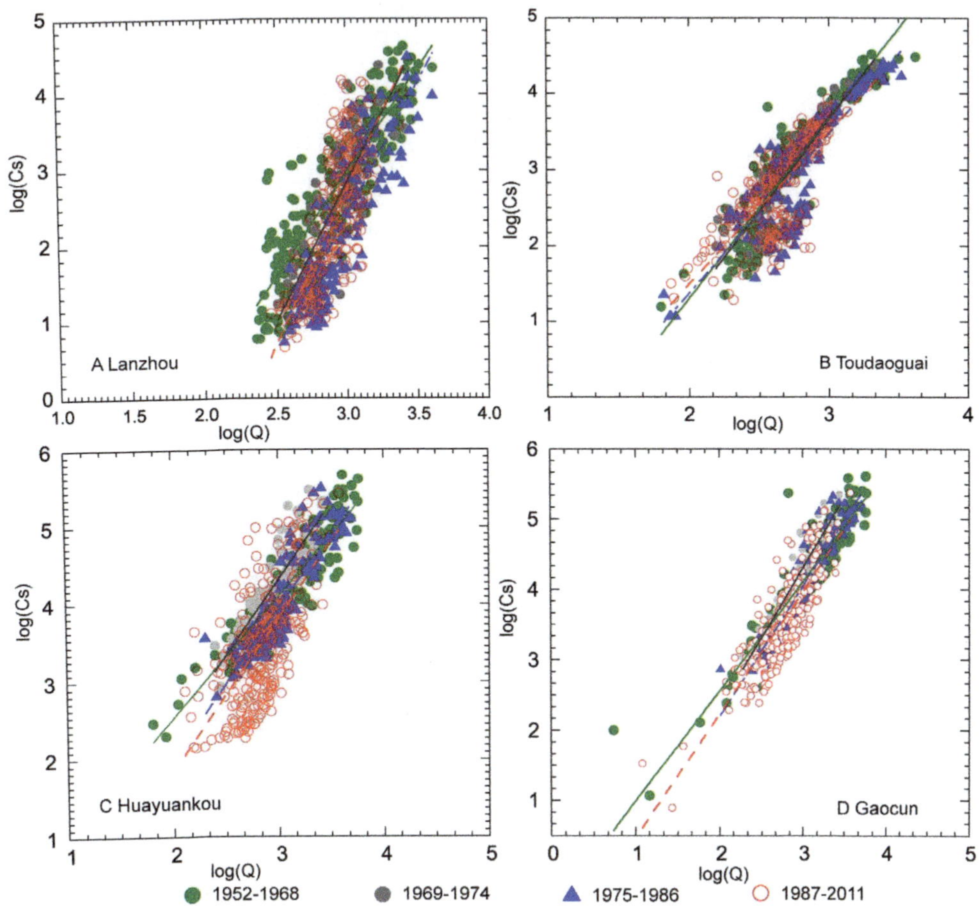

Figure 5. Sediment rating curve analysis at four gauging station along the Yellow River.

$$P_k = \frac{1-\alpha^2}{|1-\alpha e^{-2i\pi k}|^2} \quad (3)$$

where k is the Fourier frequency index that corresponds to the wavelet scale s. A specific confidence level against red noise, such as 95% confidence level (significance at 5%) can be obtained by [28]:

$$P\left(\frac{|W_n^X(s)|^2}{\delta_X^2} < p\right) = \frac{1}{2} p_k \chi_v^2(p) \quad (4)$$

where $\chi_v^2(p)$ represents the value of Chi-square distribution with two degree of freedom at the confidence level p, and v is equal to 1 for real and 2 for complex wavelet. The geophysical series have the red noise characteristics which can be modeled by a univariate lag-1 autoregressive (AR-1) process. A detailed explanation for the concept and procedure of the wavelet transform has been presented by previous studies [27,28].

Sediment Rating Curves

The sediment rating curve is defined as the statistical relationship between sediment discharge (*Qs*) or concentration (*Cs*) and water discharge (*Q*). It generally represents a functional relationship of the form [29,30]:

Table 3. Sediment rating parameters for different periods at four gauging stations.

Stations	Lanzhou			Toudaoguai			Huayuankou			Gaocun		
	log(a)	b	R^2	log(a)	b	R^2	log(a)	b	R^2	log(a)	b	R^2
1952–1968	−5.22	2.73	0.78	−3.44	2.37	0.86	−0.6	1.57	0.86	−0.57	1.56	0.88
1969–1974	−8.18	3.67	0.70	−3.60	2.42	0.68	−1.57	1.96	0.77	−1.69	1.99	0.88
1975–1986	−8.15	3.51	0.73	−2.83	2.10	0.75	−1.87	1.94	0.81	−1.52	1.86	0.88
1987–2011	−9.43	4.06	0.60	−2.52	2.01	0.60	−2.08	1.96	0.44	−1.29	1.75	0.71

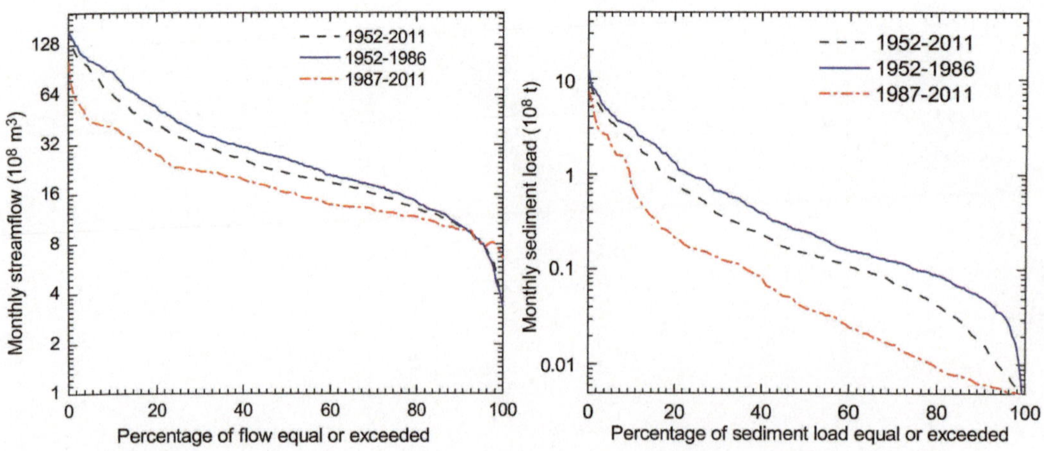

Figure 6. Flow duration curves for monthly runoff and sediment load at Huayuankou station.

$$C_s = aQ^b \tag{5}$$

$$\text{Or} \quad \log C_s = \log a + b \log Q \tag{6}$$

in which a and b are the coefficients of the sediment rating curve, and their values can be obtained from the measurement data through a linear regression between logCs and log$Q$. The sediment rating curve is regarded as a 'black box' type of model expressing the relationship between sediment and discharge as a power function. In general, the parameters a and b in the sediment rating curve do not contain particular physical meaning [31]. However, Morgan (1995) argued that the sediment rating parameter a was as an index of erosion severity in the river channel [32]. A high value of a is associated with the area, which is characterized by easily erodible materials and high loads of transported materials. The coefficient b depicts the erosive power of the river, with high values indicating strong increase in erosive power of the river. It has also been addressed that the coefficient b is related to the regional climate pattern, channel morphology, grain-size distribution of sediment and the erodibility within the river basin [31,33].

Flow Duration Curve

The flow duration curve (FDC) is a cumulative distribution function that shows the percentage of specific discharge that is equaled or exceeded. It has been considered as an important approach for assessing the frequency and magnitude of a time series. For a given time series $Q_i(q_1, q_2, \cdots, q_n)$, it can be ranked to produce an ordered record $q_{(i)}\left(q_{(1)}, q_{(2)}, \cdots, q_{(n)}\right)$, where n is the sample size, and q_1 and q_n are the largest and smallest observations, respectively. The FDC can be plotted by the measured values $q_{(i)}$ associated with probability of p_i, expressed as:

$$p_i = P\left(Q > q_{(n)}\right) = \frac{i}{n+1} \tag{7}$$

The FDC provides a simple graphical view of the overall variability for long term streamflow or sediment load [34]. It is probably one of the most informative approaches of displaying the complete range of hydrological time series.

Results

Trends Analysis for Streamflow and Sediment Load

Fig. 3 shows simple linear trends in average annual streamflow and sediment load at Lanzhou, Toudaoguai and Huayuankou stations. Overall, both streamflow and sediment load shows significant decreasing trends ($p<0.05$) at all stations. The annual streamflow and sediment load exhibits the most significant decrease at Huayuankou station ($p<0.05$), with average reduction rates of -5.44×10^8 m^3/a and -0.26×10^8 t/a, respectively, while the streamflow and sediment load at Lanzhou station display gently decrease (-1.33×10^8 m^3/a and -0.02×10^8 t/a).

Table 4. Flow duration curve analysis for different periods at four gauging stations along the Yellow River.

Stations	Lanzhou		Toudaoguai		Huayuankou		Gaocun	
	Pre$_{1986}$	Post$_{1986}$	Pre$_{1986}$	Post$_{1986}$	Pre$_{1986}$	Post$_{1986}$	Pre$_{1986}$	Post$_{1986}$
Q_{10} (10^{8}m^3)	56.25	32.92	45.26	22.50	88.13	40.44	81.96	36.69
Q_{50} (10^{8}m^3)	22.79	22.01	14.93	10.55	26.18	16.37	24.19	14.49
Q_{90} (10^{8}m^3)	12.24	11.89	6.22	6.51	10.45	9.75	7.53	8.81
Sed_{10} (10^{6}t)	20.5	8	39	9.86	308	76	274	70
Sed_{50} (10^{6}t)	0.67	0.66	354	177	25	4	26	6
Sed_{90} (10^{6}t)	0.04	0.04	25	30	5	1	4	2

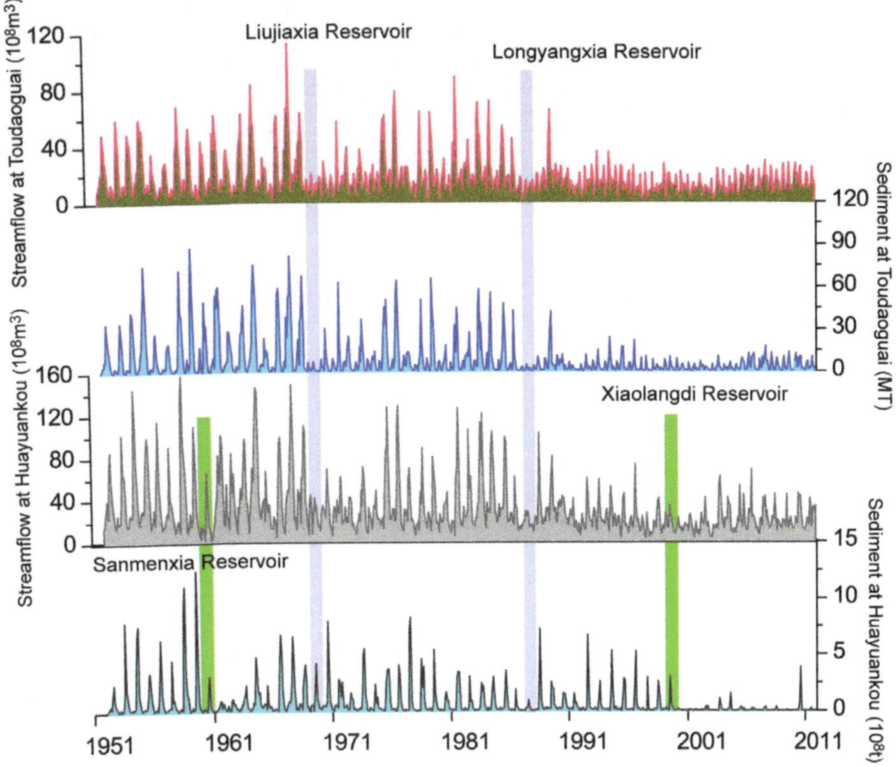

Figure 7. Monthly streamflow and sediment load at Toudaoguai and Huayuankou stations associated with reservoirs construction.

Periodicities in Monthly Streamflow and Sediment Load

The wavelet transform was applied to examine the variation of periodicities in monthly streamflow and sediment load at the four gauging stations. Fig. 4 illustrates the wavelet power spectrum of the monthly hydrological variables. It can be clearly seen that both streamflow and sediment load are in good agreement between low- and high-frequency oscillation. The wavelet power spectrums indicate intermittent periodicities of 0.5 and 1 year due to the seasonal and annual alternation of hydrological variables at all the stations. Furthermore, the 95% confidence intervals are continuously distributed in one year band during 1952–1968 and 1974–1986. Periodicity was not detected from 1969 to 1973 and after 1987. This can be attributed to the intensive human activities in the upper and middle reaches of the Yellow River during the past six decades. The data in Fig. 4 indicate that human activities have made the seasonal and annual cycles of monthly series weaker since the late 1980s, and even disappearing completely after 2000.

According to the wavelet transform results of monthly streamflow and sediment load, five periods can be distinguished to quantify the temporal changes of discharge and sediment in the Yellow River basin. Table 2 demonstrates the mean water discharge and sediment concentration at four gauging stations

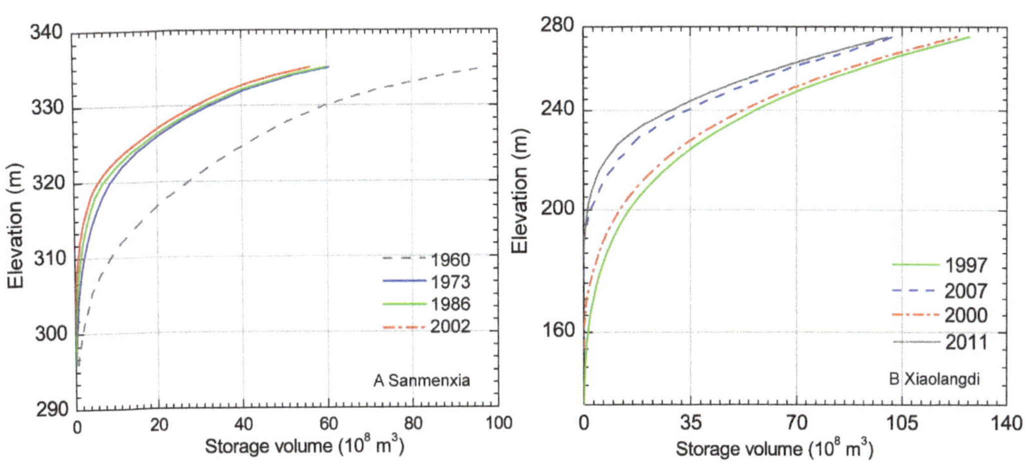

Figure 8. Historical changes of storage volumes for Sanmenxia and Xiaolangdi reservoirs.

Table 5. Statistics of soil and water conservation measures in the middle reaches of the Yellow River basin [37,39].

Soil-water conservation measures	Area controlled by various measures during different period (10^3 km^2)					
	1959	1969	1979	1989	1996	2006
Terrace	0.53	2.41	5.62	10.06	14.41	28.54
Afforestation	1.97	5.21	13.56	33.31	44.40	58.61
Grassing	0.42	0.66	1.57	5.51	6.37	14.07
Check dam	0.036	0.185	0.467	1.009	1.224	1.310
In Total	2.95	8.47	21.22	49.89	66.41	102.53

along the Yellow River during different periods. A general decrease in both water discharge and sediment concentration can be clearly detected.

Due to the length of the river basin, changing trends of mean water discharge and sediment concentration are highly variable, both spatially and temporally. In detail, the mean water discharge at Lanzhou station shows a reduction of nearly 20%, while the sediment concentration shows an evident decline from the period of 1952–1968 (3.84 g/l) to 2000–2011 (0.63 g/l). Compared to the Lanzhou station upstream, the mean water discharge decreased by nearly 50% or even more at the stations in the middle and lower reaches. A much more evident reduction (decreased by approximately 90%, Table 2) is observed for the sediment concentration at Huayuankou and Gaocun stations. The average annual sediment concentration during 2000–2011 was 3.2 g/l, only accounting for 7.3% of that from 1952 to 1968 at Huayuankou station.

Sediment Rating Curves at Four Key Stations Along the Yellow River

Sediment rating curves at four main gauging stations along the Yellow River were obtained for four distinct periods (i.e. 1952–1968, 1969–1974, 1975–1986, and 1987–2011) according to the periodicities of monthly streamflow and sediment load. All sediment rating curves at four stations in different periods generally displayed a decreasing trend in the later period (1987–2011) (Fig. 5). The steepest rating curves were found at Lanzhou, followed by Toudaoguai station, and the relatively flat rating curves were found at Huayuankou and Gaocun stations. This pattern may relate to the channel morphology of the rivers. High slope gradient and bedrock-confined V-shaped channels in the upstream of Lanzhou suggests higher unit stream power with steep rating curve. In contrast, the wide U-shape river channel in the mid-lower reaches is characterized by gentle slope gradient with lower unit stream power.

It has been mentioned that the coefficient log(a) represents the soil erodibility in the river basin. High values of log(a) denote high sediment supply. The variation of a with time for a given basin is expected to reflect the human disturbances on the hydrological processes. The exponent b represents the erosive power of the river and the influence on the sediment supply from the entire basin surface. Large values of b indicate an increased transport capacity of the river [31]. The decreasing log(a) and increasing b from 1952 to 2011 imply a decreasing sediment supply from the source region and an increased erosive power of the river at Lanzhou station (Table 3). However, the sediment rating exponent log(a) values at Toudaoguao station continuously increased from −3.44 in 1952–1968 to −2.52 in 1987–2011, and the rating coefficient b synchronously decreased from 2.37 to 2.01, suggesting increasing sediment supply and decreased transport capacity of the river due to severe siltation. The sediment rating parameters show similar changes at Huayuankou and Gaocun stations within the past six decades. Relatively high values of log(a) in 1952–1968 are induced by the abundant sediment income from the upper and middle reaches, whereas the higher values of b at lower reaches of Huayuankou and Gaocun from 1969 to 2011 might indicate the influence of the construction of the Sanmenxia Reservoir.

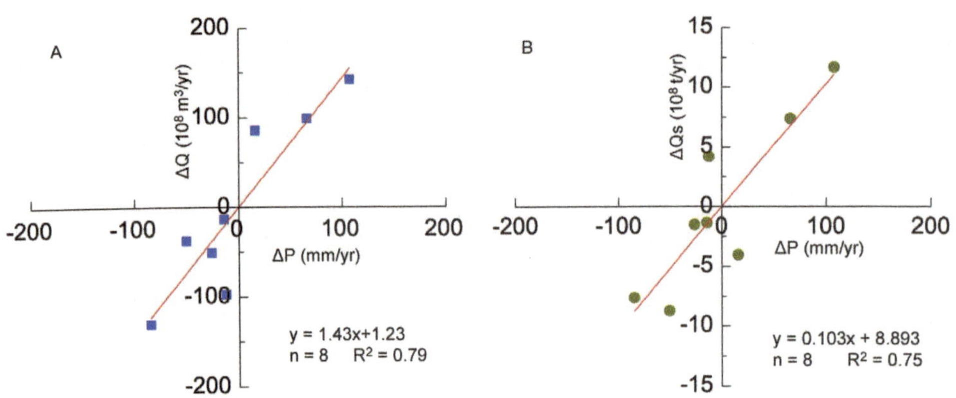

Figure 9. Relationship between the interannual variability of precipitation streamflow and sediment load at Huayuankou from 1952 to 1959.

Flow Duration Curve Analysis for Monthly Streamflow and Sediment Load

According to the wavelet analysis of monthly streamflow and sediment load, we divided the studied period into two periods, namely before and after 1986, to simply quantify the variation in monthly streamflow and sediment load. Fig. 6 shows an illustrative example of the FDCs for both monthly streamflow and sediment load at Huayuankou station during the two periods described previously. The results indicate that there was evident decline for both mean monthly streamflow and sediment load. More than 50% of the reduction was detected when comparing the high values of flow and sediment load during 1987–2011 to those of 1952–1986.

To clearly identify the temporal variation of streamflow and sediment load at the four stations along the Yellow River, six indices with different magnitudes of flow and sediment load (i.e. Q_{10}, Q_{50}, Q_{90}, Sed_{10}, Sed_{15}, and Sed_{90}) were selected for analysis during two different periods. The indices of high-flow (Q_{10}), median-flow (Q_{50}) and low-flow (Q_{90}) were defined as the specific values exceeded 10%, 50% and 90% of the monthly streamflow respectively. Similar indices were also defined for sediment load. As shown in Table 4, the magnitude of monthly streamflow and sediment load generally decreased during the period of 1987–2011 compared with that from 1952 to 1986 at all four key gauging stations. Specifically, the high-flow (Q_{10}) reduced by more than 50% at all the stations except for the upstream Lanzhou station. However, the low-flow indicated relatively gentle changes between the two periods. In contrast, the high-sediment index (Sed_{10}) within the period of 1987–2011 illustrated a drastic decline at the four stations, particularly at Toudaoguai, Huayuankou and Gaocun stations, where reductions of approximately 70% can be observed.

Discussion

Impacts of Dam Operation on Streamflow and Sediment

Regulation from dams and reservoirs is the most direct way among the human activities that drastically alters the regimes of streamflow and sediment load in the river basin. In the Yellow River basin, more than 3,150 reservoirs have been built and the total storage capacity of all the registered reservoirs reaches up to 72 km^3 [35]. This is much higher than the average annual streamflow of 24.4 km^3 between 2001 and 2011 at Huayuankou station.

Fig. 7 illustrates the regulating effect of jointly operated reservoirs on the water and sediment discharge regimes at Toudaoguai and Huayuankou stations. The abrupt decline in monthly streamflow and sediment load indicates substantial trapping effects of four major reservoirs along the Yellow River: the Longyangxia (with a total volume of 27.6 km^3), Liujiaxia (5.7 km^3), Sanmenxia (9.6 km^3) and Xiaolangdi Reservoir (12.7 km^3). In general, the magnitude and variability of monthly streamflow at Toudaoguai and Huayuankou stations has been reduced by these large reservoirs, particularly after the late 1980s. The seasonal variability in both streamflow and sediment load has decreased since 1986 when the Longyangxia Reservoir started operation. The observed streamflow in the flood seasons (June to October) accounted for approximately 68.3% of the average annual total in the 1950s, but decreased to 53.1% after 2000 at Huayuankou station. This may be largely attributed to the trapping effect of reservoirs, which dramatically reduced the flood peaks by storing the water during the rainy season and releasing the water in the dry period to meet the agricultural demands. Furthermore, the discontinuous periodicities of monthly streamflow and sediment load (Fig. 4) were consistent with the timing of dams' construction, particularly after the construction of the Longyangxia Reservoir in 1986.

The Sanmenxia reservoir started operation in 1960 and trapped large amounts of sediment during its first few years of operation due to abundant sediment discharged into the Yellow River from the Loess Plateau. As shown in Fig. 8a, the storage volume decreased significantly from 1960 to 1973, afterwards the reduction is relative gentle. Sediment retention was estimated to reach up to 79.1×10^8 t with an average of 3.13×10^8 t/a from 1960 to 1973 [25]. The sediment load at Huayuankou station was

Table 6. Relative contribution of precipitation and human activities to the decrease of streamflow and sediment load at Huayuankou within different periods.

	Streamflow (10^8 m^3/a)			Sediment load (10^8 t/a)		
Time period	1960–1986	1987–1999	2000–2011	1960–1986	1987–1999	2000–2011
Average precipitation	490–469	490–448	490–443	490–469	490–448	490–443
average streamflow and sediment	487.4–443.8	487.4–277.8	487.4–241.0	16.8–10.7	16.8–7.1	16.8–1.0
Decrease in streamflow and sediment	43.6	209.6	246.4	6.1	9.7	15.8
	Contribution to streamflow reduction (10^8 m^3/a)			**Contribution to sediment load reduction (10^8 t/a)**		
Soil-water conservation[a]	–	45.8	74.5	1.8	3.1	6.1
Trapping by Reservoirs[b]	6.4	2.1	2.4	3.1	1.5	3.99
Changes from upper reaches	–	73.2	84.3	0.10	1.06	1.10
Precipitation variation[c]	16.6	34.1	38.3	1.3	2.7	3.0
Changes in water consumption[b]	14.5	20.9	15.2	–	–	–
Total decrease	37.5	176.1	214.7	6.3	8.36	14.19

[a]Data was obtained from [41] and [42];
[b]data from YRCC [25];
[c]values are estimated from the equation shown in Fig. 9.

11.06×10^8 t/a during the period of 1960–1973, which has been reduced by nearly 35% compared to that from 1952 to 1959.

In order to reduce severe siltation in the lower reaches of the Yellow River, the Xiaolangdi reservoir was constructed between Sanmenxia reservoir and Huayuankou station (Fig. 1) and started to store water in 1999. Considerable siltation occurred with the operation of the reservoir, leading to a total sediment retention of about 32.5×10^8 t from 1997 to 2007 (Fig. 8b). However, the construction of Xiaolangdi reservoir did not resolve the severe siltation in the lower reaches of the Yellow River. To efficiently mitigate siltation and scour the sediment deposited in the channel, the Yellow River Conservancy Commission initiated the Water-Sediment Regulation Scheme in 2002. This scheme aims at the joint operation of large reservoirs, which releases large amounts of stored water to deliver sediment retained within the Xiaolangdi reservoir downstream and to scour the lower reaches in the summer season. It has been estimated that about 3.9 $\times10^8$ t of sediment have been scoured into the sea after executing this measure 13 times between 2002 to 2011 with a total released water volume of 50.91 km^3 [25].

Effects of Soil and Water Conservation on Streamflow and Sediment Load

A series of soil and water conservation practices have been carried out in the upper and middle reaches of the Yellow River basin since the late 1950s. These soil and water conservation measures include biological measures (i.e. terracing, afforestation) and engineering measures (such as check dams and reservoirs) [36]. Table 5 shows statistics of soil and water conservation measures in the middle reaches of the Yellow River basin. The measures present various increase rates during different periods. Before the late 1970s, the area controlled by soil and water conservation measures was very small due to the limited awareness and funding, and increased significantly afterwards because of considerable government-funded projects. Between the 1970s and 1996, the area controlled by various types of soil and water conservation practices has almost tripled and recent statistics indicate that nearly 0.2×10^6 km^2 of area are controlled with various soil and water conservation measures by 2006, including more than 1, 800 key hydraulic projects and 110,000 check dams for gully erosion control [37]. Thus, it can be undoubtedly concluded that the dramatically increased soil and water conservation measures are responsible for the decreasing streamflow and sediment load in those catchments. Generally, monthly streamflow and sediment load of the Yellow River demonstrated seasonal cycles before 1970s without intensively intervention from human activities. However, the discontinuous seasonal cycles on monthly streamflow and sediment load indicated that various soil and water conservation measures had greatly altered the hydrology regime of the river and reduced the discharge and sediment load (Fig. 4 and 6), particularly after the 1980s.

Associated with large number of soil and water conservation measures implementation, land use and land cover showed great changes in the whole river basin. Increasing cropland, built-up land and bare land have been detected from 1980s to 2000 according to Landsat TM interpretation, the areas of grassland and woodland decreased dramatically [16,38]. Zheng et al. (2009) indicated that land use changes were responsible for more than 70% of the streamflow reduction in the 1990s [19]. Since 1999, vegetation cover exhibited an increasing trend due to the GFGP [17]. The increasing vegetation cover may result in more water remained in the soil store for evaporation and then reduce river runoff. Zhang et al. (2008) addressed that land use/cover changes (including soil and water conservation measures) accounted for over 50% of the reduction in mean annual streamflow in most studied catchments in the middle reaches of the Yellow River basin [36].

Relative Contributions of Precipitation and Anthropological Effects

Consistent with previous studies [18,40], both wavelet transform and FDC analysis suggested that the magnitude and regime of streamflow and sediment have been greatly altered, and a more uniform runoff and sediment regime has formed. It has also been confirmed that the significant decrease in streamflow and sediment load of the Yellow River resulted mainly from the variability in precipitation and anthropogenic activities [13,14]. To quantitatively separate the contribution of precipitation and human activities on streamflow and sediment load reductions, numerous studies applied simple linear regression approaches or water balance-based methods [15,18,21]. In this study, we applied a simple linear regression method based on approximating the relationship between changes of streamflow/sediment and precipitation to estimate the impact of precipitation variability on them in the river basin.

Consistently to previous analysis and other studies [21], we assumed that streamflow and sediment load were not intensively influenced by human activities in the period of 1952 to 1959, which can therefore be considered as a reference period. Fig. 9 illustrates the relationship between the inter-annual variability of precipitation (ΔP) and the inter-annual variability of streamflow/sediment ($\Delta Q/\Delta Q_s$) at Huayuankou station. By developing the relationship shown in Fig. 9, contributions for precipitation and human activities to the decrease of streamflow and sediment load can be estimated (shown in Table 6).

The average precipitation from 1960 to 1987 was 461 mm, 4 mm lower than the reference levels. The decrease of streamflow was estimated to be 0.29×10^8 m^3/a according to the relationship between ΔQ and ΔP, and as shown in the equation in Fig. 9a (Table 6). The changes of precipitation contributed approximately 1.3×10^8 t/a to the sediment load reduction based on the equation displayed in Fig. 9b. During this period, the decreased streamflow and sediment trapped by reservoirs were 6.4×10^8 m^3/a and 3.1×10^8 t/a, with a large portion of the reduction at Huayuankou station (Table 6). However, soil and water conservation started from the late 1950s and became effective since the late 1970s [18].

The average precipitation from 1987 to 1999 was 420 mm, 45 mm below that from 1952 to 1959. According to the equation shown in Fig. 9, this results in a reduction in streamflow of 34.1×10^8 m^3/a and sediment load of 3.7×10^8 t/a. Dramatic decrease in streamflow was detected due to an evident decline of incoming water from the upper reaches and soil and water conservation measures in the middle reaches (Table 6). Sediment retention by reservoirs decreased to 1.5×10^8 t/a, accounting for 15.5% of the measured decrease at Huayuankou station. Variation in precipitation and soil and water conservation measures were two predominant factors leading to 2.7×10^8, and 3.1×10^8 t/a of sediment decrease at Huayuankou station. In summary, the total decreases of streamflow and sediment load due to these factors were 176.1×10^8 m^3/a and 8.36×10^8 t/a respectively, which were slightly lower than the measured decrease at Huayuankou station. This may be caused by the underestimation of water consumption and effects of soil and water conservation.

The average precipitation from 2000 to 2011 was 443 mm, which resulted in decrease of streamflow and sediment load 38.3×10^8 m^3/a and 3.0×10^8 t/a. In this period, soil and water conservation practices became the dominant factor, accounting for 30.2% and 38.6% of the measured decrease in streamflow and sediment load at Huayuankou station. Large reservoirs built within this period caused an increasing trapping effect and a sediment decrease with an average rate of 3.99×10^8 t/a. The total decrease of sediment load due to these factors was 14.29×10^8 t/a, which agrees relatively well with the measured decline at Huayuankou station (Table 6).

Conclusion

This study investigated the variation of monthly streamflow and sediment load at four key gauging stations (Lanzhou, Toudaoguai, Huayuankou and Gaocun) along the Yellow River from 1952 to 2011 by making use of the wavelet transform, sediment rating curve and flow duration curve methods. The findings can be summarized as follows:

Both streamflow and sediment load, demonstrate significant decreasing trends ($p<0.05$) at all the stations. The most significant decline in annual streamflow and sediment load is detected at Huayuankou station, with average reduction rates of -5.44×10^8 m^3/a and -0.26×10^8 t/a, respectively, while the streamflow and sediment load at Lanzhou station display a gentle decrease (-1.33×10^8 m^3/a and -0.02×10^8 t/a).

The wavelet transform revealed intermittent periodicities of 0.5 and 1 year within 95% confidence interval due to the seasonal and annual alternation of hydro-climatic variables. Periodicity was not detected from 1969 to 1973 and after 1986. This can be attributed to the intensive human activities in the upper and middle reaches of the Yellow River during the past six decades.

Sediment rating curves at four gauging stations generally displayed decreasing trends in response to the decline of sediment induced by human activities. The decreasing log(a) and increasing *b* imply a reduced sediment supply from the source region and an increase of the erosive power of the river at Lanzhou station. However, the Toudaoguai station showed an opposite trend with decreasing transport capacity due to severe siltation. Similar changes were found at Huayuankou and Gaocun stations. Relatively high values of log(a) in 1952–1968 were induced by the abundant sediment income from the upper and middle reaches, whereas the increasing *b* from 1969 to 2011 indicates trapping effects from large reservoirs.

Flow duration curve analysis showed that the regime and magnitude of monthly streamflow and sediment load have been completely altered from 1952 to 2011. Compared to the changes of streamflow, sediment indices illustrated a dramatic decline at the four stations, particularly in the mid- and lower reaches with reduction rates of more than 70%.

Significant reduction of streamflow and sediment load is largely caused by reservoir construction and the implementation of numerous soil and water conservation measures. Precipitation changes, incoming water from the upper reaches, water consumption and soil and water conservation measures are the dominant factors for the decrease in streamflow. Sediment retention by Sanmenxia and Xiaolangdi reservoirs is related to great trapping effects on sediment discharge since their operation, leading to an evident reduction in their total volumes. Variation in precipitation and various soil and water conservation practices also contribute a large proportion to the sediment reduction during 1986 to 2011.

Acknowledgments

We would like to appreciate the Yellow River Conservancy Committee for providing valuable data and the Loess Plateau Database Center with data collection (Http://loess.geodata.cn).

Author Contributions

Conceived and designed the experiments: GJZ XMM PT. Performed the experiments: GJZ XMM PT. Analyzed the data: GJZ PT. Contributed reagents/materials/analysis tools: GJZ AS PT. Wrote the paper: GJZ XMM.

References

1. Milliman JD, Meade RH (1983) World-wide delivery of river sediment to the oceans. Journal of Geology 91: 1–21.
2. Walling DE, Fang D (2003) Recent trends in the suspended sediment loads of the world's rivers. Global and Planetary Change 39: 111–126.
3. Cerda A (1998) The influence of aspect and vegetation on seasonal changes in erosion under rainfall simulation on a clay soil in Spain. Canadian Journal of Soil Science 78: 321–330.
4. Fan X, Pedroli B, Liu G, Liu Q, Liu H, et al. (2012) Soil salinity development in the yellow river delta in relation to groundwater dynamics. Land Degradation & Development 23: 175–189.
5. Cerda A (1998) Effect of climate on surface flow along a climatological gradient in Israel: a field rainfall simulation approach. Journal of Arid Environments 38: 145–159.
6. Syvitski JPM, Vorosmarty CJ, Kettner AJ, Green P (2005) Impact of humans on the flux of terrestrial sediment to the global coastal ocean. Science 308: 376–380.
7. Xu JX (2005) The water fluxes of the Yellow River to the sea in the past 50 years, in response to climate change and human activities. Environmental Management 35: 620–631.
8. Milliman JD, Farnsworth KL, Jones PD, Xu KH, Smith LC (2008) Climatic and anthropogenic factors affecting river discharge to the global ocean, 1951–2000. Global and Planetary Change 62: 187–194.
9. Zhao GJ, Mu XM, Hormann G, Fohrer N, Xiong M, et al. (2012) Spatial patterns and temporal variability of dryness/wetness in the Yangtze River Basin, China. Quaternary International 282: 5–13.
10. Wang HJ, Bi NS, Saito Y, Wang Y, Sun XX, et al. (2010) Recent changes in sediment delivery by the Huanghe (Yellow River) to the sea: Causes and environmental implications in its estuary. Journal of Hydrology 391: 302–313.
11. Cerdà A, Lavee H, Romero-Diaz A, Hooke J, Montanarella (2010) Soil erosion and degradataion in mediterranean type ecosystems. Land Degradation & Development 21: 71–74.
12. Cerda A (1999) Parent material and vegetation affect soil erosion in eastern Spain. Soil Science Society of America Journal 63: 362–368.
13. Xu JX (2009) Plausible causes of temporal variation in suspended sediment concentration in the upper Changjiang River and major tributaries during the second half of the 20th century. Quaternary International 208: 85–92.
14. Liu CM, Zheng HX (2004) Changes in components of the hydrological cycle in the Yellow River basin during the second half of the 20th century. Hydrological Processes 18: 2337–2345.
15. Miao CY, Ni JR, Borthwick AGL, Yang L (2011) A preliminary estimate of human and natural contributions to the changes in water discharge and sediment load in the Yellow River. Global and Planetary Change 76: 196–205.
16. Li XL, Gao J, Brierley G, Qiao YM, Zhang J, et al. (2013) Rangeland Degradation on the Qinghai-Tibet Plateau: Implications for Rehabilitation. Land Degradation & Development 24: 72–80.
17. Zhao GJ, Mu XM, Wen ZM, Wang F, Gao P (2013) Soil erosion, conservation, and eco-environment changes in the Loess Plateau of China. Land Degradation & Development 24: 499–510.
18. Zhao GJ, Mu XM, Tian P, Wang F, Gao P (2013) Climate changes and their impacts on water resources in semiarid regions: a case study of the Wei River basin, China. Hydrological Processes 27: 3852–3863.
19. Zheng HX, Zhang L, Zhu RR, Liu CM, Sato Y, et al. (2009) Responses of streamflow to climate and land surface change in the headwaters of the Yellow River Basin. Water Resources Research 45.
20. Gao P, Mu XM, Wang F, Li R (2011) Changes in streamflow and sediment discharge and the response to human activities in the middle reaches of the Yellow River. Hydrology and Earth System Sciences 15: 1–10.
21. Peng J, Chen SL, Dong P (2010) Temporal variation of sediment load in the Yellow River basin, China, and its impacts on the lower reaches and the river delta. Catena 83: 135–147.
22. Xu JH, Li XM, Zhang PD, Lin YP (1998) Delimitation of coarse sediment and study on overlap of coarse sediment and sediment abundant areas in the middle Yellow River Basin. Journal of Sediment Research 4: 36–46 (In Chinese).
23. Giordano M, Zhu ZP, Cai XM, Hong SQ, Zhang XC, et al. (2004) Water management in the Yellow River Basin: background, current critical issues and future research needs. Comprehensive Assessment of Water Management in Agriculture Research Report 3. 48.
24. Valentin C, Poesen J, Li Y (2005) Gully erosion: Impacts, factors and control. Catena 63: 132–153.
25. YRCC, Yellow River Water Conservancy Committee (2000–2011) Bulletin of the Yellow River. http://wwwyellowrivergovcn/nishagonggao.

26. Zhang Q, Xu CY, Jiang T, Wu YJ (2007) Possible influence of ENSO on annual maximum streamflow of the Yangtze River, China. Journal of Hydrology 333: 265–274.
27. Grinsted A, Moore JC, Jevrejeva S (2004) Application of the cross wavelet transform and wavelet coherence to geophysical time series. Nonlinear Processes in Geophysics 11: 561–566.
28. Torrence C, Compo GP (1998) A practical guide to wavelet analysis. Bulletin of the American Meteorological Society 79: 61–78.
29. Syvitski JP, Morehead MD, Bahr DB, Mulder T (2000) Estimating fluvial sediment transport: The rating parameters. Water Resources Research 36: 2747–2760.
30. Hu BQ, Wang HJ, Yang ZS, Sun XX (2011) Temporal and spatial variations of sediment rating curves in the Changjiang (Yangtze River) basin and their implications. Quaternary International 230: 34–43.
31. Asselman NEM (2000) Fitting and interpretation of sediment rating curves. Journal of Hydrology 234: 228–248.
32. Morgan RPC, editor (1995) Soil Erosion and Conservation. Second ed. ed. London: Longman.
33. Yang GF, Chen ZY, Yu FL, Wang ZH, Zhao YW, et al. (2007) Sediment rating parameters and their implications: Yangtze River, China. Geomorphology 85: 166–175.
34. Brown AE, Western AW, McMahon TA, Zhang L (2013) Impact of forest cover changes on annual streamflow and flow duration curves. Journal of Hydrology 483: 39–50.
35. Xu KH, Milliman JD, Xu H (2010) Temporal trend of precipitation and runoff in major Chinese Rivers since 1951. Global and Planetary Change 73: 219–232.
36. Zhang XP, Zhang L, Zhao J, Rustomji P, Hairsine P (2008) Responses of streamflow to changes in climate and land use/cover in the Loess Plateau, China. Water Resources Research 44.
37. Ran DC, Liu B, Wang H, Luo QH, Ma Y, editors (2006) Trapping effects of soil and water conservation measures on streamflow and sediment load in the Middle Reaches of Yellow River. Zhengzhou, China: Yellow River Conservancy Press.
38. Wang SY, Liu JS, Ma TB (2010) Dynamics and changes in spatial patterns of land use in Yellow River Basin, China. Land Use Policy 27: 313–323.
39. Wang G, Fan Z, editors (2002) Variation of streamflow and sediment flux in the Yellow River. Zhengzhou, China: Yellow River Conservancy Press.
40. Mu XM, Zhang XQ, Shao HB, Gao P, Wang F, et al. (2012) Dynamic Changes of Sediment Discharge and the Influencing Factors in the Yellow River, China, for the Recent 90 Years. Clean-Soil Air Water 40: 303–309.
41. Pan XD, Li Y, Zhang XH, Shen GQ, Yue DJ (2006) The Riverbed Process in the Lower Yellow River After the Construction of Sanmenxia Reservoir. Zhengzhou, China: Yellow River Conservancy.
42. Yao WY, Xu JH, Ran DC, editors (2011) Assessment of changing trends in streamflow and sediment fluxes in the Yellow River Basin. Zhengzhou: Yellow River Conservancy Press (In Chinese).

9

Spatial and Temporal Habitat Segregation of Mosquitoes in Urban Florida

Paul T. Leisnham[1*], **Shannon L. LaDeau**[2], **Steven A. Juliano**[3]

1 Department of Environmental Science and Technology, University of Maryland, College Park, Maryland, United States of America, **2** Cary Institute of Ecosystem Studies, Millbrook, New York, United States of America, **3** School of Biological Sciences, Illinois State University, Normal, Illinois, United States of America

Abstract

Understanding mechanisms fostering coexistence between invasive and resident species is important in predicting ecological, economic, or health impacts of invasive species. The non-native mosquitoes *Aedes aegypti* and *Culex quinquefasciatus* have been resident in the southeastern United States for over a century. They coexist at some urban sites with the more recent invasive *Aedes albopictus*, which is usually superior in interspecific competition. We tested predictions of temporal and spatial habitat segregation that foster coexistence of these resident species with the superior invasive competitor. We measured spatial and temporal patterns of site occupancy and abundance for all three species among standard oviposition traps in metropolitan Tampa, Florida. Consistent with the condition-specific competition hypothesis, *A. albopictus* and *A. aegypti* abundances were greater and *C. quinquefasciatus* abundance was lower late (September) versus early (June) in the rainy season, and the proportional increase of *A. albopictus* abundance was greater than that of *A. aegypti*. These results are postulated to result from greater dry-season egg mortality and associated greater rainy-season competitive superiority of larvae of *A. albopictus*, followed by *A. aegypti*, and *C. quinquefasciatus*. Spatial partitioning among landscape variables was also evident among species, with *A. albopictus* more likely to oviposit across a range of open grass landscapes whereas *A. aegypti* were mostly restricted to cemeteries. *Culex quinquefasciatus* showed a shift in abundance from cemeteries early in the rainy season to developed areas characterized by built environments with large proportions of impervious surfaces late in the rainy season, where *A. albopictus* was not in its highest abundance. These results suggest that both temporal and spatial variation, and their interaction, may contribute to local coexistence between *Aedes* and *Culex* mosquito species in urban areas.

Editor: Igor V. Sharakhov, Virginia Tech, United States of America

Funding: Field work was funded by NIAID grant R01- AI-44793 (Illinois State University subaward) and by a grant to SAJ from Illinois State University. Time for analyses and authorship for PTL was funded by the University of Maryland. The funders had no role in study design, data collection and analysis, decision to publish, or preparation of the manuscript.

Competing Interests: The authors have declared that no competing interests exist.

* E-mail: leisnham@umd.edu

These authors contributed equally to this work.

Introduction

Ecological theory and empirical work indicate that competition often results in competitive exclusion when resources are limited. However, competitive exclusion may be avoided via a number of mechanisms, including differential resource use (e.g., [1]), temporally varying condition-specific competition and the storage effect (e.g., [2]), and spatial resource partitioning (e.g., [2]). Understanding mechanisms of coexistence is particularly interesting in the context of biological invasions. Competitively superior invasive species may impact the distribution and abundance of resident species without causing their extinction over the entire introduced range (e.g., [1]). Identifying mechanisms contributing to local or spatially patterned coexistence of invasive and resident species is important for predicting future ecological, economic, or health impacts of species invasions.

The Asian Tiger mosquito, *Aedes albopictus* (Skuse), is native to Asia and has invaded North and South America, Europe, and Africa in the past three decades (see [3] and references therein). This species is well studied [4], and provides us with an opportunity to determine whether patterns of coexistence with natives in its new range are consistent with any of these coexistence mechanisms. *Aedes albopictus*, and its competitors, such as *Aedes aegypti* (L) and *Culex quinquefasciatus* Say, a member of the *Culex pipiens* complex, utilize water-holding natural (e.g., tree holes, plant axils) and artificial (e.g., tire casings, trash, bird baths) container habitats for their egg and larval stages. *Aedes aegypti* and *C. quinquefasciatus* invaded the Americas from Africa during the 15–17th centuries and the 19th century, respectively [5]. Both species have widely colonized container habitats in the southeastern United States, and are now widely considered non-native resident species with regards to the more recent *A. albopictus* invasion [4]. Biting females that emerge from these aquatic container habitats are principal vectors of arboviruses worldwide, including dengue, chikungunya, yellow fever, La Crosse encephalitis, and West Nile virus [6–9], and understanding mechanisms of coexistence among these vectors is of human health importance.

Most laboratory and field studies (see [10], [11] and references therein) convincingly show that larval *A. albopictus* are superior competitors for resources over *A. aegypti* and members of the *C. pipiens* complex, and *A. aegypti* appears to have a clear competitive advantage over *C. pipiens pipiens* [12–14]. However despite this competitive hierarchy, inferior competitors *A. aegypti* and *C. quinquefasciatus* often coexist with invading *A. albopictus* in the southeastern USA, and are sometimes the most common container mosquitoes [15], [16]. A number of hypotheses could explain the persistence of *A. aegypti* and *C. quinquefasciatus* after the invasion of *A. albopictus*. Here we use a spatially and temporally explicit field dataset to evaluate two prominent hypotheses: condition-specific competition and spatial partitioning.

Differential mortality among development stages affects R^*, the equilibrium resource abundance necessary to produce zero net population growth [1], which determines competitive advantage [1]. Abiotic conditions that differentially affect mortality, even in non-competing life-stages, such as eggs, can alter the outcome of interspecific competition by differentially changing species' R^*s [17], [18]. Condition-specific competition occurs when the outcome of competition is altered or reversed under different abiotic conditions [19]. When there are temporal or spatial fluctuations in abiotic environments, and species with environmentally resistant life-cycle stages respond differently to those environments, the competitive outcome can be altered and coexistence can result [1], [2].

Condition-specific competition between *A. albopictus* and *A. aegypti* has been observed in the laboratory [17]. Experimental manipulation of container drying regime in laboratory colonies shows that in dry conditions, *A. albopictus* suffers a greater interspecific density effect than *A. aegypti*, whereas under wetter conditions, *A. aegypti* suffers a greater interspecific density effect than *A. albopictus* [17]. Under dry conditions, *A. albopictus* eggs suffer greater mortality than do *A. aegypti* eggs [17]. Drought is a strong environmental influence where these species coexist in the southern part of Florida, which experiences a distinct cycle of rainy (June–September) and dry (October–May) seasons (National Oceanic and Atmospheric Administration: URL: http://www.ncdc.noaa.gov/cdo-web/search). Abundances of both species in field containers in south Florida support condition-specific competition, with abundances being higher late in the rainy season (i.e., September) vs. early in the rainy season (i.e., June) [20]. The proportional increase of *A. albopictus* from the early to late rainy season is greater than that of *A. aegypti* presumably due to higher dry-season egg mortality and strong rainy-season competitive superiority of larval *A. albopictus* [20]. A single field study investigating competition between these species during rainy and dry seasons did find seasonal differences in the intensity of competition, but competitive effects favoring *A. albopictus* were evident only in the rainy season [21]. Both species responded in similar ways to the dry season, and evidence for a role for desiccation induced mortality was absent [21]. More intense competition during the rainy season was likely due to differences in detritus resource availability, which was greater during the dry season. Thus, although condition-dependent coexistence of *A. albopictus* and *A. aegypti* remains plausible, there is a clear need for additional testing of this hypothesis.

Competition between *C. quinquefasciatus* and *Aedes* mosquitoes in varying wet-dry regimes has not been tested in the laboratory but we expect that competitive success of *C. quinquefasciatus* would be highly dependent on wet conditions because it oviposits by depositing floating rafts of eggs in existing aquatic habitats. Thus, unlike *A. albopictus* and *A. aegypti*, which oviposit desiccation-tolerant eggs on the insides of containers prior to flooding, ovipositing *C. quinquefasciatus* require existing aquatic habitats, and thus are more strongly limited by their availability. Additionally, habitat persistence to support *Culex* egg hatching and larval eclosion directly affects *Culex* mortality and adult production. Dry conditions are expected to cause greater mortality of *C. quinquefasciatus* eggs and larvae compared to *Aedes* eggs and larvae because *Aedes* larvae are only at risk from habitat drying if eggs hatch after a flooding event. For coexistence between *A. albopictus* and *A. aegypti* to occur under these circumstances, environments must fluctuate between conditions favoring the different species and there must be resistant life stages (e.g., dormant eggs) that persist through times when a species is at a disadvantage [2]. For *Culex* to coexist with both the *Aedes* species there must be sufficient rainfall and favorable climate conditions to allow habitats to persist long enough for both oviposition and immature development.

Poorer competitors may also escape exclusion if there is resource partitioning in space [22], [23]. Spatial partitioning is most likely to arise due to greater availability of habitats and attraction of competitors to different environmental conditions [2], [24]. Prior studies have shown differential habitat utilization of competing mosquitoes among land use types [25–27]. For example, a study of the oviposition ecology of *A. albopictus* and *A. aegypti* in Rio de Janeiro, Brazil and Boca Raton, Louisiana, USA showed that in general *A. aegypti* was most prevalent in highly urbanized areas and *A. albopictus* in rural, suburban and vegetated urban areas, but that abundances of both species were similar in suburban areas [26]. Spatial partitioning among land use types likely results from direct effects of landcover on both aquatic and terrestrial habitat quality and population success, and from behavioral habitat choice, but few studies have rigorously examined the specific relationships and mechanisms explaining the distribution, abundance and co-occurrence of mosquito species in a heterogeneous urban landscape [25], [27].

In this paper, we evaluate the importance of seasonal condition-specific competition and habitat segregation hypotheses in explaining temporal and spatial distributions and co-occurrence of immature *A. albopictus*, *A. aegypti*, and *C. quinquefasciatus* across three cities (Tampa, Bradenton, and Palmetto) in greater metropolitan Tampa, Florida, USA. Metropolitan Tampa is ideal for such tests because it has a diverse range of habitats within a few kilometers of one another, and because these species have coexisted at some sites in metropolitan Tampa for over 15 years (G.F. O'Meara, unpublished data), suggesting stability. A prior study of mosquito oviposition ecology in Tampa found evidence for habitat segregation and condition specific competition between *A. aegypti* and *A. albopictus* among broad classes of residential, industrial, and commercial land during the rainy season [20]. However, [20] had three important limitations that we address in this paper. First, [20] did not consider *C. quinquefasciatus* despite this species being the third most common mosquito collected from oviposition traps, and potentially having a different pattern of abundance between dry vs. wet conditions than *Aedes*. Second, [20] did not evaluate how habitat partitioning among species may vary between the early vs. late rainy season. Third, although past work has indicated that cemeteries containing flower vases may act as important habitat for mosquitoes in Florida (e.g., [15], [20], [28]), [20] only compared cemeteries with randomly chosen non-cemetery urban areas rather than the full range of land cover types that exist within diverse urban landscapes.

Our investigation in this paper directly builds on [20] by re-sampling the same study areas, evaluating habitat partitioning among *Aedes* and *Culex* mosquitoes across different land uses, and explicitly comparing cemeteries with other land uses. We evaluate occupancy, abundance, and co-occurrence across different land cover types at the beginning of the rainy season (June) and after several months of wet conditions (September). As with [20] and other studies that sample mosquitoes among landscapes (e.g., [25], [26], [27]), our investigation is observational and can only provide a snapshot of habitat utilization in the field. However, our hypotheses explaining coexistence among species generate testable predictions about the distribution and abundance of these mosquitoes in a heterogeneous urban landscape.

1. Condition-specific competition predicts that all species occupy a greater proportion of containers and are more abundant per container late in the rainy season but that there are greater proportional increases in *A. albopictus* and *C. quinquefasciatus* from early to late rainy season, compared to the dry season, due to high dry-season (October–May) egg mortality and loss of oviposition habitat, respectively, and strong rainy-season (June–September) competitive superiority of *A. albopictus* and greater available habitat for *C. quinquefasciatus.*
2. The spatial habitat segregation hypothesis predicts co-occurrence if species occupy and are more abundant in different habitats defined by landuse. Competitive superiority of *A. albopictus* predicts persistence of *A. aegypti* and *C. quinquefasciatus* in urban areas if they are able to exploit different habitats compared to *A. albopictus,* especially in the rainy season when *A. albopictus* is expected to be most common.

These hypotheses are not mutually exclusive and may both contribute to coexistence among *A. albopictus, A. aegypti,* and *C. quinquefasciatus.*

Materials and Methods

Three areas within metropolitan Tampa (Tampa, Hillsborough County [lat. 27.9°, long. −82.4°]; St. Petersburg, Pinellas County [lat. 27.8°, long. −82.6°; and Bradenton, Manatee County [lat. 27.5°, long. −82.4°), all separated by Tampa Bay, were selected for study. Southern Florida experiences greater rainfall and temperatures in the summer rainy season (June–September) compared to the dry season (October–May) (National Oceanic and Atmospheric Administration: URL: http://www.ncdc.noaa.gov/cdo-web/search). Mean monthly rainfall and daily average temperature for the three meteorological stations closest to our study sites reflect this pattern for both the period from 1981–2010 (Dry season: 66.4 mm, 20.5°C; Rainy season: 200.1 mm, 28.1°C), the dry season before our sampling in 2008 (55.8 mm, 21.2°C), and the rainy season in 2008 (162.0 mm, 28.1°C).

Each area consisted of 60 km^2 divided into 60 1-km^2 grid cells. These areas were selected because they have residential, commercial, and industrial land uses [29], a high density of human-created structures, diverse vegetation types, and a number of cemeteries with suitable mosquito habitats (Tampa, $n = 10$; St. Petersburg, $n = 5$; and Bradenton, $n = 4$) that were used by [20]. Within each area, we randomly selected the centers of a subset of urban cells (Tampa, $n = 34$; St. Petersburg, $n = 34$; and Bradenton, $n = 32$) and each cemetery as sampling sites to give a total of 100 urban and 19 cemetery sites. Other sites with high densities of container habitats, such as tire yards, are present in each area but preliminary observations suggested that few had long standing piles of tires outside that would fill with water and provide larval mosquito habitat. Thus, we decided not to survey tire yards in this study.

Consistent with [20], we used oviposition traps (ovitraps) to sample mosquito populations. Ovitraps allow the rigorous testing of relationships between mosquito oviposition ecology and broad landscape-scale variables by standardizing individual sampling container. Ovitraps are widely used to sample *Aedes* oviposition in time and space (e.g., [25–27], [30], [31]), and are effective at sampling *Culex* when baited with resource (hay, grass, leaves etc) infusion (e.g., [20], [32], [33]). The occurrence of eggs in ovitraps is considered a sensitive indicator for identifying the presence of many mosquito species [34], [35], and more sensitive even when compared to adult collection methods [36], [37]. The occurrence and abundances of *Aedes* from ovitraps has been positively associated with rainfall and temperature patterns [38], [39], and because of their strong sensitivity, ovitraps are considered particularly effective at sampling mosquitoes during unfavorable seasons (i.e., dry seasons and/or lower temperatures) [36]. Abundances from ovitraps can be less reliable at predicting oviposition intensity and adult densities than other methods [35], but they are more affordable, and thus more highly replicable than adult traps, and easier to sample than existing resident containers [30], making them ideal to compare broad trends in mosquito presence and abundance across large numbers of sites and between seasons, as in this study.

In the early rainy season (June) in 2008, 3 ovitraps were placed in the shade at each site. Ovitraps were placed at ground level and within 20 m of one another (357 total traps). Ovitraps consisted of black plastic cups (400 ml), with holes drilled 4 cm from the base to prevent flooding and hatching of *Aedes* eggs. Ovitraps were lined with seed germination paper (Nasco Science®), filled with 225 ml deionized water, and baited with an additional 25 ml of grass/oak leaves infusion (72 g senescent live oak (*Quercus virginiana*) leaves and 36 g Zoisa grass in 5.4 L of deionized water for 3 days). Ovitraps in alternating counties were set out over six days, with 14–20 stations being provisioned with traps on any particular day. After 7 days, ovitraps were collected and all larvae identified. *Culex* egg rafts from each ovitrap were collected and stored on water to allow eggs to hatch. Germination paper from each ovitrap was also stored in humid conditions for 10 days then immersed in nutrient broth solution to hatch eggs. Numbers of field-collected larvae and pupae and laboratory-hatched larvae were summed by species. Larvae were reared to 4th instar to facilitate identification and larvae and pupae stored in 70% ethanol for later examination in the laboratory. During the late rainy season (September) in 2008, we repeated our sampling using this same protocol. When collecting ovitraps in June we measured the remaining water in each trap and recorded if the trap had dried completely. Preliminary analyses of June data indicated no relationships between water volume and mosquito abundances among 'wet' traps; thus we only scored traps as 'dry' or 'wet' in September collections. Field collected larvae included no endangered or protected species, and no prior permissions were required to access study sites or collect mosquitoes.

The relatively small size and inconspicuous color of these ovitraps was expected to minimize chances of them being disturbed in areas of relatively high human activity, and has been used in previous oviposition surveys in urban landscapes [20], [40]. Twelve and 10 traps were disturbed across all three areas in June and September respectively, including all three traps at one site in St. Petersburg in September, thus these cups were removed from all analyses.

NLCD and Land Use Data

A subset of our sites from each city was purposely located in cemeteries. Additionally, we examined land cover around each site at two spatial scales (50-m and 200-m buffers). These distances were chosen *a priori* to represent the local development site characteristics (50 m) and characteristics of the wider landscape within the dispersal distances of all species (200 m). Data from the National Land Cover Database (NLCD, 2001) were downloaded at the USGS Landcover Institute (available online at: http://landcover.usgs.gov/). We tabulated the number of pixels in each NLCD-defined land cover class and used the standardized proportional area in each class as explanatory variables in analyses described below. Given that all of our sites were located in a metropolitan area, we selected three NLCD available land cover class categories that best described the variation seen among our sites for the analyses: High Intensity Developed (impervious surfaces >79%), Open Space (predominantly lawns and parkland), and Wetland cover (including both woody and herbaceous). A fourth category, Low Intensity Developed was also common but was significantly correlated with each of the other three and thus, was not used in this analysis.

Statistical Analyses

We evaluated the relative abundance of mosquito larvae across 119 sites in three cities from two sampling dates. We were specifically interested in examining how relative abundance and multiple species occurrence (co-occurrence at a site) varied with seasonality and with land cover. We used the species-specific mean number (rounded to nearest integer) of larvae from the replicate traps at each site as our dependent variable in the analyses described below.

Count data are generally assumed to follow a Poisson distribution, although large numbers of zeros can invalidate the Poisson assumption of equal mean and variance. We formally compared a standard Poisson linear regression with a zero-inflated Poisson (ZIP) model. In the ZIP model, zeros can be generated both due to the Poisson sampling structure (e.g., random zeros) and due to site unsuitability. We used a coupled GLM approach (e.g., [41], [42]) to estimate both the proportion of zeros that could best be attributed to site unsuitability (i.e., latent site suitability parameter) vs. the (Poisson) structural zeros and a generalized linear regression with explanatory variables describing both processes. In this model, the probability of a Poisson zero is δ_{is} for species s at a site i and occurs with probability ρ_{is}:

$$\delta_{is} : \text{Bernoulli}(\rho_{is})$$

and logit(ρ_{is}) is a species-specific function of sampling season.

When the latent 'site suitability' estimate is 1, (i.e., zeros are generated from a Poisson sampling distribution), then observed counts (Y_{is}), including zeroes, are conditionally independent with unknown parameters λ_{is} representing the true abundance of species s at site i. We assumed that each of these conditional distributions is Poisson with mean λ_{it}:

$$Y_{is} \sim Pois(\lambda_{is} * \delta_{is})$$

Then, for all $i = 1,\ldots,N$ sites and $s = 1,\ldots,S$ species,

$$\log(\lambda_{is}) = \alpha_o s X_i + \eta_{c[i],s}$$

Relative abundance is thus, a function of site suitability, the explanatory variables in matrix $\boldsymbol{X}$ (as defined below), and a hierarchical structure, $\eta_{c[i],s}$, which allows for correlation among species-specific abundances within a city beyond what is explained by components of $\boldsymbol{X}$.

We used the model framework described above to test each of our hypotheses, testing specifically the relative importance of seasonal and land use effects as predictors of population abundances and co-occurrences of the two *Aedes* species and *Culex quinquefasciatus*. We ran an initial model where the matrix X included the vector of ones, and three binary indicator variables for month and presence of each of the other two species. We ran the model again where X included a vector of ones, a binary indicator variable for month and cemetery location, and the three land cover variables described in the previous section at both 200 and 50 meter scales. Although we could have integrated the seasonal and co-occurrence variables into the land cover analysis, we chose to focus on two simpler models with rapid convergence that more directly addressed our separate hypotheses about seasonality (condition-dependent) and habitat segregation. Results in the text are shown as the mean followed by the 95% credible interval derived from the posterior distribution for each parameter. We evaluated the fit of the ZIP model relative to a simpler Poisson regression that does not include zero inflation but uses the same definition of λ_{is}.

Bayesian inference requires prior distributions be assigned to all unknown parameters and we used standard conjugate distributions in each case [43]. 1,..., P parameters (α_p) describing covariate effects in the regression equation was given a vague (flat) Gaussian prior distribution with mean 0 and large variance (1000). For each city c, $\eta_k \sim N(0,\tau_\eta^2)$ and τ_η^2 was sampled from a relatively uninformative inverse gamma distribution with parameters (0.01, 0.01). Model parameters were estimated by simulating from the joint posterior distribution of all unknown parameters using a Markov chain Monte Carlo (MCMC) algorithm implemented using the WinBUGS software [43]. We evaluated model fit in three ways. DIC is a measure of how expanding or decreasing model structure changes the prediction accuracy of the model, with lower DIC values (>5 units difference) representing the preferred model [44]. We further evaluated model fit with and without zero-inflation by comparing the proportion of zeros and overdispersion using model predictions and raw data. We used the overdispersion index: OD.i = Var(y)/E(y), where y = either observed or model predicted count data. For each of the three focal species, the zero-inflated Poisson models provided a better fit to data than a simple Poisson structure (Table 1).

Results

Aedes aegypti, *Culex quinquefasciatus*, and *A. albopictus* were the most widespread species sampled, being collected from 25.0% (173/692), 20.5% (142/692), and 12.1% (84/692) of ovitraps respectively. When pooling abundances across all ovitraps, *C. quinquefasciatus* was the most common species collected (17,844/22,360 individuals), followed by *A. aegypti* (3,086/22,360) and *A. albopictus* (1,430/22,360). *Culex nigripalpus*, *Aedes triseriatus* and *Toxorhynchites rutilus* were also collected from ovitraps but constituted less than 1% of the total number of individuals. A total of 59 ovitraps, occurring at 32.8% of sample sites (39/119) were dry in the early rainy season (June), including all ovitraps at 6 sites (5.0%). Dry ovitraps in the early rainy season were most common at sites surrounded by more high intensity developed cover (r = 0.217, t = 2.40, p = 0.017), and not related to any other land cover variables (p>0.050). The 6 sites with all dry ovitraps were removed from further analyses.

The proportion of sites *A. aegypti* and *C. quinquefasciatus* occupied decreased between the early and late rainy season, whereas the proportion of sites with *A. albopictus* increased over this same period. *Aedes albopictus* was nearly twice as likely to co-occur at sites with *A. aegypti* in the late rainy season (10.1%, 12/119) relative to the early rainy season (5.3%, 6/113), whereas its co-occurrence with *C. quinquefasciatus* was less common and did not vary (3.5%, 4/113 vs. 3.4%, 4/119) (Table 2). *Aedes aegypti* and *C. quinquefasciatus* co-occurred at 31.0% (35/113) sites in the early rainy season but only 1.7% (2/119) sites in the late rainy season (Table 2).

Table 1. Comparisons between simple Poisson structure and zero-inflated Poisson (ZIP) models for describing *Aedes albopictus, Aedes aegypti, and Culex quinquefasciatus* abundances.

	Proportion Zeros			Overdispersion Index			DIC*
	Observed	**Poisson**	**ZIP**	**Observed**	**Poisson**	**ZIP**	**Poisson**
A. albopictus	0.76	0.379	0.79	22.25	5.42	13.89	+973
A. aegypti	0.51	0.046	0.53	23.59	2.04	9.61	+1424
C. quinquefasciatus	0.59	0.001	0.58	101.60	22.21	53.57	+8233

*DIC, deviance information criterion, shown as deviation from smallest DIC for each species.

Table 2. Numbers and proportions of sites and individual oviposition traps (in parentheses) occupied by immature of *Aedes albopictus* (ALB), *A. aegypti* (AEG), and *C. quinquefasciatus* (CQ) in cemeteries and urban sites in Palmetto, St. Petersburg, and Tampa, FL, in the early (June) and late (Sept.) rainy seasons.

	Number of sites	ALB only	AEG only	CQ only	ALB+AEG	ALB+CQ	AEG+CQ	ALB+AEG+CQ	None	Proportion ALB	Proportion AEG	Proportion CQ
Early rainy season (June)												
Cemeteries	16 (37)	0 (1)	4 (7)	1 (8)	0 (0)	3 (4)	3 (5)	2 (1)	3 (11)	0.13 (0.16)	0.56 (0.35)	0.56 (0.49)
Palmetto	32 (88)	0 (0)	3 (9)	3 (14)	3 (5)	0 (1)	14 (17)	3 (1)	6 (41)	0.19 (0.08)	0.72 (0.36)	0.63 (0.38)
St. Petersburg	32 (76)	0 (1)	6 (12)	6 (16)	0 (0)	1 (1)	12 (12)	1 (0)	6 (34)	0.06 (0.03)	0.59 (0.32)	0.63 (0.38)
Tampa	33 (85)	0 (6)	5 (14)	7 (13)	3 (4)	0 (0)	6 (12)	6 (4)	6 (32)	0.27 (0.16)	0.61 (0.40)	0.58 (0.34)
Early total	113 (286)	0 (8)	18 (42)	17 (51)	6 (9)	4 (6)	35 (46)	12 (6)	21 (118)	0.19 (0.10)	0.63 (0.36)	0.60 (0.38)
Late rainy season (Sept)												
Cemeteries	19 (57)	4 (7)	4 (7)	1 (2)	2 (5)	2 (2)	0 (0)	2 (1)	4 (27)	0.53 (0.26)	0.53 (0.23)	0.26 (0.09)
Palmetto	33 (93)	2 (14)	8 (14)	5 (6)	3 (3)	0 (0)	0 (1)	3 (2)	12 (62)	0.24 (0.20)	0.42 (0.22)	0.24 (0.10)
St. Petersburg	33 (95)	2 (12)	7 (12)	7 (12)	0 (0)	2 (0)	2 (0)	0 (0)	13 (67)	0.12 (0.13)	0.27 (0.13)	0.33 (0.13)
Tampa	34 (102)	6 (15)	7 (15)	5 (6)	7 (9)	0 (0)	0 (0)	2 (1)	7 (61)	0.47 (0.25)	0.44 (0.25)	0.21 (0.07)
Late total	119 (347)	14 (48)	26 (48)	18 (26)	12 (17)	4 (2)	2 (1)	7(4)	36 (217)	0.39 (0.20)	0.31 (0.20)	0.26 (0.10)

Only totals for cemeteries among cities are shown.

Occurrence and Abundance: Seasonal Patterns

The relative abundance of *A. albopictus* and, to a lesser degree, *A. aegypti* increased from the early to the late rainy season across occupied sites, while *C. quinquefasciatus* abundances declined during this same time period (Table 3). Consistent with the data summary in Table 2, the model estimates that *A. aegypti* and *C. quinquefasciatus* occupied a larger proportion of sites across both sample periods (0.47 and 0.42, respectively) relative to *A. albopictus* (0.22, Table 3). Our analysis suggests asymmetric influence of co-occurrence on species abundances. The relative abundance of *A. aegypti* was an estimated 73.3% and 41.9% more abundant across sites where *C. quinquefasciatus* or *A. albopictus* were present, respectively (Table 3), although occurrence of *Ae. aegypti* was not predictive of either *Culex* or *A. albopictus* abundance. Similarly, *A. albopictus* was an estimated 63.2% more abundant when *C. quinquefasciatus* larvae were present (Table 3), although *C. quinquefasciatus* abundance was on average 6.8% lower at sites where *A. albopictus* was present.

Relative Importance of Land Cover

The importance of cemetery habitat varied by species and between the early and late rainy season. Cemetery habitat was positively associated with abundance of *A. albopictus* and *A. aegypti* in the late rainy season (September), whereas both species were less abundant at cemetery sites relative to other land cover categories during the early rainy season (June) (Table 4). *Culex quinquefasciatus* showed the opposite response, with cemetery sites being more important predictors of abundance in June and negatively associated with abundance in September (Table 4). There was no improvement in model fit when land cover characteristics at the 50-m scale were added to the base model that included month and cemetery indicator variables. If cemetery was excluded, Open Cover Developed area within a 50-m buffer was a positive predictor of *C. quinquefasciatus* in the early rainy season (0.427, 95% CI (0.339, 0.515) and of *A. albopictus* in the late rainy season (3.188, 95% CI (1.838, 4.708)).

In addition to cemeteries, High Intensity Developed area and Open Cover Developed area were important predictors of abundance for at least one of the species at the 200-m scale (Table 4). Wetland area was not significantly positively associated with any species' abundances but was negatively associated with *A. aegypti* and *C. quinquefasciatus* (Table 4). Both High Intensity Developed area and Open Cover Developed area were positively associated with *A. aegypti* early in the rainy season but by September, cemetery locations were the predominant habitat for this species (Table 4). Relative abundance of *A. albopictus* was not positively associated with any habitat variables in the early rainy season but was greatest at cemeteries and at sites with more Open Cover Developed area in the late rainy season (Table 4). By contrast to the two *Aedes* species, *C. quinquefasciatus* shifted from an early-season cemetery focus to greater abundances at sites with greater High Intensity Developed and Open Cover Developed area by September (Table 4).

All models included the latent site-suitability parameter and a seasonal effect was estimated. Zero counts for both *A. aegypti* and *C. quinquefasciatus* were less likely to be a random Poisson zero and more likely to be due to unsuitable habitat in late versus early season (seasonal effect: *A. aegypti* −0.862, 95% CI (−1.390, −0.343); *C. quinquefasciatus*: −1.389 95% CI (−1.946, −0.839)), while unsuitable habitat was a more likely explanation for zeros in the early season for *A. albopictus* (0.658, 95% CI (0.056, 1.279)).

Table 3. Posterior parameter estimates (with 95% CI) for models describing variation in relative abundances of *Aedes albopictus, Aedes aegypti,* and *Culex quinquefasciatus* by season, when each of the other species was present.

	Aedes albopictus			*Aedes aegypti*			*Culex quinquefasciatus*		
Parameter	**Mean**	**2.5%**	**97.5%**	**Mean**	**2.5%**	**97.5%**	**Mean**	**2.5%**	**97.5%**
Intercept	**0.99**	**0.44**	**1.43**	**1.48**	**1.25**	**1.76**	**4.27**	**4.09**	**4.49**
Month	**1.42**	**1.15**	**1.71**	**0.82**	**0.68**	**0.96**	**−0.68**	**−0.74**	**−0.61**
A. albopictus presence	–	–	–	**0.35**	**0.22**	**0.48**	**−0.07**	**−0.14**	**−0.02**
A. aegypti presence	−0.05	−0.25	0.15	–	–	–	0.03	−0.02	0.08
C. quinquefasciatus presence	**0.49**	**0.3**	**0.68**	**0.55**	**0.41**	**0.68**	–	–	–
Presence	**0.22**	**0.16**	**0.27**	**0.47**	**0.41**	**0.54**	**0.42**	**0.36**	**0.48**

Mean values less than 0.0 for Month indicate a decline in abundance between early and late rainy season. Bold-face indicates that the posterior distribution (95% CI) does not include zero.

Table 4. Posterior parameter estimates (with 95% CI) for models describing the relative importance of land cover (200-m buffer) in explaining variation in larval abundance across the rainy season.

	Aedes albopictus		*Aedes aegypti*		*Culex quinquefasciatus*	
Parameter	**Early**	**Late**	**Early**	**Late**	**Early**	**Late**
Intercept	−0.431		1.393		0.821	
	(−3.385, 1.783)		(−0.133, 2.641)		(−0.335, 1.580)	
Cemetery	**−1.665**	**2.161**	**−1.275**	**1.689**	**0.513**	**−1.128**
	(−3.053, −0.617)	**(1.053, 3.551)**	**(−1.659, −0.908)**	**(1.263, −2.143)**	**(0.416, 0.615)**	**(−1.464, −0.802)**
Wetland	0.312	2.158	**−2.286**	**−3.759**	−0.367	**−0.983**
	(−3.270, 3.436)	(−1.355, 5.856)	**(−3.777, −0.887)**	**(−6.160, −1.435)**	(−0.759, 0.073)	**(−2179, −0.061)**
High Intensity Developed	1.066	−0.039	**0.959**	**−2.051**	−0.084	**0.753**
	(−0.969, 2.938)	(−2.321, 2.329)	**(0.021, 1.915)**	**(3.223, −0.810)**	(−0.379, 0.221)	**(0.289, 1.214)**
Open Cover Developed	−0.658	**2.755**	**1.544**	**−2.289**	**−0.320**	**1.325**
	(−2.638, 1.361)	**(0.684, 4.984)**	**(1.008, 2.093)**	**(−3.000, −1.598)**	**(−0.500, −0.128)**	**(0.759, 1.914)**

Parameters describing the association with each land cover type were estimated separately for samples from early (June) and late (Sept.) rainy season. Bold-face indicates that the posterior distribution (95% CI) does not include zero.

Discussion

The results of this study show clear patterns in the seasonal and spatial distributions of *A. albopictus*, *A. aegypti*, and *C. quinquefasciatus* in metropolitan Tampa that are consistent with seasonal condition-specific competition and habitat segregation as mechanisms contributing to species coexistence. These results also highlight important interactions between these two mechanisms that can only be explored when both space and time are considered explicitly, as we have done in this study. The persistence of *A. aegypti* and *C. quinquefasciatus* in the presence of the competitively superior *A. albopictus* may be explained by seasonal condition-specific competition, if there are seasonally-related differences in survival among the species. We predicted *A. albopictus* would show a greater increase in the proportion of sites occupied and per site abundance during the rainy season than *A. aegypti*, due to its high dry season egg mortality and strong rainy-season competitive superiority. Our results were consistent with this prediction, with both *Aedes* species increasing in per site abundance from early to late season, but with *A. albopictus* abundance increasing almost twice as much on average relative to *A. aegypti* abundance (Table 3). While the proportion of occupied sites increased from early to late rainy season for *A. albopictus*, they actually decreased for *A. aegypti*, suggesting that rainy-season competition from *A. albopictus* had a particularly severe negative effect on the distribution of *A. aegypti*. While both *Aedes* species appeared to favor cemetery sites later in the rainy season (which may explain why *A. aegypti* abundances were generally greater when *A. albopictus* was also present at a given site, Table 3), *A. albopictus* was also abundant in sites characterized by greater Open Developed Cover. This NLCD category includes golf courses, open parks, lawns and would also cover cemeteries. These findings suggest that *A. albopictus* was more likely to oviposit in wet sites across the range of open grass categories, whereas *A. aegypti* was more restricted to habitat specific to cemetery sites.

Laboratory tests have shown that *A. albopictus* is a superior competitor to members of the *C. pipiens* complex [13], [45], and we found *C. quinquefasciatus* abundance to be lower at sites with *A. albopictus* than those without the invader. *Culex quinquefasciatus* is more sensitive to dry conditions than are *Aedes*. Therefore, we expected *C. quinquefasciatus* to experience high dry-season mortality resulting in low abundance early in the rainy season, and that abundances would increase during the rainy season as more water-filled container habitats become available. However, as with *A. aegypti*, *C. quinquefasciatus* occupied a greater proportion of sites early in the rainy season, and its site occupancy and abundance declined from the early to late season. These results are inconsistent with condition-specific competition as the main mechanism of *C. quinquefasciatus* coexistence with *A. albopictus* and *A. aegypti*. *Culex quinquefasciatus* did however demonstrate a seasonal shift in habitat preference that may be important for its persistence with these *Aedes* species, and that is consistent with a spatial habitat segregation mechanism of coexistence by season. The associations between *C. quinquefasciatus* and land cover characteristics were seasonally distinct from both *Aedes* species, with a shift in abundance from cemetery sites early in the rainy season to both High Intensity Developed and Open Cover Developed areas by the late rainy season (Table 4). High Density Developed areas include heavily built environments such as some commercial and residential areas where impervious cover is high.

Numerous studies have shown that the seasonal patterns of *A. aegypti* and *A. albopictus* abundances are linked with local rainfall (e.g., [31], [46], [47]), and these patterns have been interpreted as being a result of dry-season egg mortality and rainy-season competitive ability [17], [21]. The southern peninsula of Florida has predictable seasonal differences in rainfall, with high total precipitation from frequent rain showers from May through September (rainy season) and low total precipitation from infrequent rain showers from October through April (dry season). In our study areas, average monthly rainfall from May through September is over three times higher than from October through April for the period 1981–2010 (200.1 vs. 66.4 mm). Compared with these long-term averages, rainfall during our study was lower (162.0 vs. 55.8 mm), but broadly consistent between rainy (81.0% of long-term averages) and dry (84.0%) seasons, and thus likely to represent the normal effects of rainfall on mosquito seasonal patterns. Although year-to-year variation in rainfall may affect either dry-season egg mortality or rainy-season competition among mosquitoes, it is difficult to predict the outcomes of such variation due to the effects of rainfall on other environmental factors that may affect mosquito communities, including detritus resources.

There is little evidence that changes in temperature in the range experienced within and between seasons affect the outcome of competition among *A. albopictus, A. aegypti,* and *C. quinequefasiatus.* In the laboratory, [48] showed no effect of temperature in the 24–30°C range on the outcome of larval competition between *A. albopictus* and *A. aegypti.* A field study found reduced competition between *A. albopictus* and *A. aegypti* in the dry season in Florida, but this was almost certainly due to increased detritus resource inputs [21]. No studies that we are aware of have tested the effect of temperature on larval competition of *C. quinquefasciatus* with *Aedes* species [49]. A laboratory study showed greater reductions in survival and development rates from 20.0°C to 15°C for *A. aegypti* than *C. quinquefasciatus* [50]. However, while daily minimum temperatures can average below 15°C in southern Florida during the coldest months (i.e., January–March), brief periods (i.e., hours) at unfavorable temperatures are unlikely to affect the outcome of population-level competition between *C. quinquefasciatus* and *Aedes* compared to a lack of available habitat from lower rainfall. Moreover, temperatures from containers are likely to be higher than values from weather stations since containers are usually sheltered.

Although the observed greater proportional increase in *A. albopictus* from early to late rainy season is consistent with strong negative effects of interspecific competition on *A. aegypti,* we found that species co-occurrence did not reduce *A. aegypti* abundance per site. A prior field study showed *A. albopictus* and *A. aegypti* responded the same way to manipulations of egg desiccation and to seasonal differences in detritus inputs in the field, and that the effect of high detritus inputs in the dry season resulted in no detectable effects of competition in the dry season [21]. Similar responses to seasonal factors that reduce competition, such as more resources, suggest that season may contribute to coexistence as an equalizing mechanism [2] rather than a stabilizing mechanism. It is possible that *A. aegypti* can utilize specific container habitats that have relaxed interspecific competition at some sites.

Cemeteries were an important larval habitat for each species during the rainy season. While abundances of both *A. aegypti* and *A. albopictus* were lower in cemeteries during the early rainy season, cemeteries increased in importance by the late rainy season. Relative abundance of *C. quinquefasciatus* had the opposite relationship with cemeteries, which were important habitat early but not later in the rainy season. These results suggest that cemetery habitat plays a key role in mechanisms controlling the rate of increases in *Aedes* abundance and decreases in *C. quinquefasciatus* abundance from early to late season. The similar responses of *A. aegypti* and *A. albopictus* to cemetery habitat between early vs. late rainy season may be due to a number of ecological processes. Cemeteries likely provide numerous existing containers for each species. Even within one cemetery, cemetery vases vary considerably in their biotic and abiotic conditions [15], [51], and it is likely that these *Aedes* species choose to oviposit in containers with particular conditions, largely independently of the other species [2], [52]. Interspecific aggregation caused both by random processes and cuing on the environment may reduce the competitive impact of *A. albopictus* on *A. aegypti,* and facilitate the local coexistence of these species in cemetery sites [52]. Coexistence of *A. aegypti* with *A. albopictus* within the same cemetery may also be facilitated if *A. aegypti* avoids oviposition in vases already inhabited by *A. albopictus.* Some mosquitoes alter their oviposition behavior in response to conspecifics or to controphic non-mosquito larvae (e.g., [53], [54]) and conspecific eggs [55] already present in the habitat. However, we are unaware of any evidence for oviposition deterrence by other competing mosquito species, and this is an area for future research. Interspecific aggregation between *A. aegypti* and *A. albopictus* within cemeteries may be especially important for coexistence in the late rainy season when abundances of *A. aegypti* are highest.

During the late rainy season, cemetery was the only positive predictor of *A. aegypti* abundance at the 200-m scale; all other land cover variables were negative predictors of *A. aegypti.* The proportion of sites where *A. aegypti* was sampled (Table 2) declined 50% between the early and late rainy season, but this was predominantly due to loss of site-level occupancy in the urban areas, whereas the proportion occupancy in cemetery sites was relatively unchanged. However, the proportion of the replicate ovitraps where *A. aegypti* were present (Table 2, in parentheses) declined across all sites between early and late rainy season. [20] compared *A. albopictus* and *A. aegypti* abundances from oviposition traps placed in the same cemeteries vs. traps in the intervening urban matrix. [20] proposed that *A. aegypti* would be superior to *A. albopictus* at colonizing vacant cemeteries. In a metapopulation setting, this superior colonization ability would result in escape from regional exclusion via competition from *A. albopictus* [22], [23], but [20] found no difference in abundances of *A. aegypti* and *A. albopictus* between cemeteries and areas in the intervening urban matrix. The present study suggests that cemeteries may be important habitat for both of these *Aedes* species, and may act as patches of ideal habitat that support these species as metapopulations. Out of the 19 total cemeteries sampled in the late rainy season, and ignoring the presence of *C. quinquefasciatus,* ten were colonized by either *A. albopictus* or *A. aegypti,* four had both species, and five were unoccupied by these *Aedes* (Table 2). Future research may be needed to further investigate the utilization of cemeteries by these *Aedes* species, and the role of metapopulation dynamics among cemeteries for maintaining species coexistence, especially in the late rainy season.

Opposite to *A. albopictus* and *A. aegypti, C. quinquefasciatus* were positively predicted by cemeteries in the early rainy season and negatively predicted by cemeteries in the late rainy season. These field results are consistent with the competitive exclusion of *C. quinquefasciatus* from cemeteries as *A. albopictus* increasingly utilized cemeteries during the rainy season. However, seasonal responses of *C. quinquefasciatus* to rainfall and interspecific competition may be more complex and difficult to understand than with *Aedes.* In addition to above ground containers, *C. quinquefasciatus* commonly utilize a wide variety of above- and below-ground storm water treatment devices, such as wetlands, retention basins, and catch basins [56–58]. The regional coexistence of *C. quinquefasciatus* with the competitively superior *A. albopictus* and *A. aegypti* may depend on utilization of these ground habitats by *C. quinquefasciatus,* with container habitats acting as secondary low quality habitats, where these *Aedes* species are less likely to oviposit. However, wetland cover was negatively associated with *C. quinquefasciatus* abundances in this study, suggesting that that these habitats may not be an important source of individuals collected in our traps. *Culex quinquefasciatus* abundance was associated with High Intensity Developed and Open Cover Developed area, both of which likely harbor below ground catch basins that may be a source for individuals collected in our traps.

Local habitat segregation in oviposition between *A. albopictus* and *A. aegypti* has been well documented along rural to urban gradients, with *A. albopictus* generally positively related to rural variables and *A. aegypti* related to urban variables (e.g., [40], [59]). Habitat segregation between *C. quinquefasciatus* and *Aedes* species has been mainly studied among individual containers [45]. To our knowledge only [20] has examined spatial patterns of competing mosquitoes at the within-city scale and related them across different land uses in an entirely urban environment. [20] did not find *A. albopictus* and *A. aegypti* densities to be specifically associated with cemeteries, but instead differed among broad land use categories, with *A. aegypti* more abundant in ovitraps in residential areas compared to industrial and commercial areas. Patchiness of urban landscapes can contribute to the coexistence of *A. albopictus* with the inferior competitors *A. aegypti* and *C. quinquefasciatus*, especially given that *A. aegypti* and *C. quinquefasciatus* start the rainy season with the numerical advantage. *Aedes aegypti* and *A. albopictus* abundances were similarly associated with cemetery habitats and not predicted by other landscape variables at the 50-m scale, suggesting that *Aedes* coexistence is unlikely to be due to spatial partitioning at the scale of land use types used in this study. Rather *A. aegypti* may be able to escape competitive exclusion by ovipositing in uninhabited vases within cemeteries. Investigating species' utilization among individual resident vases or other container habitats within cemeteries or other land uses was beyond the scope of this study, although past studies have found evidence that interspecific aggregation is likely be important for *A. aegypti* coexistence with *A. albopicitus* in Florida [52].

The results of this study are generally consistent with those of [20], a prior study in the same locations, as well as studies in other urban areas that have shown spatial and temporal differences among *A. albopictus* and *A. aegypti* (e.g., [23], [25], [60]). However, this study builds on prior studies in two important ways. First, it shows that seasonal patterns of *C. quinquefasciatus* oviposition into small containers show a trend opposite to that of co-occurring *Aedes* species. Second, this study examined specific land cover variables within each city and quantified their relative importance for explaining species abundances, as well as how the effects of seasonality on mosquito ecology varies in conjunction with land use.

Acknowledgments

We thank the Mosquito and Aquatic Weed Control Center of Hillsborough County, FL for their use of facilities and resources.

Author Contributions

Conceived and designed the experiments: PTL SAJ. Performed the experiments: PTL. Analyzed the data: PTL SL SAJ. Contributed reagents/materials/analysis tools: PTL SL SAJ. Wrote the paper: PTL SL SAJ.

References

1. Tilman D (1982) Resource competition and community structure. Princeton, New Jersey, USA: Princeton University Press.
2. Chesson P (2000) Mechanisms of maintenance of species diversity. Annual Review of Ecology and Systematics 31: 343–366.
3. Juliano SA, Lounibos LP (2005) Ecology of invasive mosquitoes: effects on resident species and on human health. Ecology Letters 8: 558–574.
4. Lounibos LP (2002) Invasions by insect vectors of human disease. Annual Review of Entomology 47: 233–266.
5. Juliano SA, Lounibos LP (2005) Ecology of invasive mosquitoes: Effects on resident species and on human health. Ecology Letters 8: 558–574.
6. Gerhardt RR, Gottfried KL, Apperson CS, Davis BS, Erwin PC, et al. (2001) The first isolation of La Crosse virus from naturally occurring infected *Aedes albopictus*. Emerging Infectious Diseases 7: 807–811.
7. Ibañez-Berñal SB, Briseño JP, Mutebi EA, Rodriguez G (1997) First record in America of *Aedes albopictus* naturally infected with dengue virus during the 1995 outbreak at Reynosa, Mexico. Medical and Veterinary Entomology 11: 305–309.
8. Turell MJ, Dohm DJ, Sardelis MR, O'Guinn ML, Andreadis TG, et al. (2005) An update on the potential of North American mosquitoes (Diptera: Culicidae) to transmit West Nile virus. Journal of Medical Entomology 42: 57–62.
9. Paupy C, Ollomo B, Kamgang B, Moutailler S, Rousset D, et al. (2010) Comparative role of *Aedes albopictus* and *Aedes aegypti* in the emergence of Dengue and Chikungunya in central Africa. Vector Borne Zoonotic Diseases 10: 259–266.
10. Juliano SA (2010) Coexistence, exclusion, or neutrality? a meta-analysis of competition between *Aedes albopictus* and resident mosquitoes. Israel Journal of Ecology & Evolution 56: 325–351.
11. Juliano SA (2009) Species interactions among larval mosquitoes: Context dependence across habitat gradients. Annual Review of Entomology 54: 37–56.
12. Peters T, Chevone B, Callahan R (1969) Interactions between larvae of *Aedes aegypti* (L.) and *Culex pipiens* L. in mixed experimental populations. Mosquito News 29: 435–438.
13. Murrell EG, Juliano SA (2012) Competitive Abilities in Experimental Microcosms Are Accurately Predicted by a Demographic Index for R*. Plos One 7: 9.
14. Costanzo KS, Mormann K, Juliano SA (2005) Asymmetrical competition and patterns of abundance of *Aedes albopictus* and *Culex pipiens* (Diptera: Culicidae). Journal of Medical Entomology 42: 559–570.
15. O'Meara GF, Evans Jr LF, Gettman AD, Cuda JP (1995) Spread of *Aedes albopictus* and decline of *Ae. aegypti* (Diptera: Culicidae) in Florida. Journal of Medical Entomology 32: 554–562.
16. Juliano SA, Lounibos LP, O'Meara GF (2004) A field test for competitive effects of *Aedes albopictus* on *A. aegypti* in South Florida: Differences between sites of coexistence exclusion? Oecologia 139: 583–593.
17. Costanzo KS, Kesavaraju B, Juliano SA (2005) Condition specific competition in container mosquitoes: The role of non-competing life-history stages. Ecology 86: 3289–3295.
18. Chase JM, Leibold MA (2002) Spatial scale dictates the productivity-biodiversity relationship. Nature 416: 427–430.
19. Taniguchi Y, Nakano S (2000) Condition-dependent competition: Implications for the distributions of stream fishes. Ecology 81 2027–2039.
20. Leisnham PT, Juliano SA (2009) Spatial and temporal patterns of coexistence between competing *Aedes* mosquitoes in urban Florida. Oecologia 160: 343–352.
21. O'Neal PA, Juliano SA (2013) Seasonal variation in competition and coexistence of *Aedes* mosquitoes: Stabilizing effects of egg mortality or equalizing effects of resources? Journal of Animal Ecology 82: 256–265.
22. Amarasekare P, Possingham H (2001) Patch dynamics and metapopulation theory: The case of successional species. Journal of Theoretical Biology 209: 333–344.
23. Calcagno V, Mouquet N, Jarne P, David P (2006) Coexistence in a metacommunity: The competition-colonization trade-off is not dead. Ecology Letters 9: 897–907.
24. Hartley S, Shorrocks B (2002) A general framework for the aggregation model of coexistence. The Journal of Animal Ecology 71: 651.
25. Rey JR, Nishimura N, Wagner B, Braks MAH, O'Connell SM, et al. (2006) Habitat segregation of mosquito arbovirus vectors in south Florida. Journal of Medical Entomology 43: 1134–1141.
26. Braks MAH, Honório NA, Lourenco-De-Oliveira R, Juliano SA, Lounibos LP (2003) Convergent habitat segregation of *Aedes aegypti* and *Aedes albopictus* (Diptera: Culicidae) in southeastern Brazil and Florida. Journal of Medical Entomology 40: 785–794.
27. Carbajo AE, Curto SI, Schweigman NJ (2006) Spatial distribution pattern of oviposition in the mosquito *Aedes aegypti* in relation to urbanization in Buenos Aires: Southern fringe bionomics of an introduced vector. Medical and Veterinary Entomology 20: 209–218.
28. Vezzani D (2007) Review: Artificial container-breeding mosquitoes and cemeteries: A perfect match. Tropical Medicine and International Health 12: 299–313.
29. Geographic Mapping Section (1999) Florida land use, cover and forms classification system handbook. Tallahassee, FL Florida Department of Transportation Surveying and Mapping Office.
30. Silver JB (2008) Mosquito Ecology: Field Sampling Methods. New York: Springer.
31. Richards SL, Apperson CS, Ghosh SK, Cheshire HM, Zeichner BC (2006) Spatial analysis of *Aedes albopictus* (Diptera: Culicidae) oviposition in suburban neighborhoods of a piedmont community in North Carolina. Journal of Medical Entomology 43: 976–989.
32. Lee JH, Kokas JE (2004) Field evaluation of CDC gravid trap attractants to primary West Nile virus vectors, *Culex* mosquitoes in New York State. Journal of the American Mosquito Control Association 20: 248–253.
33. Rawlins SC, Martinez R, Wiltshire S, Legall G (1998) A comparison of surveillance systems for the dengue vector *Aedes aegypti* in Port of Spain, Trinidad. Journal of the American Mosquito Control Association 14: 131–136.
34. Morato VCG, Teixeira MD, Gomes AC, Bergamaschi DP, Barreto ML (2005) Infestation of *Aedes aegypti* estimated by oviposition traps in Brazil. Revista De Saude Publica 39: 553–558.
35. Focks DA (2003) A review of entomological sampling methods and indicators for dengue vectors. Gainsville, FL: World Health Organization. 40 p.
36. Schultz GW (1993) Seasonal abundance of dengue vectors in Manila, Republic of the Philippines. Southeast Asian Journal of Tropical Medicine and Public Health 24: 369–375.
37. Romero-Vivas CME, Falconar AKI (2005) Investigation of relationships between *Aedes aegypti* egg, larvae, pupae, and adult density indices where their

main breeding sites were located indoors. Journal of the American Mosquito Control Association 21: 15–21.
38. Dibo MR, Chierotti AP, Ferrari MS, Mendonca AL, Neto FC (2008) Study of the relationship between *Aedes* (Stegomyia) *aegypti* egg and adult densities, dengue fever and climate in Mirassol, state of Sao Paulo, Brazil. Memorias Do Instituto Oswaldo Cruz 103: 554–560.
39. Hoeck PAE, Ramberg FB, Merrill SA, Moll C, Hagedorn HH (2003) Population and parity levels of *Aedes aegypti* collected in Tucson. Journal of Vector Ecology 28: 65–73.
40. Rey JR, Nishimura N, Wagner B, Braks MAH, O'Connell SM, et al. (2006) Habitat segregation of mosquito arbovirus vectors in south Florida. Journal of Medical Entomology 1134–1141.
41. Ghosh SK, Mukhopadhyay P, Lu JC (2006) Bayesian analysis of zero-inflated regression models. Journal of Statistical Planning and Inference 136: 1360–1375.
42. Kery M (2010) Introduction to WinBUGS for ecologists: A Bayesian approach to regression, Anova, mixed models and related analyses. Burlington, MA: Academic Press. 302 p.
43. Lunn DJ, Thomas A, Best N, Spiegelhalter D (2000) WinBUGS – a Bayesian modelling framework: concepts, structure, and extensibility. Statistics and Computing 10: 325–337.
44. Spiegelhalter DJ, Best NG, Carlin BR, van der Linde A (2002) Bayesian measures of model complexity and fit. Journal of the Royal Statistical Society Series B-Statistical Methodology 64: 583–616.
45. Carrieri M, Bacchi M, Bellini R, Maini S (2003) On the competition occurring between *Aedes albopictus* and *Culex pipiens* (Diptera: Culicidae) in Italy. Environmental Entomology 32: 1313–1321.
46. Serpa LLN, Costa K, Voltolini JC, Kakitani I (2006) Seasonal variation of *Aedes aegypti* and *Aedes albopictus* in a city of Southeastern Brazil. Revista de Saude Publica 40: 1101–1105.
47. Barrera R, Amador M, Clark GG (2006) Ecological factors influencing *Aedes aegypti* (Diptera: Culicidae) productivity in artificial containers in Salinas, Puerto Rico. Journal of Medical Entomology 43: 484–492.
48. Lounibos LP, Suárez S, Menéndez Z, Nishimura N, Escher RL, et al. (2002) Does temperature affect the outcome of larval competition between *Aedes aegypti* and *Aedes albopictus*? Journal of Vector Ecology 27: 86–95.
49. Juliano SA (2008) Species interactions among mosquitoes: context dependence across habitat gradients. Annual Review of Entomology 54.
50. Rueda LM, Patel KJ, Axtell RC, Stinner RE (1990) Temperature-dependent development and survival rates of *Culex quinquefasciatus* and *Aedes aegypti* (Diptera: Culicidae). Journal of Medical Entomology 27: 892–898.
51. Murrell EG, Damal K, Lounibos LP, Juliano SA (2011) Distributions of competing container mosquitoes depend on detritus types, nutrient ratios, and food availability. Annals of the Entomological Society of America 104: 688–698.
52. Fader JE, Juliano SA (2013) An empirical test of the aggregation model of coexistence and consequences for competing container-dwelling mosquitoes. Ecology 94: 478–488.
53. Blaustein L, Kotler BP (1993) Oviposition habitat selection by the mosquito, *Culiseta longiareolara*: effects of conspecifics, food and green toad tadpoles. Ecological Entomology 18: 104–108.
54. Mokany A, Shine R (2003) Oviposition site selection by mosquitoes is affected by cues from conspecific larvae and anuran tadpoles. Austral Ecology 28: 33–37.
55. Chadee DD, Corbet PS, Greenwood JJD (2002) Egg-laying yellow fever mosquitoes avoid sites containing eggs laid by themselves or by conspecifics. Entomologia Experimentalis et Applicata 57: 1990.
56. Reiter ME, LaPointe DA (2009) Larval habitat for the avian malaria vector *Culex quinquefasciatus* (Diptera: Culicidae) in altered mid-elevation mesic-dry forests in Hawai'i. Journal of Vector Ecology 34: 208–216.
57. Workman PD, Walton WE (2003) Larval behavior of four *Culex* (Diptera: Culicidae) associated with treatment wetlands in the southwestern United States. Journal of Vector Ecology 28: 213–228.
58. Harbison JE, Metzger ME, Hu RJ (2010) Association between *Culex quinquefasciatus* (Diptera: Culicidae) oviposition and structural features of belowground stormwater treatment devices. Journal of Medical Entomology 47: 67–73.
59. Braks MAH, Honorio NA, Lourenço-de-Oliveira R, Juliano SA, Lounibos LP (2003) Convergent habitat segregation of *Aedes aegypti* and *Aedes albopictus* (Diptera: Culicidae) in southeastern Brazil and Florida, USA. Journal of Medical Entomology 40: 785–794.
60. Braks MA, Honorio NA, Lourencqo-De-Oliveira R, Juliano SA, Lounibos LP (2003) Convergent habitat segregation of *Aedes aegypti* and *Aedes albopictus* (Diptera: Culicidae) in southeastern Brazil and Florida. J Med Entomol 40: 785–794.

10

The Push and Pull of Land Use Policy: Reconstructing 150 Years of Development and Conservation Land Acquisition

Maria João Santos[1*¤], Terry Watt[2], Stephanie Pincetl[3]

1 Center for Spatial and Textual Analysis, Spatial History Project, and Bill Lane Center for the American West, Stanford University, Stanford, California, United States of America, 2 Terry Watt Planning Associates, San Francisco, California, United States of America, 3 California Center for Sustainable Communities, Institute of the Environment University of California Los Angeles, Los Angeles, California, United States of America

Abstract

The growth of human populations and their resource needs have stressed the conservation of natural land resources. Many policies and programs have been implemented to address the pressures on land resources and notwithstanding this pressure, significant acquisition of land for conservation has occurred throughout history in the U.S., and internationally. Here we assess the on-the-ground result of the evolution of land use policies in California as a pioneer forerunner, in the form of acquisition of land for conservation (i.e. Open Space), and its impact on the rest of the U.S. and beyond. To this end we describe the timeline and spatial representation of the growth of California's conservation network over the last 150 years, and link it to the history of land use policies. We then assess whether conservation land acquisition has consistently grown through time or occurred in specific decades. About ¼ of the state is now designated Open Space. Fewer and larger areas conserved and acquired at the beginning of the 20th century; the conservation network was complemented with a larger number of smaller sized properties. Despite acquisition of land in every decade, the process was uneven (E = 0.3 for California, E = 0.14±0.08 average for the state's counties), mostly due to the large acquisitions and land set asides in the 1900s, followed by 1930s and 1940s. This process was a result of a comprehensive set of legislation that evolved through time, and resulted from the competing needs for development and conservation. Even with the impressive 174,000 km^2 of public lands in California, the future of California's natural infrastructure and natural heritage cannot rely solely on these public lands, nor public agencies and their resources. Critically a future course of land preservation relying on the purchase of new lands – in California and beyond – for conservation is tremendously expensive.

Editor: Rob Slotow, University of KwaZulu-Natal, South Africa

Funding: MJS was funded by a postdoctoral fellowship from the Bill Lane Center for the American West and the Spatial History project at Stanford University. The funders had no role in study design, data collection and analysis, decision to publish, or preparation of the manuscript.

Competing Interests: The authors have declared that no competing interests exist.

* Email: mjsantos@stanford.edu

¤ Current address: Department of Innovation, Environmental and Energy Sciences, Utrecht University, The Netherlands

Introduction

The growth of human populations and their resource needs have stressed the conservation of natural land resources. Conservation planning has focused on moving beyond opportunism [1], and is using strategic planning to identify the best spatial solutions to meet conservation goals, and implement such solutions on-the-ground [2]. This is evident in the recent approaches to conservation of the Nature Conservancy, for example, identifying the most valuable and endangered conservation hotspots across the globe, and investing funds in those places to preserve land. However, temporal aspects have been somewhat neglected (but see [3]), only currently coming to play as changes in land cover/land use and climate may threaten the persistence of the already existing conservation networks [4]. This, of course raises difficult scientific and ethical issues beyond the scope of this paper, about moving fauna and flora to locations where they will survive under climate change conditions and other types of interventions. At the same time, the creation of a conservation network can in itself induces change [5], and this change can be tied with historical legacies or periods of intense action [3]. These periods can be identified in the timeline of protected area establishment and in the creation of key conservation legislation, but integration of both timelines is yet to be conducted. While the former is easily assessed worldwide, the latter is affected by local, regional, national and international contexts. Here we tie both timelines, to understand the push-and-pull of land use policies over the last 150 years, using California as case study to demonstrate the importance of the integration of both timelines and to suggest such research be conducted in other places. We focus on California for its preeminence in this area, and its international significance. California is arguably the state that led the implementation of the concept of resource reserves since the 19th century, in the country and elsewhere [3].

Forty four percent of the Earth's plant species and 35% of the Earth's terrestrial vertebrates are endemic to 25 hotspots, despite the fact these hotspots make up just 1.4% of the world's land cover [6]. These biodiversity hotspots have immense implications for conservation strategies to protect large number of species living in relatively small regions of land. They also represent lands that have been and are highly threatened by land cover and climate changes. The California Floristic Province is one of the world's 25 most biodiverse areas on the planet [6]. For California itself, 44% of its plant and vertebrate species are endemic to the state [7]. Today, ¼ of California are properties designated as Open Space. Open Space refers to "*lands protected through title ownership by a public agency or non-profit land conservation organization*" [8]. Open Spaces includes a wide array of types of properties from City Parks to National Parks, and some heavily managed land, such as that managed by the federal Bureau of Land Management (BLM). Fee title ownership by the public exists for all of these Open Spaces. However, Open Space does not include private land in easement contracts. Moreover, lands in public ownership – aside from the National Park lands – are subject to the Federal Land Policy and Management Act, FLPMA (1976) that allows multiple-use of public lands and for the most part does not recognize conservation as an exclusive use except on lands with designations such as wilderness.

Already in the later decades of the nineteenth century, preservation of California's remarkable natural resources was underway. In 1864 Yosemite Valley and the Mariposa Grove of Giant Sequoia trees were protected by Abraham Lincoln, with a grant of 80.9km^2 of federal land. At the time, the federal government maintained ownership of vast lands, part of the public domain, that had not been sold through the Homestead Act (1865) and other acts that distributed the lands the federal government inherited with its new territories. Much of California's early preserved land was land that belonged to the federal government and was "withdrawn" from sale to the public, and instead, conserved as part of the public domain. This eventually became the backbone of the lands of the National Park Service, and the National Forest Service. Similar patterns of land reservation occurred in many states of the American West. Nevada, for example, is 80% public lands, or federal ownership. In southern California, the San Gabriel and San Bernardino Forest reserves were created in 1892 and 1893 respectively, from the public domain by President Harrison, some of the first land reservations from sale from the public lands in the nation. California is in many ways the state that led the implementation of the concept of resource reserves since the 19th century [9], lobbying Congress and the President about the public interest in the public lands, and that they should be conserved. State interest in land conservation was reflected in 1927 by the then recently created State Parks Commission that instructed Frederick Law Olmsted Jr. "*to make a survey to determine what lands are suitable and desirable for the ultimate development of a comprehensive, well-balanced state park system, and to define the relation of such a system to other means of conserving and utilizing the scenic and recreational resources of the state*"[10]. The survey was conducted, and set the vision, for example, for the creation of the state's Redwood Parks. Birth home of the Sierra Club, there is no question that interest in protecting the state's scenic beauty has been important for nearly a century and a half, an idea exported to other parts of the country and the world to establish their own conservation timelines [3,9].

In this paper we review the history of land conservation in California from a land use and governance perspective, and assess the on-the-ground result of the evolution of land use policies in the state in the form of acquisition of land for conservation. To do so we reconstruct the conservation history of California, that is, the growth of the conservation network per decade over the last 150 years, and link it to the timeline of land use policies as an example for further research of this type. We assess whether conservation land acquisition has taken place relatively evenly over time or has been concentrated in specific decades. We then review the challenges of current land use regulations, including fiscal constraints and regulatory requirements, which create a complex terrain for further land conservation. New tools are emerging for protection but structural constraints that we will discuss, remain in place as well, and they will differ from state to state and from nation to nation. In this discussion we develop a method that can be replicated to better help understanding both the ways in which lands have been conserved and the potential pathways for further conservation given the structural conditions that exist in California.

Materials and Methods

Open Space land

To obtain geospatial data for all Open Spaces in the state, we used the California Protected Area Database (CPAD) as a base database [8]. CPAD was developed by GreenInfo Network as a GIS inventory of all fee-protected Open Space properties in California. We selected this database and not the World Database on Protected Areas [3,11] because CPAD has a more comprehensive representation of the Open Space properties in the state, from City Parks to National Parks. It does not include all public land, nor private owners, or properties protected through the use of easements. Neither WDPA nor CPAD included information on the acquisition and establishment dates of each and every Open Space property in California or other protected lands across the globe [8,11]. To complement this information gap, and reconstruct a conservation history for the state, we contacted federal and state agencies and Non-Governmental Organizations that manage Open Space properties to request information. We asked for acquisition date, that is, when the fee title was purchased, and establishment date – when the property was open to the public. Below we present statistics for data acquisition for the great majority of the properties and area of the state designated as Open Space. In the remaining analysis we use date of acquisition for our timeline of conservation land acquisition. To understand the conditions under which land were protected, knowing the timelines of establishment is important, as it allows better insights into the reasons why such acquisitions might have occurred [3].

Land use policies

To complement, compare and contrast the Open space acquisition timeline we developed the land use history timeline. We reviewed the legal documents that detailed land use policies in California. These included primary source legislation and published literature on examples, applications and assessments of the impacts of such legislation (for example see http://ceres.ca.gov/planning/state.html). Key literature was searched for the date of the legislation, its description, intended targets, and when possible implications of the legislation for land use planning and Open Space acquisition.

Data analysis

The timeline of conservation land acquisition describes the cumulative area acquired per decade in the state. To assess whether each decade contribution to the conservation network was higher or lower than the state's average, we calculated the slope of the regression line of all the decades and compared it with the slope of the regression for consecutive decades. Significantly different decades were illustrated in the state's timeline. We then tested whether the acquisition of land was even throughout time for the state and how it compared across counties. We calculated the Simpson evenness index (E) which varies from 0 (uneven

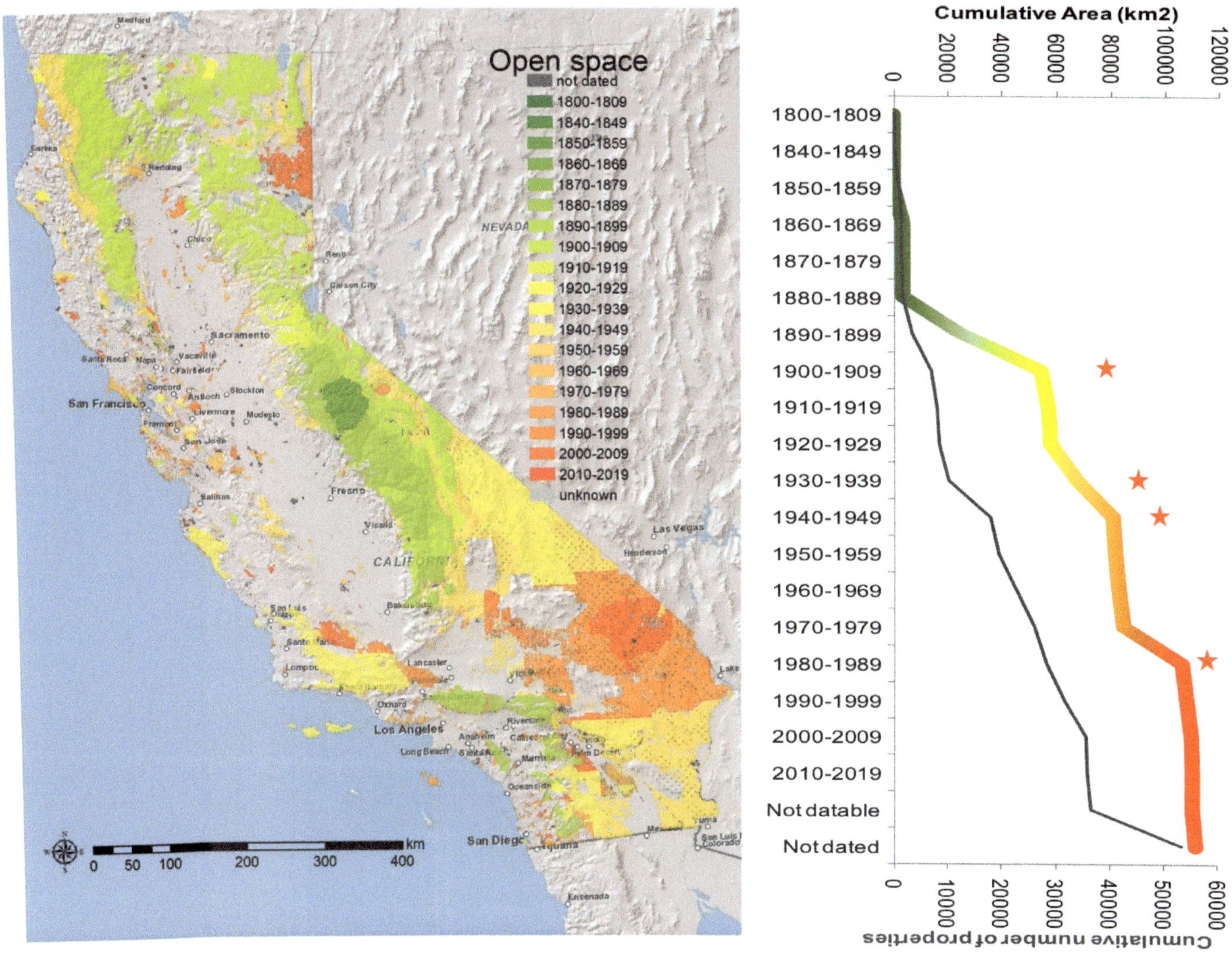

Figure 1. Timeline of Open Space acquisition and establishment in California over the last 150 years. Left panel – spatial representation, Right panel – timeline (red stars represent the decades where the conservation land acquisition was higher than the average over all the decades).

distribution of conservation action through the decades) to 1 (equal distribution of conservation action through the decades; [12]). We calculated evenness for the state and for each county, and determined whether counties differed from the state values. We then linked the timeline of conservation land acquisition with the history of land use policy in the state. Because of the link between funding for land acquisition and development, we analyzed in detail the most populous counties in the beginning of the 20th century.

Results

California has 16,000 Open Space properties in the form of 53,337 parcels that were acquired throughout the last 150 years. These properties cover ¼ of the state's area (112,156km^2). We were able to retrieve information on acquisition date for a total of 35,807 properties (67%), corresponding to 110,300km^2 (98.5%) of the state's Open Space. Merced, Orange, San Joaquin, Stanislaus, and Sutter were the counties from which it was harder to collect information. City Parks were the properties that had the least data about date acquired and/or created.

History of conservation land acquisition

Large tracts of land were set aside from the remaining public domain at the end of the 19th century. These were the mountain forest lands to be managed by the subsequently created United States Forest Service (Figure 1). Land acquisitions in the 1880s and 1890s contributed with an area of 30,000km^2, 30% of the current Open Space land. With the close of the public domain and into the twentieth century, additional land had to then be acquired for parks (mostly by federal agencies) as it was privately owned. Geographically, earlier land preservation focused on mountain regions, mostly in the Central and Southern Sierras, and the mountains surrounding Los Angeles (Figure 1). Land set asides and early acquisitions were predominantly in the northern California mountain ranges until 1920s (Figure 1). In the 1930's and 1940's, large tracts in southern California were purchased, as for example the lands for Death Valley and Joshua Tree National Parks. At this time the California State Parks further contributed to land acquisition, a pattern that continued into the 1950's (Figure 1). These acquisitions were also greatly aided by regulations passed in the 1940's that allowed Special Districts (a type of government unit that is formed to provide a specific service, such as schools, sewers, street lighting, parks or other) to acquire

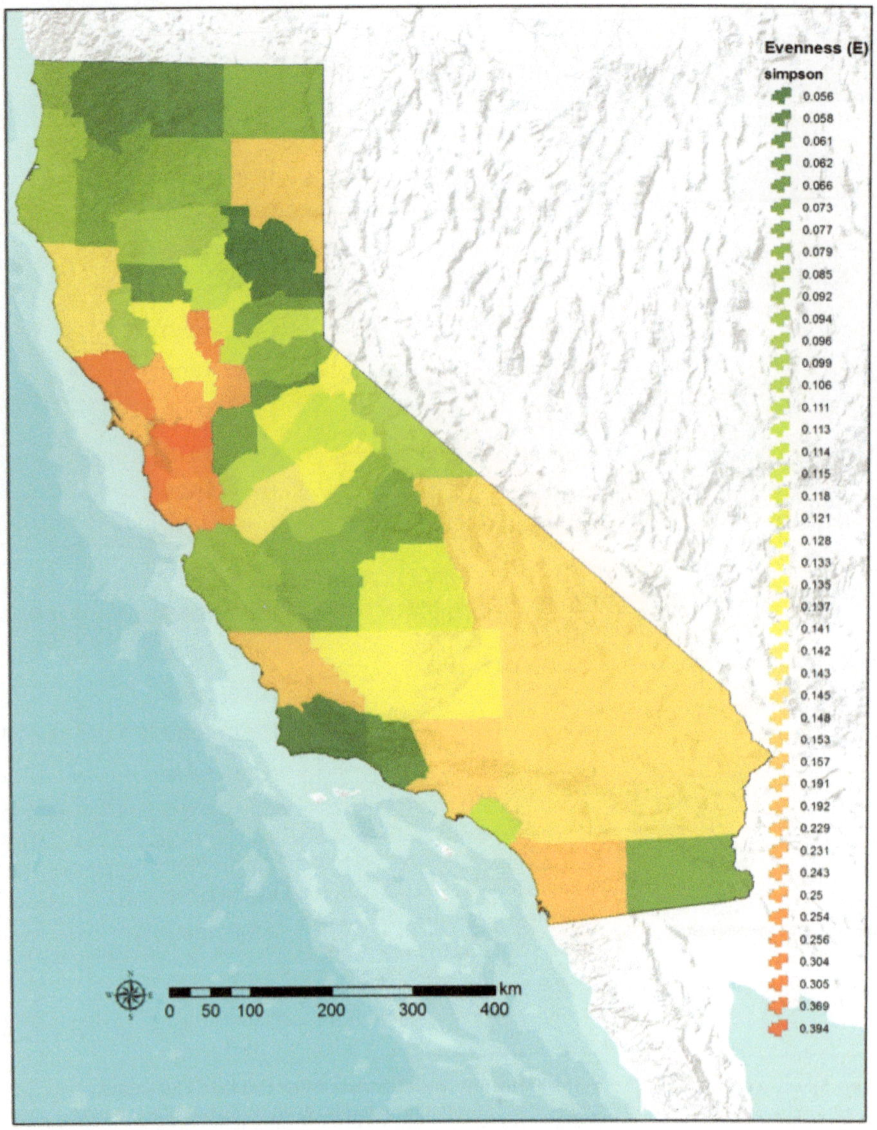

Figure 2. Per county evenness of the conservation timeline. Lighter colors represent higher evenness. The higher evenness is achieved around the San Francisco Bay area, and most of the counties show an evenness half of that of the state ($E_{California} = 0.3$).

conservation lands. From 1950 to 1980 more scattered and smaller sized properties filled in the conservation network, as the larger tracts of land were already public. However, this pattern shifted from the 1980's to today with the acquisition of large tracts of desert lands (Figure 1).

The rate of land acquisition (slope of the regression line for the state conservation acquisition) was 6874. 8 km^2 per decade (Figure 1). This rate was significantly higher in five decades (1890–1900, 1900–1910, 1920–1930, 1930–1940 and 1970–1980, indicated in Figure 1 with red stars), with acquisition rates ranging from 9697km^2 in 1930–1930, to 29799km^2 in 1900–1910, a very steep acquisition slope 5 times higher than the state's rate (Figure 1). The rate of conservation land acquisition in California over the last 150 years was also spatially uneven. The state's evenness value was 0.3, and 0.15 as the counties average (Figure 2). Plumas, Siskiyou and Santa Barbara counties were very uneven in their conservation land acquisition (Figure 2), while the Bay Area counties had a more even acquisition rate over time, and greater or equal to that of state. This is because of the greater amount of Open Space area acquired or set aside in the 1900's, followed by the 1930's and 1940's, with rates up to 5-times greater than the state's average rate of land acquisition (red stars in Figure 1).

Of the most populous counties in 1910, San Francisco engaged in some of the earliest acquisitions and then continued a constant rate of acquisition throughout the twentieth century (Figure 3). Los Angeles conservation land grew more slowly at the beginning of the twentieth century, showed a stepwise pattern. High federal involvement at the turn of the twentieth century and again in 1970–1990, was supplemented by land acquisitions by nonprofit organizations. San Diego shows a big increase in land acquisition in 1920's followed by a smooth rate thereafter; this is a similar pattern to Alameda, San Mateo and Santa Clara, but the peak of Open Space acquisition in these counties starts in 1950's. The most populous counties show, however, a very different proportion of conservation land; the largest being in San Diego and the smallest in San Francisco and Sacramento (Table 1).

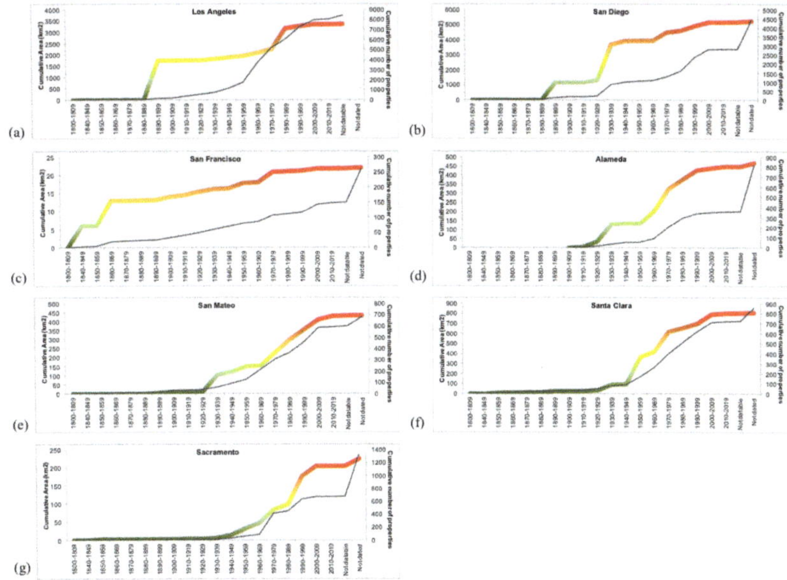

Figure 3. Timeline of conservation for the most populous counties in California in 1910.

History of land use policies

Local control over land use was instituted very early on to ensure autonomy of localities from state interference. This local independence was staunchly solidified by 1914 state legislation authorizing charter cities to make and enforce all laws and regulations in respect to municipal affairs ([13]; Table 2). This includes by 1937, the requirement to adopt comprehensive plans, which are guiding documents that set out a general vision for how the land in the jurisdiction of the city or unincorporated in the county should be utilized. Extensive land development patterns were facilitated by urban and state policies including uniform subdivision regulations in 1929 [14,15,16,17,18], and attempts to establish and enforce General Plans to guide development were overwhelmed by the tenacity of local developers and the proliferation of new municipalities, each with its own planning authority [14]. Most of the growth between 1950 and 1960 was located in suburban communities [19], aided by the recently built highways. By the early 1960s, state policy makers and Governor Pat Brown were concerned about growing fragmentation at the local level, including the increasing numbers of special districts formed to service urban growth. The state legislature created a state Planning Office in 1959. Governor Pat Brown appointed the Coordinating Council on Urban Policy to come up with policy proposals to improve planning and coordination of growth. The Coordinating Council's report urged the creation of one multi-purpose district in each of the state's metropolitan areas to deal with regional issues such as air pollution, water supply, sewage, parks and more. These proposals were roundly resisted by the Chamber of Commerce, cities and counties. In 1963, the state legislature, instead, created watered down county-level Local Agency Formation Commissions who were empowered to approve or disapprove any petition for incorporation, special-district formation or dissolution, and annexation in each county.

Frustrated by the lack of leadership at the state level, movements to attempt to curb growth and to protect the state's natural resources mobilized (Table 2). The attempt by the state Highway Commission to improve Highway 101 by cutting through portions of the Prairie Creek Redwoods and Jedediah Smith Redwoods State Parks, was defeated by the Sierra Club that then became a watch dog over projects proposed by the Highway Commission that often targeted land in state ownership, thus a less costly strategy that the purchasing of privately held lands. Another important example was the creation of California Tomorrow in 1962, a nonprofit educational institution that developed influential proposals for reforming state government and for the creation of

Table 1. Proportion of county land in Open Space.

County	County area (km^2)	Open Space area (km^2)	Area of county in Open Space (%)
Los Angeles	10242.41	3375.847	32.96
San Diego	10973.57	5286.147	48.17
San Francisco	277.02	22.12	7.98
Alameda	2126.87	469.39	22.07
San Mateo	1430.55	435.12	30.42
Santa Clara	3378.21	811.63	24.03
Sacramento	2580.21	225.94	8.76

Selected counties represent the most populous counties as in 1910s.

Table 2. The push-and-pull of development and Open space land use policies in the state.

The push and pull of development and natural resources legislation		
Natural resources		*For development*
	1914 – Charter cities	Authorizes the making and enforcement of all laws and regulations for municipal affairs
	1937 – Comprehensive (General) Plans	Sets out a general vision for the land in the jurisdiction of the city or county, and how it should be utilized
	1947 – State Highway Act	Authorize and promote the development of highways
	1956 and 1962 – Federal Highway Act	Authorize and promote the development of highways
Sets out a general vision for the water in the jurisdiction(s) and how it should be utilized	**1957 – California Water Plan**	
	1962 – California Tomorrow	Regional government to manage growth in the state
Stop the infill of the San Francisco Bay	**1965 – Bay Area Conservation and Development Commission**	
Allows contracts between the state and farmers to preserve agriculture lands	**1965 – Williamson Act**	
Developers to set aside a portion of their subdivisions as Parks or Open Space, or pay fees for parkland acquisition	**1965 – Quimby Act**	
	1970 - Open Space in General Plans	Requires that General Plans include Open Space elements
Protocol of analysis and public disclosure of environmental impacts of proposed projects, and measures for mitigation	**1970 – California Environmental Quality Act**	
	1971 – General Plans	Become required, not voluntary
For water protection	**1972 – California Wild and Scenic Rivers Act**	
For coastal area protection	**1972 – Coastal Protection Act**	
For protection of endangered species and their habitat	**1972 – Endangered Species Act**	
	1978 – Proposition 13	Property tax reduction initiative
	1978 – AB 857	Curb urban sprawl and directing new urban growth to existing cities and suburbs – infill
For protection of endangered species habitat	**1982 – Habitat Conservation Plans**	
For protection of natural communities (endangered or not)	**1991 – Natural Communities Conservation Planning**	
Infrastructure Planning	**2001 – AB857 Farmland Protection Act**	
Plan and manage natural communities for global climate change	**2006 – AB32 The Global Warming Solutions Act**	
Greenhouse Gas emissions regulation	**2007 – SB97 CEQA and Greenhouse Gas Emissions**	
	2008 – SB The Sustainable Communities and climate Protection Act	Adaptation and vulnerability for human populations

regional government to manage growth in the state. These environmental concerns contributed to the passage of the 1965 California Land Conservation Act (Williamson Act). The Williamson Act allowed farmers and ranchers to qualify for lower property tax rates if they entered into contracts keeping their lands in agriculture for a minimum of ten years. The 1965 Quimby Act was also passed. It allowed local governments to require developers to set aside a portion of their subdivisions as parks or Open Space, or pay fees for parkland acquisition and maintenance. The 1969 Santa Barbara oil spill had a jarring effect on the state, prompting a number of legislative initiatives, including the passage of the California Environmental Quality Act (CEQA) in 1970. Modeled on the federal National Environmental Policy Act (NEPA), it requires state and local agencies within California to follow a protocol of analysis and public disclosure of environmental impacts of proposed projects, and to adopt feasible measures for mitigating those impacts. It is a mandatory part of every California state and local agency process.

In 1972 state voters created the Coastal Commission, an entity appointed by the Governor, in charge of protecting California's coast from over development. This was one of the strongest measures to emerge from the first slow growth era [20], and a clear reaction to the increased pace of urbanization that was occurring. Other counteracting effect was that developments were being held up by the federal Endangered Species Act (ESA), passed in 1972 with the aim to preserve charismatic species. Environmental organizations used the ESA to challenge development with some success and developers found themselves stymied and slowed down. ESA challenges were based on a species by species endangerment threat, and it became clear that policies based on protecting individual species would scarcely achieve the goal of preserving species as they relied on habitats that needed to be preserved as well. Habitat conservation plans (HCP's) were added to the ESA by Congress in 1982 in an effort to address the need to conserve the habitat for imperiled species. Congress viewed HCPs as a win-win situation for imperiled species; HCPs took habitat into consideration, allowing a more encompassing approach, and they created Incidental Take Permits (ITPs), which allowed – for the first time – the taking of a limited amount of habitat in exchange for a commitment to an HCP to protect and manage other habitat areas, ensuring the species' overall recovery chances [20]. But HCPs still were created to protect one species at time. In response, California, under Governor Wilson, created Multiple Species Habitat Conservation Planning (MSHCP) and the 1991 Natural Communities Conservation Planning (NCCP) in an effort to encompass entire ecosystems and their processes. The NCCP process was intended to bring all stakeholders to the table in order to set aside coherent, regional habitat preserves [18,21]. NCCPs were endorsed by Secretary of the Interior, Bruce Babbitt as a legitimate implementation of the ESA that would allow the protection of endangered species, but not halt development. This is perhaps the impetus behind the increase in conservation land acquisition in San Diego (Figure 3), one of the precursor counties to develop a NCCP.

All in all, the 1970s were a period of active initiatives to preserve land at the local level with concurrent initiatives aimed to restrict or slow growth throughout the state, but in 1978 state voters passed Proposition 13, the property tax reduction initiative. Local government revenues plummeted from about $10 billion just prior to the passage of Proposition 13 to approximately half of that shortly thereafter. Growth management efforts were hampered by the need for revenues and most counties and cities adopted land use policies most likely to replenish their depleted budgets [14,18,22,23,24]. The impacts of Proposition 13 likely explain the asymptote of land acquisition rate from 1980s onwards (Figure 1). Some exceptions, however, emerge for counties like Sacramento, where the cumulative rate of conservation land acquisition still increased after 1980s (Figure 3). This is perhaps due to the emergence of creative solutions for raising funds for the acquisition of conservation land. Yet, in addition to Proposition 13 and Proposition 4, voters also passed Propositions 62 and 218 in 1996 extending the supermajority approval requirement of Proposition 13 to virtually all types of assessments, fees or taxes used by local government. This further constricted local government's flexibility to raise funds for necessary services, let alone to purchase additional lands for conservation purposes (Table 2).

With the decline in local government's ability to raise money for land preservation through tax dollars, new mechanisms began to be invented. One such new strategy was developed successfully by the Planning and Conservation League (PCL), a statewide environmental organization, which developed a set of inventive campaign funding tactics and ballot initiatives to finance the acquisition and development of park and recreation areas [25]. Nongovernmental organizations and quasi-governmental agencies emerged from this period as new stewards of public spaces, including habitat. State conservancies with budgets funded by grants, donations and foundation funding were created. Recent innovations include the Regional Advanced Mitigation Plan (RAMP), which is an effort to develop a more comprehensive and strategic approach to mitigating biological resource impacts caused by major infrastructure projects, before projects are constructed. RAMPs typically build on a conservation plan that validates and matches the mitigation actions and cumulative impacts. Another emerging concept is the inclusion of Regional Greenprints as an integral part of local General Plans and Sustainable Community Strategies. Greenprints would offer a new way to improve conservation planning by providing a process to map a region's important Open Space for a full range of ecosystem services including habitat, farmland, recreation, water resources and more. Conservation planning could provide a baseline on which land use plans are developed, and could be linked to CO_2 sequestration, a new and popular idea. For example, Preservation Ranch, a 7890ha property in Sonoma County, was just recently purchased with carbon credits funds (http://baynature.org/articles/preservation-ranch-big-conservation-thanks-to-carbon-credits/).

With the advent of climate change, priorities have expanded to include renewable energy and land use that reduces production of greenhouse gas emissions. All of these concerns are embodied in adopted state law and policy. Examples include the 1978 Urban Strategy, AB 857 (Wiggins codified at Section 65041.1 of the Government Code), the 2006 Assembly Bill 32 (California Global Warming Solutions Act), and the 2008 SB 97 (CEQA and Greenhouse Gas Emissions), and Senate Bill 375 (Sustainable Communities and Climate Protection Act).

Discussion

It can be argued that over the past 150 years a great deal of California's most scenic landscapes has been preserved. The fundamental dilemma of conserving land is the necessity for land to be purchased. Under the Fifth Amendment of the U.S. Constitution, government must purchase land at its market value, it cannot take land away from land owners, nor force them to sell below the land's market value. Thus for additional conservation to occur, as the state's population and cities grow into previously undeveloped areas that have important ecosystems, this land must be purchased from its owners. Moreover, since land development provides localities their revenues, there is a push to continue to zone land for urbanization. Such dynamics are typical throughout the U.S. but in addition, in California, for new developments, fees are assessed on units to create funds to acquire the conservation lands, near or adjacent to the new development – mitigation land. This leads to a paradox – for there to be funds to acquire conservation lands, there needs to be development [26]. In turn this quest for growth limits the land available for conservation – one of the original goals for which growth was planned for.

California's voluntary approach to meet conservation needs has been a reactive, and accretionary approach that creates obvious mixed messages between local power and regional planning efforts, and mismatches between policy goals and instruments for their implementation. Mixed messages are due to the autonomy of land use designations of cities while conservation needs and habitats do not observe political boundaries [27]. Land use planning is by definition based on the parcelizing of land for development, which is done by cities. This parcelizing then creates sections of land that can be bought and sold individually, allows the land market to work at the municipal and county scale, but this approach fragments the landscape and its governance. While there have been numerous attempts to create regional governments with regulatory authority, and to encourage greater collaboration among local governments [28]; they still have narrow single issue mandates (for example, regional "special districts" for air quality management and flood control). In contrast, conservation needs and habitats do not observe political boundaries, they are regional and even statewide. Preservation of ecosystem requires a landscape or regional approach, keeping parcels together to provide connectivity and or enough space for ecosystems to function well, and to preserve watersheds. Further, conservation lands require revenue to purchase them. Funding streams are often predicated on growth with development fees and various impact fees to finance conservation acquisitions [26]. Thus the state has a paradox: to conserve, there must be development.

There has been a century-long series of efforts to create effective institutions to help guide metropolitan growth and development in California [14,19,27]. While early conservation land acquisition tended to occur in the mountain regions, and away from the rapidly growing cities; the push-and-pull of land use policies lead to striking differences in the timeline of the 1910s most populous counties. The early most populous counties took different interpretations of the regulations, and the growth of the conservation network took different shapes. The most steady growth rate is found for San Francisco. San Francisco, along with Alameda, San Mateo, Santa Clara and 5 other counties form the Bay Area, the region that showed the higher evenness in conservation land acquisition. This has been a key region in the state and even an example of conservation success in a growing metropolitan area [28,29]. In 1965 the Bay Area Conservation and Development Commission was created, a single purpose regional planning agency, to supervise development around San Francisco Bay, and to stop Bay infill projects. The southern California counties show a stepwise growth of the conservation network, associated with the protection of the mountain areas around Los Angeles and desert areas around San Diego.

There are also mismatches between policy goals and instruments for their implementation. California has long recognized the costs of sprawl, and prioritized specific land use outcomes including orderly and efficient land use, infill, farmland protection and conservation of resource lands, but has developed weak planning tools and rules to strongly protect habitat. With the advent of climate change, priorities have expanded to include renewable energy and land use that reduces production of greenhouse gas emissions. These measures aim to curb urban sprawl and directing new urban growth to existing cities and suburbs – infill, revitalizing central cities and neighborhoods, and protecting resource lands, in alignment with the proposals put forward by California Tomorrow and others. While the most recent SB 375 statute is possibly the closest California has come to regional planning and encourages urban and suburban infill, clustered development, mixed land uses, transit oriented development to reduce GHGs [30], it does not change land use law and local control over local land use.

At the same time, other activities are occurring that will also challenge conservation. Paradoxically, concern about GHG emissions is fueling the Governor Jerry Brown's drive to have a high speed rail line built to link Southern California to the Bay Area and Sacramento. Its route, as currently designed, goes through some of the state's prime agricultural land, and is argued that it will support even more sub-urbanization in the already urbanizing Central Valley. Governor Jerry Brown is also pursuing more energy resources, including natural gas fracking. A great deal of the state's natural gas and petroleum resources are also found the Central Valley and will also compete for water, adding another demand in addition to water for agriculture, urban areas and for places of origin. Protected federal lands are not exempt either in California with large scale electricity generation projects in the high desert of the Mohave having been given the green light by the Department of the Interior, despite documented impacts on the desert tortoise among other species.

California has been fortunate to have had magnificent habitats preserved starting in the late 19th century. This is quite different from what has happened in the country and in the world [3,31]. At the same time, California has had tremendous population growth and urbanization as well. This has led to the push-and-pull in land use policies, creating conflicts between habitat preservation and development. Urban growth is still seen as the key to the prosperity of localities, so there is little incentive to build denser more contained cities and to collaborate across jurisdictions over infrastructure and/or land use. The state legislature, and voters through the ballot initiative process, have passed a number of laws, and created new agencies, to attempt to improve conservation planning and to protect conservation lands. But with the reduction of funding since the early 1980s, conserving lands has been more difficult. Any land purchased by government must be bought at market value, and since land in California is expensive, it means conserving more lands is costly. Innovative approaches to gather conservation acquisition funds are now being proposed and implemented.

Lost perhaps is the bigger vision of how we live on the land over time. Planning law, open space protection processes, finding funding for conservation through development and CO2 markets, are instrumental approaches to an issue that involves much a longer term vision about how people live on the land. Protected areas are not one more land use type – they are a result of a (land use) policy. While protected areas have distinct geographic boundaries they are part of a conservation network, which grows through a dynamic process [32,33]. U.S. land conservation policies, enabled by the vast area in public ownership, was seen as a model to emulate in many parts of the globe. For example, the International Union for the Conservation of Nature has been at the forefront of land conservation for habitat protection in Africa and increasingly in other parts of the world. Initially this was to simply preserve wildlife, now the mission includes attempting to anticipate the conservation needs due to climate change impacts. In Europe too there are now National Parks, based on the U.S. template, but often include working landscapes [11]. The reconstruction of the history of conservation can then describe the process of growth of a conservation network. This process is yet not well documented [3]; however, the creation of such baseline knowledge can help strategize the maintenance and future additions to the conservation network [33]. Further it can allow us to understand the feedback between land use policies and conservation action on meeting and maintaining conservation goals in California and elsewhere in the world [5,30,34].

Perhaps the most important value of historic analyses is what they indicate about the future. Policies could redirect future growth and conservation. As we observed, multiple funding sources contributed to the highest rates of land acquisition in different times in history (federal in 1890–1910, state in 1920–1940, and non-governmental organizations in 1970–1980). We also observed that reduction of development fee funding lead to two outcomes: creativity on methods to raise funds, and strategic acquisitions, perhaps because of the concurrent emergence of systematic conservation planning [1,2]. Perhaps the reduction of the funding stream imposed by Proposition 13 sparked innovative funding strategies, arguably more sustainable and less paradoxical. As future housing growth depends heavily on economic conditions [35], so do development fees. Innovative funding approaches that do not depend on development fees may be more sustainable in the long run. Linking histories of policies and conservation land acquisition allows reflections on the results of such policies but also on their perpetuity. Over the last 150 years we have seen a push-and-pull between development and conservation policies, some policies remained and some got replaced. The same applies to conservation lands, especially those that are not conserved in perpetuity. The future of these lands holds new challenges, with the onset of climate induced changes. Dynamic reserve systems are being proposed [4], which may take advantage of demoting some of these currently non-in-perpetuity conserved lands, or others, and it also presents challenges on funding sources. Perhaps a RAMP-like strategy can be applied to meet the conservation network dynamics needed to meet climate change-suited conservation. While solutions for the future of conservation are contextual and geographically restricted, the issues of fund raising, land acquisition, development versus conservation, and persistence of the conservation network operate at regional, national and global scales. Only through merging information at such scales can guarantee that the benefits of conservation can be enjoyed by the future generations.

Acknowledgments

We would like to thank Harold Mooney for connecting the authors in the paper.

Author Contributions

Conceived and designed the experiments: MJS TW SP. Performed the experiments: MJS TW SP. Analyzed the data: MJS TW SP. Contributed reagents/materials/analysis tools: MJS TW SP. Wrote the paper: MJS TW SP. Collected and analyzed the data on California conservation history: MJS. Contributed the history of land use policy: SP. Contributed the legislative timeline and provided oversight over the legislative components: TW.

References

1. Pressey RL, Humphries CJ, Margules CR, Vane-Wright RI, Williams PH (1993) Beyond opportunism: key principles for systematic reserve selection. Trends in Ecology and Evolution 8: 124–128.
2. Balmford A (2003) Conservation planning in the real world: South Africa shows the way. Trends in Ecology and Evolution 18: 435–439.
3. Radeloff VC, Beaudry F, Brooks TM, Butsic V, Dubinin M, et al (2013) Hot moments for biodiversity conservation. Conservation Letters 6: 58–66.
4. Pressey RL, Cabeza M, Watts ME, Cowling RM, Wilson KA (2007) Conservation planning in a changing world. Trends in Ecology and Evolution 22: 583–592.
5. Lovejoy TE (2006) Protected areas: a prism for a changing world. Trends in Ecology and Evolution 21: 329–333.
6. Brooks TM, Mittermeier RA, Mittermeier CG, da Fonseca GAB, Rylands AB, et al (2002) Habitat loss and extinction in the hotspots of biodiversity. Conservation Biology 16: 909–923.
7. Calsbeek R, Thompson JN, Richardson JE (2003) Patterns of molecular evolution and diversification in a biodiversity hotspot: the California Floristic Province. Molecular Ecology 12: 1021–1029.
8. GreenInfo Network (2013) California Protected Areas Database – CPAD. http://www.openspacecouncil.org/programs/index.php?program = 6. Accessed 2013 October 11.
9. Barton GA (2000) Empire forestry and American Environmentalism. Environment and History 6: 187–203.
10. Olmsted FL (1929) Report of State Parks Survey of California. California State Park Commission 106pp.
11. IUCN and UNEP-WCMC (2013) The World Database on Protected Areas (WDPA) [On-line]. Cambridge, UK: UNEP-WCMC. Available at: www.protectedplanet.net. Accessed 2013 October 11.
12. Simpson EH (1949) Measurement of diversity. Nature 163: 688.
13. Silva FJ, Barbour E (1999) The state-local fiscal relationship in California: A changing balance of power. Public Policy Institute of California, San Francisco.
14. Pincetl S (1999a) Transforming California, the Political History of Land Use in the State. Johns Hopkins University Press, Baltimore.
15. Fogelson RM (1967) The Fragmented Metropolis Los Angeles, 1850–1930. University of California Press, Berkeley.
16. Dear M (1996) In the City, Time Becomes Visible: Intentionality and Urbanism in Los Angeles, 1781–1991, in Scott AJ, Soja EW, eds., The City: Los Angeles and Urban Theory at the End of the Twentieth Century, University of California Press, Berkeley.
17. Hise G (1997) Magnetic Los Angeles; Planning the Twentieth Century Metropolis. Johns Hopkins University Press, Baltimore.
18. Pincetl S (1999b) The politics of influence: Democracy and the growth machine in Orange County, U.S. in Jonas A, Wilson D (eds.). The Urban Growth Machine: Critical Perspectives, Two Decades Later. State University of New York Press, Buffalo NY.
19. Barbour E (2002) Metropolitan Growth Planning in California 1900–2000. Public Policy Institute of California, San Francisco.
20. Press D (2002) Saving Open Space, the Politics of Local Preservation in California. University of California Press, Berkeley.
21. Fulton W (1997) The Reluctant Metropolis: The Politics of Urban Growth in Los Angeles. Solano Press Books, Point Arena.
22. Pincetl S (2004) The preservation of nature at the urban fringe. In Wolch J, Pastor M, Drier P (Eds.), Up Against the Sprawl, Public Policy and the Making of Southern California. University of Minnesota Press, Minneapolis, MN.
23. Pincetl S (2006) Conservation planning in the West, Problems, new strategies and entrenched obstacles. Geoforum 37: 246–255.
24. Wolch J, Pastor M, Drier P (2004) Up Against the Sprawl, Public Policy and the Making of Southern California. Minneapolis, MN: University of Minnesota Press.
25. Pincetl S (2003) Nonprofits and park provision in Los Angeles: an exploration of the rise of governance approaches to the provision of local services. Social Science Quarterly 84: 989–1001.
26. Pincetl S, Jonas AE, Sullivan J (2011) Political ecology and habitat conservation for endangered species planning in Southern California: Region, places and ecological governance. GeoForum 42: 427–438.
27. Schrag P (1998) Paradise Lost, California's Experience, America's Future. University of California Press, Berkeley.
28. Walker R (2007) The country in the city: the greening of the San Francisco Bay. University of Washington Press, Seattle, USA. 430pp.
29. Thorne JH, Santos MJ, Bjorkman J (2013) Regional assessment of urban impacts on landcover and Open space finds a smart urban growth policy performs little better than business as usual. PlosONE. DOI:10.1371/journal.pone.0065258.
30. California Transportation Commission, Regional Transportation Guidelines (2010) http://www.catc.ca.gov/programs/rtp/2010_RTP_Guidelines.pdf. Accessed 2013 March 12.
31. Moilanen A, Arponen A (2011) Administrative regions in conservation: balancing local priorities with regional to global preferences in spatial planning. Biological Conservation 144: 1719–1726.
32. Bishop K, Phillips A, Warren L (1995) Protected for ever? Factors shaping the future of protected areas policy. Land Use Policy 4: 291–305.
33. Wätzold F, Schwerdtner K (2005) Why be wasteful when preserving a valuable resource: review article on the cost-effectiveness of European biodiversity conservation policy. Biological Conservation 123: 327–338.
34. McDonald RI, Boucher TM (2011) Global development and the future of protected area strategy. Biological Conservation 144: 383–392.
35. Radeloff VC, Stewart SI, Hawbaker TJ, Gimmi U, Pidgeon AM, et al (2010) Housing growth in and near United States protected areas limits their conservation value. Proceedings of the National Academy of Sciences 107: 940–945.

11

Estimating Demand for Industrial and Commercial Land Use Given Economic Forecasts

Filipe Batista e Silva[1,2]*, Eric Koomen[3], Vasco Diogo[3], Carlo Lavalle[1]

1 Institute for Environment and Sustainability, Joint Research Centre, Ispra, Italy, 2 Centro de Estudos de Geografia e Ordenamento do Território, Porto, Portugal, 3 Department of Spatial Economics/SPINlab, VU University Amsterdam, Amsterdam, The Netherlands

Abstract

Current developments in the field of land use modelling point towards greater level of spatial and thematic resolution and the possibility to model large geographical extents. Improvements are taking place as computational capabilities increase and socioeconomic and environmental data are produced with sufficient detail. Integrated approaches to land use modelling rely on the development of interfaces with specialized models from fields like economy, hydrology, and agriculture. Impact assessment of scenarios/policies at various geographical scales can particularly benefit from these advances. A comprehensive land use modelling framework includes necessarily both the estimation of the quantity and the spatial allocation of land uses within a given timeframe. In this paper, we seek to establish straightforward methods to estimate demand for industrial and commercial land uses that can be used in the context of land use modelling, in particular for applications at continental scale, where the unavailability of data is often a major constraint. We propose a set of approaches based on 'land use intensity' measures indicating the amount of economic output per existing areal unit of land use. A base model was designed to estimate land demand based on regional-specific land use intensities; in addition, variants accounting for sectoral differences in land use intensity were introduced. A validation was carried out for a set of European countries by estimating land use for 2006 and comparing it to observations. The models' results were compared with estimations generated using the 'null model' (no land use change) and simple trend extrapolations. Results indicate that the proposed approaches clearly outperformed the 'null model', but did not consistently outperform the linear extrapolation. An uncertainty analysis further revealed that the models' performances are particularly sensitive to the quality of the input land use data. In addition, unknown future trends of regional land use intensity widen considerably the uncertainty bands of the predictions.

Editor: Rodrigo Huerta-Quintanilla, Cinvestav-Merida, Mexico

Funding: No current external funding sources for this study. The study was carried out in the scope of the ordinary internal research activities of the Joint Research Centre of the European Commission, and no specific funding was allocated to it. This study was conducted by the sole initiative of the authors.

Competing Interests: The authors have declared that no competing interests exist.

* E-mail: filipe.batista@jrc.ec.europa.eu

Introduction

The expansion of industrial and commercial land is poorly understood. Much research focuses on sector-specific dynamics and aspects such as industry location, productivity and employment [1–3]. The relation with land use change, however, is hardly studied. Yet this is an important aspect of the potential impact of economic development on the landscape and other environmental conditions. The development of certain economic activities requires the conversion of land from natural/semi-natural to artificial covers, often irreversibly. These dynamics are difficult to grasp: they relate to global technological and economic processes (e.g. deindustrialization of developed countries, outsourcing of production to cheap-labour countries, increased importance of information and communication technologies) as well as regional and local dynamics reflected in, for example, regional competitiveness and specialization, agglomeration economies and the performance of individual firms [4–9].

The understanding of the land dynamics related to industrial and commercial activities is particularly relevant for land use models that try to assess the potential future of the landscape. Land use modelling can be used to identify the drivers of land use change, and explain how these drivers and local and spatial factors interact to produce the observed landscapes. By understanding these mechanisms, past landscapes can be reconstructed given known historical records and future landscapes can be envisaged under different scenarios (assumptions on socioeconomic changes and policy alternatives). As a consequence of these capabilities, land use models have become an important element in integrated ex-ante impact assessment of policies at a wide range of spatial scales [10,11]. Land use models, as part of a broad range of land system models, have an important role in supporting future land use policy, and may provide input for planning processes [12,13].

In practice, land use models are used to make simulations of land use change in terms of quantity and/or location [14]. Non-spatial land use models are specialized in estimating the amount of change per land use type as country or regional aggregates, while spatially-explicit models are also able to reproduce where land use changes are likely to occur, and which local land use conversions (from one land use type to another) are expected to take place [15]. Typically, spatially explicit land use models involve the use of techniques broadly classified as cellular automata. In such models, space is represented by matrices of regular sized cells. Each cell may have a finite number of states (land uses), and these may swap over time according to a predefined set of rules regarding local and neighbour characteristics [16,17]. On the other hand, non-spatial land use models may utilize a range of techniques, from econometric to system dynamic and agent-based approaches.

A complete land use modelling approach requires the two aspects to be integrated into a coherent framework: the estimation of the quantity and the spatial allocation of land uses for a given timeframe [18]. In most approaches, land use demand projections are computed externally and then are fed into spatial-explicit land use models for the allocation. As Verburg and Overmars [11] mentioned, this 'top-down' approach is necessary especially when land demand is mainly determined by "forces that are exogenous to the land allocation". The allocation of the required land is then simulated by the geographical model considering two main dimensions: local suitability and neighbourhood interactions between the different land uses [19]. The Land Use Modelling Platform is an example of a structured platform able to integrate the two essential components of land use modelling: quantity of change and spatial allocation [20]. This platform was designed for territorial impact assessment of European policies, and can be configured project-wise, as the work by Mubareka et al. demonstrates [21].

Changes in land use quantity are often influenced by dynamics that occur at larger spatial and temporal frames and involving macro-economic and demographic changes. Therefore, the prediction of changes in land use require adequate economic context [14]. Economic models "provide a structure to represent the competition among different sectors, changes in management and technology and demand shifts due to trade or policy interventions" and are thus an important and representative input to quantify some of the drivers of land use demand [19]. As Rounsevell et al. [13] have recognized,

> *innovative coupling of a range of models would allow for the consistent analysis of the land system and its interactions as a whole. The multi-model approach makes use of the strengths of existing, individual land system models and, at the same time, avoids the development of an unmanageably complex model with which to represent the whole system.*

At the continental scale, most recent land use models are now able to simultaneously simulate the allocation of more than one land use type, allowing land use competition to be represented. This means that, besides the demand for each modelled land use type, specific local suitability and the spatial interactions between land uses must be known. In models such as the CLUE-S [22] and EU-CLUE-Scanner [20], the latter aspects are addressed through a combination of empirical-statistical and rule-based approaches.

The thematic detail of models applied to small scale/large extent areas is usually restricted to just a few major groupings of land use types. Typically, artificial cover (often named 'urban' or 'built-up') is modelled as a single class, lumping together uses as distinct as residential, industrial, commercial, services, while other 'artificial' land uses, like transport facilities (networks, ports, airports) or green urban areas and sports and leisure facilities are kept static (i.e., not modelled) due to difficulties in mimicking the spatial dynamics underpinning such land uses through current pixel-based models. Also lacking are approaches to model industrial and commercial areas as a separate class. Having sensible methods to estimate land demands for these uses is probably the first condition to allow a split of the 'artificial' class into residential and industrial/commercial/services.

This paper explores and tests methodically various alternatives of estimating land demand for industrial, commercial and services land uses (from here on, we may refer to this grouping simply as 'industrial and commercial land use'). The main focus is put into developing and testing approaches that transfer economic projections into potential land demand for a large area comprising several European countries. By studying the links between economic performance and land use dynamics, we hope also to contribute to wider and tighter integration of geography-based and economic-oriented models. Quantitative and empirical approaches will be used to explain recent expansion of industrial and commercial land, thus in line with Rounsevell et al. [13] who stated that

> *empirical analysis of past and present land use change has an important role in providing insights into the socio-economic and ecological processes that shape land use transitions.* (...) For this, *quantitative data and spatial information* (...) *are necessary to detect and assess land system change, enable up-scaling of results, cross-regional comparisons and longitudinal analysis.*

In the following section, we review a selection of existing approaches to model demand for urban and/or industrial and commercial land. The subsequent sections postulate the methodology applied in this study, present a validation of the results and discuss the sensitivity to various forms of inputs and potential sources of uncertainty. The last section of this article wraps up and discusses the main findings and their implications.

Estimation of Land Demand: Review of Existing Approaches

The estimation of future built-up area demand is usually done in the context of scenario studies but rarely the implications of the selected approaches are analysed or even discussed [23]. The appropriateness of the approaches is likely to vary according to the study area, spatial resolution and temporal scope of each application. Hoymann [23] identified three approaches to calculate demand for built-up area: trend extrapolation, regression models and density measures. In the same study, a validation exercise of the three approaches to calculate future built-up area was implemented for Germany and Czech Republic at two different spatial scales. Different models were calibrated with historical data and simulations were made for current built-up areas and then compared with observed data. One striking conclusion was that different approaches to determine future urban demand could lead to different outcomes, thus highlighting the uncertainty associated particularly with long-term projections.

Trend extrapolations apply observed growth rates of built-up area to estimate future land demand. This approach does not take into account any driving force, and simply assumes that past trends will remain constant in the future. This assumption may hold for short-term projections. However, the accuracy is expected to deteriorate with time as no causal factors are taken into account. On the contrary, regression models integrate explanatory variables to drive land use changes. Interactions between different drivers of land use change can be combined in multiple regression models. The selection of the variables to integrate the regression models are either subject to theory or automatic selection through exploratory analysis.

In the work of Reginster and Rounsevell [24] population and gross domestic product (GDP) were used as predictors of urban land use by means of a regression approach. The coefficients for the two independent variables of the regression were estimated using data from the year 2000 for Europe. Then, these coefficients were applied to different projected population and GDP in order to calculate future urban land use demand for different scenarios for the temporal range 2000–2080. However, a formal validation of the regression model was not performed. Seto et al. [25] studied the urban expansion between 1970 and 2000 at the global level by collecting and analysing results from the literature, summing a total of 326 case studies of 292 unique geographical locations distributed across the world. The annual growth rate or urban expansion was regressed against several socio-economic and locational variables. It was found that population and GDP growth rates positively influenced urban expansion, while farm subsidies, by increasing returns of agriculture land, hindered urban expansion. Results also suggested that urban growth is more sensitive to GDP growth in higher income countries than in lower income countries, and that in India and Africa, population growth is the main driver of urban expansion. The model also showed that low elevation coastal areas are more prone to higher expansion rates. Regression models to explain industrial land use have also been tested. In the work of Beckers and Schuur [26], a set of regression models establishing a relationship between employment and industrial land use in the Netherlands have been critically assessed. Their empirical findings suggest that regression models based on sectorial employment as the only predictors are insufficient to explain industrial land use, even when time lag effects and different scales of analysis are taken into account. De Vor [9], on the other hand, studied the impact of spatial factors on the location choice of industrial sites, concluding that high accessibility and economies of scale (translated in high land value and size of the working age population) are positively related to the observed supply of industrial sites in a Dutch region.

In regions where land is scarce and there is pressure to develop available land, there is a growing concern regarding the conflicts between the development of land and the protection of other assets like forested areas, farm land, landscapes and ecosystems [27]. Particularly in China, high demographic and economic growth rates of urbanizing regions have led to significant land consumption in the last two decades, thus increasing concerns about deterioration and depletion of land. Moreover, the availability of land for industrial development at low prices by local governments does not encourage an efficient use of land, leading to extensive land use and increased loss of agriculture land [28]. For these reasons, in China, as well as in other densely populated parts of the world like Japan and the Netherlands, the promotion of land use efficiency is becoming an important aspect in sustainable spatial planning [28–31]. In this context, measures of land use efficiency are being used by land use researchers and planners. Studies like those from Meng et al. [30] and from Huang et al. [28] have measured land use intensity/efficiency for the industrial land use in two different areas in Beijing, China. In its most simple form, land use intensity is measured as the economic output (in monetary terms) per unit of land surface. Empirical evidence collected and presented in both studies showed that land use intensity varies greatly across sectors, and that these differences may reach a factor of 40 between the least and the most land efficient sectors. In a study carried out for the whole of the Netherlands, it was also found that the average land use intensity of industrial land varies considerably across regions, and that those variations are mainly related to the sector composition of each region [31]. In line with what has been diagnosed for China, Louw et al. [31] also argued that, in the Dutch case, the supply of generous amounts of low-priced industrial land by municipalities (as a way to foster local economy) does not encourage the land use efficiency.

Intensity measures of the land use can also be used in the context of the estimation of future land demand, as proposed by Hoymann [23]. The general principle of the approach is formulated below:

$$LUI_{u,r,t} = \frac{V_{r,t}}{A_{u,r,t}} \quad (1)$$

where *LUI* stands for Land Use Intensity of the land use *u*, in a region *r* in the year *t*. *V* refers to a socio-economic variable related to the amount of land *A*. Consequently, the estimated land demand for land use type *u* for a year *t+n*, given *V* for *t+n* is calculated as follows:

$$A_{u,r,t+n} = \frac{V_{r,t+n}}{LUI_{u,r,t}} \quad (2)$$

To illustrate, say that, for a given region in a given time in the past, both the GDP and the industrial and commercial land use are measured in monetary and surface area terms, respectively. GDP can be obtained from official national or regional statistics, while the industrial and commercial land can be obtained either from land use maps or statistical registries. The land use intensity is obtained by dividing the GDP by the total industrial and commercial area, and expressed as units of currency per hectare of industrial and commercial use (eq. 1). Given a projected GDP for a year in the future, the amount of industrial and commercial land required to 'support' the expected GDP can be calculated (eq. 2). This approach assumes a stationary land use intensity over time. Yet, a dynamic land use intensity could be inferred from time-series analysis or estimated through regression techniques.

Another family of approaches to estimate future urban growth is the System Dynamics. System dynamic modelling was first introduced in the mid-1950's by Forrester and was initially applied to solve engineering problems related to control systems in industry [32]. Soon after, the precursor of this family of models realized the potential for application to a wide range of social and economic problems, and dedicated a publication to urban dynamics [33]. This framework is suited for resolving non-linear and complex problems, allowing a representation of the behaviour of dynamic systems over time and the feedbacks between the various elements. The use of system dynamic techniques in the context of land use change modelling has become particularly popular among researchers from China. For example, Luo et al. [34] and Zheng et al. [35] have used the system dynamics approach to compute urban land demand, and then used the CLUE-S model to allocate the demands in the spatial dimension. In both studies, the demand was the result of a complex system where demography, economy and land use were interrelated. In addition, the work of Wu et al. [36] is an example of how system dynamic models can be used to make ex-ante evaluation of the impacts of scenarios of different urban land use policies. Even more recently, Lauf et al. [37] were inspired by the system dynamic principles to address the problem of modelling urban systems where growth and shrinkage occur simultaneously within the same city-region due to contradictory factors (declining population; changing of population/household structure; changing of housing preferences).

Fragkias and Geoghegan [38] developed a spatially explicit model for a county in the United States focusing on industrial and commercial land use change. The objective of this study was not so much concerned with the estimation of aggregated land demands for industrial and commercial areas, but more related to understanding the local factors affecting the discrete choices of land conversions. The underlying model is mainly econometric and land use changes are function of individual decisions to convert undeveloped into developed land parcels (residential, industrial or commercial). Two main assumptions are present in the econometric model. First, landowners seek the maximization of their earnings with respect to the net expected returns of a variety of possible conversions. Second, each land parcel has characteristics that influence both the one-time net return of the land conversion and the returns related to earnings of the land in its undeveloped state. Distance to urban centres and transportation networks/nodes, neighbourhood, environmental conditions, planning and regulations are among such characteristics. In sum, this approach could be described as purely bottom-up, whereby the land demand for industrial or commercial use is not calculated a priori, but rather the result of individual decisions over time. This exhaustive modelling approach, though, requires a wealth of detailed data which is often not available for entire countries let alone for larger regions.

In the scope of the European SENSOR project (http://www.sensor-ip.eu), an integrated approach to calculate demand for different land uses was proposed [39]. One component of the entire modelling framework was the NEMESIS econometric model, which was adapted in order to calculate endogenously demand for different land uses: agriculture, forestry, tourism, transport infrastructures, natural areas and urban. The latter is further differentiated in housing and commercial/industrial built-up. The investment in commercial and industrial buildings is computed for each given moment in time as a negative function of rental price of buildings, a positive function of production and a negative function of technical progress. The net investment in buildings for a given time *t* corresponds to the building stock, in Euros, in time *t* minus the building stock in time *t-1* times a parameter reflecting the rate of decay of buildings. The net investment is determined for over 30 economic sectors represented in the model, and then summed to obtain the total net investment in commercial and industrial buildings. A 'technical coefficient' transforms the total net investment, expressed in monetary terms, in actual land requirements. A similar approach is used to obtain demand for new residential land. However, the investment in housing buildings is, instead, a positive function of real disposable income, and negative functions of the real interest rates and building prices. Calibrated with data from 2000, the model is able to make estimations on demand for housing and commercial and industrial areas by first determining the net investment in buildings for any given year. The land demands are computed at the country level and then passed on to the CLUE-S model for spatial allocation at the resolution of 1 km^2 [39]. An important characteristic of this approach is the linkage between the economic dynamics and its consequences in terms of potential land uptake, as well as the consideration of feedbacks in the process. As a result, the land claims of the different sectors "[are] price elastic to the extent that they will respond negatively to any increase in building price" [40].

In a nutshell, and despite all the progress made so far in the field of land use change, specific focus on the industrial and commercial areas has been limited so far. Models addressing dynamics of industrial and commercial land are usually applied at local scales and/or require data inputs inaccessible or even non-existent at continental scales. In addition, formal validations of the various approaches are lacking. Yet, nowadays, policy support at supra-national level demands more integrated assessments together with more spatial and thematic detail.

Methods to Estimate Demand for Industrial and Commercial Land

The objective of this study is to develop and validate approaches to estimate demand for industrial and commercial land. The approaches should be relatively straightforward so that they can be easily replicable and applicable to large spatial regions (e.g. countries and continents) in the context of land use modelling. We propose to explore in particular those approaches based on land use intensity measures. The main reason for this choice relies on the fact that, as reviewed in the previous section, these measures are not especially intensive to calculate, requiring only a few aggregate variables, characteristics which become relevant when working at very large spatial extents. Still, intensity measures are informative and conceptually easy to interpret. Moreover, they link to sector-specific processes of economic development that are expected to be relevant for the land uptake of industrial and commercial use. Using intensities allows linking land use simulations to regional economic projections.

Below in this section, several variants of the land use intensity approach are formally introduced. By definition, intensity measures integrate one driver of land use change at a time. In this study, sector gross value added (GVA) is used as a proxy for sector economic output. In addition to the land use intensity approaches, trend extrapolations are also tested. Trend extrapolations can be seen as the simplest way to make estimations because they do not specifically address drivers of land use change, but rather apply observed growth trends to describe possible future conditions. The main reason to consider trend extrapolations in this study is to create an adequate term of comparison for the estimations based on intensity measures.

Once introduced, the models will be applied to estimate the demand for industrial and commercial land for a set of countries in Europe.

Models Description

Trend extrapolation (models 1 & 2). Two trend extrapolation methods are considered: a linear extrapolation (model 1) and an exponential extrapolation (model 2). The linear extrapolation is formulated in equation 3:

$$A_{t_2} = A_{t_1} + \left(\left(\frac{A_{t_1} - A_{t0}}{t_1 - t_0} \right) * (t_2 - t_1) \right) + \varepsilon \quad (3)$$

where *A* refers to the industrial and commercial area, t_0 and t_1 correspond to the starting and ending years of the calibration, respectively, and ε is the error term. In this method, the average yearly absolute growth of the calibration period (t_0 to t_1) is multiplied by the total forecasting years (t_2–t_1) to obtain the estimate for desired year t_2. In the exponential extrapolation, the average yearly growth rate *G* observed between t_0 and t_1 is firstly obtained through equation 4, and it is then applied to estimate the industrial and commercial land in t_2 (equation 5). The graph in Figure 1 shows the application of both models to a hypothetical region with 200 ha of industrial and commercial land in 1990, and 300 ha in 2000.

$$G_{t_0:t_1} = \left(\frac{A_{t_1}}{A_{t_0}} \right)^{\left(\frac{1}{t_1 - t_2} \right)} - 1 \quad (4)$$

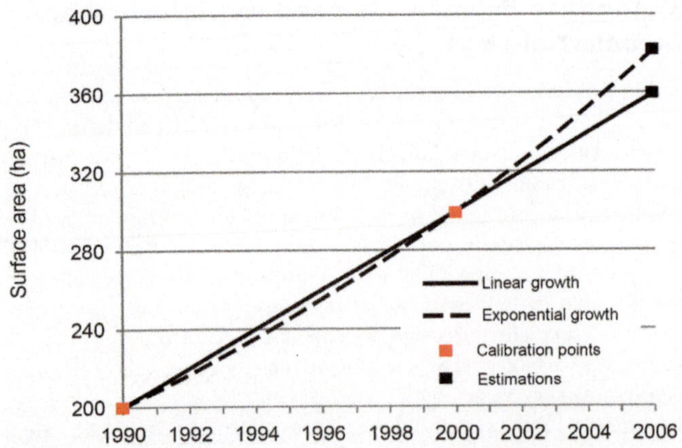

Figure 1. Extrapolation models for hypothetical region with 200 and 300 hectares of industrial and commercial land in 1990 and 2000.

$$A_{t_2} = \sqrt[\left(\frac{1}{t_2-t_1}\right)]{G+1} * A_{t_1} + \varepsilon \quad (5)$$

Region specific land use intensity (models 3 & 4). The economic product and surface area of commercial and industrial uses are consistently highly correlated at the regional level. If we take the sum of the regional gross value added of the industrial, commercial and service sectors and relate it to the respective surface area as reported by the CORINE Land Cover (CLC) datasets, correlation coefficients ranging between 0.74 and 0.76 can be found for the years 1990, 2000 and 2006, in Europe. This suggests that, in general, the higher the economic product of a region, the more physical infrastructure is required to support the economic activity.

Models 3 and 4 are characterized by using economic output or product P of regions as the driver of development of industrial and commercial areas. In both models, a land use intensity approach is used to relate the economic product with the respective area of industrial and commercial units. In model 3, the land use intensity LUI is computed for the year t_1 and measured as economic output per hectare of industrial and commercial land (eq. 6). Then, assuming a stable land use intensity in time, and knowing the product P for t_2, the total industrial and commercial land is predicted (eq. 7).

$$LUI_{t_1} = \frac{P_{t_1}}{A_{t_1}} \quad (6)$$

$$A_{t_2} = \frac{P_{t_2}}{LUI_{t_1}} + \varepsilon \quad (7)$$

This model takes the whole regional product and the whole existing area of industry and commerce per region in t_1 to compute the land use intensity. However, the total amount of industrial land is strongly related to historic developments and only partly dependent on current economic performance. In fact, as existing industrial and commercial land is likely to remain (with or without actual economic activity), this inertia is not captured by a single and static snapshot of the land use intensity. So we should perhaps focus especially on changes in economic development and related changes in the amount of land needed. This implies that the land use intensity of new developments is important in order to capture shifts in the production structure. Model 4 builds upon this idea. It measures land use intensity only of the industrial and commercial land developed during the calibration period t_0 and t_1 (eq. 8). The 'land use intensity of the recently developed land' is then used to estimate the extra land related to the growth of the product in the subsequent period (t_1:t_2) (eq. 8). Contrary to the model 3, this approach ignores the land use intensity of the industrial and commercial land developed prior to t_0.

$$LUI_{t_0:t_1} = \frac{P_{t_1} - P_{t_0}}{A_{t_1} - A_{t_0}} \quad (8)$$

$$A_{t_2} = A_{t_1} + \left(\frac{P_{t_2} - P_{t_1}}{LUI_{t_0:t_1}}\right) \quad (9)$$

We call these approaches 'region specific' because the intensity measures described above can be computed separately for any set of regions composing the whole of any area of interest. Consequently, regional differences in land use intensity (which are underpinned by differences in productivity and production structure) are captured.

Region and sector specific land use intensity (models 5 & 6). Industrial and commercial land is a rather broad and heterogeneous land use class. For example, in the CORINE Land Cover nomenclature, the homonymous land use class includes factories of all different kinds of industries, facilities for energy production and telecommunication networks, facilities related to defence and security, shopping malls and exposition sites, and a wide range of facilities related to public or private services likes schools, university and research campuses and hospitals [41]. Trying to model such a heterogeneous class as a whole poses obvious limitations. Most obvious of all, land use intensities vary considerably among industries [28,30,31], let alone the differences between the various economic sectors.

To address this limitation, one could think of making the land use intensity measures both region and sector specific. In this case, the economic product of a given sector s would be related to the land area A known to be used by sector s in year t_1 (eq. 10). At this point, it would be possible to estimate the aggregated industrial and commercial land for a given t_2 (eq. 11). Conceptually, this formulation is more robust than models 3 and 4 because it allows the integration of land use intensities specific to n number of sectors (model 5). In addition, a factor ω could be used to transform the land use intensities when calculating A in t_2, as a function of the observed changes in LUI between t_0 and t_1, that is $\omega_s = f(\Delta LUI_{s,t0,t1})$. In this study, we let $\omega = 1$ for all sectors.

$$LUI_{s,t_1} = \frac{P_{s,t_1}}{A_{s,t_1}}, \text{ with } s \in [1,n] \quad (10)$$

$$A_{t_2} = \sum_{s=1}^{n} \left(\frac{P_{s,t_2}}{LUI_{s,t_1} * \omega_s} \right) + \varepsilon \tag{11}$$

This model can be also combined with the concept of 'land use intensity of the recently developed land', as introduced in model 4. This is done by applying equations 12 and 13 (model 6):

$$LUI_{s,t_0:t_1} = \frac{P_{s,t_1} - P_{s,t_0}}{A_{s,t_1} - A_{s,t_0}} \tag{12}$$

$$A_{t_2} = A_{t_1} + \sum_{s=1}^{n} \left(\frac{P_{s,t_2} - P_{s,t_1}}{LUI_{s,t_0:t_1}} \right) + \varepsilon \tag{13}$$

Case Study: Estimation of Industrial and Commercial Land Use in Europe

Overview of the Case Study Set-Up

The case study consists of estimating the amount of industrial and commercial land use in Europe and comparing the estimates against reference land use data. Estimates will be produced by each of the six models described in the previous section and listed in table 1, for a set of Central and Western European countries. Countries from Scandinavia, the Balkans Eastern Europe and the UK were not included in the analysis due to incomplete land use and/or economic data time-series. The spatial unit of analysis used was the NUTS2 regions. The NUTS, or Nomenclature of Territorial Units for Statistics, is the European Union's official regional subdivision of member states, comprising three hierarchical levels (NUTS1, NUTS2 and NUTS3).

In order to measure the predictive power of land change models, Pontius et al. [42] and Pontius and Malanson [43] recommended that the calibration and validation should be separated processes, and that the modelling results should be compared to a 'null model'. The null model predicts pure persistence, i.e., no change during the modelling time span. In line with these recommendations, each of our models is calibrated using historical data for two points in time, t_0 and t_1, and is then used to estimate industrial and commercial land for a third point in time, t_2. Finally, all models (including the null model), are compared in terms of their ability to predict the actual total industrial and commercial land for t_2 as reported in a reference data source.

The following subsection will focus on the data used to feed the models and generate the estimates. Finally, the indicators used to measure the model performances are presented, and the results are reported and commented.

Data

Economic data. Gross value added per sector of activity was collected from Eurostat's online database (http://epp.eurostat.ec.europa.eu) at regional level (NUTS3). A time-series comprising the period 1985 to 2009 was compiled. All values were initially collected in current prices in Euros. The existing gaps were filled by using United Nations (UN) data (http://data.un.org), which were available in current US Dollars at country level only. For the missing years in the Eurostat database, annual growth rates were derived from the UN data, and then applied to generate country level data in Euros. Finally, the country values were regionally disaggregated using the regional shares of the closest available year in Eurostat. For the specific purpose of the case study, two additional procedures were applied. The values in current prices in Euros were transformed to constant prices as of 2005. The economic output expressed in constant prices is more suitable for time-series analysis because the effect of inflation is removed, thus reflecting the actual economic growth. Finally, the NUTS3 values were aggregated to NUTS2 to match with the spatial unit of analysis used in this study. GVA from three main categories of economic activity was used: industry; commerce and private services; and public services and administration.

Land use data. The source for 'industrial and commercial land' is the CORINE Land Cover (CLC). The three available editions (1990, 2000, 2006) were used. In the context of this case study, we considered 1990 as t_0, 2000 as t_1 and 2006 as t_2. The maps produced in the context of the CLC project are the only datasets providing a time-series of land use change that are consistent across European countries [44,45] because common nomenclature and standard methodology guidelines were used in its elaboration [41,46]. However, one major disadvantage of CORINE Land Cover is related to the thematic detail of its nomenclature. As mentioned earlier, the CLC class 'industrial and commercial' land use class aggregates a broad range of land use sub-categories that are not distinguishable by any further breakdown. As a result, CLC alone does not provide the minimum necessary sectorial detail to implement models 5 and 6.

To address this limitation, we focused on two countries in more detail. These countries were selected based on their different economic structure and the availability of detailed land use

Table 1. Main model characteristics.

Model nr.	Family of approach	Driver of land use change	Calibration years	Recent land use intensity	Sector specific LUI	Equations
M1	Trend extrapolation (linear)	None	1990, 2000	Not applicable	Not applicable	3
M2	Trend extrapolation (exponential)	None	1990, 2000	Not applicable	Not applicable	4 & 5
M3	Land use intensity measures	Gross Value Added	2000	No	No	6 & 7
M4	Land use intensity measures	Gross Value Added	1990, 2000	Yes	No	8 & 9
M5	Land use intensity measures	Gross Value Added	2000	No	Yes	10 & 11
M6	Land use intensity measures	Gross Value Added	1990, 2000	Yes	Yes	12 & 13

datasets and comprise of Spain (Sistema de Información de Ocupacióm del Suelo en España, http://www.siose.es/siose) and the Netherlands (Statistics Netherlands, http://www.cbs.nl). To calculate sector specific land use intensities, we first had to correspond the broad economic sectors with land use classes found in the Spanish and Dutch land use maps (see table 2). This correspondence allowed us to compute the land use intensities for the three broad economic sectors s for the Spanish and Dutch NUTS2 regions, using equation 10 where the economic product P was represented by the GVA. The land use intensities were computed for $t=2005$ in the case of Spain and for $t=2006$ in the case of The Netherlands. These years were chosen in order to match the reference dates of the national land use data sources.

We found that $LUI_{commerce}>LUI_{services}>LUI_{industry}$ for all regions, i.e., the highest economic output (GVA) per unit of land occurs in the commerce sector, followed by the services sector and by the industry sector. This relationship can also be interpreted as the area necessary to produce the same monetary unit, which is highest in the 'industry' sector and lowest in the 'commerce' sector. To find whether these patterns were consistent across regions, the coefficient of variation CV was computed for each sector s and country c, with σ_{LUI} and μ_{LUI} being the standard deviation and the average of the land use intensity, respectively (see eq. 14).

$$CV_{s,c} = \frac{\sigma_{LUIs,c}}{\mu_{LUIs,c}} \quad (14)$$

Table 3 presents the average land use intensities per sector and per country and the respective coefficient of variation. The results show a relatively small variance of the land use intensities of each sector within each country (CV <1 for all sectors in both countries). In addition, considering Spain and Netherlands altogether, we could infer that the commerce and service sectors are, on average, 27.6 and 6.7 times more land use intense than the industry sector (see values between brackets in table 3). These values can be interpreted as 'land use weights'. The higher the 'weight' the higher the land use intensity and, therefore, the less land required to produce one monetary unit of GVA.

The above empirical findings can be used to make an estimation of the sector composition of the 'industrial and commercial' class of CLC, i.e., how much of the whole land classified as 'industrial and commercial' in CLC refers to the generic sectors 'industry', 'commerce' and 'services' individually. At first, we weighted the sector product (GVA) by the sector land use weights w_s (eq. 15). The area of each sector is then calculated by multiplying the whole 'industrial and commercial' area (as reported in the CLC) by the estimated share (eq. 16). By definition, $A_t=\sum_s A_{s,t}$. This procedure was applied to 'disaggregate' the CLC class 'industrial and commercial' and thus obtain estimates of its sectorial composition for the calibration years t_0 and t_1 for all regions covered in this study.

$$WP_{s,t} = \frac{P_{s,t}}{w_s} \quad (15)$$

$$A_{s,t} = A_t * \left(\frac{WP_{s,t}}{\sum_{s=1}^{3} WP_{s,t}}\right), \quad (16)$$
$$\text{with } s \in \{1=\text{industry}, 2=\text{commerce}, 3=\text{services}\}$$

Validation Results

All models were calibrated with data up to the year 2000 and were then applied to estimate the industrial and commercial land in 2006. The validation is done by comparing each model's estimates with the actual amount of industrial and commercial land as reported by the CLC 2006, which is the nearest available to ground truth for the whole study area. The indicators used to measure the performance of the models are summarized in table 4 and the results can be consulted in tables 5 and 6 and in Figure 2.

Results reported in table 5 show that all models performed better than the null model which tell us that modelling demand for industrial and commercial areas appears to be a worthwhile exercise. However, all models have more or less underestimated the total amount of industrial and commercial land in 2006. While models 2 and 3 have best approximated the absolute expansion of land use for the whole study area, it can also be concluded that no model sufficiently reproduced the actual observed growth for the period 2000–2006. Nonetheless, most of the underestimations fall in a relatively narrow range, from −0.56% to −3.36%, and −7.06% in the worst case.

Overall, model 1 seems to be the best performer, as it scored best for AAE and TAE, and also showing one of the lowest relative differences. In addition, this model also shows the narrowest distribution of errors (see Figure 2). This indicates that the models that incorporated the economic output as driver for land use change were not able to perform better than trend extrapolations. Among the models that use GVA as driver of land use change, model 3 stands out, as its predictions are overall closer to the known land use in 2006 than predictions from the others. Models 4 and 6, which used the land use intensity of the land developed in the period 1990–2000 to estimate the land developed in 2006 performed worse than models 3 and 5, respectively, the latter using

Table 2. Correspondences between broad economic sectors and land use nomenclatures (SIOSE and CBS).

Broad sector label	Land use classes (SIOSE, Spain)	Land use classes (CBS, Netherlands)
Industry	Industry (821, 822, 823); mining and quarrying (833); energy (891, 892, 893, 894, 895, 896); water supply (911, 913)	Business estates (24); mining area (33)
Commerce and private services	Commerce and offices (841); hotels (842); recreation parks (843); camping (844)	Retail and catering (21)
Public services and administration	Public administration (851); health (852); education (854); penitentiary (855)	Public facilities (22); socio-cultural facilities (23)

Between brackets are the respective class codes of both Spanish and Dutch land use maps.

Table 3. Land use intensities and coefficient of variation (CV) per sector of main economic activity and per country.

	Industry		Commerce and private services		Public services and administration	
Country	**LUI (M€/Yr*ha)**	**CV LUI**	**LUI (M€/Yr*ha)**	**CV LUI**	**LUI (M€/Yr*ha)**	**CV LUI**
Spain	0.53 (1.0)	0.58	14.47 (27.1)	0.46	4.83 (9.0)	0.28
Netherlands	1.16 (1.0)	0.43	33.92 (29.1)	0.42	4.14 (3.6)	0.33
Spain+Netherlands	**0.67** (1.0)	**0.70**	**18.60 (27.6)**	**0.69**	**4.53 (6.7)**	**0.32**

Note: Values between brackets correspond to each sector's land use intensity in respect to the industry's land use intensity ($LUI_s/LUI_{industry}$).

the overall land use intensity of all existing industrial and commercial land use. Finally, the models which integrated both regional and sectorial specific land use intensities (models 5 and 6) did not outperform the models which relied only on an overall land use intensity per region.

Performances vary significantly country wise (table 6). Countries like Belgium, Denmark, France, Germany and Spain show fairly low errors for most modelling approaches, whereas the estimations for Austria, Ireland and Luxembourg were overall much worse, with all models severely underestimating the observed land use expansion. The negative outlier points identified in the box-whisker plot (Figure 2) correspond to NUTS2 regions of the latter countries. Even though the linear extrapolation (model 1) showed the lowest overall estimation errors, model 3 performed best for Austria, Denmark, Spain, Ireland and Luxembourg. In addition, models 4 and 5 obtained second best estimations for a number of countries.

Sensitivity to land use data. As all models strongly depend on land use data, we can expect the final results to be very sensitive to the accuracy of such input. Biases and inaccuracies in the reporting of observed land use propagate to the land use intensity measures which thus influence the final land use demand estimation. Regional industrial and commercial land use areas were derived from CLC data, which covers all Europe with a time-series comprising the years 1990, 2000, and 2006. Despite the common nomenclature and mapping methods, temporal and spatial inconsistencies have been reported [45]. Moreover, the large minimum mapping unit (MMU) of 25 hectares may create mapping artefacts. For example, land use patches smaller than the MMU in t_0 are 'hidden' within the dominant surrounding land use patch of another land use class. If the former patch expands to an area above the MMU in t_1, the patch is then mapped, thus giving the impression of an overestimated land use expansion between t_0 and t_1.

To test whether the results obtained in the validation were influence by CLC data issues, we have applied the same six models to the Dutch regions using finer land use time-series data from a national source (Statistics Netherlands, http://www.cbs.nl). The economic data was kept the same, as well as specifications for all six models. A comparison of modelling errors for the Netherlands using CLC and CBS land use data is reported in table 7. Two major conclusions can be drawn from the figures presented. First, there is a clear improvement of performances of all models when using finer land use data from a national source. Moreover, the proportion of performance improvement to the null model is higher when using finer land use data. Second, the consistent underestimation of land use demand when using CLC data is not observed when using finer data. This leads us to infer that issues directly related to the land use data partly explain the consistent model underestimation for the European case study. More specifically, these results may indicate that CLC underestimated the amount of industrial and commercial land use in 1990 and 2000 in relation to 2006, thus contributing to an overestimation of the land use intensities in 1990 and 2000.

In addition to these aspects, we must acknowledge temporal nonstationarities that might be present in the real world but which are not captured by any of the models. For example, in certain regions, spatial planning policies driven by expected economic growth may have led to oversupply of business estates that remain empty, thus decreasing the land use intensity in 2006. Other economic dynamics, such as changes in economic structure (e.g. shifts from labour intensive to capital intensive industries), can lead to appreciable changes in the land use intensity of regions over time. The uncertainties related to the future trends in

Table 4. Validation indicators computed for each model.

Indicator name	Short description	Formula
Relative difference (RD)	Relative difference between the estimated and the observed industrial and commercial area for the whole study area. It shows the magnitude of the aggregated deviation as well as the sign of the deviation. Negative and positive values mean under and overestimation, respectively. Expressed as percentage.	$RD=\left(\left(\frac{\sum_r A'_r}{\sum_r A_r}\right)-1\right)*100$
Average Absolute Error (AAE)	Average of all absolute regional deviations. It is always positive. Expressed in hectares.	$AAE=\frac{\sum_r \lvert A'_r - A_r \rvert}{n}$
Total Absolute Error (TAE)	Sum of all absolute regional deviations. It is expressed as percentage of the total known industrial and commercial land in 2006. It is always positive.	$TAE=\left(\frac{\sum_r \lvert A'_r - A_r \rvert}{\sum_r A_r}\right)*100$

Note: A – Known industrial and commercial area in 2006 (as reported in CLC2006); A' – Estimated industrial and commercial area for 2006; r – NUTS2 region; n – total number of NUTS2 regions.

Table 5. Validation results.

Model nr.	RD (%)	AAE (ha)	TAE (%)
Null	−11.68	1033	11.75
M1	−2.70	501	5.70
M2	−0.56	563	6.40
M3	−2.03	571	6.49
M4	−3.36	631	7.18
M5	−7.06	700	7.97
M6	−2.55	854	9.72

regional land use intensities will be addressed in the following chapter.

Analysis of Land Use Uncertainty

It has been noted that "when model parameters are fit by calibration to historical data, additional uncertainty is introduced due to the inherent temporal nonstationarity of processes" [47].

In this section we explore the main sources of uncertainty related to our modelling approach and their implications in terms of predicted land use demand. As explained in chapter 3, we proposed a deterministic method to estimate future demand for industrial and commercial land uses. The method relies on a single parameter, the land use intensity, which is calculated based on regional data on land use and economy. The land use intensity is expressed in terms of gross value added per hectare of industrial and commercial land in a given year in each region. This parameter is then assumed to remain constant in time and it can be used to predict future demand for industrial and commercial land given regional economic projections. The total uncertainty of the resulting land use demand predictions includes both the uncertainty of the land use intensity parameter and the uncertainty of the economic projections.

In this paper we focus only on the uncertainty of the land use intensity parameter. The uncertainty of economic projections is beyond the scope of this paper, as it is a field of research in its own. Moreover, the primary concern of this study is to design and test methods to translate given economic projections into future demand for industrial and commercial land use. The economic projections are herein dealt as exogenous assumptions, whose uncertainties shall be estimated in the appropriate framework of the economic modelling. Under this premise, the uncertainty of the estimates of future land use demand boils down to the land use intensity parameter. The uncertainty related to this parameter exists in two forms: first, the uncertainty of the measurement itself for a given moment in the past; second, the uncertainty regarding its future evolution. The sections below will focus on each of these two aspects of uncertainty.

Accounting for Land Use Mapping Errors

The land use intensity of a region is determined by dividing the regional GVA by the regional land use acreage. The latter is normally described in the form of spatially explicit land use maps, while the former is typically reported by governmental agencies according to international conventions. In Europe, Eurostat – the official statistical body of the European Commission – ensures standardization and discloses GVA data for all European countries and regions. Figures about the state of the economy can be subject to various distortions, such as measurement errors, intentional biases from reporting entities, exclusion of the parallel (not officially registered) economy. The uncertainty of the national accounts figures is, however, not communicated and thus cannot be included in this analysis. Therefore, we focus our uncertainty analysis on the errors associated with the land use maps we apply and that relate to aspects such as classification errors, and minimum mapping unit. The latter issue is particularly critical when using CLC data, as demonstrated previously.

Notwithstanding the importance of CLC as the sole European-wide land use/cover map, there has been limited reporting on its quality. Only the 2000 version of this dataset was subject to an extensive and systematic validation [44]. The thematic accuracy of CLC 2000 was assessed by comparing its classification with a classification derived from a field survey carried in the year 2000, the land use/cover area frame survey, better known as LUCAS. However, the validation yielded statistically inconclusive results for a number of land use classes for which the sample size was particularly small. This was the case of the land use class 'industrial and commercial units', for which only 34 points were controlled in all Europe.

To obtain an idea about the mapping errors in CLC we, therefore, rely on a statistically sound validation of CLC 2006 that was performed for one specific country [48]. In this validation effort for Portugal, a stratified random sampling scheme was adopted, with 100 sample points randomly selected for each land use class, in order to guarantee "a representative and meaningful basis for accuracy assessment" [48]. For each sample point, the mapped land use class was compared with visual observations of land use, enabling the construction of a 'contingency table' (or 'confusion matrix') which allows map accuracy indicators to be calculated [49]. The contingency table for the Portuguese CLC 2006 is reported in the work of Caetano et al. [48].

The binomial distribution is often applied to discrete land use classifications because each land use class can either occur or be absent at each location. When the sample size is large enough, it can be assumed that the proportion of errors of a land use class with the other land use classes is normally distributed. This allows the confidence intervals of land use accuracy assessments to be estimated through the use of a normal approximation of the binomial distribution. Given the large sample size, this approach is recommended by Cochran [50], and was adopted, for instance, by Card [49], EEA [44], and Carrão et al. [51]. By applying the detailed formulas presented by Carrão et al. [51] to the validation figures reported in the contingency table from Caetano et al. [48], we were able to estimate the true total area of the industrial and commercial land use for each Portuguese region, and the respective variances. The unbiased estimate of the true total land use and the respective estimated variance for each region depends upon the confusion between land uses presented in the contingency table and the abundance in each region of land use classes with which the industrial and commercial classes are confused with. After estimating the variance, confidence intervals can be drawn for each region around the estimated true total area of the industrial and commercial land use. Because a normal approximation of the binomial distribution was adopted, the confidence interval associated with each estimated parameter (i.e. true industrial and commercial land use in each region) is symmetric.

The results for Portugal are depicted in Figure 3A. For each NUTS2 region, it shows the amount of industrial and commercial sites mapped in CLC, the estimates of the true area of industrial and commercial land use, and its 90% confidence interval. If we take as example the region PT16, we find that the amount of industrial and commercial sites reported in CLC approximates the

Table 6. TAE per country (%).

	Models						
Country	**Null**	**M1**	**M2**	**M3**	**M4**	**M5**	**M6**
Austria	35.5	31.9	31.4	27.7	30.4	29.5	31.4
Belgium	5.7	6.8	9.2	8.0	6.1	2.8	2.0
Germany	8.9	4.5	5.0	5.1	4.9	6.9	11.9
Denmark	9.6	3.6	3.7	2.3	3.7	6.2	5.7
Spain	17.7	7.1	7.8	6.3	8.7	10.9	13.0
France	8.4	3.3	3.3	4.2	4.0	3.7	4.5
Ireland	37.1	23.3	17.1	15.9	23.3	31.8	34.6
Italy	13.2	7.2	7.8	8.0	11.1	10.4	11.8
Luxembourg	23.7	18.0	17.3	3.0	17.9	15.3	17.4
Malta	4.3	11.3	12.0	19.6	7.9	13.7	8.3
Netherlands	17.2	5.4	7.3	9.4	7.2	11.5	11.0
Portugal	15.0	5.1	12.4	8.5	10.4	10.6	11.8
Slovenia	0.9	0.3	0.3	25.9	0.1	21.5	0.5

estimated true value, and that it is 90% likely to find the true value between 10.2 and 13.7 thousand hectares (based on the validation sample).

The uncertainty in the accounting of the industrial and commercial land use propagates to the land use intensity parameter. As such – and assuming that the GVA figures are correct, as discussed earlier –, the distribution of the land use

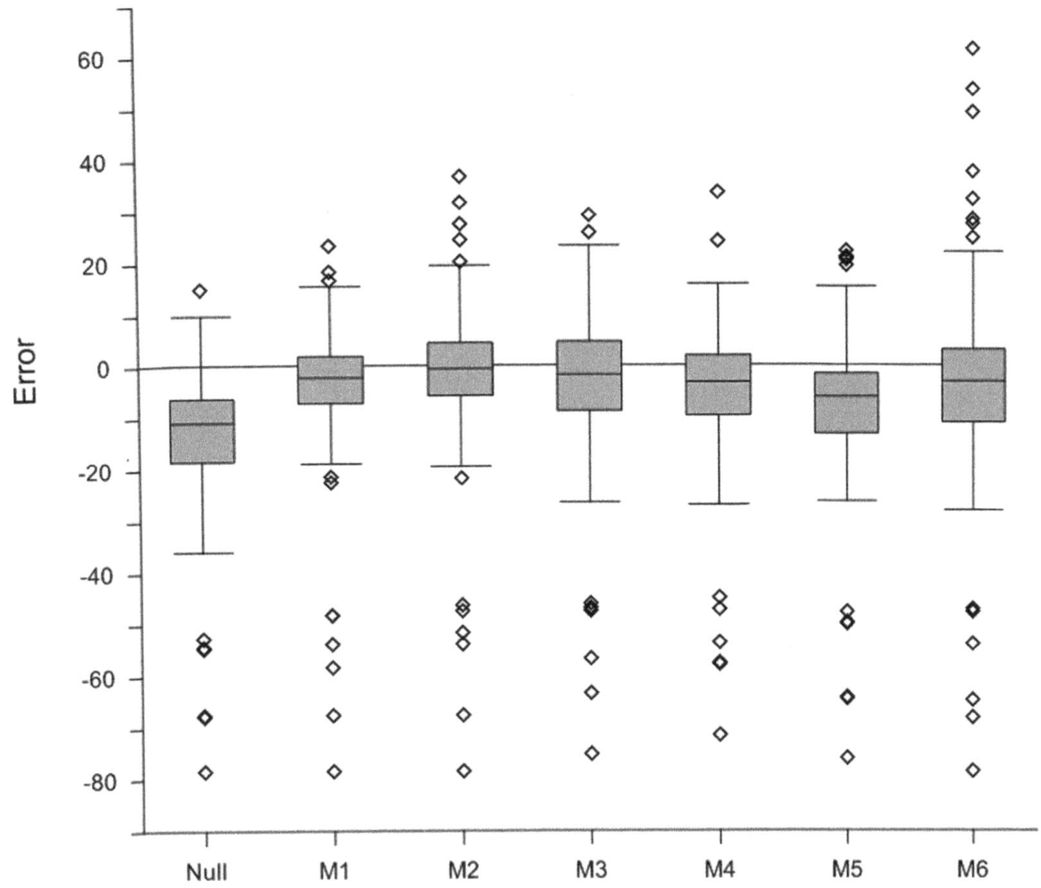

Figure 2. Distribution of the errors for each model (%).

Table 7. Validation results for Netherlands using different land use sources (CLC and CBS).

Model nr.	RD (%)		AAE (ha)		TAE (%)	
	CLC data	CBS data	CLC data	CBS data	CLC data	CBS data
Null	−17.16%	−6.93	1163	734	17.16	6.93
M1	−2.58%	1.44	364	214	5.38	2.02
M2	3.87%	2.24	495	282	7.31	2.66
M3	−9.13%	1.42	637	261	9.41	2.47
M4	−5.87%	−3.75	490	397	7.24	3.75
M5	−9.12%	−1.13	657	369	9.70	3.48
M6	−0.24%	−6.14	742	751	10.95	7.09

intensities for the Portuguese regions is depicted in Figure 3B. The regions PT11 and PT15 are particularly sensitive to the uncertainties regarding the true amount of industrial and commercial land use, while PT16, PT17 and PT18 are considerably less sensitive.

Summing up, in this section we looked at the uncertainty of the land use intensity in an indirect way: we first assessed the uncertainty of the industrial and commercial acreage reported in the land use map (figure 3A), and then looked at the impact of such uncertainty in the actual measurement of the land use intensity (figure 3B). Higher relative uncertainties of industrial and commercial acreage cause higher uncertainties on the true land use intensity. In the next section, we will combine the uncertainty of the land use intensity parameter (as just discussed) with the uncertainty of its future evolution.

Accounting for Nonstationarity in Land Use Intensity

To account for nonstationarity in land use intensity we analysed temporal changes in observed intensity figures for the Portuguese regions. We focus on this country as it allows us to assess the impact of nonstationarity in relation to the mapping error addressed in the previous section. Figure 3C shows the measured land use intensity of the Portuguese regions for the years 1990, 2000 and 2006, using data from Eurostat and the CLC time-series. We can observe an increase in intensity from 1990 to 2000, followed by a slight decrease between 2000 and 2006.

In studies by Chen et al. [29], Meng et al. [30], and Louw et al. [31], increases in land use intensity over time were observed for different study areas. Regional differences in land use intensity have been attributed to agglomeration economies, differences in economic structure and production characteristics. In addition, Louw et al. [31] argued that policy-related factors influence land use intensity, in particular by interfering with the supply of industrial land. The existing research, however, does not provide a solid framework to anticipate future changes in land use intensity for the European regions. To account for this uncertainty, we constructed two extreme trends for the future evolution of the land use intensity. One trend assumes that the land use intensity will continue to increase as observed in the period 1990–2006 for each respective region. The other assumes that the land use intensity will decrease as observed in the period 2000–2006. These two variants for the unknown evolution of the land use intensity are seen as bracketing the likely future values, thus providing a worst-case scenario for the uncertainty range of future land use intensity.

To illustrate the potential variance in estimates of future land use demand, we apply our demand model 3 to a scenario in which the economic output of regions is assumed to grow linearly until 2020, as observed in the period 1995–2008. Based on these premises, we constructed five possible trends of future industrial and commercial land use demand:

- Trend 1 (central estimation, constant land use intensity). The estimated true land use intensity for 2006 and for each region remains constant in time.
- Trend 2 (central estimation, increasing land use intensity). The estimated true land use intensity for 2006 increases in the same pace as observed in the period 1990–2006 in each region.
- Trend 3 (central estimation, decreasing land use intensity). The estimated true land use intensity for 2006 decreases in the same pace as observed in the period 2000–2006 in each region.
- Trend 4 (maximum estimation, decreasing land use intensity). The upper endpoint of the 90% confidence interval for the land use intensity for 2006 decreases in the same pace as observed in the period 2000–2006 in each region.
- Trend 5 (minimum estimation, increasing land use intensity). The lower endpoint of the 90% confidence interval for the land use intensity for 2006 increases in the same pace as observed in the period 1990–2006 in each region.

Figure 3D shows the resulting aggregated demand for industrial and commercial land use for Portugal. Trend 1 is a typical central and deterministic trajectory, which does not incorporate the uncertainties related to the land use mapping and assumes a stationary land use intensity over time. Trends 2 and 3 incorporate the uncertainty related to the future evolution of the land use intensity, thus providing lower and upper bounds for the future land use demand assuming we are certain about the measured land use in 2006. In 2020, the estimated value in trend 3 is circa 1.7 times higher than the one estimated in trend 2. Finally, trends 4 and 5 incorporate the uncertainty related to the future evolution of the land use intensity plus the uncertainty regarding the true acreage of industrial and commercial in 2006. In 2020, the estimated value in trend 4 is circa 2.7 times higher than the one estimated in trend 5.

These trends were constructed in order to translate the likely maximum variance of future demand for industrial and commercial land use, avoiding any underestimation of uncertainties. These results indicate that the uncertainty band for the projected land use is rather large, which is not surprising given the coarse resolution of the CLC and the unforeseen trajectories of future land use intensity.

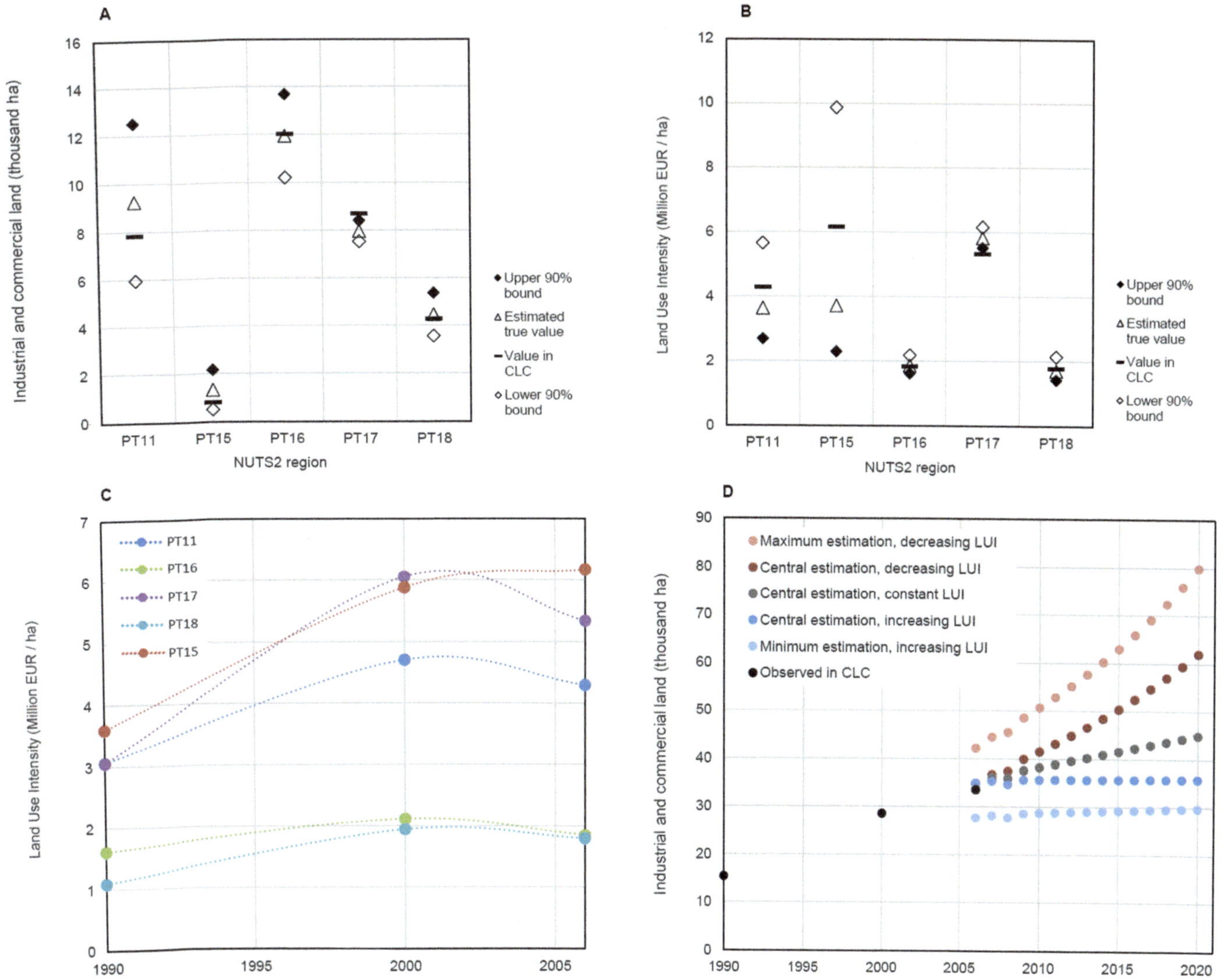

Figure 3. A: Industrial and commercial land use in 2006 per region, with 90% confidence interval. B: Land use intensity in 2006 per region, with 90% confidence interval. C: Land use intensities 1990–2006 per region. D: Scenarios of future demand for industrial and commercial land use (sum of all Portuguese regions).

Discussion and Conclusions

Estimating demand for future industrial and commercial land use is a challenging exercise. Very few attempts are known in the literature, and, when attempted, demands are estimated for small study areas, with detailed input data on land use and economy. In this study we aimed at developing and testing straightforward methods to estimate demand for new industrial and commercial land at continental scale with sub-national detail. The main difficulties concerned the input land use data, the CORINE Land Cover, which provides low spatial detail (minimum mapping unit of 25 hectares) and low thematic detail (the industrial, commercial and services land uses are all lumped together in one single land use class). The CLC, however, has key advantages: it provides a times-series (1990, 2000, 2006) and was designed for temporal and spatial consistency. Time series of consistent, more detailed national land use maps are scarce, the presented Dutch case being an exception. In effect, detailed national land use datasets are often not comparable (between years and between countries) due to different nomenclatures and mapping protocols. On the other hand, the drivers of the development of industrial, commercial and services land uses are difficult to grasp. Does economic development lead to spatial impacts or the other way around? Or, is there a more complex non-linear interaction between land use change and regional economic performance? These questions, although legitimate and pertinent, were not the focus of this study.

Given the high correlation between the economic output of a region and its total industrial, commercial and services land use, we started out assuming a direct and linear relationship between those two variables. Four methods based on land use intensity measures were devised and compared to simple trend extrapolation techniques in a case study for South and Western European countries. All models were, in addition, compared with the null model, which assumes no land use change during the validation time interval. The models were calibrated using information for the period 1990–2000 and then used to estimate the observed industrial and commercial land use in 2006, as reported in the CLC.

All models performed substantially better than the null model, which indicates that any of the devised modelling approaches is better than not modelling at all. However, none of the land use intensity approaches consistently outperformed the linear trend. Results seem to indicate that simpler assumptions to estimate industrial and commercial land return overall higher accuracies at least for short-term estimations. Nonetheless, by analysing the validation results at the country level we cannot discard that approaches based on land use intensity measures yield superior accuracies for many regions and even whole countries (see table 6). It seems, indeed, that there is not a one best approach for the entire set of tested countries.

The linear extrapolation has slightly produced less error dispersion when compared to other methods, despite the fact that it does not integrate any actual independent driver of land use change. As the estimation period extends, the high performances of the linear extrapolation model may phase out quicker than those of any model which relies on the economic activity as a proxy for land use changes. While for short-term estimations the linear extrapolation might actually produce very plausible overall estimations, for the medium and long term estimations, simple linear trends are thus conceptually unacceptable.

It is worth exploring the reasons why approaches based on intensity measures did not work as well as one could have expected at the start of this study. Even though GVA and total industrial and commercial land use are highly correlated at NUTS2 level, it cannot be inferred that change in GVA and change in land use are equally correlated. Changes in the economic structure will certainly lead to changes in the ratio between economic output and required land. On top of that we can expect time-lags between economic developments and their spatial impact and vice-versa. Old industrial sites (brownfields) may wait decades for redevelopment and thus remain present in the landscape long after their economic activities cease. While on the other hand expected economic development may lead to the construction of new offices and business estates that may remain empty for years. As mentioned earlier, complex issues like these call for more in depth study of the spatial development of industrial and commercial land use and its interaction with the underpinning economic drivers. Finally, the methods based on regional and sector specific land use intensity did not perform better than those based only on regional specific land use intensity. The failure to obtain better results for these particular methods may be at least partially explained by the weaknesses in the correspondences between a) the NACE classification, in which the GVA data is based, b) the CLC nomenclature, and c) the nomenclature of the two national land use datasets. All these three elements were required, first to disaggregate the sectorial composition of CLC class 'industrial and commercial units', and second, to estimate regional and sector-specific land use intensities. In fact, the assumptions made when coupling different nomenclatures may have led to uncertainties and errors that propagated to the final results of models 5 and 6.

The use of static land use intensities, as measured in the calibration years, was yet another drawback. In fact, the assumption of stable land use intensities has contributed to the overall error of these approaches. For the model 5 in particular, a factor ω was introduced, allowing for change in the land use intensities. However, in this case study, the factor was set to 1 in order to have a neutral effect on the land use intensities. Later, by making ω_s a function of the observed changes in the sectorial land use intensities between 1990 and 2000, we observed a substantial increase in the accuracy of the model 5. The total relative difference decreased to −4.78%, the average absolute error decreased to 572 ha, and the total absolute error dropped to 6.51%, thus making M3 and M5 very close in terms of overall accuracy. This demonstrates how important it is to account for temporal changes in the land use intensities, rather than keeping them static. However, the study of the changes in land use intensities, their trends and drivers, is yet to be made. Unfortunately, the available data at European level is yet insufficient for an appropriate assessment. Finally, even if more detailed land use maps are available for some European countries, consistent time-series are still lacking.

Another intriguing aspect stood out from the validation results. All models have underestimated the amount of industrial and commercial land use in 2006 (table 5). We tested the hypothesis that the observed consistent underestimation was at least partially a consequence of issues related to the input land use data from CLC. By applying all six methods to the Dutch regions using a more detailed national land use source, substantial reduction of deviations was observed for all models, and the resulting relative differences between known and estimated land use were much less biased, with models 1, 2 and 3 actually producing slight overestimations (see table 7). These results, although referring to a small portion of the entire case study, demonstrate how sensitive the methods are to the detail and accuracy of the input land use data.

Finally, to illustrate the practical application of the land use intensity-based methods, we used model 3 (best performer among all the tested models) to project industrial and commercial land use demand, given a hypothetical scenario of linear economic growth up to 2020 for the Portuguese NUTS2 regions. While implementing this 'forecasting' exercise, we identified the main sources of uncertainty related to the model used. The land use intensity parameter itself, which was measured for the year 2006 using the CORINE Land Cover and economic statistics from Eurostat, was found to be uncertain due to inaccuracies in the land use mapping. Based on a statistically sound validation of the CLC 2006 for Portugal, it was possible to draw a 90% confidence interval around the land use intensity for each region. In addition, we proposed two extreme scenarios for the evolution of the regional land use intensities, based on past trends. From one single hypothetical economic scenario, we arrived to 5 possible trajectories of future industrial and commercial land use demand, confirming that the uncertainties can be substantial.

Despite the limitations herein summarized, we argue that straightforward approaches, such as the ones based on land use intensity measures were lacking, and are relevant and suitable for large study areas, where data are limited. Whereas the uncertainties of these methods could be narrowed in part by using more detailed and consistent land use time series data, the uncertainties related to economic forecasts will remain and are intrinsic to the source of the forecast (i.e. economic models). The likely future evolution of regional land use intensities remains unknown for the most part, and more detailed studies are needed to grasp the underpinning factors. However, the dynamic change in land use intensities could be addressed, for instance, as proposed in equation 11, or through any other suitable variant. The ω factor can be the result of a calibration process or used as a 'policy parameter' in the context of scenario analyses and ex-ante impact assessment of policies. In sum, the proposed methods allow the generation of sensible and scenario-dependent results, on developments of industrial and commercial land uses across regions, which are linked to macro-economic models. As for the limitations and uncertainties, they should be acknowledged and dealt with transparency, as in any other modelling exercise.

Acknowledgments

The authors would like to thank Hugo Carrão for the valuable advice regarding the estimation of unbiased area of industrial and commercial land use and the respective estimated variance, based on CORINE Land Cover data. The authors are also very grateful for the constructive comments received from the anonymous referees, who considerably contributed to expand and improve the manuscript.

Author Contributions

Conceived and designed the experiments: FBS EK VD CL. Performed the experiments: FBS VD. Analyzed the data: FBS EK VD. Wrote the paper: FBS. Reviewed the manuscript: EK VD CL.

References

1. Esteban J (2000) Regional convergence in Europe and the industry mix: A shift-share analysis. Reg Sci Urban Econ 30: 353–364.
2. Brülhart M, Traeger R (2005) An Account Of Geographic Concentration Patterns In Europe. Reg Sci Urban Econ 35: 597–624.
3. Ezcurra R, Pascual P (2007) Spatial disparities in productivity in Central and Eastern Europe. East Europ Econ 45: 5–32.
4. Brülhart M (2001) Evolving Geographical Concentration of European Manufacturing Industries. Review of World Economics 137: 215–243.
5. Cohen J, Paul C (2005) Agglomeration economies and industry location decisions: the impacts of spatial and industrial spillovers. Reg Sci Urban Econ 35: 215–237.
6. Tregenna F (2009) Characterising deindustrialisation: An analysis of changes in manufacturing employment and output internationally. Cambridge J Econ 33: 433–466.
7. Mulatu A, Gerlagh R, Rigby D, Wossink A (2010) Environmental Regulation and Industry Location in Europe. Environ Resour Econ 45: 459–479.
8. Raspe O, van Oort F (2011) Growth of new firms and spatially bounded knowledge externalities. Ann Reg Sci 46: 495–518.
9. De Vor F (2011) The Impact and Performance of Industrial Sites: Evidence from the Netherlands. Doctoral thesis. Book no. 509. Tinbergen Institute Research Series. Thela Thesis and Tinbergen Institute.
10. Koomen E, Rietveld P, De Nijs T (2008) Modelling land-use change for spatial planning support. Ann Reg Sci 42: 1–10.
11. Verburg PH, Overmars KP (2009) Combining top-down and bottom-up dynamics in land use modeling: Exploring the future of abandoned farmlands in Europe with the Dyna-CLUE model. Landsc Ecol 24: 1167–1181.
12. Koomen E, Borsboom-van Beurden J (2011) Land-use modeling in planning practice. Geojournal library 101. Heidelberg: Springer. 214 p.
13. Rounsevell MDA, Pedroli B, Erb KH, Gramberger M, Busck AG, et al. (2012) Challenges for land system science. Land Use Policy 29: 899–910.
14. Veldkamp A, Lambin EF (2001) Predicting land-use change. Agric Ecosyst Environ 85: 1–6.
15. Koomen E, Stillwell J, (2007) Modeling land-use change. In: Koomen E., Stilwell J, Bakema A, Scholten H, editors. Modelling Land-Use Change: progress and applications. Dordrecht: Springer. 1–21.
16. Batty M (1997) Cellular Automata and Urban Form: A Primer. J Am Plann Assoc 63: 266–274.
17. Batty M, Xie Y, Sun Z (1999) Modeling urban dynamics through GIS-based cellular automata. Comput Environ Urban Syst 23: 205–233.
18. Verburg P, Kok K, Pontius GR, Veldkamp A (2006) Modeling Land-Use and Land-Cover Change. In: Lambin E, Geist H, editors. Land-Use and Land-Cover Change: Local Processes and Global Impacts. The IGBP Series, Heidelberg: Springer-Verlag. 117–135.
19. Heistermann M, Müller C, Ronneberger K (2006) Land in sight? Achievements, deficits and potentials of continental to global scale land-use modeling. Agric Ecosyst Environ 114: 141–158.
20. Lavalle C, Baranzelli C, Batista e Silva F, Mubareka S, Rocha Gomes C, et al. (2011) A high resolution land use/cover modelling framework for Europe. ICCSA 2011, Part I, Lect Notes Comput Sc 6782: 60–75.
21. Mubareka S, Estreguil C, Baranzelli C, Rocha Gomes C, Lavalle C (2013) A land-use based modelling chain to assess of the impacts of Natural Water Retention Measures on Europe's Green Infrastructure. Int J Geogr Inf Sci 27: 1740–1763.
22. Verburg PH, Eickhout B, Meijl H (2008) A multi-scale, multi-model approach for analyzing the future dynamics of European land use. Ann Reg Sci 42: 57–77.
23. Hoymann J (2012) Quantifying demand for built-up area – a comparison of approaches and application to regions with stagnating population. J Land Use Sci 7: 67–87.
24. Reginster I, Rounsevell M (2006) Scenarios of future urban land use in Europe. Environ Plann B Plann Des 33: 619–636.
25. Seto KC, Fragkias M, Güneralp B, Reilly MK (2011) A Meta-Analysis of Global Urban Land Expansion. PLoS One 6: e23777.
26. Beckers P, Schuur J (2013) The Future Demand for Industrial Sites in the Netherlands: Is Employment a Good Predictor? Reg Stud: in press.
27. Lambin E, Meyfroidt P (2011) Global land use change, economic globalization, and the looming land scarcity. Proc Natl Acad Sci USA 108: 3465–3472.
28. Huang D, Wan W, Dai T, Liang J (2011) Assessment of Industrial Land Use Intensity: A Case Study of Beijing Economic-technological Development Area. Chin Geogr Sci 21: 222–229.
29. Chen SY, Liu YL, Chen CF (2007) Evaluation of Land-Use Efficiency Based on Regional Scale: A Case Study in Zhanjiang, Guangdong Province. Journal of China University of Mining and Technology 17: 215–219.
30. Meng Y, Zhang FR, An PL, Dong ML, Wang ZY, et al. (2008) Industrial land-use efficiency and planning in Shunyi, Beijing. Landsc Urban Plan 85: 40–48.
31. Louw E, van der Krabben E, van Amsterdam H (2012) The Spatial Productivity of Industrial Land. Reg Stud 46: 137–147.
32. Forrester JW (1961) Industrial dynamics. Pegasus Communications. 479 p.
33. Forrester JW (1969) Urban Dynamics. Pegasus Communications. 299 p.
34. Luo G, Yin C, Chen X, Xu W, Lu L (2010) Combining system dynamic model and CLUE-S model to improve land use scenario analyses at regional scale: A case study of Sangong watershed in Xinjiang, China. Ecological Complexity 7: 198–207.
35. Zheng XQ, Zhao L, Xiang WN, Li N, Lv LN, et al. (2012) A coupled model for simulating spatio-temporal dynamics of land-use change: A case study in Changqing, Jinan, China. Landsc Urban Plan 106: 51–61.
36. Wu Y, Zhang X, Shen L (2011) The impact of urbanization policy on land use change: A scenario analysis. Cities 28: 147–159.
37. Lauf S, Haase D, Seppelt R, Schwarz N, (2012) Simulating demography and housing demand in an urban region under scenarios of growth and shrinkage. Environ Plann B Plann Des 39: 229–246.
38. Fragkias M, Geoghegan J (2010) Commercial and industrial land use change, job decentralization and growth controls: a spatially explicit analysis. J Land Use Sci 5: 45–66.
39. Jansson T, Bakker M, Boitier B, Fougeyrollas A, Helming J, et al. (2008) Cross sector land use modeling framework. In: Helming K, Pérez-Soba M, Tabbush P, editors. Sustainability Impact Assessment of Land Use Change. Heidelberg: Springer. 159–180.
40. Le Mouël P, Boitier B, Zang AN, Chevallier C, Zagamé P, et al. (2009) NEMESIS adapted to SENSOR sectors, extension to EU-25, and inclusion of land supply module; forecast simulation of baseline scenarios and policy cases. In: Helming K, Wiggering H, editors. SENSOR Report Series 2009. 80 p.
41. Büttner G, Feranec G, Jaffrain G (2006) CORINE Land Cover Nomenclature Illustrated Guide – Addendum 2006. Unpublished Report. 84 p.
42. Pontius RG, Huffaker D, Denman K (2004) Useful techniques of validation for spatially explicit land-change models. Ecol Model 179: 445–461.
43. Pontius GR, Malanson J (2005) Comparison of the structure and accuracy of two land change models. Int J Geogr Inf Sci 19: 243–265.
44. European Environment Agency (2006) The Thematic Accuracy of CORINE Land Cover 2000: Assessment Using LUCAS (Land Use/Cover Area Frame Statistical Survey). Technical Report 7/2006. Copenhagen: European Environment Agency. 85 p.
45. Batista e Silva F, Lavalle C, Koomen E (2013) A procedure to obtain a refined European land use/cover map. J Land Use Sci 8: 255–283.
46. European Environment Agency (2007) CLC2006 Technical Guidelines. Technical Report 17/2007. Copenhagen: European Environment Agency. 67 pp.
47. National Research Council (2013) Advancing Land Change Modeling: Opportunities and Research Requirements. Washington DC: The National Academies Press. 135 p.
48. Caetano M, Araújo A, Nunes A, Nunes V, Pereira M (2009) Accuracy assessment of the CORINE Land Cover 2006 map of Continental Portugal. Technical report. Lisbon: Instituto Geográfico Português. 31 p.
49. Card D (1982) Using known map category marginal frequencies to improve estimates of thematic map accuracy. Photogramm Eng Remote Sensing 48: 431–439.
50. Cochran WG (1977) Sampling Techniques. New York: John Wiley. 448 p.
51. Carrão H, Araújo A, Gonçalves P, Caetano M (2010) Multitemporal MERIS images for land-cover mapping at a national scale: a case study of Portugal. Int J Remote Sens 31: 2063–2082.

12

Land Use and Land Cover Change Dynamics across the Brazilian Amazon: Insights from Extensive Time-Series Analysis of Remote Sensing Data

João M. B. Carreiras[1,2]*, Joshua Jones[3], Richard M. Lucas[3,4], Cristina Gabriel[1]

1 Tropical Research Institute (IICT), Lisboa, Portugal, **2** National Centre for Earth Observation (NCEO), Centre for Terrestrial Carbon Dynamics (CTCD), University of Sheffield, Sheffield, South Yorkshire, United Kingdom, **3** Department of Geography and Earth Sciences, Aberystwyth University, Aberystwyth, Ceredigion, United Kingdom, **4** Centre for Ecosystem Science, School of Biological, Earth and Environmental Sciences, University of New South Wales, Sydney, New South Wales, Australia

Abstract

Throughout the Amazon region, the age of forests regenerating on previously deforested land is determined, in part, by the periods of active land use prior to abandonment and the frequency of reclearance of regrowth, both of which can be quantified by comparing time-series of Landsat sensor data. Using these time-series of near annual data from 1973–2011 for an area north of Manaus (in Amazonas state), from 1984–2010 for south of Santarém (Pará state) and 1984–2011 near Machadinho d'Oeste (Rondônia state), the changes in the area of primary forest, non-forest and secondary forest were documented from which the age of regenerating forests, periods of active land use and the frequency of forest reclearance were derived. At Manaus, and at the end of the time-series, over 50% of regenerating forests were older than 16 years, whilst at Santarém and Machadinho d'Oeste, 57% and 41% of forests respectively were aged 6–15 years, with the remainder being mostly younger forests. These differences were attributed to the time since deforestation commenced but also the greater frequencies of reclearance of forests at the latter two sites with short periods of use in the intervening periods. The majority of clearance for agriculture was also found outside of protected areas. The study suggested that a) the history of clearance and land use should be taken into account when protecting deforested land for the purpose of restoring both tree species diversity and biomass through natural regeneration and b) a greater proportion of the forested landscape should be placed under protection, including areas of regrowth.

Editor: Bruno Hérault, Cirad, France

Funding: This study was supported by the "REGROWTH-BR - Remote sensing of regenerating tropical forests in Brazil: mapping and retrieving biophysical parameters" project (Ref. PTDC/AGR-CFL/114908/2009), funded by the Foundation for Science and Technology (FCT, http://www.fct.pt/index.phtml.en), Portugal. J.M.B. Carreiras' work was partially funded by NERC National Centre for Earth Observation (Ref: R8/H12/82). The funders had no role in study design, data collection and analysis, decision to publish, or preparation of the manuscript.

Competing Interests: The authors have declared that no competing interests exist.

* Email: j.carreiras@sheffield.ac.uk

Introduction

Humans are increasingly changing the state and dynamics of the Earth system, affecting processes within and between the biosphere, hydrosphere and atmosphere. Climate change is a recognized consequence of such changes and would be occurring at a faster rate if several ecosystem services at the global scale were absent [1]. For example, less than half of the total amount of carbon dioxide (CO_2) released to the atmosphere each year remains there because of removal by terrestrial and ocean carbon (C) sinks [2]. Losses of biodiversity from terrestrial ecosystems have also been substantial.

The role of terrestrial ecosystems in the global C cycle had been widely recognized in the literature (e.g., [3]), with particular recognition given to the pan-tropical belt [4,5]. In the past two decades, two main international conventions have sought to establish mechanisms aimed at stabilizing greenhouse gases (GHG) concentrations in the atmosphere (the United Nations Framework Convention on Climate Change; UNFCCC) and protecting biological diversity (Convention on Biological Diversity; CBD). Specifically, the alarmingly high rate of tropical land use and land cover change (LULCC) [6] and resulting biodiversity loss with further severe consequences for ecosystem function and structure [7] has driven the UNFCCC to establish several investment mechanisms and market based C transactions. These are related to the enhancement of forest C stocks and the decrease of deforestation and forest degradation, while promoting sustainable development in developing countries (UNFCCC Non-Annex I countries). Such mechanisms include the UNFCCC Clean Development Mechanism (CDM) initiative and the post-Kyoto Protocol Reduced Emissions from Deforestation and Degradation (REDD+) program (e.g., [8]). These efforts are proving successful in many regions leading to greater conservation of the intact forests of the Neotropical, Afrotropical, Australasia and Indo-Malay biogeographical regions [9] and the carbon and biodiversity they contain. However, many of the deforested and degraded lands of the tropics are also capable of supporting forests and hence, in addition, there is potential to restore some of the ecosystem values (e.g., carbon amounts and biodiversity) lost through previous disturbances [10,11]. This capacity depends in part, however, on the history of forest clearance and land use that has occurred and hence a clearer understanding of the potential of

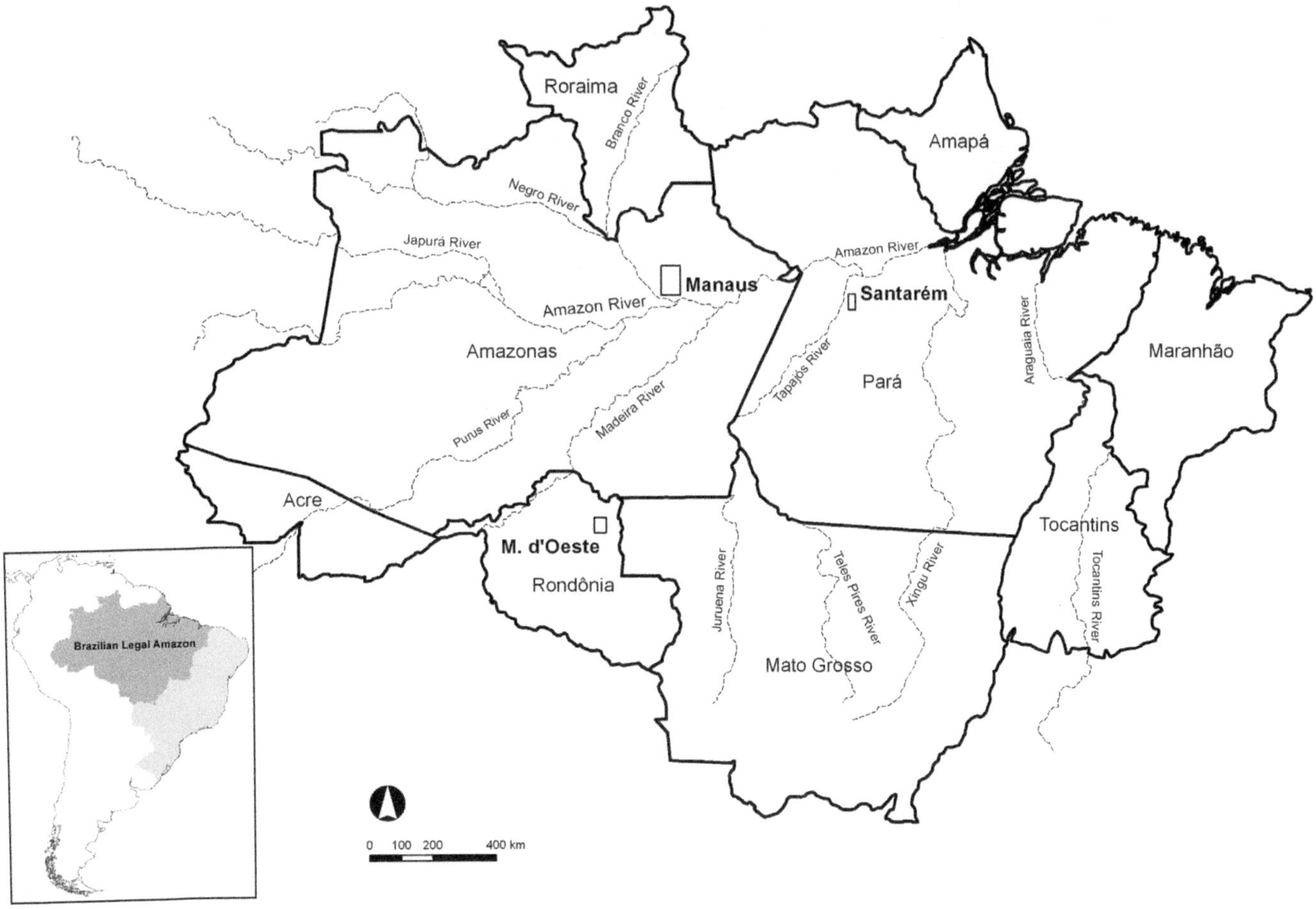

Figure 1. Location of the three study sites. Location of the three study sites: Manaus, Santarém and Machadinho d'Oeste (M. d'Oeste), and the insertion (bottom left) of the Brazilian Legal Amazon (dark grey) in Brazil (light grey) and South American continent. Also shown are the major rivers included in the Amazon basin.

deforested areas to recover ecosystems is needed. This study addresses these issues by focusing on sites in the Brazilian Legal Amazon (BLA) that have experienced deforestation and variable histories of subsequent land use.

The BLA covers an area of approximately 5,000,000 km^2 and consists primarily of primary tropical forest (rain and seasonal forest) [12]. Brazil's National Institute for Space Research (Instituto Nacional de Pesquisas Espaciais, INPE) has, since 1988, undertaken annual mapping of deforestation within the BLA using remote sensing data [13], which has been highly variable across the region. For example, an average deforestation rate of 21,050 km^2 yr^{-1} was reported between 1977 and 1988, which decreased subsequently to 11,030 km^2 yr^{-1} up to 1991; from 1991 to 2004 the deforestation rate increased to 27,772 km^2 yr^{-1} (reaching a record high of 29,059 km^2 yr^{-1} from 1994 to 1995) and then decreased to a record low of 4,571 km^2 yr^{-1} in 2012 [13]. Currently, much of the deforested area is under agricultural use (following disturbance mainly with slash-and-burn practices) but in many areas, the land has been abandoned and frequently supports forests at different stages of regeneration [14].

Several estimates of the deforested land occupied with regeneration have been generated, mainly from interpretation of remote sensing data, but also using transition modeling. Data from the National Oceanic and Atmospheric Administration (NOAA) Advanced Very High Resolution Radiometer (AVHRR) from 1988–1991 were used by Stone et al., 1994 [15] to generate a land cover map of South America, and subsequently Schroeder and Winjum, 1995 [16] used this dataset to estimate the extent of forest regeneration in the BLA at $\sim 151 \times 10^3$ km^2. Fearnside, 1996 [17], using a matrix of annual transition probabilities, estimated that approximately 48% ($\sim 195 \times 10^3$ km^2) of the deforested landscape in 1990 supported forest regeneration. Lucas et al., 2000 [18] using NOAA AVHRRR data estimated that $\sim 158 \times 10^3$ km^2 supported some type of forest regeneration in the period 1991–1994. Cardille and Foley, 2003 [19] used a matrix of annual transition probabilities to estimate that 36% ($\sim 91 \times 10^3$ km^2) of the area deforested between 1980 and 1995 in the entire Amazon river drainage basin was in some stage of secondary succession forest. Carreiras et al., 2006 [20] exploited a time-series of 12 monthly composite images of the year 2000, derived from the *Satellite Pour l'Observation de la Terre* (SPOT-4) VEGETATION sensor and estimated that, in 2000, $\sim 140 \times 10^3$ km^2 of land was occupied by regenerating forests. Neeff et al., 2006 [21] used large-area land cover maps derived from remote sensing datasets to generate estimates of regrowth extent within the BLA and concluded that regrowth increased from $\sim 29 \times 10^3$ km^2 in 1978 to $\sim 161 \times 10^3$ km^2 in 2002. Whilst providing information on the extent of regenerating forests at certain points in time, the dynamics of regenerating forests and the land on which they occupy has only been attempted using remote sensing by a few studies (e.g., [22]). However, such information is needed across large areas as regenerating forests may represent one of the prime

Table 1. Landsat Multi-spectral Scanner (MSS), Thematic Mapper (TM) and Enhanced TM (ETM+) data available for the three Amazonian sites.

Manaus (Path 231, Row 62)		Santarém (Path 227, Row 62)		Machadinho d'Oeste (Path 231, Row 67)	
Date[1]	Sensor	Date[1]	Sensor	Date[1]	Sensor
19730707	MSS	19840824	TM	19840617	TM
19770731	MSS	19850726	TM	19860810	TM
19780822	MSS	19860729	TM	19870712	TM
19790703	MSS	19870716	TM	19890717	TM
19830709	MSS	19880803	TM	19900618	TM
19850604	TM	19890822	TM	19910925	TM
19880815	TM	19900809	TM	19940816	TM
19890802	TM	19910711	TM	19950803	TM
19910808	TM	19931020	TM	19960704	TM
19920607	TM	19951010	TM	19970723	TM
19941019	TM	19960825	TM	19980624	TM
19950920	TM	19970727	TM	19990729	TM
19960720	TM	19980815	TM	20010803	TM
19990713	TM	19990903	TM	20030724	TM
20010827	ETM+	20000905	TM	20050713	TM
20020830	ETM+	20010916	ETM+	20060716	TM
20030809	TM	20030829	TM	20070703	TM
20060716	TM	20050701	TM	20080806	TM
20070804	TM	20060805	TM	20090809	TM
20080806	TM	20070621	TM	20100625	TM
20090910	TM	20081130	TM	20110612	TM
20100727	TM	20090712	TM		
20110831	TM	20100629	TM		

[1]Date format is yyyymmdd.

mechanisms by which biomass, biological diversity and ecosystem services can be restored [23–25].

The aim of this study was to establish the extent to which the age class distribution of regenerating forests was determined by the history of deforestation and subsequent use and management of the land, thereby giving insight into the potential for forest restoration and conservation into the future. This was achieved by comparing Landsat sensor data classified into the broad amalgamated categories of mature forest, non-forest (i.e., agriculture, including stock pasture) and secondary forest over decadal periods and time-separated by a maximum of four years. Three study areas were chosen to perform the analysis: Manaus (Amazonas state), Santarém (Pará state) and Machadinho d'Oeste (Rondônia state). Several measures were used to compare the three areas since the inception of widespread agricultural practices in the region (1970s for Manaus and Santarém and 1980s for Machadinho d'Oeste) up to the present: Deforestation (conversion from mature forest to non-forest or from secondary forest to non-

Table 2. Error matrix obtained from the accuracy assessment of the 2007 land cover map of the Manaus site (MF – mature forest, NF – non-forest, SF – secondary forest).

	Classified				
Reference	**MF**	**NF**	**SF**	**Total**	**Omission error**
MF	186	4	10	200	0.070
NF	8	185	7	200	0.075
SF	11	2	187	200	0.065
Total	205	191	204	600	
Commission error	0.093	0.031	0.083		

The overall accuracy was 0.930.

Table 3. Error matrix obtained from the accuracy assessment of the 2010 land cover map of the Santarém site (MF – mature forest, NF – non-forest, SF – secondary forest).

	Classified				
Reference	**MF**	**NF**	**SF**	**Total**	**Omission error**
MF	196	0	4	200	0.020
NF	0	198	2	200	0.010
SF	23	11	166	200	0.170
Total	219	209	172	600	
Commission error	0.105	0.053	0.035		

The overall accuracy was 0.93.

forest) and regrowth rates (conversion from non-forest to regeneration), the age of secondary forest, the period of active land use prior to abandonment to regeneration, and the frequency of clearance. The study builds on that of Prates-Clarke et al., 2009 [38] in that the time-series was extended from 2003 to 2011 for Manaus and Santarém and a new time-series was generated for Machadinho d'Oeste.

Study Areas

For all three sites (Manaus, Santarém and Machadinho d'Oeste; Figure 1), the natural vegetation prior to disturbance was primary (undisturbed) *terra firme* (i.e., non-flooded) forest. In each case, the canopy heights averaged between 25 and 35 m, with emergent trees exceeding 50 m at the more productive sites (e.g., Manaus). Species diversity is also high, often exceeding 225 species ha^{-1} [26], with an estimated mean above ground biomass of approximately 270 Mg ha^{-1} [27].

The first site, north of Manaus, the capital of Amazonas state, occupied an area of 5,042 km^2 (2°33′11″ S, 60°5′7″ W, Figure 1) and included the Instituto Nacional de Pesquisas da Amazônia (INPA) and the Smithsonian Institution (SI) Biological Dynamics of Forest Fragments Project (BDFFP) research sites (established in 1979) [28], the Adolfo Ducke and Walter Egler forest reserves, and other state and federal environmental protection areas. Significant deforestation began in the area following construction in the early 1970s of the BR-174 highway connecting Manaus with Boa Vista in Roraima. Most deforestation activity occurred either side of the highway, with this fuelled primarily by agricultural expansion. Recognition of the opportunity to study the impacts of clearance and forest fragmentation on biodiversity as well as ecosystem services spurred the preservation of fragments of various dimensions (typically less than 2 km^2) prior to felling of the surrounding tropical forest. Many of these fragments were, however, reconnected because of rapid regrowth of forests following abandonment of many clearances from the mid-1980s onwards. Up until the present, a substantial amount of research has been conducted, with these informing future management strategies across the Amazon [28].

The second site (3°10′5″ S, 54°55′42″ W) covered an area of 1,118 km^2 and was located approximately 80 km to the south of Santarém, the second main city in the Pará state (Figure 1). This study area is within the Tapajós National Forest, between the Tapajós River and the BR-163 highway connecting Santarém to the state capital of Mato Grosso, Cuiabá. This conservation unit was created in 1974 and has been recognized as a model for sustainable forest management, including logging activities, at levels ranging from individuals to communities (e.g., [29,30]). Compared to Manaus, forests are less dense with a mixed arrangement of broadleaves and palm trees such as Açaí (*Euterpe oleracea*) and Babaçu (*Orbignya phalerata*) being commonplace [31].

The third site (9°32′56″ S and 62°6′27″ W) occupied an area of 1,780 km^2 and was located mostly within the Machadinho d'Oeste municipality, Rondônia state. This municipality originated from the former Machadinho settlement project, deployed by the Brazilian federal government (through the *Instituto Nacional de Colonização e Reforma Agrária*, INCRA) in 1982 as part of the POLONORDESTE Program [32]. As such, the majority of its inhabitants are dependent on agriculture for subsistence [33]. Open rain forest is the dominant vegetation type in the municipality [32]. According to data from the Brazilian Ministry

Table 4. Error matrix obtained from the accuracy assessment of the 2010 land cover map of the Machadinho d'Oeste site (MF – mature forest, NF – non-forest, SF – secondary forest).

	Classified				
Reference	**MF**	**NF**	**SF**	**Total**	**Omission error**
MF	200	0	0	200	0.000
NF	0	197	3	200	0.015
SF	19	37	144	200	0.280
Total	219	234	147	600	
Commission error	0.087	0.158	0.020		

The overall accuracy was 0.90.

Table 5. Area (ha) and relative incidence (RI, %) of perennial (PC) and tree (TC) crops with the mapped area of mature forest (MF), non-forest (NF) and secondary forest (SF).

	Manaus (2007)		Machadinho d'Oeste (2010)			
	TC		PC		TC	
Classified	Area (ha)	RI (%)	Area (ha)	RI (%)	Area (ha)	RI (%)
MF	139	0.0	31	0.1	0	0.0
NF	398	1.0	540	0.9	14	0.0
SF	4,237	4.4	1,323	2.1	3	0.0
Total	4,774		1,894		17	

Estimates were based on the land cover map closest to the date that was used to delineate crop types using very high resolution (VHR) imagery.

of the Environment (MMA), this site includes several conservation units, mainly extractive reserves, which were established in and after 1995.

The area of the Manaus site was approximately three to four times the size of the other two study areas for several reasons. Deforestation in Manaus started in the early 1970s and land conversion from primary forest to agriculture/pasture was much more scattered. By comparison, deforestation patterns were more concentrated at Santarém (either side of and in proximity to a highway) and Machadinho d'Oeste (a planned settlement). A larger area was also selected to encompass older deforestation within the BDFFP study area and more recent clearing towards the city of Manaus, thereby allowing a wide range of LULCC processes to be captured.

At all three sites, much of the history of deforestation and land use has been captured by optical (primarily Landsat) sensors during the dry season, when cloud cover is minimal. In Manaus, a moderately strong dry season occurs between June and October [28], in Santarém between May and October [34], and in Machadinho d'Oeste between April and November [32]. Rainfall is generally higher at Manaus (between 1,900 and 2,500 mm annually) and lower at Santarém and Machadinho d'Oeste with an annual average of 2000 mm. The mean annual temperatures at the sites range from 25°C to 26°C [32,34,35]. The distribution of forests types is determined in part by the topography and geology. The topography is moderately flat (up to 160 m elevation) north of Manaus and is divided by a large number of waterlines. Soils are nutrient-poor (i.e., Oxisols), with these being typical of many across the Amazon basin [36]. The area south of Santarém is also moderately flat, ranging from approximately 50 to 240 m, and consists of nutrient poor soils (i.e., Oxisols and Ultisols) [34,37]. The terrain is moderately undulated at Machadinho d'Oeste, ranging from approximately 90 to 370 m, and the dominant soil types are Alfisols, Oxisols, Ultisols, and Alluvial [32].

Data

For Manaus, Landsat Multi-spectral Scanner (MSS), Thematic Mapper (TM) and Enhanced Thematic Mapper Plus (ETM+) data were acquired between 1973 and 2011 (path 231, row 62; Table 1). The Landsat MSS data acquired between 1973 and 1983 were only available as hard-copy prints [38]. The remaining Landsat TM and ETM+ scenes (1985 onwards) were provided in digital format from both INPE and the United States Geological Service (USGS).

In the previous study of Prates-Clark et al., 2009 [38], scenes acquired between 1985 and 2003 were used to study forest regeneration in this area. In this study, the time-series was extended to 2011 using six additional annual scenes, which were relatively cloud free (<20%), or cloud was distributed in only certain areas of the image. In total, the time-series extended over an approximate 40-year period from 1973 (when the deforested area was minimal) to 2011. For Santarém, Prates-Clark et al., 2009 [38] again used Landsat TM and ETM+ imagery acquired between 1984 and 2003 (path 227, row 62), with most scenes unaffected by substantive cloud cover. The time-series was extended using Landsat TM data acquired for each year from 2005 to 2010. Coverage for 2011 was actively sought but no data with minimal cloud coverage were available. Whilst most changes at Santarém were associated with deforestation for agriculture and pasture, extensive wildfires damaged the surrounding forests in 1992 and 1997. The last fire episode was associated with the 1997–1998 El Niño Southern Oscillation (ENSO) event. The time-series classification for Machadinho d'Oeste had not been undertaken previously and so 21 new time-series digital images acquired between 1984 and 2011 were obtained from the USGS. Gaps in the time-series ranged from between one (70%) to approximately three years (5%). Cloud cover was minimal on all but two dates, although the vast majority of the deforested area was visible within these.

Methods

4.1. Image pre-processing

Landsat images acquired by the USGS were orthorectified and calibrated to units of spectral radiance (W m^{-2} sr^{-1} μm^{-1}) and then calibrated to top of atmosphere (TOA) reflectance using calibration factors and equations provided by Chander et al., 2009 [39]. Images acquired by INPE were processed to TOA reflectance and geometrically corrected using Environment for Visualizing Images (ENVI) software (Exelis Visual Information Solutions, Boulder, CO, USA), a third order polynomial and a nearest neighbor transformation [38]. Each Landsat image was then subsetted to encompass the main area of deforestation and the intersection of all images in the time series.

4.2. Image classification

As with Prates-Clark et al., 2009 [38], each digital image within the time-series for each site was classified into mature forest (MF), non-forest (NF), and secondary forest (SF), with the second category including crops (herbaceous and woody) and pastures. Areas of open water were also mapped and a common mask was applied to all dates in the time-series. Several classification approaches were utilized with the intention of providing the best

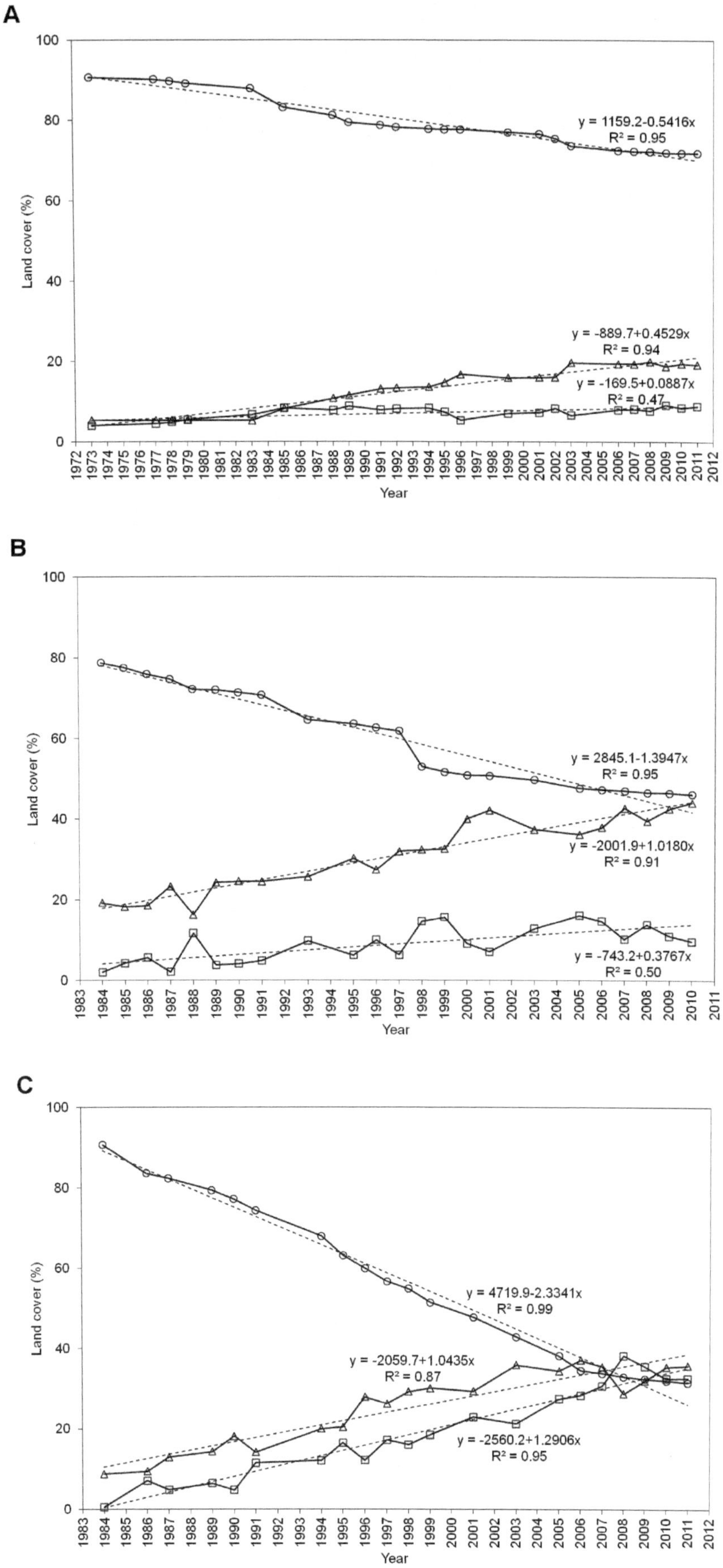
A
Land cover (%)
Year
y = 1159.2-0.5416x
R² = 0.95
y = -889.7+0.4529x
R² = 0.94
y = -169.5+0.0887x
R² = 0.47
B
Land cover (%)
Year
y = 2845.1-1.3947x
R² = 0.95
y = -2001.9+1.0180x
R² = 0.91
y = -743.2+0.3767x
R² = 0.50
C
Land cover (%)
Year
y = 4719.9-2.3341x
R² = 0.99
y = -2059.7+1.0435x
R² = 0.87
y = -2560.2+1.2906x
R² = 0.95

Figure 2. Land cover change (%) across the three selected sites. Land cover change (%) across the three selected sites; A) Manaus (1973–2011), B) Santarém (1984–2010), and C) Machadinho d'Oeste (1984–2011). Each dashed line represent the linear fit of the proportion of each land cover class as a function of the year; also showing are the corresponding equations and coefficient of determination (R^2).(○ - mature forest, □ - non-forest, Δ- secondary forest).

classification of the desired categories. At Manaus, and for digital images acquired before 2006, supervised classifications based on the minimum distance or maximum likelihood algorithms were applied (e.g., [40]), with regions of interest representing the main cover types defined around a number of target areas described on the basis of field observations or through reference to very high resolution data. For the Landsat TM images acquired between 2006 and 2011, an object oriented classification (e.g., [41]) was followed and involved a decision-rule classification applied within the eCognition software (Trimble Geospatial Imaging, Munich, Germany). The rule base method used data from the available Landsat sensor bands as well as data layers derived from spectral indices, namely the Normalized Difference Vegetation Index (NDVI) and Normalized Difference Water Index (NDWI). The classification of MF was refined with a cloud-free mask of the MF area obtained through classification of the most-cloud-free and recent image in the time-series. Images acquired on the 31st August, 2011, 29th June, 2010 and 12th June 2011 were used to generate the MF mask, on the assumption that the land cover type had remained as MF for the whole of the time-series. If MF was covered by cloud in earlier images of the time-series, this was subsequently classified as MF. At Manaus, SF were older compared to the other sites and the structure of the upper canopy of the regenerating forests led to a spectral response in the near and shortwave infrared channels that was similar to MF; hence, reference to images earlier in the time-series was necessary [18,22,42]. At Santarém, the Landsat data acquired between 1984 and 2003 were classified using a fuzzy logic approach that was applied to the original Landsat bands as well as derived fractional images (shade/moisture, vegetation and soil). Each pixel of the Landsat data acquired between 2005 and 2010 was classified into one of the three main land cover classes (described above) using the random forests machine learning classification algorithm (e.g., [43]) and training data. This algorithm has a number of options (e.g., [43]) that were tested for each date to obtain the best possible classifier. As with Santarém, all Landsat TM dates for Machadinho d'Oeste were classified using the random forests algorithm. In all cases, the objective was to obtain the best classification of NF, MF and SF for each year of observation.

4.3. Post-classification

The comparison of the time-series of classified images (as MF, NF, or SF) identified several cases where some pixels classified as SF or NF in a given date were classified as MF in the following date. As this sequence is not plausible, in-house Interactive Data Language (IDL) code (Exelis Visual Information Solutions, Boulder, CO, USA) was written to identify these cases by comparing the classification over two consecutive dates. Where SF or NF areas were classified as such in one year and as MF in the next, the MF class was reallocated to SF. These areas were not reclassified as NF because it was assumed that spectral confusion between NF and MF was unlikely. This procedure was similar to that undertaken by Roberts et al., 2002 [44], who also considered "disallowed transitions" between cover types.

4.4. Deforestation and regrowth rates

Annual rates of change from MF (or SF) to NF (deforestation) and from NF to SF (regeneration) were calculated for consecutive dates in the time-series using equations (1) and (2) [45]

$$r = \frac{1}{t_2 - t_1} \ln \frac{A_2}{A_1} \tag{1}$$

$$R = \frac{A_1 - A_2}{t_2 - t_1} \tag{2}$$

where r and R are the annual rate of change expressed in percentage per year and hectares per year respectively; A_1 and A_2 are the MF (or SF, in the case of deforestation occurring in regeneration areas) or NF cover areas (hectares) at time t_1 and t_2 respectively (time period). These rates were used to characterize the temporal evolution of change across the three sites, to detect events (primarily clearance) and describe general trends.

4.5. Age of secondary forest, period of active land use, and frequency of clearance

Algorithms were written and implemented in IDL to compare the classifications of MF, NF, and SF between dates. Subsequently, datasets relating to the history and dynamics of land use, namely the age of secondary forest (ASF), the period of active land use (PALU) prior to abandonment, and the frequency of clearance (FC) were generated for each site and for each year in which an image had been acquired. The ASF was estimated by summing the time (in years) that each pixel was occupied by SF since the last clearance event. However, when SF occurs in the first date of the time series, and it is not subsequently cleared, the ASF can only be considered as a minimum age as the exact date of land abandonment is not known. The PALU was defined as the difference (in years) between the time of initial forest clearance and the onset of regeneration. However, where reclearance of regenerating forest had occurred, the PALU was calculated by summing the period since the last reclearance event until the forest cover had re-established. The FC was estimated by summing the frequency of transitions from MF (or SF) to NF. For both the Manaus and Santarém study areas, fire scars were evident within some clearances by the lower near and shortwave infrared reflectance compared to the original vegetated surface [46]. However, many burned areas acquired a vegetated cover quite rapidly [47], and so only a partial fire history could be retained.

To better spatially represent and discuss the main results, classes of ASF, PALU, and FC were generated and used to assess their temporal evolution. Each metric was classified into three classes as to identify a lower, middle and upper interval range. However, these intervals do have some degree of subjectivity. ASF classes were defined as initial (≤5 years), intermediate (6–15 years) and advanced (≥16 years), adapted from Lucas et al., 2000 [18]. PALU classes were defined as short (≤2 years), medium (3–4 years) and long (≥5 years), with these representing various different crop or pasture cycles. FC classes were defined as low (1 time), medium (2 times) and high (≥3 times), representing several land use patterns in terms of the frequency of the deforestation -> agriculture/pasture -> abandonment -> re-growth temporal sequence.

Table 6. Land cover proportion (%) of mature forest (MF), non-forest (NF) and secondary forest (SF) in each year of the time-series at Manaus, Santarém and Machadinho d'Oeste.

Manaus				Santarém				Machadinho d'Oeste			
Year	**MF**	**NF**	**SF**	**Year**	**MF**	**NF**	**SF**	**Year**	**MF**	**NF**	**SF**
1973	91	4	5	1984	79	2	19	1984	91	1	9
1977	90	4	5	1985	78	4	18	1986	84	7	9
1978	90	5	5	1986	76	6	19	1987	82	5	13
1979	89	6	5	1987	75	2	23	1989	79	6	14
1983	88	7	5	1988	72	12	16	1990	77	5	18
1985	83	8	8	1989	72	4	24	1991	74	11	14
1988	81	8	11	1990	71	4	25	1994	68	12	20
1989	79	9	12	1991	71	5	24	1995	63	16	20
1991	79	8	13	1993	65	10	26	1996	60	12	28
1992	78	8	13	1995	64	6	30	1997	57	17	26
1994	78	8	14	1996	63	10	27	1998	55	16	29
1995	78	7	15	1997	62	6	32	1999	51	18	30
1996	78	5	17	1998	53	15	32	2001	48	23	29
1999	77	7	16	1999	52	16	33	2003	43	21	36
2001	77	7	16	2000	51	9	40	2005	38	27	34
2002	76	8	16	2001	51	7	42	2006	35	28	37
2003	74	7	20	2003	50	13	37	2007	34	31	35
2006	73	8	19	2005	48	16	36	2008	33	38	29
2007	72	8	19	2006	47	15	38	2009	32	36	32
2008	72	8	20	2007	47	10	43	2010	32	33	35
2009	72	9	19	2008	47	14	40	2011	32	33	36
2010	72	9	19	2009	46	11	43				
2011	72	9	19	2010	46	10	44				

Table 7. Summary of deforestation and regrowth rates between consecutive dates in the Manaus time-series (MF – mature forest; SF – secondary forest).

	MF annual rate of deforestation		SF annual rate of deforestation		Annual rate of deforestation		Annual rate of regrowth	
	(% yr^{-1})	(ha yr^{-1})	(% yr^{-1})	(ha yr^{-1})	(% yr^{-1})	(ha yr^{-1})	(% yr^{-1})	(ha yr^{-1})
Minimum	0.1	277	0.0	0	0.1	691	0.0	0
Maximum	2.9	13,451	27.7	8,453	3.8	17,936	133.9	24,738
Mean	0.7	3,123	6.4	3,231	1.3	6,355	34.1	6,017
Median	0.4	1,591	5.0	2,725	0.9	4,513	24.8	4,793
Inter-quartile range	0.4	2,172	5.1	3,006	1.3	5,881	31.9	6,156
Standard deviation	0.8	3,726	6.5	2,699	1.1	4,905	33.7	5,756

Table 8. Summary of deforestation and regrowth rates between consecutive dates in the Santarém time-series (MF – mature forest; SF – secondary forest).

	MF annual rate of deforestation		SF annual rate of deforestation		Annual rate of deforestation		Annual rate of regrowth	
	(% yr^{-1})	(ha yr^{-1})	(% yr^{-1})	(ha yr^{-1})	(% yr^{-1})	(ha yr^{-1})	(% yr^{-1})	(ha yr^{-1})
Minimum	0.4	235	0.5	194	0.5	448	10.5	832
Maximum	15.6	11,046	82.3	11,352	17.7	16,952	261.5	12,257
Mean	2.1	1,598	16.7	3,189	4.9	4,787	78.5	4,675
Median	1.2	963	10.7	2,769	4.0	4,236	71.3	3,361
Inter-quartile range	1.3	1,085	16.6	2,350	3.0	2,831	79.6	4,031
Standard deviation	3.2	2,292	18.7	2,849	4.4	4,188	61.5	3,332

Table 9. Summary of deforestation and regrowth rates between consecutive dates in the Machadinho d'Oeste time-series (MF – mature forest; SF – secondary forest).

	MF annual rate of deforestation		SF annual rate of deforestation		Annual rate of deforestation		Annual rate of regrowth	
	(% yr^{-1})	(ha yr^{-1})	(% yr^{-1})	(ha yr^{-1})	(% yr^{-1})	(ha yr^{-1})	(% yr^{-1})	(ha yr^{-1})
Minimum	1.2	748	0.0	17	1.5	2,386	3.7	103
Maximum	9.9	9,223	88.1	16,711	17.7	18,202	84.2	17,389
Mean	3.9	4,142	19.5	5,289	7.2	9,286	25.8	6,804
Median	3.5	4,171	14.3	4,273	6.0	7,436	19.1	6,172
Inter-quartile range	3.4	4,057	13.5	7,114	6.7	9,571	18.0	4,609
Standard deviation	2.3	2,394	19.5	4,467	4.4	5,248	21.5	4,141

4.6. Accuracy assessment and area calibration of the time-series of classified images

The most recent very high resolution (VHR) imagery was used to carry out the accuracy assessment of the time-series classifications at each site. As VHR data were not available to perform the accuracy assessment for all the classified images in the time-series, the accuracy in the classification of the image that was closest in time was assumed to be similar for the remaining images. To quantify the accuracy of the 2010 classification for Santarém, reference was made to 5 m SPOT-5 panchromatic data from 2009 and 2011 whilst the 2007 and 2010 classifications for Manaus and Machadinho d'Oeste respectively were validated using GeoEye imagery available on Google Earth.

At each site, 200 random points located at the center of larger polygons (greater than 6 ha) were generated for each class. The points on the classification were compared with the VHR imagery from the same date. At Santarém, it was assumed that if the same land cover was present in both the 2011 and 2009 image, the land cover in 2010 would be the same. The accuracy assessment was reported as a standard error matrix, including the overall accuracy and omission and commission errors (e.g., [48]). From the same VHR imagery and past field studies of the study sites (e.g., [32,38]), woody agriculture crops had established on some of the deforested areas. At Manaus, a large area of tree crops (TC, mainly oil palm) was progressively planted in the north west of the site from the 1983 to 1989, whilst at Machadinho d'Oeste, perennial crops (PC, mainly coffee plantations) were more common. To assess how well the areas of MF, NF and SF were discriminated using the classification, these areas were delineated manually from the VHR imagery and used in the generation of the error matrices.

The most straightforward way of estimating total areas for each class was to count the number of pixels in the MF, NF and SF classes with respect to the ground truth data. This is called "naïve estimation" in the remote sensing literature [49]. As a matter of fact, areal estimates obtained this way are biased (e.g., [50]). Some post-classification methods have been used to improve these estimates, namely the so-called calibration techniques that are divided into classical and inverse methods. The information contained in the error matrix was used to correct for misclassification bias [50]. Walsh and Burk, 1993 [50] carried out a simulation study and concluded that the inverse method consistently performed better than the classical method and lead to unbiased estimates for total areas. The idea behind this calibration technique was to use the misclassification probabilities among classes to revise the proportions given by pixel counting (p_i) [51]. According to Tenenbein, 1972 [51] and Walsh and Burk, 1993 [50], the revised (calibrated) proportion of the total area occupied by class i (π_i) can be estimated with equation (3)

$$\pi_i = \sum_{j=1}^{k} \left[(f_j + n_{.j})/N \right] \left(n_{ij} / n_{.j} \right) \quad (3)$$

where i represents the observed class, j the predicted class, k the number of classes, n the number of observations in the validation sample, N the combined number of observations in the validation and satellite datasets, f_j the (N-n) in class j, n_{ij} the number of observations in class i that were classified as j, and $n_{.j}$ the sum of all training observations predicted as class j. This method was applied to calibrate the area estimates obtained from each date of classified images using the error matrices that were produced for the three sites.

Table 10. An assessment of the dynamics of land use at Manaus, Santarém and Machadinho d'Oeste (figures in brackets represent the number of years).

		Sites		
Indicators	**Impact**	**Manaus**	**Santarém**	**Machadinho d'Oeste**
a) Clearance of MF only with little or no deforestation over SF	High pressure for new land	1973–1985 (12)	Pre-1984	1984–1987 (3)
b) Area of MF cleared>SF cleared.	Pressure for new land with some contained re-use of the existing deforested area (through reclearance)	1985–1989 (4)	1986–1987 (1)	1987–1990 (3)
		2002–2003 (1)	1991–1993 (2)	1991–1996 (5)
			1997–1998 (1)	1997–1998 (1)
			2006–2007 (1)	2001–2003 (2)
			2008–2009 (1)	
c) Area of MF cleared<SF cleared	Greater re-use of existing deforested land but still requirement for more land	1989–2002 (13)	1984–1986 (2)	1990–1991 (1)
		2003–2011 (11)	1987–1991 (4)	1996–1997 (1)
			1993–1997 (4)	1998–2001 (3)
			1998–2006 (8)	2003–2011 (8)
			2007–2008 (1)	
			2009–2010 (1)	
d) Area of MF and SF cleared>area of SF regenerating		1973–1985 (12)	1984–1986 (2)	1984–1986 (2)
		1988–1989 (1)	1987–1988 (1)	1987–1989 (2)
		1991–1994 (3)	1989–1993 (4)	1990–1995 (5)
		1996–2002 (6)	1995–1996 (1)	1996–1997 (1)
		2003–2007 (4)	1997–1999 (2)	1998–2001 (3)
		2008–2009 (1)	2001–2005 (4)	2003–2008 (5)
		2010–2011 (1)	2007–2008 (1)	2010–2011 (1)
e) Reclearance of SF but no clearance of MF	Contained re-use of area already deforested	No years	No years	No years
f) Area of SF regeneration>MF and SF deforestation	Net abandonment of land to regenerating forests	1985–1988 (3)	1986–1987 (1)	1986–1987 (1)
		1989–1991 (2)	1988–1989 (1)	1989–1990 (1)
		1994–1996 (2)	1993–1995 (2)	1995–1996 (1)
		2002–2003 (1)	1996–1997 (1)	1997–1998 (1)
		2007–2008 (1)	1999–2001 (2)	2001–2003 (2)
		2009–2010 (1)	2005–2007 (2)	2008–2010 (2)
			2008–2010 (2)	

Results

5.1. Accuracy assessment of land cover classification

The error matrices resulting from the accuracy assessment over Manaus, Santarém, and Machadinho d'Oeste are presented in Table 2, 3 and 4 respectively. For all three sites, the overall accuracy was high (above 0.90), although some relatively high omission and commission errors were identified, especially in Santarém and Machadinho d'Oeste. For Manaus, the omission and commission errors were always below 10%. For Santarém, a higher omission error in the SF class was detected (17.0%), mainly due to misclassification as MF (11.5%), which reflects also in the high commission error in the MF class (10.5%). At Machadinho d'Oeste, a major source of error also comes from a high omission error in the SF class (28%), with this mainly being a consequence of misclassification as NF (18.5%) and to a lesser extent as MF (9.5%); a high commission error was also observed in the NF class (15.8%), with this being due to misclassification with SF.

In Manaus, TC (4,774 ha) was mostly incorrectly classified as SF (89%), illustrating the similarity in their spectral signatures, but accounted only for 4.4% of the SF area in 2007 (Table 5). However, most of the TC classified as SF in Manaus was associated with the oil palm plantation situated in the north west of the study site. The majority of TC in Machadinho d'Oeste was correctly classified as NF, although the overall area was very small (17 ha). Conversely, 70% of the PC (mainly coffee plantations) at Machadinho d'Oeste was misclassified as SF, although this only accounted for 2.1% the entire SF area in 2010.

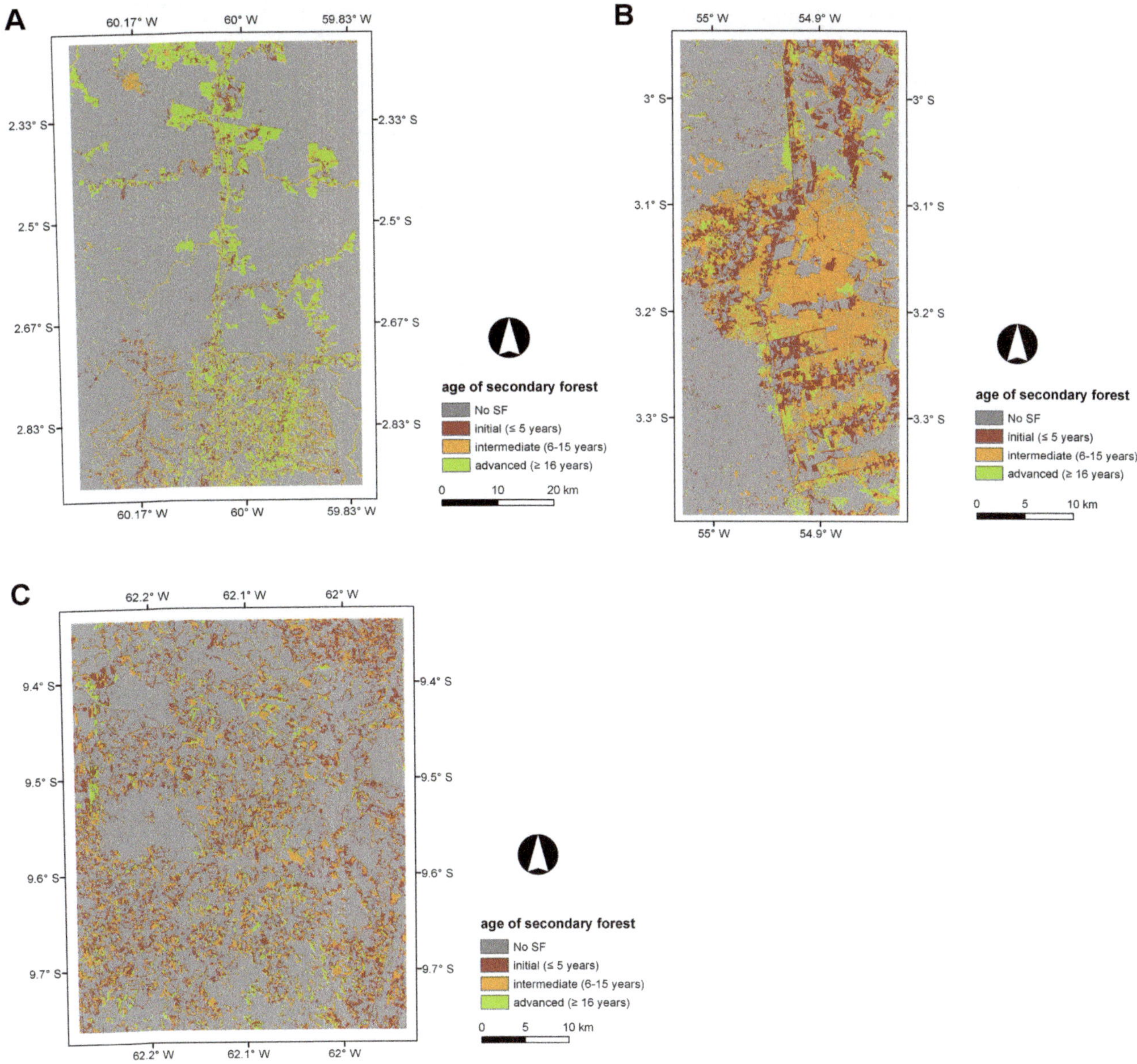

Figure 3. Maps displaying the age of secondary forest (ASF, years) across the three selected sites. Maps displaying the age of secondary forest (ASF) across the three selected sites for areas undergoing secondary forest (SF) in the last year of the corresponding time-series; A) Manaus (2011), B) Santarém (2010), and C) Machadinho d'Oeste (2011).

5.2. Land cover change

The evolution of land cover change in Manaus, Santarém and Machadinho d'Oeste is depicted in Figure 2A, 2B and 2C respectively and in Table 6, which shows that the most rapid change occurred in Machadinho d'Oeste, followed by Santarém and Manaus. For all sites, there was a strong correspondence between the land cover percentage and time (year) for the MF class, with the rate decreasing by an average of 0.54% in Manaus to 1.39% in Santarém and 2.33% in Machadinho d'Oeste. For all sites, a corresponding increase in the area deforested occurred, with the land cover alternating between NF and SF in all cases. The change rate in the NF class increased from 0.09% (Manaus) and 0.38% (Santarém) to 1.29% (Machadinho d'Oeste) whilst the SF change rate progressively increased from Manaus (0.45%) to Santarém (1.02%) and Machadinho d'Oeste (1.04%). The relationships between the proportion (%) of MF, NF and SF and time (year) were significant at all sites ($P<0.001$).

5.3. Deforestation and regrowth rates

Tables depicting relative and absolute deforestation and regeneration rates between consecutive dates for Manaus, Santarém and Machadinho d'Oeste are presented as Supporting Information (Table S1, S2 and S3 respectively in File S1) and a summary is given in Tables 7, 8 and 9 respectively. The maximum relative annual rate of deforestation over MF was observed at Santarém, with a 15.6% yr^{-1} loss between 1997 and 1998, with this associated with the extensive wildfires in 1997, and the highest average annual rate of deforestation was in Machadinho d'Oeste (3.9% yr^{-1}). The maximum relative rates of SF clearance were in Machadinho d'Oeste (88.1% yr^{-1}, 1990–1991) and Santarém

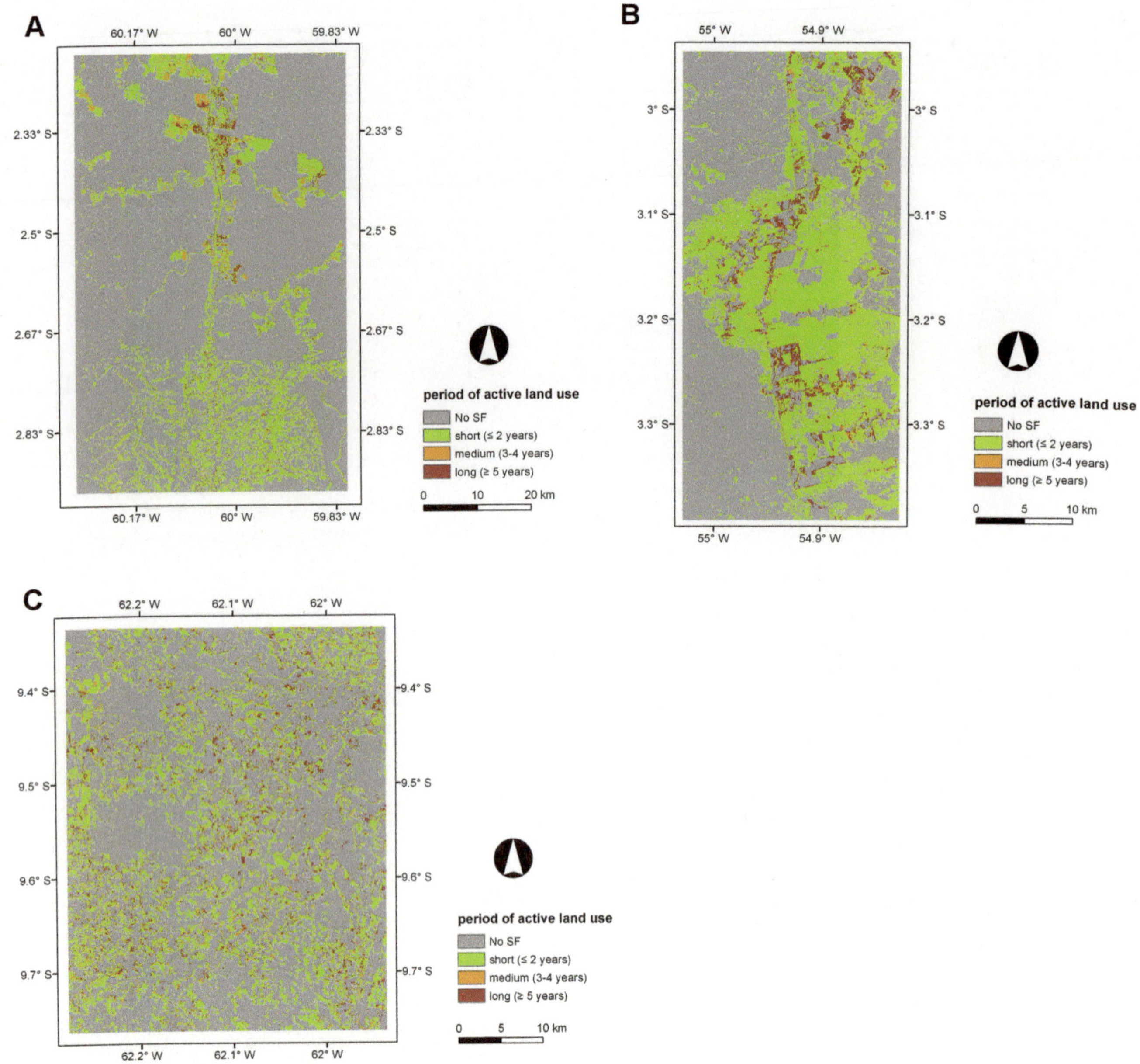

Figure 4. Maps displaying the period of active land use (PALU, years) across the three selected sites. Maps displaying the period of active land use (PALU, years) across the three selected sites for areas undergoing secondary forest (SF) in the last year of the corresponding time-series; A) Manaus (2011), B) Santarém (2010), and C) Machadinho d'Oeste (2011). An area of 667 ha (1.5%) with a PALU of zero in the Santarém site was included in the short PALU class (≤2 years) and corresponds to areas that were SF from 1984 to 2010.

(82.3% yr^{-1}, 1987–1988), with average rates being 19.5% and 16.7% for these sites respectively. The maximum and the highest average relative rates of regeneration were observed at Santarém, with 261.5% yr^{-1} (1997–1998) and 78.5% yr^{-1} respectively. The lowest relative rates of regrowth were at Machadinho d'Oeste, with these averaging 25.8% yr^{-1}. The average overall relative deforestation rates (combined MF and SF) were greatest for Machadinho d'Oeste (7.2% yr^{-1}), followed by Santarém (4.9% yr^{-1}) and Manaus (1.4% yr^{-1}). On average, deforestation rates over SF were higher than those over MF, indicating reclearance at all sites and particularly Machadinho d'Oeste (averaging 19.5% yr^{-1}) and Santarém (averaging 16.7% yr^{-1}).

Absolute areas (annual or based on periods of observation) are useful when comparing deforestation and regeneration rates between sites and the processes of change. Several key indicators are then listed in Table 10 together with the dates for each site to which these apply. For all sites, a trend from indicators a) to f) was observed with the majority being cleared initially, regenerating forests progressively establishing with these then recleared to varying degrees, and, of these, several are then abandoned to regrowth again.

5.4. Age of secondary forest (ASF), period of active land use (PALU), and frequency of clearance (FC)

Maps of the ASF, PALU, and FC for the last date of the time-series are presented in Figures 3, 4 and 5 for Manaus (2011), Santarém (2010) and Machadinho d'Oeste (2011) respectively and the proportion of areas with respect to these classes is indicated in

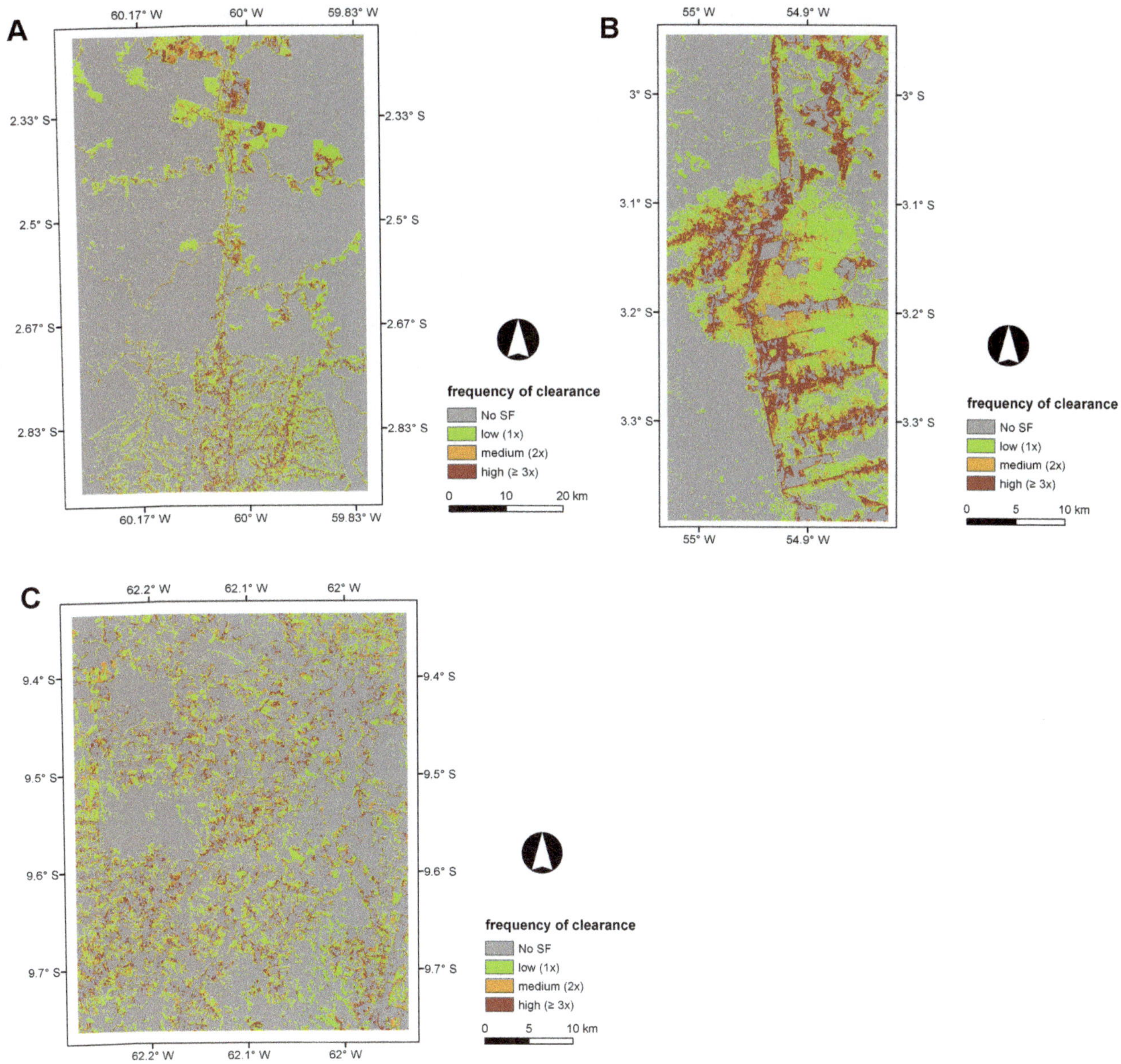

Figure 5. Maps displaying the frequency of clearance (FC) across the three selected sites. Maps displaying the frequency of clearance (FC) across the three selected sites for areas undergoing secondary forest (SF) in the last year of the corresponding time-series; A) Manaus (2011), B) Santarém (2010), and C) Machadinho d'Oeste (2011). In Manaus, Santarém and Machadinho d'Oeste an area of 103 ha (0.1%), 726 ha (1.6%), and 55 ha (0.1%), respectively, with a PALU of zero was included in the low FC class (1 time) and corresponded to areas that were non-forest (NF) in the first date of the time-series and no clearance has occurred in the secondary forest (SF) that persisted until the end of the time-series. In Santarém, the 726 ha also included areas that were already SF in 1984 that persisted until 2010.

Figure 6. At Manaus, regrowth forests were comparatively older, with 50% occupied by forests ≥16 years (Figure 6A). At Santarém, 57% of forests were aged 6–15 years, and at Machadinho d'Oeste, the majority of the area occupied by SF aged ≤5 years (46%). Hence, the three sites contain different distributions of age classes, with this attributable to differences in the time since deforestation but also the PALU and the frequency of clearance events. At all sites, the majority of land has been used actively for ≤2 years (Figure 6B), either following the initial clearance of the primary forest or subsequent reclearance events, but the deforestation had commenced at different times. At Manaus, the PALU was also ≤2 years for over 64% of the area and many of the forests had been able to regenerate undisturbed, with some approaching 40 years in 2011, and had been cleared mainly on one (65%) occasion (Figure 6C). At Santarém, 88% of the area under regrowth had been used actively for ≤2 years prior to the last abandonment and the majority (53%) cleared only once. At Machadinho d'Oeste, the PALU was typically ≤2 years (75%), and again, forests were relatively young because most of the clearance occurred only on one (57%) occasion. Hence, different typologies of SF were observed at each site.

As a result of combining the 3-class PALU and FC maps for the last date of the time-series, a proxy for land use intensity in each site was generated (Table 11). At Manaus, of the 65% of land occupied by SF in 2011 that was subjected to only one clearing event, 41% had a short PALU but a significant proportion (16%

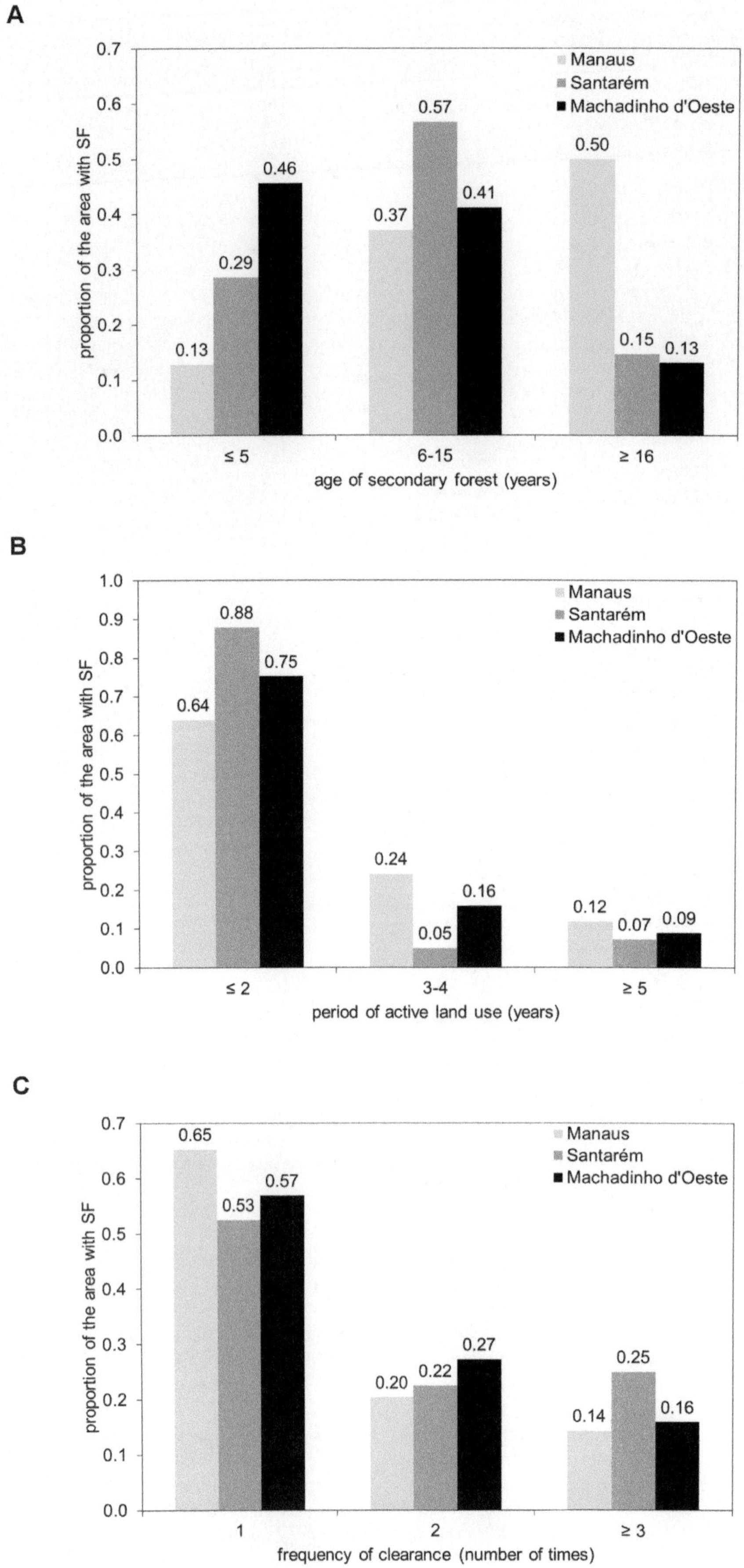
A
Manaus
Santarém
Machadinho d'Oeste
proportion of the area with SF
0.13
0.29
0.46
0.37
0.57
0.41
0.50
0.15
0.13
≤ 5
6-15
≥ 16
age of secondary forest (years)
B
Manaus
Santarém
Machadinho d'Oeste
proportion of the area with SF
0.64
0.88
0.75
0.24
0.05
0.16
0.12
0.07
0.09
≤ 2
3-4
≥ 5
period of active land use (years)
C
Manaus
Santarém
Machadinho d'Oeste
proportion of the area with SF
0.65
0.53
0.57
0.20
0.22
0.27
0.14
0.25
0.16
1
2
≥ 3
frequency of clearance (number of times)

Figure 6. Classes of age of secondary forest, period of active land use, and frequency of clearance for the three sites. Classes of A) age of secondary forest (ASF), B) period of active land use (PALU), and C) frequency of clearance (FC) for the three sites, as a proportion of the area with secondary forest (SF) in the last date of the time-series. The proportion of the area with SF in the first class of PALU (≤2 years) in the Santarém site includes a small proportion (1.5%) corresponding to areas of SF persisting in all dates of the time-series. The proportion of the area with SF in the first class of FC (1x) in Manaus, Santarém and Machadinho d'Oeste includes a small percentage (0.1%, 1.6%, 0.1%, respectively) corresponding to those cases where non-forest (NF) was observed in the first date of the time-series and no clearance has occurred (i.e., the only transition was from NF to SF).

and 9%) had a medium and long PALU respectively. Santarém is paradigmatic, as almost all land with SF in 2010 that was cleared only once was subjected to a short PALU. Assuming that a high FC and short and medium PALU will represent a higher land use intensity, then Manaus had 13% of the area undergoing SF in 2011 in these conditions, while Santarém and Machadinho d'Oeste had 20% and 15% in 2010 and 2011 respectively.

To investigate whether these typologies were dependent on the last date of the time-series, the same analysis was performed for all the dates in the time-series. At Manaus, the proportion of the area with SF in the ≤5 year ASF class decreased irregularly with the onset of regeneration up to approximately 15% in the last date of the time-series (Figure 7A), with this pattern being mirrored for the 6–15 year ASF class, and the ≥16 year ASF class steadily increasing up to around 50% in 2011. At Santarém (Figure 7B), the proportion of the area with SF in the initial ASF class decreased to around 30% in 2010, again mirrored by an overall increase to about 60% in the 6–15 year ASF class. As with Manaus, the proportion of the area with SF in the advanced ASF class increased from 15 years following the onset of regeneration to cover approximately 15% in 2010. In Machadinho d'Oeste (Figure 7C), the proportion of SF in the initial ASF class systematically decreased from the onset of regeneration leveling at around 45% in 2011, and an opposite pattern was observed in the intermediate ASF class, with a systematic increase since the onset of regeneration up to approximately 40% in 2011. As in Manaus and Santarém, a steady increase in the proportion of area with regeneration in the advanced ASF class was observed, from 17 years after the onset of regeneration up to around 15% in 2011. By combining the initial and intermediate ASF classes, a more consistent but asymptotic increase in the area of SF was observed, with this compensating for the interplay between these ASF classes, which occurs because of reclearance. The general trend across all sites is a decrease in the proportion of the area with SF in the initial ASF class up to the present, with a simultaneous increase in the proportion of the intermediate and advanced ASF classes. No asymptotic pattern is visible in the relationship between the proportion of the area with SF and the number of years following the onset of regeneration for this variable in the three selected sites.

At Manaus, the proportion of areas with a short PALU was initially high (100%) but, over time, this area decreased to about 40% but then increased subsequently to 60%, where it remained constant 32 years from the onset of regeneration (Figure 8A). This decrease was associated with an increase in the area with a medium and long PALU (14–16 years after the onset of SF regeneration). At Santarém (Figure 8B), the proportion with short PALU remained high, with this suggesting clearance of SF was common practice. The areas with medium and long PALU remained relatively low (<5–10%) but increased slightly towards the end of the time-series. At Machadinho d'Oeste, the temporal pattern was similar to that observed at Manaus; after 12–14 years, the proportion of short, medium and long PALU stabilized around 75%, 15% and 10% respectively.

The FC over areas undergoing SF also varied between sites (Figure 9). At Manaus, the SF had been cleared only once until about 14 years after the onset of SF, but this proportion decreased, stabilizing at approximately 65% in 2011. This occurred because SF began to be cleared again 15 years after the onset of regeneration and again after 19 years (i.e., 3 FC events), with these areas stabilizing at approximately 20% and 15% respectively. At

Table 11. Percentage of area with SF in the last date of the time-series, per combination of PALU and FC classes.

PALU (classes)	FC (classes)			Total
Manaus	**Low (1x)**	**Medium (2x)**	**High (≥3x)**	
Short (≤2 years)	41	13	10	64
Medium (3–4 years)	16	5	3	24
Long (≥5 years)	9	2	1	12
Total	66	20	14	
Santarém	**Low (1x)**	**Medium (2x)**	**High (≥3x)**	
Short (≤2 years)	51	20	17	88
Medium (3–4 years)	1	1	3	5
Long (≥5 years)	0	2	5	7
Total	52	23	25	
Machadinho d'Oeste	**Low (1x)**	**Medium (2x)**	**High (≥3x)**	
Short (≤2 years)	48	17	11	75
Medium (3–4 years)	5	7	4	16
Long (≥5 years)	4	4	1	9
Total	57	28	16	

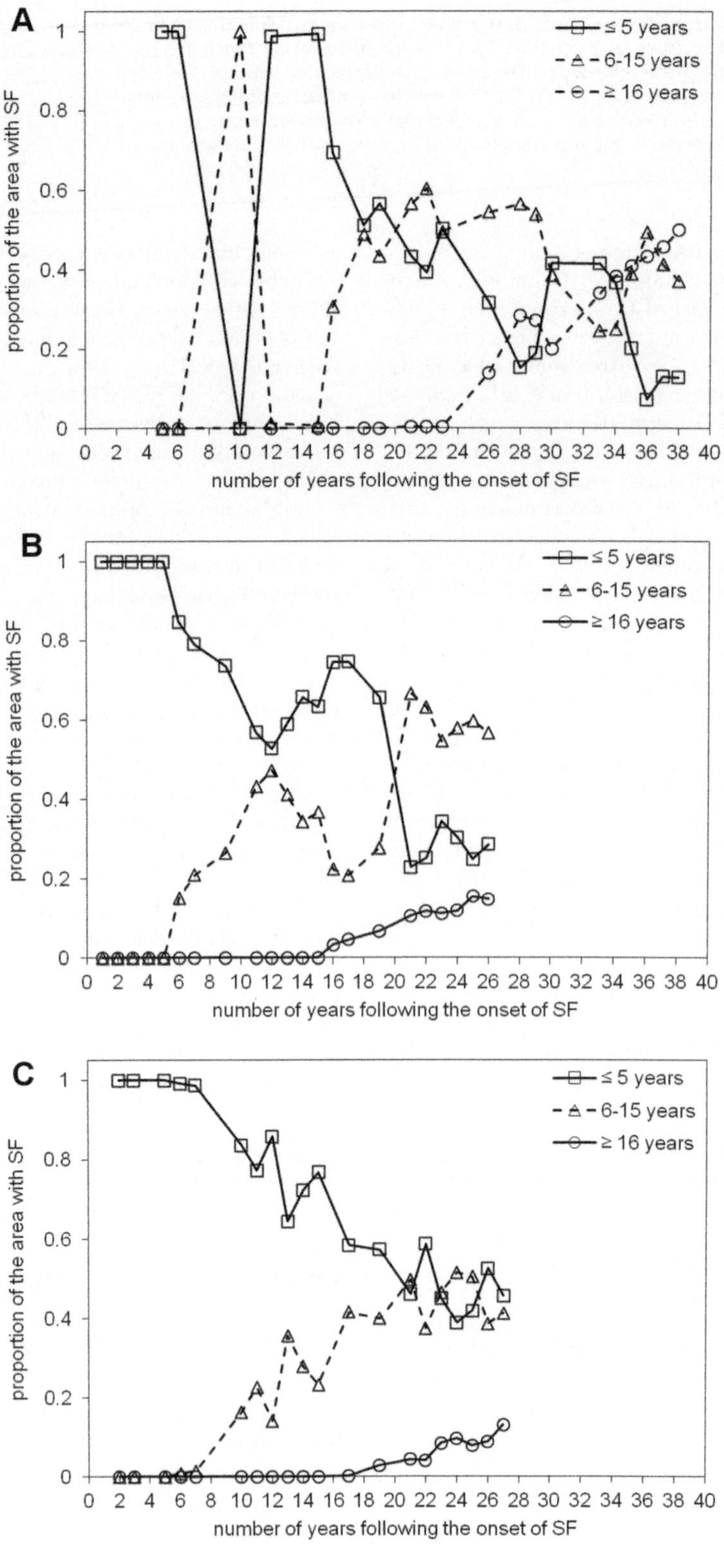

Figure 7. Proportion of the area with secondary forest (SF) as a function of the number of years following the onset of SF for the three classes of age of secondary forest. Proportion of the area with secondary forest (SF) as a function of the number of years following the onset of SF for the three classes of age of secondary forest (ASF) in A) Manaus, B) Santarém, and C) Machadinho d'Oeste: initial, ≤5 years; intermediate, 6–15 years; advanced, ≥16 years.

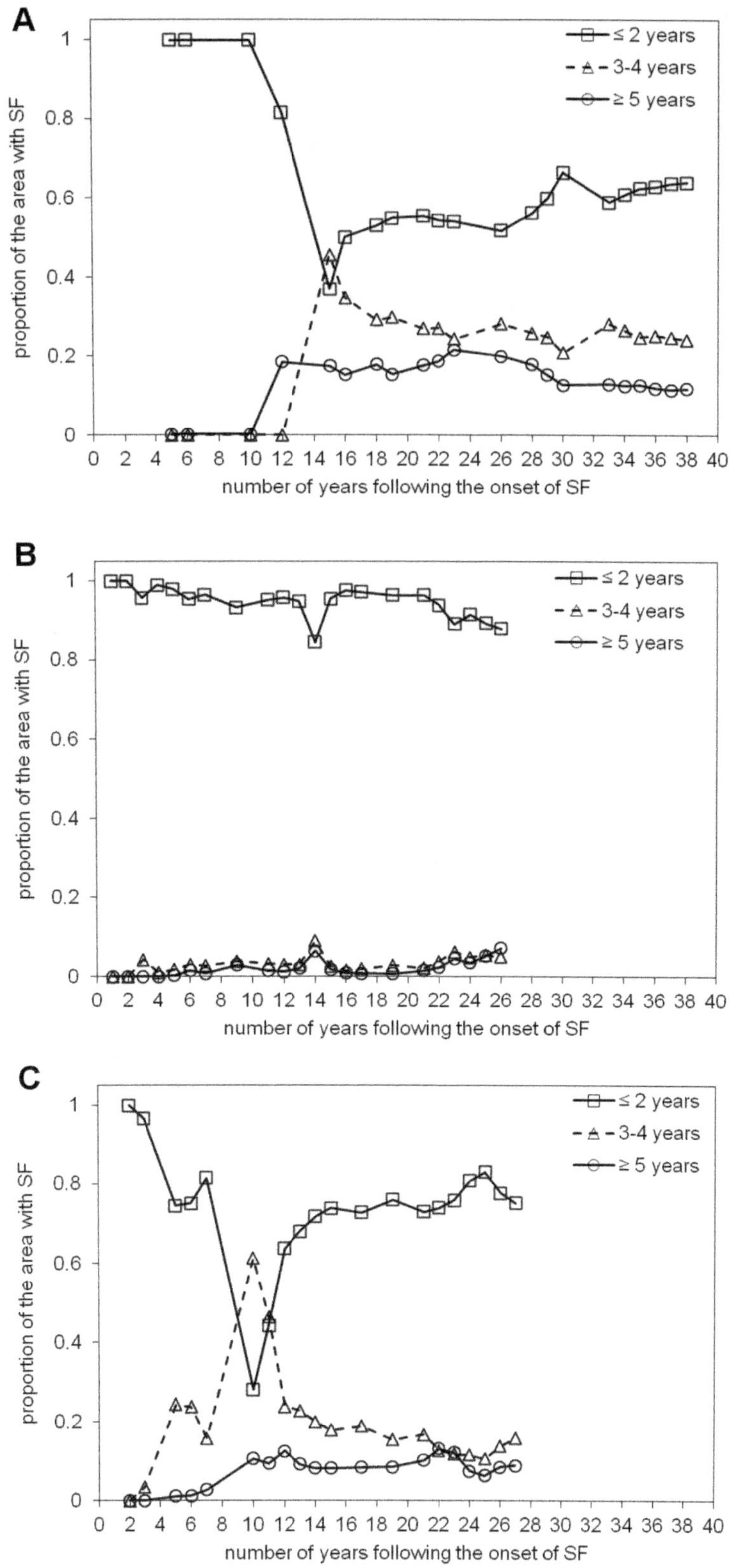

Figure 8. Proportion of the area with secondary forest (SF) as a function of the number of years following the onset of SF for the three classes of period of active land use. Proportion of the area with secondary forest (SF) as a function of the number of years following the onset of SF for the three classes of period of active land use (PALU) in A) Manaus, B) Santarém, and C) Machadinho d'Oeste: short, ≤2 years; medium, 3–4 years; long, ≥5 years.

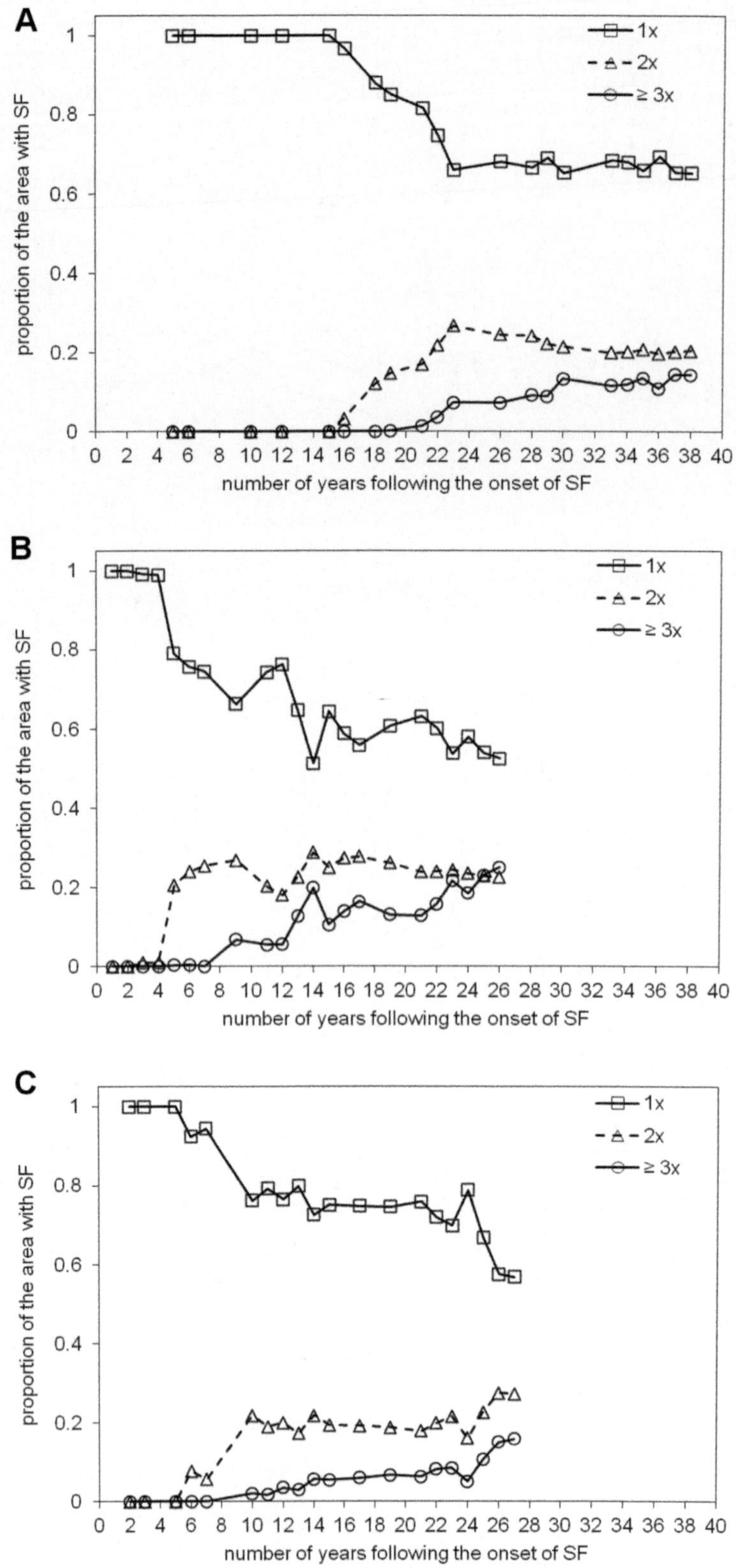

Figure 9. Proportion of the area with secondary forest (SF) as a function of the number of years following the onset of SF for the three classes of frequency of clearance. Proportion of the area with secondary forest (SF) as a function of the number of years following the onset of SF for the three classes of frequency of clearance (FC) in A) Manaus, B) Santarém, and C) Machadinho d'Oeste: low, 1 time; medium, 2 times; high, ≥3 times.

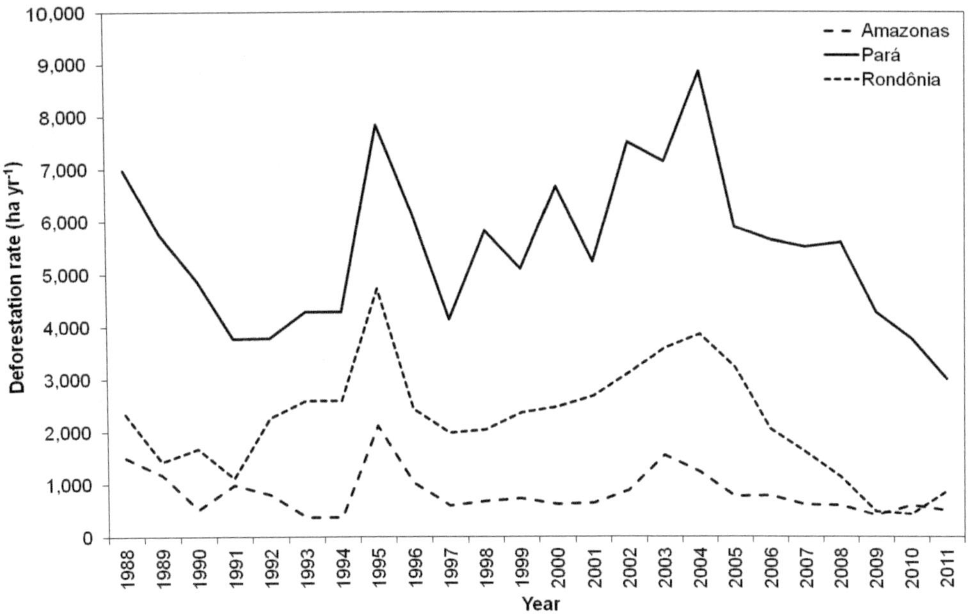

Figure 10. Deforestation rates in the Amazonas, Pará and Rondônia states according to INPE, 2013 [13]. Deforestation rates since 1988 (ha yr^{-1}) in the Amazonas, Pará and Rondônia states according to the data reported by INPE, 2013 [13].

Santarém, SF was cleared within 2–4 years from the onset of regeneration and continually thereafter. This was reflected in the increase in the proportion of SF areas cleared twice (after 4 years) and three times (after 7 years) with these both stabilizing at approximately 20%. At Machadinho d'Oeste, the proportion of areas with only a clearing event decreased to 55% 28 years after the onset of regeneration. The medium and high FC classes commenced after 4 years and 7 years respectively with the proportions of these remaining relatively constant (25% and 15%).

Discussion

6.1. Accuracy assessment of the land cover time-series

Several studies carried out in the Amazon have acknowledged some issues in discriminating SF from MF and, sometimes, NF. Lucas et al., 2000 [18] reported omission and commission errors ranging from 55–75% and 51–76%, respectively, associated with a greater confusion between NF and SF rather than between MF and SF. Carreiras et al., 2006 [20] reported an omission and commission error in the SF class of 66% and 43%, respectively, due to confusion between the SF, MF and NF classes. These authors concluded that misclassification among SF, MF, and NF was understandable, since from a spectral standpoint, SF is a transitional class between NF (i.e., agriculture/pasture) and MF. The initial stages of regeneration are spectrally similar to agriculture/pasture. Conversely, final stages of regrowth are more related to MF. The values reported in the two previous studies are much higher than those from this study, and that could be associated with the spatial resolution of the datasets that were used: ~1-km SPOT 4 VEGETATION and NOAA AVHRR vs. the 30-m spatial resolution Landsat TM and ETM+ data. The spatial arrangement of SF in the BLA, dominated by small previously abandoned areas, especially in Rondônia (e.g., [52]), creates an additional difficulty when trying to discriminate these from the surrounding land cover classes with coarse spatial resolution data. Kimes et al., 1999 [53] used 20-m spatial resolution SPOT High Resolution Visible (HRV) data between 1986 and 1994 to discriminate MF, NF and SF in Rondônia and reported an accuracy of 95% in the SF class with misclassification happening mostly with the MF class. Metzger, 2002 [54] used Landsat TM data over the Bragantina region (Pará state) in 1996 to map various land cover classes and reported omission and commission errors in SF age classes ranging from 8–11% and 7–10%, respectively, with misclassification occurring mainly from confusion with MF. Vieira et al., 2003 [55] used Landsat 7 ETM+ data to map different stages of SF in São Francisco do Pará (northeastern Pará state) and reported omission and commission errors between 17–25% and 22–40% respectively, due mainly to misclassification among SF stages. Kuplich, 2006 [56] used Landsat TM and Synthetic Aperture Radar (SAR) data to discriminate various land cover classes in Manaus (included the BDFFP) and reported omission and commission errors in the SF age classes of 10–99% and 32–80% respectively, but those errors decreased when the SF age classes were aggregated into a unique class (78% and 25%, respectively). Although errors are still high in some studies (e.g., [18,20,56]), others have been able to discriminate SF from MF and NF with a high degree of accuracy (e.g., [53–55]).

Prates-Clark et al., 2009 [38] validated the 1995 and the 2001 land cover maps over Manaus and Santarém respectively using field-based reference data. For the validation of the 1995 land cover map over Manaus, field data collected in 1995 were used [22,42] and the authors reported an overall accuracy of ~99%, with omission errors in the MF, NF and SF classes of 0.3%, 3.5% and 1.9%, respectively, and commission errors of 0.5%, 7.0% and 9.5%, respectively. The major source of error in the SF class (higher commission error) was misclassification as MF class. At Santarém, field data collected in 2002 [57] were used to validate the 2001 land cover map. The overall accuracy was 88%, the omission errors were 2.2%, 1.1% and 4.2% for the MF, NF and SF classes respectively, and the commission errors were 0.6%, 28.8% and 71.9% respectively. The particularly high commission error in the SF class was essentially the consequence of misclassification as MF. The NF class displayed a high commission error that was related to misclassification as both MF and SF. Higher errors indicated in Prates-Clark et al., 2009 [38] for the Santarém site were also depicted in our accuracy assessment of the 2010 land cover map, although of much lower magnitude.

A major limitation of the accuracy assessment carried out in this study it that only one date (classification) of each time-series per study area was validated with very high spatial resolution data. Nevertheless, a 10-fold cross-validation procedure was implemented for those dates classified with the random forests algorithm, i.e., the 21 Landsat TM images of the Machadinho d'Oeste time-series (1984–2011) and the post 2003 Santarém Landsat TM images (2005–2010). In Machadinho d'Oeste, the average overall accuracy was ~98%, the omission errors of the MF, NF and SF classes ranged from 0.0–4.4%, 0.0–2.1% and 0.7–7.6% respectively, and the commission errors were between 0.0–4.2%, 0.0–0.3% and 1.7–7.1% respectively. At Santarém, the 2005–2010 time-series of classified Landsat TM data resulted in an average overall accuracy of 95%, the omission errors of the MF, NF and SF classes ranged between 1.8–6.7%, 0.0–0.9% and 8.1–16.7%, respectively, and the commission errors were between 3.7–7.5%, 0.0–0.6% and 4.1–13.2% respectively. Both the omission and commission errors of the NF class were very low, thus indicating that the major source of error of the SF class was misclassification as MF.

6.2. Differences of land cover change patterns among sites

Extensive deforestation across the BLA started in the 1970s and was concentrated along the southern and eastern rims of the region [58]. The main deforestation drivers were conversion of forest to cropland and/or cattle ranching, which were carried out both by small farmers and large landholders [59]. According to the data provided by INPE, 2013 [13], the states where the three sites are located displayed different deforestation rates (Figure 10). Deforestation rates were always higher in Pará, followed by Rondônia and Amazonas. However, since 2003–2005, deforestation rates have been decreasing, with Rondônia experiencing a rapid decrease leading to values closer to those observed in the Amazonas state. Several factors contributed to this decline, namely the improvement of market-driven environmental governance [60]. Pará experienced the highest number of settled families by agrarian reform in the nine states composing the BLA (~31,000 settled families per year in the period 2003–2006), and the number has been increasing since the 1960s [61]. Although the vast majority of deforestation in the BLA can be tracked to large landholders occupying the land for cattle ranching, the implementation of planned settlements is not negligible at all, especially in Rondônia, which is known by its small farmers' radial, fishbone and watershed deforestation patterns (e.g., [58]). For example, up to the mid-1990s, the number of settled families in Rondônia was only second to Pará, with 1,423 settled families per year against 1,462 settled families per year in Pará [61].

The three study areas have different socio-economic drivers leading to deforestation, different land uses and subsequent regeneration. The Manaus site is mostly included in the Manaus municipality, although 24% is also included in the Rio Preto da Eva municipality. This study area basically encompasses several environmental protection areas. As such, it seems that these conservation areas have been critical at preserving most of the forest in the region. The conservation units (CUs) limits accessed from the Brazilian Ministry of the Environment (MMA, available at http://mapas.mma.gov.br/i3geo/datadownload.htm) were used to estimate the ratio of relative incidence of deforestation (NF and SF) outside and inside the CUs in the last date of the time-series for each site. The three sites had a significant proportion of the area included in the CUs: 43% in Manaus, 37% in Santarém and 23% in Machadinho d'Oeste. In Manaus, the relative incidence of deforestation outside the CUs was twice that inside CUs and in Santarém and Machadinho d'Oeste that value increased to five times. In fact, several studies have concluded that protected areas in the Amazon region are indeed effective at reducing deforestation rates (e.g., [60,62,63]). Figure 2A illustrates the lower amount of land cover change in Manaus, with approximately 20% of MF lost between 1973 and 2011, when compared with Santarém (more than 30% of MF lost between 1984 and 2010) and Machadinho d'Oeste (~40% of MF lost between 1984 and 2011). Deforestation in the Manaus study area started in the 1970s, with the construction of the BR-174 highway connecting Manaus and Boa Vista. Early deforestation is traceable along this highway, but many areas were abandoned as a result of poor soil fertility; indeed, many small farmers moved to Roraima, which was considered to have younger fertile soils [64]. The area of forest in 2011 was approximately 69% and 81% of the territory of the Manaus and Rio Preto da Eva municipalities respectively [13]. Figure 2A shows that the proportion of MF in the Manaus site decreased from approximately 90% in 1973 to 71% in 2011, which is similar to the value reported by INPE, 2013 [13] for the Manaus municipality.

The Santarém site is located south of the city of Santarém, in the Belterra municipality, along the BR-163 highway, and including part of the Tapajós National Forest. This highway links Santarém and Cuiabá in the south. According to Scatena et al., 1996 [65] and Brondizio and Moran, 2012 [66], this region has a tradition of agriculture and agroforestry since the early 1900s, which has further escalated as a consequence of the construction of the BR-163 in the 1970s. In fact, of the three sites Santarém had the lowest proportion of MF at the beginning of the time-series (~80%) and the highest proportion of SF (~20%), thus suggesting that deforestation was commonplace in the area in the early 1980s. According to INPE, 2013 [13], approximately 68% of the forest area in the Belterra municipality was remaining in 2011, whereas we report a proportion of MF in 2010 of 46%. A vast majority of the Belterra municipality is included in the Tapajós National Forest, mainly the part that is west of the BR-163 highway. So, it is not surprising that the proportion of remaining forest in this municipality is higher than the proportion of remaining forest in the study area that includes the fraction east of the BR-163 highway that is not a protected area.

The Machadinho d'Oeste site (Rondônia state) is included mostly in the Machadinho d'Oeste municipality, although 20% overlaps the Vale do Anari municipality. Of the three sites, this was the one with the highest degree of land cover change, with a reduction of ~60% in the MF cover from 1984 (91%) to 2011 (32%). Deforestation in this region started in the 1980s with the implementation of settlements by INCRA, of which Machadinho d'Oeste and Vale do Anari are just two examples [33]. According to Batistella and Moran, 2005 [33], the proportion of MF in 1988 and 1998 in Machadinho d'Oeste was around 80% and 66% respectively, while our estimates were 82% in 1987 (1988 was not mapped) and 55% in 1998. INPE, 2013 [13] reports an area of MF in the Machadinho d'Oeste municipality in 2011 of approximately 61%, and our study estimated the percentage of MF in the Machadinho d'Oeste study area as 32%. Google Earth data from 2013 shows that the areas of the Machadinho d'Oeste municipality to the east and north of the study area have much less deforestation and so this is a possible explanation for the higher proportion of MF detected by Batistella and Moran, 2005 [33] in 1998 and INPE, 2013 [13] up to 2011. Also, Batistella and Moran,

2005 [33] identified the land cover types occurring in the deforested areas of Machadinho d'Oeste. In 1988, approximately 2% was covered by SF, the value increasing to approximately 13% in 1998; our study mapped SF in 1987 and 1998 and the proportion was ~13% and ~29% respectively. Again, when considering the entire Machadinho d'Oeste municipality, there are extensive areas to the north and east of the study area that remained intact, with these covered by MF. Nevertheless, there is a coincident temporal trend that is the increase of SF from 1988 to 1998.

6.3. Land abandonment and the emergence of secondary forest

Land abandonment in the BLA is a consequence of a range of factors that have changed across the last decades since the inception of large scale deforestation in the region. Namely these are; reduced crop productivity as a consequence of poor soil fertility; lack of financial incentives, migratory patterns; non-traditional land uses and market fluctuations [67]. At Manaus, in 2011, land abandonment resulted in SF that was comprised mainly of areas with advanced ASF (50% ≥16 years), although large areas of intermediate ASF (37% between 6–15 years) were also present (Figure 3A and Figure 6). This suggests that the conservation areas were not only effective at preserving MF but also SF areas. As mentioned before, deforested areas in this study area were abandoned mainly because of poor soil fertility, with settlers moving to the nearby state of Roraima [64]. It is clear from Figure 3A that the higher proportion of advanced ASF in the study area is in the northern half of the site. These areas undergoing regeneration in 2011 had mostly short PALU (64%) as a consequence of intensification of land use and requirement for new land, and 65% was deforested only once. In comparison, the Santarém site, which included parts of the Tapajós National Forest, had regrowth dominated by intermediate ASF in 2010. Those areas experienced essentially short PALU (88% ≤2 years) and low FC (52% deforested once). According to Brondizio and Moran, 2012 [66], land abandonment in the region, and associated regeneration, was high until 1999. At this point, large-scale soybean cultivation started, mainly because of the decay of primary and secondary roads, lack of social services, and limited access to water. On the other hand, land abandonment in Machadinho d'Oeste resulted in SF that was mainly in the initial (46%) and intermediate (41%) ASF classes, with these areas experiencing mostly (75%) short PALU and low FC (57%). This was a planned settlement and most of the deforested area is under agriculture and or pasture use, suggesting that a vast majority of the area undergoing SF could be indeed forest fallow that might be under cattle ranching use or being subjected to subsistence agriculture in a crop/fallow cycle [66].

One important aspect worth mentioning is the large proportion of SF in the Santarém site that resulted from the 1997 wildfires. This area of regeneration was not the consequence of land abandonment but a natural process following fire-induced disturbance. This wildfire was just one among many others that occurred in the BLA as a consequence of the 1997–1998 *El Niño* Southern Oscillation (ENSO) event (e.g., [68]). In fact, the proportion of NF in 1997 (prior to the wildfire) was around 6% and in 1998 had increased to approximately 15%, mainly as a consequence of that large wildfire, perfectly visible in Figure 3B as the orange irregular zone of intermediate ASF in the centre of the study area. As a consequence of the natural regeneration process, the SF proportion increased to approximately 44% in 2000 (~32% in 1997). Therefore, approximately 20% of the area of SF in 2010 was estimated as resulting from natural regeneration following that wildfire event.

6.4. Implications of prior land use for biomass accumulation and biodiversity restoration in secondary forests

Regenerating forests in the three selected sites have been classified in terms of ASF, PALU prior to abandonment and FC. Prior LULCC practices have a fundamental influence on the vegetation regenerating following land abandonment in the Amazon [10,22,69–75]. Vegetation structure, species composition and dominance are just some of the parameters of the regenerating vegetation that have been studied to identify the impact of prior land use practices (e.g., [76]). As such, these have an impact on the capability of these forests to accumulate biomass (and to act as a C sink at a greater or lesser extent) and to restore biodiversity (e.g., [77]).

Uhl et al., 1988 [71] studied areas undergoing secondary succession following pasture abandonment in Pará (Paragominas municipality) and identified major patterns related to prior land use intensity; light use sites, characterized by a lower PALU (0–4 years) and with no reclearance or just one slash-and-burn episode, had high species richness dominated by pioneer species (e.g., *Cecropia* sp. and *Solanum* sp.), and a ~13 m closed canopy with an AGB accumulation rate of 10 Mg ha^{-1} yr^{-1} (8-year old sites); comparatively, moderate use sites, having higher PALU (6–12 years) and with 1 to 5 clearing episodes, had lower species richness dominated by a 7–8 m *Vismia* sp. partially developed canopy and an AGB accumulation rate of 5 Mg ha^{-1} yr^{-1}; heavy use sites, subjected to large-scale mechanized operations, had poor species richness (*Solanum* sp. and *Cecropia* sp.), mainly composed by scattered trees and an AGB accumulation rate of only 0.6 Mg ha^{-1} yr^{-1}. Mesquita et al., 2001 [75] studied regeneration pathways in a region north of Manaus that is included in our Manaus site (BDFFP) and identified major differences related to prior land use. Basically, areas that were deforested and abandoned immediately were dominated by *Cecropia* spp. trees, whereas those that were abandoned after some years of pasture use were dominated by *Vismia* spp. trees. Lucas et al., 2002 [22], also identified two distinct regeneration pathways in a region north of Manaus and in the same way as Mesquita et al. [75]; the authors identified *Vismia* spp. dominated regeneration associated to sites with more intensive land use and those dominated by *Cecropia* spp. to less intensive land use. According to Chazdon et al., 2007 [74], in terms of AGB accumulation, tropical secondary forests are characterized by a rapid growth rates in the first years after land (agriculture or pasture) abandonment. Furthermore, Zarin et al., 2001 [78] showed that there is a strong correlation between the AGB of secondary forests and the number of years following abandonment (i.e., ASF), soil texture and climate data in the Amazon region. On this pretext, the secondary forests in Manaus have the potential to accumulate more AGB, followed by Santarém and Machadinho d'Oeste. In 2011 (the last date of the time-series), Machadinho d'Oeste had approximately 45% of the area of SF with less than five years of age, and around 40% with less than three years of age. Machadinho d'Oeste was initially a settlement project implemented by the Brazilian government and today most of its people are dependent on subsistence agriculture [32]. Therefore, it is possible that some of the area mapped as SF was indeed forest fallow, part of a crop/fallow system of subsistence agriculture, which have been acknowledged by several studies (e.g., [79,80]).

Concluding Remarks

An accurate assessment of large-scale land use and land cover change dynamics over remote areas (e.g., Amazon) can only be carried out with an adequate monitoring system. Remote sensing data, in this case annual or quasi-annual time-series of high spatial resolution optical data (Landsat program), has proven its ability to accurately discriminate MF, SF and NF over three sites in the BLA experiencing several decades of deforestation. Lower deforestation rates and a greater proportion of intermediate and advanced ASF classes were characteristic of the Manaus site. On the other hand, Machadinho d'Oeste had the highest deforestation rates and lower regrowth rates, with most of its regeneration occurring in areas abandoned over the past 5 years. LULCC at Santarém displayed an intermediate behavior, with intermediate deforestation rates and the highest regrowth rates, which lead to regeneration dominated by areas of intermediate ASF. Conservation units were effective at reducing deforestation at all three sites. The temporal evolution of the spatial arrangement of the various parcels of land identified as one of the three classes could provide new insights about the fragmentation patterns in the region (Ewers et al. 2013 [81]).

Although several generalizations about the type and composition of secondary forest occurring in each of the three sites can be made, a correct assessment can only be made after an *in situ* assessment. Therefore, the research presented here will provide the opportunity to assess the influence of previous LULCC dynamics on the biomass accumulation and biodiversity restoration and to further investigate hypothesis related to differences among the three sites. A field campaign is foreseen for 2014 to collect forest inventory data that will be used to study the influence of prior LULCC on the capability of regrowth areas to accumulate AGB and to restore biodiversity. In the end, a framework related to the influence of prior LULCC on the biomass accumulation and biodiversity will be built and could be used to inform land management policies in the region.

Acknowledgments

The United States Geological Survey (USGS) and Brazil's National Institute for Space Research (INPE) are particularly thanked for providing free access to the Landsat sensor data used in this study. Nathan Thomas (Department of Geography and Earth Sciences, Aberystwyth University, Aberystwyth, United Kingdom) is particularly acknowledged for reviewing earlier versions of the manuscript.

Author Contributions

Conceived and designed the experiments: JC RL. Performed the experiments: JC JJ RL CG. Analyzed the data: JC JJ RL. Contributed reagents/materials/analysis tools: JC JJ RL CG. Wrote the paper: JC JJ RL.

References

1. Falkowski P, Scholes RJ, Boyle E, Canadell J, Canfield D, et al. (2000) The global carbon cycle: A test of our knowledge of earth as a system. Science 290: 291–296.
2. Raupach M (2011) Pinning down the land carbon sink. Nature Climate Change 1: 148–149.
3. Heimann M, Reichstein M (2008) Terrestrial ecosystem carbon dynamics and climate feedbacks. Nature 451: 289–292.
4. Pan Y, Birdsey R, Fang J, Houghton R, Kauppi P, et al. (2011) A Large and Persistent Carbon Sink in the World's Forests. Science 333: 988–993.
5. Gibson L, Lee T, Koh L, Brook B, Gardner T, et al. (2011) Primary forests are irreplaceable for sustaining tropical biodiversity. Nature 478: 378–381.
6. Achard F, Eva H, Stibig H, Mayaux P, Gallego J, et al. (2002) Determination of deforestation rates of the world's humid tropical forests. Science 297: 999–1002.
7. Hooper D, Adair E, Cardinale B, Byrnes J, Hungate B, et al. (2012) A global synthesis reveals biodiversity loss as a major driver of ecosystem change. Nature 486: 105–108.
8. Lederer M (2011) From CDM to REDD+ – What do we know for setting up effective and legitimate carbon governance? Ecological Economics 70: 1900–1907.
9. Olson D, Dinerstein E, Wikramanayake E, Burgess N, Powell G, et al. (2001) Terrestrial ecoregions of the worlds: A new map of life on Earth. Bioscience 51: 933–938.
10. Chazdon RL (2003) Tropical forest recovery: legacies of human impact and natural disturbances. Perspectives in Plant Ecology Evolution and Systematics 6: 51–71.
11. Cardinale B, Duffy J, Gonzalez A, Hooper D, Perrings C, et al. (2012) Biodiversity loss and its impact on humanity. Nature 486: 59–67.
12. Goulding M, Barthem R, Ferreira EJG (2003) The Smithsonian atlas of the Amazon. Washington, D.C.: Smithsonian Books. 253 p.
13. INPE (2013) Projeto PRODES: Monitoramento da Floresta Amazônica Brasileira por Satélite. Instituto Nacional de Pesquisas Espaciais. Available: http://www.obt.inpe.br/prodes/index.php. Accessed 11 March 2014.
14. Davidson EA, de Araujo AC, Artaxo P, Balch JK, Brown IF, et al. (2012) The Amazon basin in transition. Nature 481: 321–328.
15. Stone TA, Schlesinger P, Houghton RA, Woodwell GM (1994) A map of the vegetation of South America based on satellite imagery. Photogrammetric Engineering and Remote Sensing 60: 541–551.
16. Schroeder PE, Winjum JK (1995) Assessing Brazil carbon budget: II. Biotic fluxes and net carbon balance. Forest Ecology and Management 75: 87–99.
17. Fearnside P (1996) Amazonian deforestation and global warming: Carbon stocks in vegetation replacing Brazil's Amazon forest. Forest Ecology and Management 80: 21–34.
18. Lucas RM, Honzak M, Curran PJ, Foody GM, Milne R, et al. (2000) Mapping the regional extent of tropical forest regeneration stages in the Brazilian Legal Amazon using NOAA AVHRR data. International Journal of Remote Sensing 21: 2855–2881.
19. Cardille J, Foley J (2003) Agricultural land-use change in Brazilian Amazonia between 1980 and 1995: Evidence from integrated satellite and census data. Remote Sensing of Environment 87: 551–562.
20. Carreiras JMB, Pereira JMC, Campagnolo ML, Shimabukuro YE (2006) Assessing the extent of agriculture/pasture and secondary succession forest in the Brazilian Legal Amazon using SPOT VEGETATION data. Remote Sensing of Environment 101: 283–298.
21. Neeff T, Lucas RM, dos Santos JR, Brondizio ES, Freitas CC (2006) Area and age of secondary forests in Brazilian Amazonia 1978–2002: An empirical estimate. Ecosystems 9: 609–623.
22. Lucas RM, Honzak M, Amaral ID, Curran PJ, Foody GM (2002) Forest regeneration on abandoned clearances in central Amazonia. International Journal of Remote Sensing 23: 965–988.
23. Fearnside PM (2000) Global warming and tropical land-use change: Greenhouse gas emissions from biomass burning, decomposition and soils in forest conversion, shifting cultivation and secondary vegetation. Climatic Change 46: 115–158.
24. Houghton RA (2005) Aboveground forest biomass and the global carbon balance. Global Change Biology 11: 945–958.
25. Silva JMN, Carreiras JMB, Rosa I, Pereira JMC (2011) Greenhouse gas emissions from shifting cultivation in the tropics, including uncertainty and sensitivity analysis. Journal of Geophysical Research-Atmospheres 116. doi:10.1029/2011JD016056.
26. Laurance S, Laurance W, Andrade A, Fearnside P, Harms K, et al. (2010) Influence of soils and topography on Amazonian tree diversity: a landscape-scale study. Journal of Vegetation Science 21: 96–106.
27. Saatchi S, Houghton R, Alvala R, Soares J, Yu Y (2007) Distribution of aboveground live biomass in the Amazon basin. Global Change Biology 13: 816–837.
28. Laurance WF, Camargo JLC, Luizao RCC, Laurance SG, Pimm SL, et al. (2011) The fate of Amazonian forest fragments: A 32-year investigation. Biological Conservation 144: 56–67.
29. van Gardingen PR, Valle D, Thompson I (2006) Evaluation of yield regulation options for primary forest in Tapajos National Forest, Brazil. Forest Ecology and Management 231: 184–195.
30. Bacha CJC, Rodriguez LCE (2007) Profitability and social impacts of reduced impact logging in the Tapajos National Forest, Brazil - A case study. Ecological Economics 63: 70–77.

31. Espírito-Santo FDB, Shimabukuro YE, Aragão LEOC, Machado ELM (2005) Analysis of the floristic and phytosociologic composition of Tapajós national forest with geographic support of satellite images. Acta Amazonica 35: 155–173.
32. Miranda EE (2013) Sustentabilidade Agrícola na Amazonia - 23 anos de monitoramento da agricultura em Machadinho d'Oeste (RO). Embrapa Monitoramento por Satélite. Available: http://www.machadinho.cnpm.embrapa.br/. Accessed 11 March 2014.
33. Batistella M, Moran EF (2005) Dimensões humanas do uso e cobertura das terras na Amazônia: uma contribuição do LBA. Acta Amazonica 35: 239–247.
34. Silver WL, Neff J, McGroddy M, Veldkamp E, Keller M, et al. (2000) Effects of soil texture on belowground carbon and nutrient storage in a lowland Amazonian forest ecosystem. Ecosystems 3: 193–209.
35. Bierregaard RO (2001) Lessons from Amazonia: the ecology and conservation of a fragmented forest. New Haven, USA: Yale University Press. 478 p.
36. Laurance WF, Fearnside PM, Laurance SG, Delamonica P, Lovejoy TE, et al. (1999) Relationship between soils and Amazon forest biomass: A landscape-scale study. Forest Ecology and Management 118: 127–138.
37. Keller M, Varner R, Dias JD, Silva H, Crill P, et al. (2005) Soil-atmosphere exchange of nitrous oxide, nitric oxide, methane, and carbon dioxide in logged and undisturbed forest in the Tapajos National Forest, Brazil. Earth Interactions 9: 1–28.
38. Prates-Clark C, Lucas R, dos Santos J (2009) Implications of land-use history for forest regeneration in the Brazilian Amazon. Canadian Journal of Remote Sensing 35: 534–553.
39. Chander G, Markham B, Helder D (2009) Summary of current radiometric calibration coefficients for Landsat MSS, TM, ETM+, and EO-1 ALI sensors. Remote Sensing of Environment 113: 893–903.
40. Lillesand TM, Kiefer RW (2000) Remote sensing and image interpretation. New York, USA: John Wiley & Sons. 724 p.
41. Navulur K (2007) Multispectral image analysis using the object-oriented paradigm. Boca Raton, USA: CRC Press/Taylor & Francis. 165 p.
42. Lucas R, Xiao X, Hagen S, Frolking S (2002) Evaluating TERRA-1 MODIS data for discrimination of tropical secondary forest regeneration stages in the Brazilian Legal Amazon. Geophysical Research Letters 29: 42-1-42-4.
43. Hastie T, Tibshirani R, Friedman JH (2009) The elements of statistical learning: data mining, inference, and prediction. New York (USA): Springer. 745 p.
44. Roberts D, Numata I, Holmes K, Batista G, Krug T, et al. (2002) Large area mapping of land-cover change in Rondonia using multitemporal spectral mixture analysis and decision tree classifiers. Journal of Geophysical Research-Atmospheres 107: LBA 40-1-LBA 40-18.
45. Puyravaud J (2003) Standardizing the calculation of the annual rate of deforestation. Forest Ecology and Management 177: 593–596.
46. Pereira MC, Setzer AW (1993) Spectral characteristics of fire scars in Landsat-5 TM images of Amazonia. International Journal of Remote Sensing 14: 2061–2078.
47. Cochrane M, Souza C (1998) Linear mixture model classification of burned forests in the Eastern Amazon. International Journal of Remote Sensing 19: 3433–3440.
48. Foody GM (2009) Sample size determination for image classification accuracy assessment and comparison. International Journal of Remote Sensing 30: 5273–5291.
49. Gallego FJ (2004) Remote sensing and land cover area estimation. International Journal of Remote Sensing 25: 3019–3047.
50. Walsh TA, Burk TE (1993) Calibration of satellite classifications of land area. Remote Sensing of Environment 46: 281–290.
51. Tenenbein A (1972) A double sampling scheme for estimating from misclassified multinomial data with applications to sampling inspection. Technometrics 14: 187–202.
52. Ferraz SFD, Vettorazzi CA, Theobald DM, Ballester MVR (2005) Landscape dynamics of Amazonian deforestation between 1984 and 2002 in central Rondonia, Brazil: assessment and future scenarios. Forest Ecology and Management 204: 67–83.
53. Kimes DS, Nelson RF, Salas WA, Skole DL (1999) Mapping secondary tropical forest and forest age from SPOT HRV data. International Journal of Remote Sensing 20: 3625–3640.
54. Metzger JP (2002) Landscape dynamics and equilibrium in areas of slash-and-burn agriculture with short and long fallow period (Bragantina region, NE Brazilian Amazon). Landscape Ecology 17: 419–431.
55. Vieira ICG, de Almeida AS, Davidson EA, Stone TA, de Carvalho CJR, et al. (2003) Classifying successional forests using Landsat spectral properties and ecological characteristics in eastern Amazonia. Remote Sensing of Environment 87: 470–481.
56. Kuplich TM (2006) Classifying regenerating forest stages in Amazonia using remotely sensed images and a neural network. Forest Ecology and Management 234: 1–9.
57. Prates-Clark CC (2004) Remote sensing of tropical regenerating forests in the Brazilian Amazon. PhD dissertation. Aberystwyth, Wales, UK: University of Wales.
58. Fearnside PM (2005) Deforestation in Brazilian Amazonia: History, rates, and consequences. Conservation Biology 19: 680–688.
59. Kirby KR, Laurance WF, Albernaz AK, Schroth G, Fearnside PM, et al. (2006) The future of deforestation in the Brazilian Amazon. Futures 38: 432–453.
60. Nepstad DC, McGrath DG, Soares B (2011) Systemic Conservation, REDD, and the Future of the Amazon Basin. Conservation Biology 25: 1113–1116.
61. Pacheco P (2009) Agrarian Reform in the Brazilian Amazon: Its Implications for Land Distribution and Deforestation. World Development 37: 1337–1347.
62. Rosa IMD, Purves D, Souza C Jr, Ewers RM (2013) Predictive Modelling of Contagious Deforestation in the Brazilian Amazon. PLoS ONE 8(10): e77231. doi:10.1371/journal.pone.0077231.
63. Nolte C, Agrawal A, Silvius KM, Soares-Filho BS (2013) Governance regime and location influence avoided deforestation success of protected areas in the Brazilian Amazon. Proceedings of the National Academy of Sciences of the United States of America 110: 4956–4961.
64. Fearnside PM, Graca P (2006) BR-319: Brazil's manaus-porto velho highway and the potential impact of linking the arc of deforestation to central Amazonia. Environmental Management 38: 705–716.
65. Scatena FN, Walker RT, Homma AKO, deConto AJ, Ferreira CAP, et al. (1996) Cropping and fallowing sequences of small farms in the "terra firme" landscape of the Brazilian Amazon: A case study from Santarem, Para. Ecological Economics 18: 29–40.
66. Brondizio ES, Moran EF (2012) Level-dependent deforestation trajectories in the Brazilian Amazon from 1970 to 2001. Population and Environment 34: 69–85.
67. Perz SG, Skole DL (2003) Social determinants of secondary forests in the Brazilian Amazon. Social Science Research 32: 25–60.
68. Nepstad DC, Verissimo A, Alencar A, Nobre C, Lima E, et al. (1999) Large-scale impoverishment of Amazonian forests by logging and fire. Nature 398: 505–508.
69. Uhl C (1987) Factors controlling succession following slash-and-burn agriculture in Amazonia. Journal of Ecology 75: 377–407.
70. Buschbacher R, Uhl C, Serrao EAS (1988) Abandoned pastures in eastern Amazonia. II. Nutrient stocks in the soil and vegetation. Journal of Ecology 76: 682–699.
71. Uhl C, Buschbacher R, Serrao EAS (1988) Abandoned pastures in eastern Amazonia. I. Patterns of plant succession. Journal of Ecology 76: 663–681.
72. Saldarriaga JG, West DC, Tharp ML, Uhl C (1988) Long-term chronosequence of forest succession in the upper Rio Negro of Colombia and Venezuela. Journal of Ecology 76: 938–958.
73. Uhl C, Jordan CF (1984) Succession and nutrient dynamics following forest cutting and burning in Amazonia. Ecology 65: 1476–1490.
74. Chazdon RL, Letcher SG, van Breugel M, Martinez-Ramos M, Bongers F, et al. (2007) Rates of change in tree communities of secondary Neotropical forests following major disturbances. Philosophical Transactions of the Royal Society B-Biological Sciences 362: 273–289.
75. Mesquita R, Ickes K, Ganade G, Williamson G (2001) Alternative successional pathways in the Amazon Basin. Journal of Ecology 89: 528–537.
76. Uhl C, Clark K, Clark H, Murphy P (1981) Early plant succession after cutting and burning in the upper Rio Negro region of the Amazon basin. Journal of Ecology 69: 631–649.
77. Chazdon RL (2008) Beyond deforestation: Restoring forests and ecosystem services on degraded lands. Science 320: 1458–1460.
78. Zarin DJ, Ducey MJ, Tucker JM, Salas WA (2001) Potential biomass accumulation in Amazonian regrowth forests. Ecosystems 4: 658–668.
79. Brown S, Lugo A (1990) Tropical Secondary forests. Journal of Tropical Ecology 6: 1–32.
80. Lu DS, Moran E, Batistella M (2003) Linear mixture model applied to Amazonian vegetation classification. Remote Sensing of Environment 87: 456–469.
81. Ewers RM, Didham RK, Pearse WD, Lefebvre V, Rosa IMD, et al. (2013) Using landscape history to predict biodiversity patterns in fragmented landscapes. Ecology Letters 16: 1221–1233.

13

Soil Bacterial Community Response to Differences in Agricultural Management along with Seasonal Changes in a Mediterranean Region

Annamaria Bevivino[1]*, Patrizia Paganin[1], Giovanni Bacci[2,3], Alessandro Florio[2], Maite Sampedro Pellicer[1], Maria Cristiana Papaleo[3], Alessio Mengoni[3], Luigi Ledda[4], Renato Fani[3], Anna Benedetti[2], Claudia Dalmastri[1]

1 ENEA (Italian National Agency for New Technologies, Energy and Sustainable Economic Development) Casaccia Research Center, Technical Unit for Sustainable Development and Innovation of Agro-Industrial System, Rome, Italy, **2** Consiglio per la Ricerca e la Sperimentazione in Agricoltura - Research Centre for the Soil-Plant System, Rome, Italy, **3** Laboratory of Microbial and Molecular Evolution, Department of Biology, University of Florence, Florence, Italy, **4** Dipartimento di Agraria, University of Sassari, Sassari, Italy

Abstract

Land-use change is considered likely to be one of main drivers of biodiversity changes in grassland ecosystems. To gain insight into the impact of land use on the underlying soil bacterial communities, we aimed at determining the effects of agricultural management, along with seasonal variations, on soil bacterial community in a Mediterranean ecosystem where different land-use and plant cover types led to the creation of a soil and vegetation gradient. A set of soils subjected to different anthropogenic impact in a typical Mediterranean landscape, dominated by *Quercus suber* L., was examined in spring and autumn: a natural cork-oak forest, a pasture, a managed meadow, and two vineyards (ploughed and grass covered). Land uses affected the chemical and structural composition of the most stabilised fractions of soil organic matter and reduced soil C stocks and labile organic matter at both sampling season. A significant effect of land uses on bacterial community structure as well as an interaction effect between land uses and season was revealed by the EP index. Cluster analysis of culture-dependent DGGE patterns showed a different seasonal distribution of soil bacterial populations with subgroups associated to different land uses, in agreement with culture-independent T-RFLP results. Soils subjected to low human inputs (cork-oak forest and pasture) showed a more stable bacterial community than those with high human input (vineyards and managed meadow). Phylogenetic analysis revealed the predominance of *Proteobacteria*, *Actinobacteria*, *Bacteroidetes*, and *Firmicutes* phyla with differences in class composition across the site, suggesting that the microbial composition changes in response to land uses. Taken altogether, our data suggest that soil bacterial communities were seasonally distinct and exhibited compositional shifts that tracked with changes in land use and soil management. These findings may contribute to future searches for bacterial bio-indicators of soil health and sustainable productivity.

Editor: Jack Anthony Gilbert, Argonne National Laboratory, United States of America

Funding: This research was funded by MIUR (Integrated Special Fund for Research - FISR) in the frame of the Italian National Project SOILSINK "Climate change and agro-forestry systems, impacts on soil carbon sink and microbial diversity", and partially supported by MIUR (Research Department of Italian Government) in the framework of the Agreement Program ENEA-CNR (Articolo 2, comma 44, Legge 23.12.2009 n. 191 - Legge Finanziaria 2010). The funders had no role in study design, data collection and analysis, decision to publish, or preparation of the manuscript.

Competing Interests: The authors have declared that no competing interests exist.

* Email: annamaria.bevivino@enea.it

Introduction

Soil microorganisms play an important role as regulators of major biogeochemical cycles and can significantly affect the ecosystem functioning [1], being involved in organic matter dynamics, nutrient cycling and decomposition processes [2]. The anthropogenic activities affect the diversity of natural habitats modifying the number of species occurring in the environment at the landscape scale. Soil management strongly influences soil biodiversity in agricultural ecosystems. Different practices can alter the below-ground ecosystem, often leading to depletion of soil carbon and loss of biodiversity, and thus affecting the structure of the resident microbial communities [3]. Therefore, characterizing genetic and functional diversity of soil bacterial communities in response to agricultural practices and/or climate is fundamental to better understand and manage the ecosystem processes.

The Mediterranean area is one of the most important biodiversity hotspots in the world and is increasingly threatened by intensive land use [4]. The high environmental diversity that characterizes the Mediterranean region is related to the integration of natural ecosystems and traditional human activities such as the agroforestry practices [5]. The collapse of the traditional agro-silvo-pastoral system that occurred during the past century has led to major changes in the extension of woodlands dominated by typical Mediterranean species, i.e. cork oak (*Quercus suber*) and/or holm oak (*Quercus ilex*) woodlands [6], [7]. Research on the influence of management practices on the biodiversity of these agro-silvo-pastoral systems is increasing but it has focused mostly on plants [8] and vertebrates [5], [9]. In the frame of the Italian Project SOILSINK (Climatic changes and agricultural and forest systems: impact on C reservoirs and on soil microbial diversity), a hilly basin in Gallura (Berchidda site, Sardinia, Italy) was selected as a reference Mediterranean site for studying the influence of land-use changes on diversity, function and seasonal variations of soil microbial communities [10], [11]. The site is within an area of about 1,450 ha and is characterised by extensive agro-silvo-pastoral systems, typical of north-eastern Sardinia (Italy) and similar areas of the Mediterranean basin [12]. The chosen site represents a sustainable balance between human activities and natural resources that have created a landscape of high heterogeneity and cultural value, whose importance has been recognized at the European level [13], [14]. Indeed, it is considered climatically (Mediterranean zone) and pedologically homogeneous with vegetation patterns similar to those called *dehesas* or *montados* of the south-western Iberian Peninsula [15], [16], [17]. In the past, this area was covered by cork-oak forests, which gradually were subjected to increasing under-storey grazing and usage for the extraction of cork. Today, there are different land-use and plant cover types that lead to a soil and vegetation gradient with an ecological progression: from a cork-oak forest undergoing minimum disturbance to managed vineyards with an intensive agricultural practice (grass covered and ploughed), passing through areas with temporary grassland, and pasture.

The different land uses altered soil potential, making possible to discriminate the role of human management on soil functioning. When forests are converted to grasslands, and grasslands turned into agricultural lands, a sharp switch from one type of soil microbial community to another one occurs. Since the ability of an ecosystem to withstand serious disturbances may partly depend on its microbial component(s), characterizing bacterial community composition and/or structure might help to better understand and manipulate ecosystem processes. The aim of the present study was to investigate the effects of soil characteristics and different agricultural managements on soil bacterial community in two seasons. Sampling was carried out in spring and autumn 2007, when the plant cover-growing season usually starts and ends. A combination of culture-based and molecular techniques along with statistical analysis of data obtained was applied to interrogate the diversity, function, and ecology of soil bacterial communities. The results obtained in the present study, along with the other ones obtained within the SOILSINK Project [11], [18], [19], [20], provide useful data on the impact of soil type, cover vegetation, and human activities on the distribution of the bacterial genetic resources in soil communities for this Mediterranean region. Our results confirm that the environments with low inputs (cork-oak forest and pasture) show a more stable soil microbial community than those subjected to increasing human input (vineyards and managed meadow) and suggest that soil bacterial communities are seasonally distinct with compositional shifts that track with changes in land use and soil management.

Materials and Methods

Ethics Statement

We carried out the study on the hilly basin in Gallura (Olbia-Tempio municipalities, Sardinia, Italy). Five soils uses were identified in private farms within Berchidda site (40°49′ 15″N, 9°17′ 32″ E) and were obtained. The soil sampling was carried out in the frame of a national research project (SOILSINK Project) and soils used in this study were collected under consent of the landowners. The responsible of the study site was Prof. P.P. Roggero (University of Sassari, Italy). We confirm that our study did not harm the environment and did not involve endangered or protected species. Specific geographic coordinates (referred to World Geodetic System, 1984) of our study area are reported in Table S1.

Sampling site

The study area (Olbia-Tempio) is representative of the climate, vegetation type and management of some of the most common agro-forestry systems in the Mediterranean basin [16].

The Berchidda site is made up of hydromorphic and granitic soil with a loamy sand texture. The altitude ranges from 275 m to 300 msl. This area is referred to as a meso-thermo Mediterranean, subhumid phytoclimatic belt with a mean annual rainfall of 862 mm and the mean annual temperature of 13.8°C [10]. In the

Figure 1. Long-term effects of different land-use with increasing level of intensification in spring and autumn. Both pasture and managed meadow included spotted cork oak trees, which are key components of the Dehesatype landscape typical of this area of Sardinia. The cork-oak formation, pasture, and managed meadow have been converted to the current use and maintained unchanged for more than 30 years, whereas the non-tilled cover cropped vineyard and the tilled one were planted in 1985 and 1994, respectively. From the left to right: cork-oak forest (CO), hayland pasture rotation (PA), managed meadow (MM), grass covered vineyard (CV), tilled vineyard (TV).

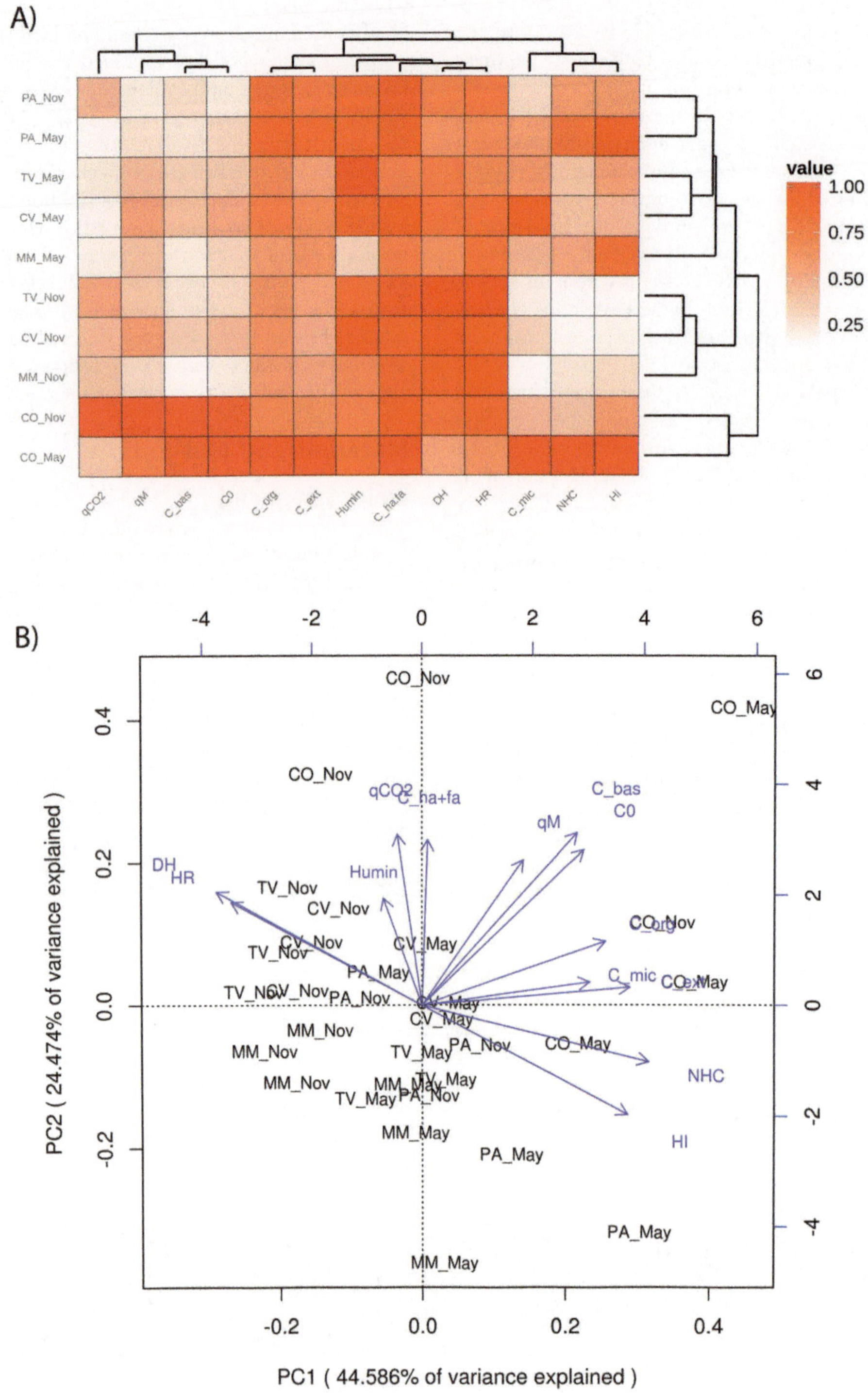

Figure 2. Effect of land-use and season on soil physical-chemical and biological parameters. A) Heat map with hierarchal clustering of physical-chemical and biological parameters across the five Sardinia soils with different land uses at the two different sampling time points (May and November). The heat map was constructed using a maximum-minimum normalization of the data in order to represent each value in a range between 0 and 1. Higher values are represented by darker colors whereas lower ones are represented by lighter colors. CO = cork-oak forest; PA = hayland-pasture rotation; MM = managed meadow; TV = tilled vineyard; CV = grass covered vineyard. B) PCA ordination of data (axes 1 and 2) generated from physical-chemical and biological properties of the different types of land use in May and November.

past, the Berchidda area was covered by cork-oak forests (dominated by *Quercus Suber* L.), which were subjected to intense usage for the extraction of cork. Today, there are five main different land-use units close together: cork-oak forest (CO), hayland-pasture rotation (PA), managed meadow (MM), and tilled (TV) and grass covered vineyard (CV). The cork-oak formation, pasture, and managed meadow have been converted to the current use and maintained unchanged for more than 30 years,

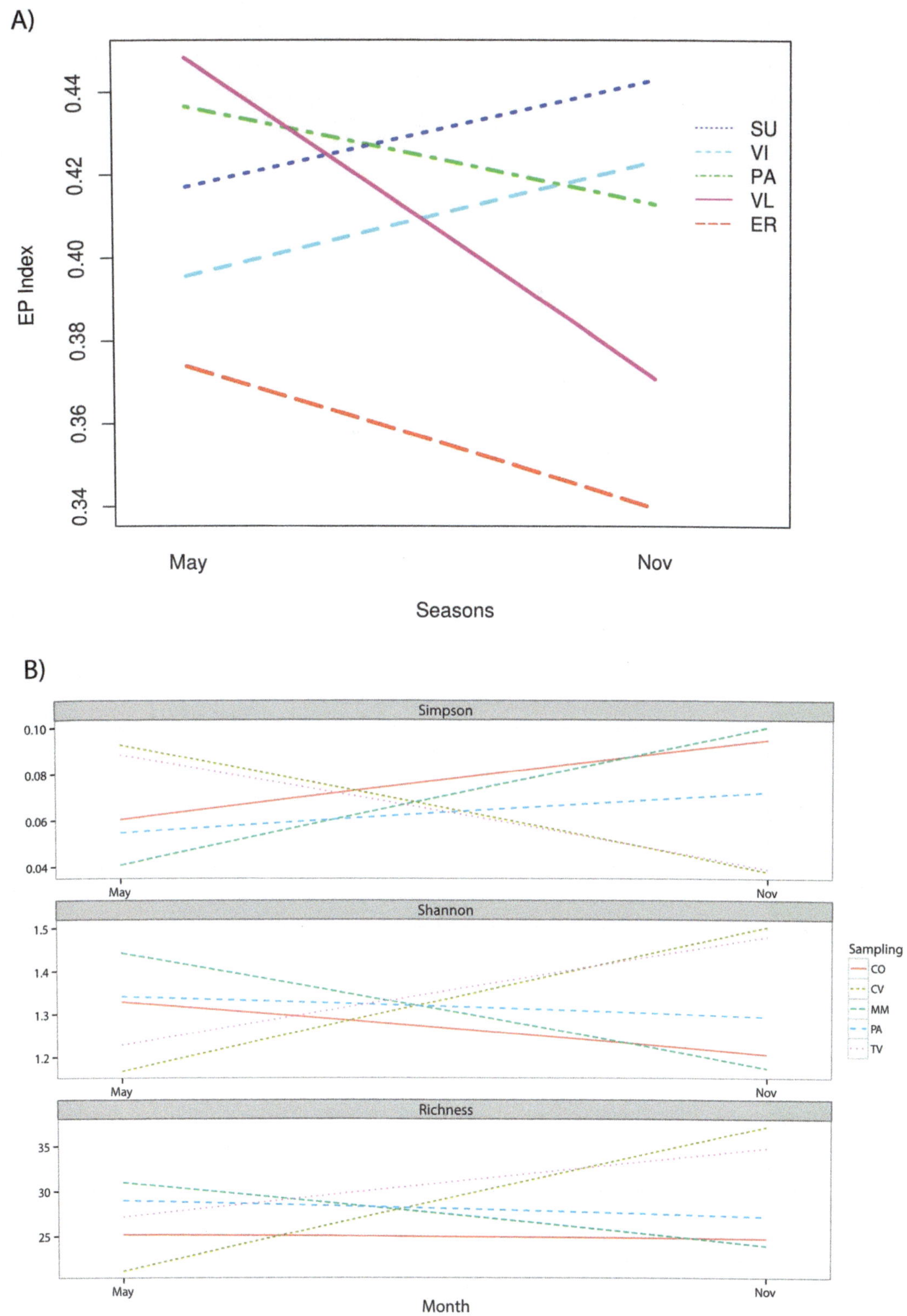

Figure 3. Effect of land-use and season on Eco-Physiological (EP) index of culturable bacteria (A) and diversity indices from CD-DGGE profiles (B).

whereas the non-tilled cover cropped vineyard and the tilled one were planted in 1985 and 1994, respectively. The five soils are located inside an area of 161.5 km^2. Detailed characteristic and management of the five soils have been previously described [12], [20], [21], [22], [23].

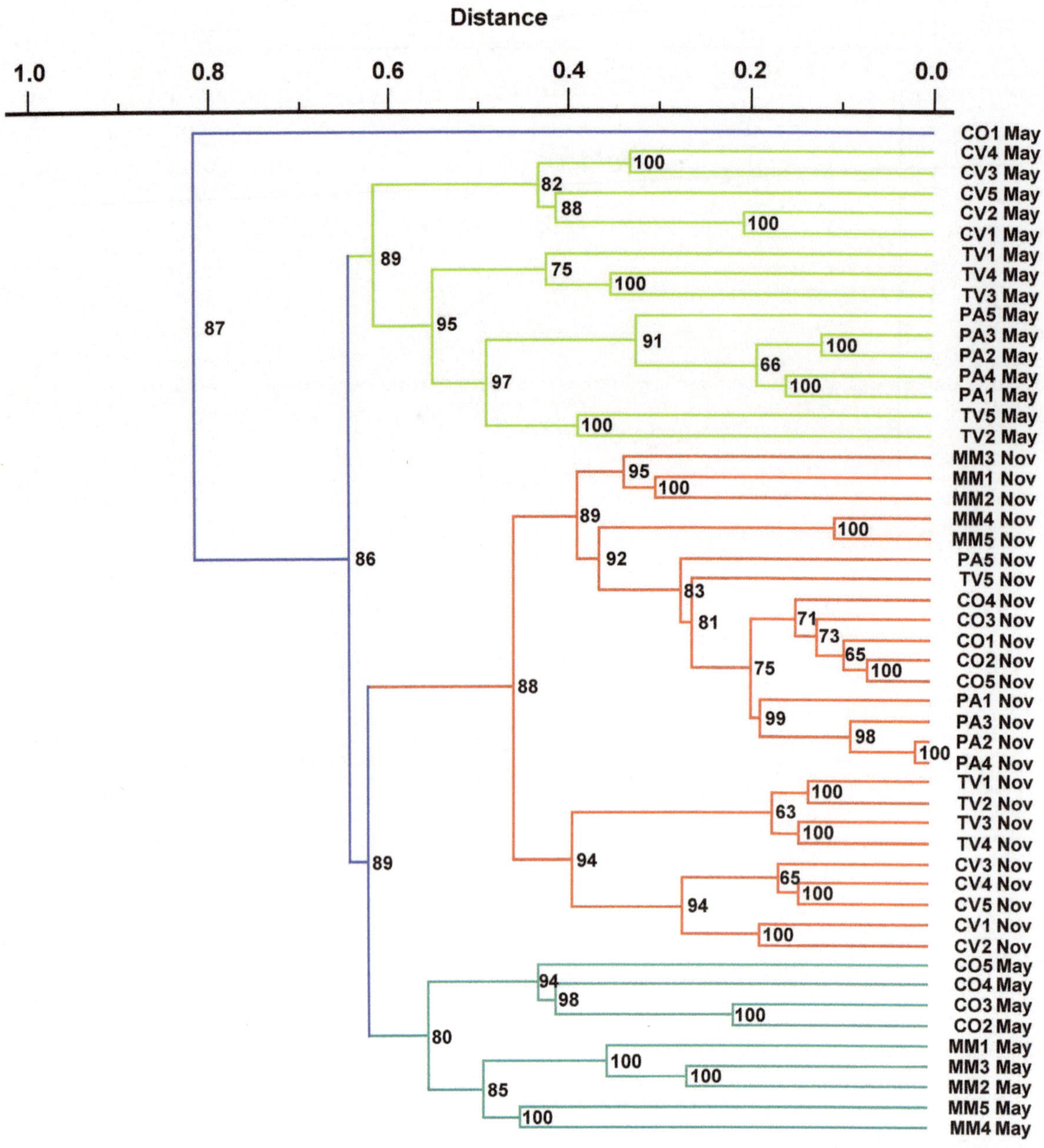

Figure 4. UPGMA dendrogram of DGGE profiles of amplified 16S rDNA of bacterial communities recovered in CO, PA, MM, CV and TV, in spring and autumn, generated using Phoretix ID advanced analysis package. Designation of samples is the same as in Fig. 1. The scale bar represents dissimilarity among samples. Consistency of each cluster was measured by the Cophenetic correlation coefficient shown at each node.

Pedological characterization of the study area

The pedogenic substrate of the study area consists of medium-grained granite, affected by localized presence of veins of quartz and porphyry. The morphology of the Berchidda area varies from flat to undulating. The processes of soil erosion by water channeled are evident only in the short-term forage crops made on soil with more than 15% of slope. All profiles have a horizons sequence of the type A-Bw-C or A-Bw-BC-C or more rarely, A-Bw-C-R (Table S1). The power of the profiles, limited to horizons A and Bw, varies from a minimum of 38 cm to a maximum of 100 cm. In three soil profiles in TV, the sequence between the horizon A and Bw and the substrate is gradually altered by the presence of a horizon BC, characterised by coarse texture, the power of which varies from about 30 cm to 90 cm. Direct contact with the unaltered rock, R horizon, was observed only in one profile in CV. The prevailing textural classes are sandy and sandy loam. The content of organic substance in the horizons A is never very high, with average value around to about 3%. The maximum values of 11.8% and 8.6% were observed on the A horizons in two soil profiles in CO. The exchange complex, in agreement with the reduced content in clays, is never high. The degree of base saturation is predominantly less than 60% (Dystric conditions of the USDA Soil Taxonomy). The profiles with the exchange complex with a degree of base saturation below 60% in all horizons between 25 and 75 cm (TV, CV, MM, and PA) were

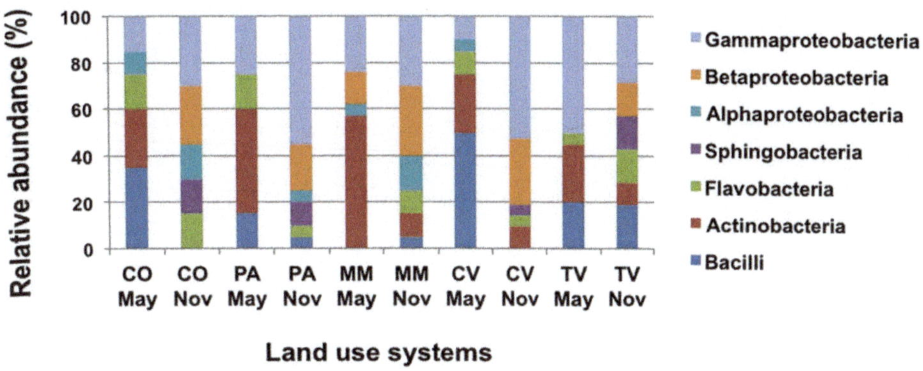

Figure 5. Relative abundances of major taxonomic groups across land use systems in spring and autumn. Detailed data of each class are listed in Additional file 1: Table S6. CO = cork-oak forest; PA = hayland-pasture rotation; MM = managed meadow; TV = tilled vineyard; CV = grass covered vineyard. Values presented are the mean percent.

classified as Typic Dystroxerepts [24]. The other one (CO) was classified as Lithic Xerorthents.

Soil sampling

Five soil replicates were collected from bulk soil of the five different managements (CO, PA, MM, CV, and TV) (Fig. 1). Soils were collected in May and November 2007. After removal of litter layer, soil core samples (50 to 100 g; diameter, 5 cm) were taken from each of the five locations, using a 5-on-dice sampling pattern with ca. 70 m distance between each sampling point. Sampling was performed at 20 cm depth, where most microbial activity is known to occur [25], [26]. In the vineyard soils, samples from along the rows and between the rows were pooled together to form a field replicate. The other soil samples (CO, PA, MM) were collected out of trees influence. At each season and for each soil type, five randomly field replicates were collected for a total of 25 soil samples (5 replicates × 5 land uses), each one being a composite sample of five soil cores. A total of 50 composite samples were taken for the two seasons. Soil samples were immediately sieved (<2 mm) to remove fine roots and large organic debris, air dried, and transported to the labs for microbiological analysis. The moisture content was adjusted to 60% of their water holding capacity (WHC) and soil samples were then left to equilibrate at room temperature in the dark for one day prior to analyses, in order to restore, within limits [27], the microbial activity of air- dried soils to that of soils in the field.

Chemical and biochemical analyses of soil samples

The chemical and biochemical analyses were performed on three replicates for each land use for a total of 15 soil samples (3 replicates × 5 land uses), each one being a composite sample of five soil cores. Total organic carbon (C_{org}) was estimated after oxidation with $K_2Cr_2O_7$ and subsequent titration of unreduced $Cr_2O_7^{2-}$ with $Fe(NH_4)_2(SO_4)_2$, as reported by Springer and Klee [28]. Soil Organic Matter (SOM) was determined by C_{org} multiplied by 1.724 van Belem coefficient. The C_{org} fractionation was set up as reported by Ciavatta and co-workers [29]. In particular, solid samples were extracted at 65°C for 24 h using 0.1 mol l^{-1} NaOH plus 0.1 mol l^{-1} $Na_4P_2O_7$ solution (1:50, solid:liquid ratio). The samples were then centrifuged at 5000 ×*g* and the supernatants were filtered through a 0.20-µm Millipore filter (Millipore, Billerica, MA) (total extractable C, C_{ext}). The humic-like acid (HA) fraction was separated from the fulvic-like acid (FA) and the non-humified carbon fractions (NHC) by precipitation after acidification of the alkaline solution (supernatant) to pH<2. Chromatography on a column of polyvinylpyrrolidone (PVP, Aldrich, Germany) was used to separate the NHC from the FA. The FA was then combined with the HA to obtain total humified fraction (HA+FA). Total extractable C (C_{ext} %), humic and fluvic acid C (C_{HA+FA} %) were determined by the dichromate oxidation method. The non-humified carbon (C_{NH}) was determined as the difference between C_{ext} and C_{HA+FA}. Humification indexes HI, DH, and HR were determined according to previous works [29], [30].

Microbial biomass C (C_{mic}) was determined by the fumigation-extraction method of Vance and co-workers [31] with some slight modifications. The measurements were performed on air-dried soils, pre-conditioned by a 10-d incubation in open glass jars, at −33 kPa water tension, and 30°C. The incubation was employed for restoring, within limits [27], the microbial activity of air-dried soils to that of soils in the field. Four replicates of each soil sample were used. Average values are given in mg C kg^{-1} of soil. For measuring microbial respiration, 20 g (oven-dry basis) of moist sample were placed in 1 L stoppered glass jars. The CO_2 evolved was trapped, after 1, 2, 4, 7, 10, 14, 21, 28 days of incubation, in 2 ml 1 M NaOH and determined by titration of the excess NaOH with 0.1 M HCl [32].

Non-linear least square regression analysis was used to calculate parameters affecting C mineralization from daily CO_2 evolution data (Stat Win 4.0 for Windows). The best fit was obtained with the exponential model of CO_2-C accumulation according to the negative exponential decay model:

$$C_m = C_0\left(1 - e^{-kt}\right),$$

where C_m is the cumulative value of mineralized C during *t* days, *k* is the rate constant, and C_o is the potentially mineralizable C [33]. The CO_2 emitted in 28 days of incubation was used as cumulative respiration (C_{cum}). The CO_2 evolved during the 28th day of incubation was used as the basal respiration value (C_{bas}). Microbial indices were calculated as follows [34], [35], [36]:

$$qCO_2 = [(\mu g\, CO_2 - C_{bas} \times \mu g^{-1} C_{mic})h^{-1}]10^3,$$
$$qM_{cum} = (\mu g\, CO_2 - C_0 \times \mu g^{-1} C_{org}),$$
$$qM_{bas} = (\mu g\, CO_2 - C_{bas} \times \mu g^{-1} C_{org})$$

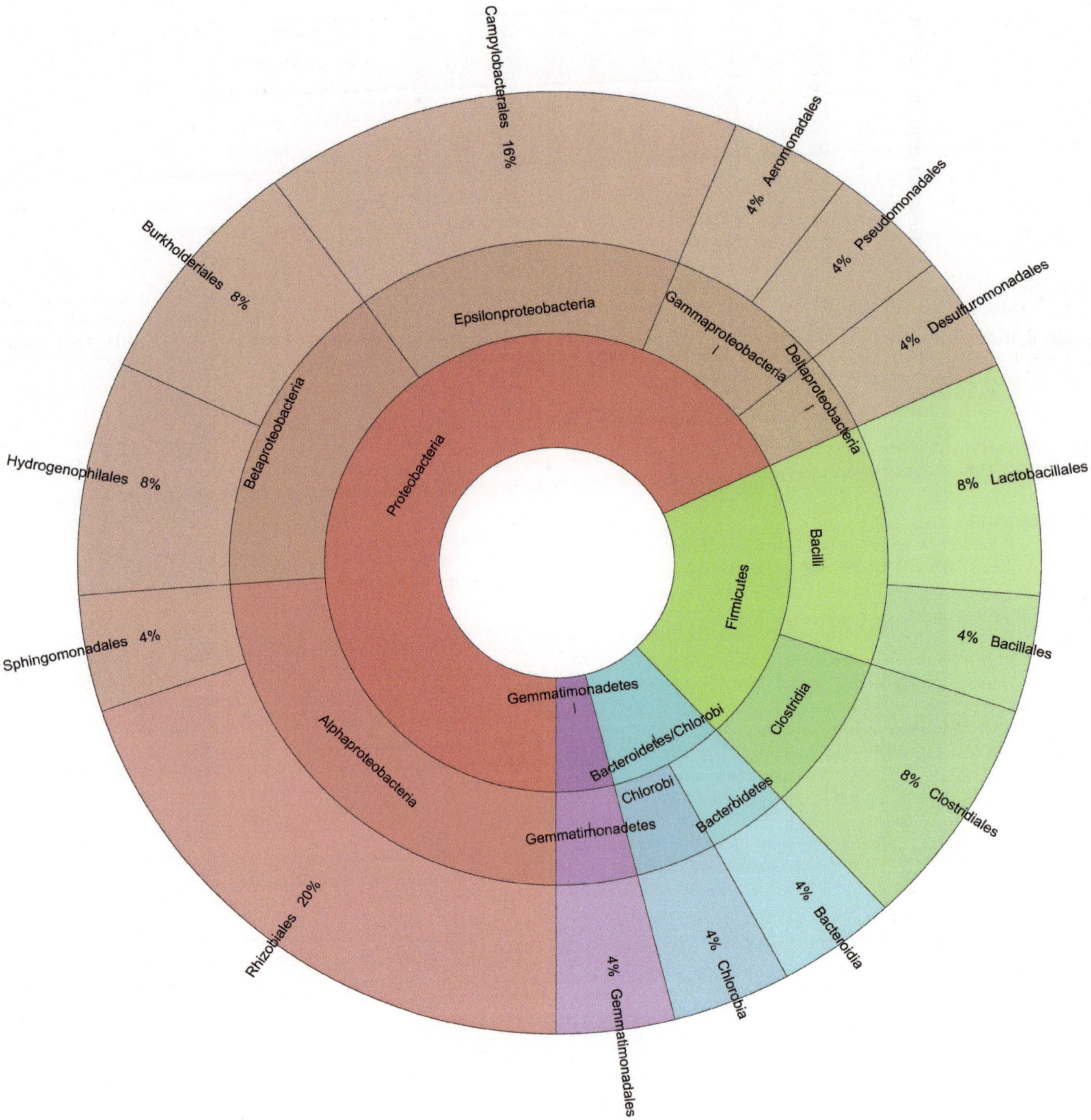

Figure 6. Plot of the taxonomic composition of total bacterial community as inferred from taxonomic interpretation of T-RFLP profiles.

Recovery of cultured bacterial cells from soil samples

Soil bacterial community analysis was performed on five soil replicates for each land use, as described above. Bacterial cell extraction was performed according to the recommendations of Smalla et al. [37] with minor modifications. Briefly, 1 g of soil was placed in a sterile 15 ml plastic tube containing 10 ml of phosphate buffered saline (PBS, pH 7.0). This mixture was homogenized for 30″ at low speed by using the Ultra-Turrax Thyristor Regle 50 (Janke & Kunkel IKA-Labortechnik). After homogenization, suspension was placed into a Erlenmeyer flask (100 ml) containing 10 g of glass beads (0,2 mm) and shaken for 1 h at 180 r.p.m. and 28°C to disperse bacteria. The flasks and glass beads were autoclaved for 20 min at 121°C before use. The soil suspension was transferred into a sterile 15 ml plastic tube and serially diluted with sterile saline solution (9 g l^{-1}NaCl) from 10^{-1} up to 10^{-7}. The, 100 μl aliquots of serially diluted soil suspensions were plated in triplicate on 0.1 Triptic Soy Broth (TSB; Difco) amended with 15 g l^{-1} agar (0.1 TSA) and 100 g μl^{-1} of cycloheximide (Sigma) to inhibit fungal growth. Plates were incubated at 28°C for 6 days.

Growth strategy and total bacterial populations

To determine the changes in the structure of culturable fraction of soil bacteria, the r/K-strategy concept proposed by De Leij and co-workers [38] was used. Bacterial colonies appearing within 48 h were designated as r-strategists, and the remaining as K-strategists. Colonies were enumerated at 1, 2 and 6 days of growth on 0.1 TSA; in this way, three counts (or classes) were generated per sample. Plates containing between 30 and 300 colonies were then selected for enumeration. Total bacterial counts obtained were expressed as colony forming units (CFU) per gram of soil. Distribution of bacteria in each class as a percentage of the total counts gave insight into the distribution of r- and K-strategists in each sample.

To evaluate the changes in the biodiversity of bacterial populations in soils, the eco-physiological (EP) index [38] was used. The EP index of each soils tested was calculated using the equation:

$$H' = -\sum(P_i)(log_e P_i),$$

where P_i represents the CFU on each day (1, 2 and 6 days of incubation) as a proportion of the total CFU in that sample after 6 days incubation, i.e., the proportion of colonies appearing on counting day i ($i = 1, 2, 6$) with $EP_{min} = 0$. Higher values of EP index imply a more even distribution of proportions of bacteria developing on different days (i.e., different classes of bacteria).

Terminal-Restriction Fragment Length Polymorphism (T-RFLP)

DNA was extracted from soil samples by using the FastDNA SPIN Kit for Soil (QBiogene). Terminal-Restriction Fragment Length Polymorphism (T-RFLP) was performed on 16SrRNA genes amplified from extracted DNA with primer pairs P0 and P6 as previously reported [39], [40]. Purified amplification products were digested separately with restriction enzymes *Rsa*I and *Msp*I and digestions were resolved by capillary electrophoresis on an ABI310 Genetic Analyzer (Applied Biosystems, Foster City, CA, USA) using LIZ 500 (Applied Biosystems) as size standard. T-RFLP analysis was performed as previously reported [41]. Diversity indices were calculated with PAST software [42] as previously reported [19], taking into account peak intensities of T-RFLP fragments. Taxonomic interpretation of T-RFLP profiles was performed by querying the Ribosomal Database Project Database by using MiCA3 web tool (http://mica.ibest.uidaho.edu/), as previously described [40].

Culture-dependent DGGE (CD DGGE) analysis

Culture-dependent DGGE (CD DGGE) fingerprinting of 16S rRNA gene was used to characterize mixed bacterial communities recovered on agar plates.

Collection of cultured bacterial communities for DGGE analysis. Cultured bacterial communities were collected following procedure proposed by Duineveld and co-workers [43] with minor modifications. Briefly, after one week of incubation at 28°C, colonies were removed from plates containing between 100 and 1000 colonies by adding 3.0 ml sterile physiological solution (0.9% NaCl) on each plate and scraping off all grown colonies with a sterile Drigalski spatula. The cell suspensions thus obtained were aliquoted into 1.5 ml Eppendorf tubes and centrifuged at 8,000 r.p.m. for 10 minutes. The pellets were stored at −80°C for subsequent DNA extraction and PCR-DGGE analysis.

DNA extraction and PCR amplification of 16S rRNA genes. Genomic DNA was extracted with sodium dodecyl sulfate-proteinase K lysis buffer, followed by a treatment with cetyl-trimethylammonium bromide (CTAB) as described in *Current Protocols for Molecular Biology* [44]. Briefly, the pellet was resuspended in 567 µl of TE [10 mM Tris-HCl - 1 mM EDTA (pH 8.0)] buffer, and glass beads 0.3 mm in diameter (250 mg) were added, followed by bead beating for 20 s. Then, 30 µl of 10% sodium dodecyl sulfate and 3 µl proteinase K 20 mg/ml (Sigma) were added and samples were incubated at 37°C for 1 h. Glass beads were removed by centrifugation 2 min at 2,800 rpm and samples were then incubated at 65°C for 10′ with100 µl of 5 M NaCl prepared with sterile water and 80 µl of CTAB/NaCl (10% CTAB in 0.7 M NaCl). Following incubation, extracts were purified by using phenol/phenol-chloroform/isoamyl alcohol (49.5:49.5:1) extraction and DNA was recovered by isopropanol precipitation at 4°C o/n. Pelleted DNA was washed twice with cold 70% ethanol, allowed to air dry, and re-suspended in 50 µl of sterile water. Quantity and purity of DNA were checked by NanoDrop (NanoDrop Technologies, USA) and gel electrophoresis. The DNA samples were stored at −20°C until required for use.

The 16S rRNA gene was amplified using 20 ng of lysate suspension and the universal bacterial primers P0 and P6 [45]. Dilutions 1:100 (2 µl) of the1450 bp PCR products were then used as template for the second PCR amplification with the forward primer 63F (5′- AGGCCTAACACATGCAAGTC -3′), with a GC clamp (5′-CGCCCGCCGCGCGCGGCGGGCGGGGCG-GGGGCACGGGGGG -3′) incorporated at the 5′ end, and the reverse primer 518R (5′-ATTACCGCGGCTGCTGG-3′), to produce 495 bp fragments suitable for DGGE analysis [46]. Both PCR reactions were performed in Qiagen Taq buffer (10X) containing 1.5 mM $MgCl_2$, with 150 ng of each primer, 250 µM (each) deoxynucleoside triphosphates, and 0.5 U of *Taq* DNA polymerase (Qiagen, Hilden, Germany) in a 25 µl reaction volume. Cycle parameters for PCR with the primer pairs P0–P6 and 63F-GC and 518R were previously described by Di Cello and co-workers [45] and El Fantroussi and co-workers [46], respectively.

Denaturing Gradient Gel Electrophoresis. 16S rRNA gene amplicons were separated by double gradient denaturing gradient gel electrophoresis (DG-DGGE) as described by Cremonesi et al. [47], in a DCode universal mutation detection system (Bio-Rad, CA, USA). Separation of purified PCR products (700 ng) was achieved in6%–12% polyacrylamide (acrylamide: N,N-methylenebisacrylamide, 37.5:1) gels containing an increasing linear gradient of denaturants ranging from 30% to 60% (100% denaturant corresponds to 7 M urea and40% deionized formamide). Each gel also included marker lanes represented by DGGE profiles containing a large number of discrete bands spanning the entire gradient, suitable for within- and between-gel alignment. Electrophoresis were carried out for 16 h at 75 mV in 1X TAE buffer at 60°C, stained with 50 µg/ml ethidium bromide for 30 min, destained in water and photographed with the UVIpro Platinum Gel Documentation System (GAS7500/7510; Eppendorf, Cambridge,UK).

Cluster analysis and diversity indices

Quantity One software package (Bio-Rad) and Phoretix 1D PRO software (Phoretix International, Newcastle upon Tyne, United Kingdom) were used for CD-DGGE profile analysis. The cluster analysis and dendrogram generation were carried out by using the Phoretix 1D Pro software according to the manufacturer's instructions (Phoretix International, Newcastle upon Tyne,

United Kingdom). Bands of CD-DGGE patterns were aligned and normalized using reference lanes. Background noise was subtracted by rolling ball algorithm with a radius of 50 pixels; the automatic band detection was performed with a minimum slope of 200 and a noise reduction of 10, and peaks smaller than 2% of the maximum peak were discarded. Bands were manually corrected and matched to create an absent/present binary matrix. The similarity between the band patterns was calculated using the Dice coefficient and the clustering analysis was performed with the unweighted pair group method with arithmetic averages (UPGMA) to generate a dendrogram by using mathematic averages algorithm programs integral to the Phoretix 1D Pro software. Coefficient of cophenetic correlation was used to measure the consistency of clusters.

DGGE banding data were used to estimate three diversity indices by treating each band as an individual operational taxonomic unit (OTU). The number of DGGE bands present in each sample was used to measure the Richness index (*R*). The Shannon-Weaver index of general diversity (H') [48] and the Simpson index of dominance (*D*) [49] were calculated from the number of bands present and the relative intensities of each band (P_i) in each lane. Relative signal intensities of detected bands, in each gel track, were determined by using the Quantity One software package (Bio-Rad) and calculated from the peak area of the densitometric curves.

The Shannon-Weaver diversity (H') was calculated using the following equation:

$$H' = -\sum(P_i)(log_e P_i),$$

where P_i (the proportion of abundances of the ith band) is measured as:

$$P_i = n_i/N,$$

where n_i is the peak height of a band, and N is the sum of all peak heights in the densitometry profiles.

The Simpson index (D) was calculated with the formula:

$$D = \sum(P_i x P_i);$$

it measures the strength of dominance because it weights towards the abundance of the OTUs and varies inversely with species diversity [50].

Bacterial isolation and identification

A total of 100 bacterial colonies were randomly picked up for each soil sample from 0.1 TSA plates (containing approximately 50 to 500 colonies), previously used for the determination of the CFU counts and EPI-index, and repeatedly streaked onto 0.1 TSA fresh plates to obtain pure cultures. Isolated colonies were then grown overnight (o/n) in TSB medium at 28°C and 200 r.p.m., and stored at −80°C in 30% glycerol until further analysis. From all five soil samples, 500 colonies were isolated in each season, for a total of 1000 bacterial colonies. A total of 203 colonies with different morphologies (about 20 colonies per each sample) were taken up to investigate their taxonomic affiliation.

Genomic DNA and PCR amplification of the 16S rRNA gene were performed as described above. Sequencing reactions were prepared from PCR products using an Applied Biosystem Big Dye Terminator sequencing kit version 3.1, according to the manufacturer's instructions and analysed using a 3730 DNA Analyzer Applied Biosystem apparatus. The sequences were compared with those in the GenBank databases by using the BLAST program and Seqmatch tool of the RDP (http://www.ncbi.nlm.nih.gov/BLAST/and http://rdp.cme.msu.edu/, respectively) and aligned with the closest relatives with the Clustal W function of the BioEdit package [51]. Bacterial identification by 16S rRNA gene sequences assignment was performed using the RDP Classification Algorithm (http://rdp.cme.msu.edu/classifier/classifier.jsp).

Statistical analysis

Bacterial population data (CFU/g of soil) were log transformed and subsequently analysed by one-way ANOVA (STATISTICA, Release 3.0b). Percentage data of EP index value were logit-transformed, as follows:

$$Logit\,(p) = log[p/(1-p)]$$

for the proportion *p*, and compared using one-way ANOVA (STATISTICA, Release3.0b).

Analyses of variance (ANOVA) on biodiversity indexes (Shannon-Weaver, Richness and Simpson), principal component analysis (PCA) and clustering analysis on biochemical data were performed using R packages "stats" and "vegan" (http://cran.r-project.org/and http://cran.r-project.org/web/packages/vegan/index.html). All data clustering were performed using the "UPGMA" algorithm implemented in the "hclust" function of the R "stats" package. Distances among samples were calculated using "Bray-Curtis" distance implemented in "vegan" package as:

$$i = sample\text{ "}i\text{"};\ j = sample\text{ "}j\text{"},\ BC_{ij} = 2C_{ij}/\left(S_i + S_j\right),$$

where C_{ij} = sum of the smaller value for species in common between samples i and j, and S_i and S_j = total number of species in samples i and j, respectively [52]. Variation of biodiversity indexes of cultured bacteria was inspected using ANOVA analysis. Biodiversity indexes were first divided into groups depending on sampling season and different managements of soils and then the analysis was performed. Clustering analysis (UPGMA) on biochemical parameters was performed. Each parameter was first divided in groups, in the same way of previous ANOVA analysis, and averaged. Then, each result obtained was normalized using the maximum-minimum normalization technique in order to make the data comparable. PCA analysis using each biochemical data was performed.

Results and Discussion

Effect of land use on soil chemical and biochemical properties

Soil organic matter (SOM) represents a dynamic system influenced by several factors, including climate, clay content, mineralogy and soil management, which all affect the processes of organic matter transformation and evolution in soil [53], [54]. Both soil fertility and stability are related to the organic matter content of soil. Many functions of SOM are due to its more stabilised fractions, humified materials and balance between the labile and the stabilised fractions [55]. Changes in SOM content are related to changes in microbial biomass turnover, because they reflect the balance between rates of microbial organic matter accumulation and rates of organic matter degradation. The extent of organic matter's organization not only impacts the amount of carbon mineralized but also the type of carbon that is consumed by the microorganisms.

In this study two categories of soil quality indicators were used: organic matter quality indicators and microbial biomass activity indicators (Tables S2 and S3). Our results (Fig. S1) revealed that both land use and sampling season affect the chemical and structural composition of the most stabilised fractions of SOM. A higher content of C_{org} occurred in CO soil in May, and C_{ext} and C_{HA+FA} showed the same pattern, whereas the average values of these parameters were slightly higher in PA soil in November. These results were reflected in the humification parameters, where the humification rate (HR%) can provide quantitative information about the humic substances content normalised with respect to total SOM, while the degree of humification (DH%) provides the amount of the humified carbon in the extracted organic fraction and the humification index (HI) can be considered as an index of soil humification activity as well as of availability of non humified labile fractions [29]. Overall, land use change reduced soil C stocks and labile organic matter at both sampling times except for PA in November. Pasture has a great potential soil organic C stock and, in the long term, grass management systems have nearly equivalent potential to store soil organic C as forest [56]. Potentially mineralizable C (C_0), which indicates the amount of C in the labile fraction of soil organic matter, decreased over sampling time in all soils, and C_{mic} similarly declined. Cultivated soils are characterised by low microbial activity, mainly due to the disappearance of easily decomposable organic compounds through tillage and soil disturbance. As previously found for chemical properties, the metabolic activity responds to the different land uses; in both May and November samplings, qCO_2 was higher in CO soil when compared to the others, resulting in increases stress. Hence, unfavourable conditions result in a decrease in the size of the microbial biomass and the efficiency of C substrates degradation, conducting to an increase in respiration rate per unit of microbial biomass [57].

As beneficial and negative effects of soil management practices are strongly linked to microbial activities and regulate soil quality and functioning [58] the relationship existing between soil management practices and variation in chemical and biochemical parameters was evaluated, by performing a clustering analysis (UPGMA) as reported in Tables S2 and S3. The dendrogram obtained was linked to a heat-map representing all the biochemical parameters variation (Fig. 2A). Data analysis revealed that chemical and biochemical parameters clustered in two different groups each of which corresponding to one of the two parameters analyzed (chemical and biochemical). In both sampling times, TV and CV clustered together, suggesting that seasonal change rather than management regime was a major driving force contributing to vineyard soil fertility. Furthermore, according to the UPGMA clustering, CO formed a separate cluster from that of the other soil uses (Fig. 2A). Interestingly, the chemical and biochemical parameters clustered together in both CO and PA, regardless of the season, in contrast to MM, CV and TV, where they clustered separately in relation to the sampling season. This finding suggests that soils subjected to low human inputs (pasture and cork-oak forest) showed a more stable chemical and biochemical soil composition than those with high human input (managed meadow and vineyard).

The different positions of the variables in the plane of the first two principal components, as revealed by PCA analysis (Fig. 2B), indicated that chemical and biochemical parameters were differentially affected by the various land use types. The first and the second principal components (PC1 and PC2) accounted for 44.59% and 24.47% of the total variance in the data, respectively (Fig. 2B). Both chemical and biochemical parameters were positively affected by PC1 except for HR and DH, whereas PC2 was able to discriminate between the two variable sets, with the most of chemical and biochemical parameters being positively and negatively affected, respectively. The above analyses revealed an important effect of land uses on both chemical and biochemical; on the contrary, only biochemical parameters were affected by seasons.

Influence of land use on soil bacterial community

Community structure analysis by EP-index and r-K strategy. Environmental conditions select organisms that either grow rapidly in uncrowded, nutrient-rich conditions (r-strategists), or can efficiently exploit resources in crowded conditions (K-strategists). In this work, we applied the method developed by De Leij et al. [38], who used the r/K-strategy concept for the characterization of soil bacterial communities. First, we compared the microbial community structures found in the different land uses during each of the two seasons. Since sampling was performed at 20 cm depth, where most microbial activity is known to occur [26], the main changes are expected through conversion from one soil management system to another one [59]. Microbial community structure found in the different land uses during each of the two seasons was investigated by means of EP index that is a measure of both richness (i.e. total number of species in the community) and evenness (i.e. how evenly individuals in the community are distributed over the different species) of groups of microorganisms with similar developmental characteristics.

The cultured bacteria belonging to fast-growing organisms, especially the r-strategists detected on day 2, dominated in all land uses and in both seasons (Table S4). The MM soil exhibited a lower EP index and a lower percentage of r-strategists in both seasons, possibly due to amensalism from a dominant bacterial group in the community. Variation of EP index of cultured bacteria was inspected using an ANOVA analysis of variance (Fig. 3A). Interestingly, statistical analysis revealed a significant effect of land uses on bacterial community structure ($P<0.001$) as well as an interaction effect between land uses and season change ($P<0.001$) on EP index. Variation of EP index due to seasonal changes was significant in the soil with higher human impact (TV) when compared with the other soils. Total bacterial concentrations varied significantly in respect of season ($P<0.001$) being higher in spring than in autumn (Table S2) in all but CO soils, with significant differences between CO and PA in spring, and between CO and MM, and MM and PA in autumn.

Overall results suggest that land uses affected cultured bacterial communities. Soils with low human impact (cork-oak forest) have a more stable bacterial density and show a less variation of bacterial community structure across seasons than soils subjected to high human impact such as tilled vineyard.

Bacterial community profiling by CD-DGGE and T-RFLP analyses. Culture dependent DGGE (CD-DGGE) fingerprinting of the 16S rRNA was used to characterize mixed bacterial communities recovered on agar plates. CD-DGGE represents a useful technique to follow the dynamics of distinct culturable fractions of the soil bacterial community in relation to physical, chemical and biological changes in the soil environment [60]. Since culture-dependent and culture-independent methods likely profile distinct fractions of the soil bacterial community with unique ecological roles [61], culturable bacteria may provide an ecologically relevant complement to culture-independent community characterization [62].

By pooling the bacterial cells growing on individual agar plates, we obtained a culture-dependent bacterial community. The analysis of DGGE profiles revealed clear banding patterns for each land-use and plant-cover types of sufficient complexity to investigate differences in soil microbial communities and identified stable communities with highly reproducible profiles (Fig. S2). A different community composition among land-use types was found as evidenced by the presence of different dominant signals; this finding indicated a compositional shift among soils examined in spring and autumn and subjected to different anthropogenic impact. Bacterial diversity was investigated through richness (*R*), Shannon-Weaver (*H'*), and Simpson (*D*) indices. Analysis of variance (ANOVA) confirmed the differences in the distribution of bacterial species due to land uses and sampling seasons ($P<0.001$). Both *R* and *H'* indices decreased from May to November in MM samples and increased in TV and CV samples, and the complementary opposite trend was observed for *D* index (Fig. 3B). Otherwise, cultured bacterial community in CO and PA soils did not vary significantly over seasons. Most likely, these results reflect the impact of both land-use and vegetation type and coverage on soil microbial communities. In particular, the low shift of biodiversity indices observed in CO and PA seems to be correlated to a higher stability of bacterial populations in natural habitats with low human impact, whereas populations inhabiting more anthropogenic areas tend to be more variable. Land-use type and, in particular, differences in vegetation dynamics may have a large role in modulating the temporal variability in soil bacterial communities. As observed by Lauber and co-workers [3], soils from the different land-use types did not exhibit identical temporal dynamics even though all the soils were located in close proximity and exposed to the same climatic conditions. Diversity indices obtained from T-RFLP analysis performed on total bacterial DNA (Fig. S3) partially confirmed the trends of diversity shown by CD-DGGE, though observed differences were not statistically significant (Table S5).

The unweighted pair-group method using arithmetic averages (UPGMA) of bacterial community profiling by CD-DGGE revealed a high diversity of the bacterial communities in each land-use and plant-cover soils. The similarity between the DGGE patterns of the soil bacterial communities revealed three distinct clusters (Fig. 4). Samples collected in autumn grouped into a separate cluster with about 52% similarity, whereas samples collected in spring grouped into two separate clusters, each of which composed by samples sharing about 40% similarity. The highest similarity values were shown by clusters based on soil land uses, suggesting a low bacterial variation within each soil type. This finding is consistent with previous reports showing that land-use type was the most important factor in determining the composition of soil microbial community [3].

The clustering results based on CD-DGGE were in agreement with culture-independent T-RFLP analysis that confirmed a different seasonal distribution of soil bacterial populations with subgroups associated to different land uses. Most of samples retrieved from the same soil (CO, PA, and MM) clustered together, whereas samples retrieved from TV and CV were often intermixed in the UPGMA dendrogram (Fig. S4). As culture-dependent and culture-independent profiles can separately resolve unique, diverse, and equally complex fractions of the soil bacterial community [60], our combined results from both CD-DGGE and T-RFLP methods, applied on culturable and total fractions, respectively, revealed an interaction effect between land uses and season change in affecting soil bacterial communities.

Taxa responses to land use determined by phylogenetic affiliation of soil bacterial isolates. Taxonomic affiliation was investigated on a total of 203 bacterial isolates. In detail, 20–21 colonies recovered from each soil sample at each season and showing different morphologies (for a total of 101 bacterial isolates in spring and 102 in autumn) were subjected to DNA extraction and PCR amplification of 16S rRNA gene. An amplicon of the expected size (about 1500 bp) was obtained from each isolate, and its nucleotide sequence was determined and submitted to GenBank (Table S6). Although it is generally accepted that not all bacteria, including types of soil bacteria, are culturable, the isolation of bacteria by agar plate cultivation and subsequent phylogenetic analysis permit to isolate and identify previously uncultured representatives or even new members of certain bacterial species for further analysis of their metabolic function. Therefore, even if the metagenome sequencing is becoming the most powerful tool to investigate microbial communities, the ability to isolate indigenous strains actually remains the unique way to further characterize and select soil bacteria showing interesting properties. Additionally, standard cultivation techniques have been shown to be able to capture members of the soil rare biosphere which could not be detected by metagenome sequencing [63].

The 203 sequences obtained were compared with those present in the GenBank databases by using the BLAST [64] program. Results showed that most of 16S rRNA gene sequences matched NCBI database sequences at 99–100% of similarity at the genus level, with *Arthrobacter*, *Bacillus*, *Stenotrophomonas*, *Pseudomonas* and *Burkholderia* as the most representative genera. So, identification at the genus level was achieved in 103 isolates and at the species level in 72 isolates, while 12 isolates were only affiliated to taxa level higher than genus and 16 remained unidentified (Table S6). Further comparison with GenBank databases by using the Seqmatch tool of the RDP indicated that 16S rRNA gene sequences were affiliated to four phyla: *Proteobacteria* (classes α, β and γ), *Bacteroidetes* (classes *Flavobacteria* and *Sphingobacteria*), *Actinobacteria* and *Firmicutes*. Even though the RDP analysis does not permit to affiliate a bacterial isolate to a given species, it allowed to assess the genus of almost all isolates, including those not identified through the BLAST search. In fact, a boostrap value of at least 80% is satisfactory for RDP requirement. Almost all our sequences gave rise to 100% boostrap, 12 ranged from 89% to 99%, and only two of them only resulted in low values (24 and 54% respectively, referred to two spring isolates).

Data obtained from both approaches allowed to assess that Berchidda soil is colonized by bacteria included in the classes of γ-*Proteobacteria* (with a prevalence of *Pseudomonas* and *Stenotrophomonas* genera), *Actinobacteria* (in particular, the genus *Arthrobacter*), β-*Proteobacteria* (with a prevalence of *Burkholderia* genus), *Bacilli* (with the genus *Bacillus* as the dominant one), Flavobacteria and Sphingobacteria. The relative abundances of the different classes identified by RDP across the different samples related to soil uses and season are represented in Figure 5. A putative taxonomic description of total bacterial community, derived from the interpretation of T-RFLP data (Fig. S3) is reported in Figure 6, in which the largest fraction of the community is represented by members of *Proteobacteria*, with α-*Proteobacteria* as the most abundant class.

Differences in class composition across the site were observed suggesting that the microbial composition changes in response to land uses. In fact, all the seven classes (α, β and γ *Proteobacteria*, *Sphingobacteria Flavobacteria*, *Actinobacteria* and *Bacilli*) were present in CV, PA, and CO soils, while all classes but α-*Proteobacteria* in TV and all classes but *Sphingobacteria* in MM were found. The observed large diffusion of *Proteobacteria* in all soil-uses is in agreement with previously reported data concerning soil bacterial communities in different land-use systems [65]. Within *Proteobacteria*, the majority of the isolates fell into the gamma subgroup, which showed higher relative abundance compared to that of any other taxa at the class level while a few were alfa proteobacteria, especially in vineyards and pasture. In all soil-uses, were also found *Actinobacteria*, in agreement with previous work [65], and β-*Proteobacteria*, like *Burkholderia* sp., previously detected by Pastorelli and co-workers [19] who analysed the denitrifying bacterial communities present in the same Berchidda soil samples. Among Bacteroidetes, it has to be noted the relative abundance of the genera *Chryseobacterium* sp. (in all soil uses) and *Flavobacterium* sp. (in all soil uses, but MM), which include isolates already observed in Korea soils [66], and bacteria with plant growth promoting properties recovered in Iran soil [67], respectively. As already pointed out by Fierer and co-workers [68], the β-*Proteobacteria* and *Bacteroidetes* follow copiotrophic lifestyles and their relative abundance were highest in soils with high C availability. In general, copiotrophic bacteria should have higher growth rates and traditional culturing methods are likely to select for microorganisms that can grow rapidly in high resource environments.

Strong differences in class composition were observed in each soil when sampled in the two seasons: for instance, *Bacilli* and *Actinobacteria* dominate in spring (particularly, *Bacilli* in CV and CO, while *Actinobacteria* in MM and PA), whilst β-*Proteobacteria* tend to dominate in autumn. Six out of the seven identified classes (β and γ-*Proteobacteria*, *Flavobacteria*, *Actinobacteria* and *Bacilli*) were recovered in both seasons whilst *Sphingobacteria* were recovered only in autumn. In all soils, *Actinobacteria* were prevalent in spring, β *Proteobacteria* predominated in autumn, while β and γ-*Proteobacteria* in both spring and autumn. When genera composition was used to infer the diversity indices of cultured bacteria (Table S7) an increase of diversity (estimated as Shannon H and Evenness) from May to November was present for all soils, but CO, where Shannon H was slightly higher in May than in November. In particular, MM soil showed the highest increases for most indices, especially for alpha diversity (*i.e.* the species richness and evenness within a sample) that has often been correlated with ecosystem stability and functionality [69]. It must be noted that cultivated bacterial populations did not fluctuate with seasonal changes in soils with low human input (cork-oak forest). The bacterial communities of the cork-oak forest soil were similar in richness and composition, furthering the point that in a community with moderate disturbance, new individuals and groups could be introduced in a manner that promotes competition and diversity of the community, thus establishing a more stable community [65].

These data suggest that shifts from forest to managed meadow and vineyard result in changes of bacterial communities composition. Previous studies have also shown that shifts from forest to grassland soil [70], from cultivated system to pasture [71] and from grazed pine forest to cultivated crop and grazed pasture [65] resulted in significant changes in bacterial community composition. Our results suggest that the use of culture-dependent 16S rRNA gene sequencing along with traditional analysis of soil physiochemical properties may provide insight into the ecological relevance of soil bacterial taxa.

Conclusion

Overall, data obtained in this work revealed an important effect of land uses on both chemical and biochemical soil parameters. Soil bacterial communities were seasonally distinct and exhibited compositional shifts that tracked with changes in land use and soil management. This study, combining the pedological and biochemical data with microbiological and molecular analysis, furnishes a good methodological approach to describe the influence of different soil managements on soil microbial community structure. In fact, the results demonstrate that, in the same pedological conditions, long-term soil management influence the community structure; i.e., soils subjected to low human inputs (cork-oak forest and pasture) showed a more stable chemical and biochemical soil composition as well as bacterial community than those with high human input (vineyards and managed meadow). Further research is required to determine whether the observed shifts in bacterial community composition produce parallel changes in the functional attributes of these communities across soil types under different long-term management regimes. The use of culture-independent approaches, like metabarcoding and metagenome sequencing, will make it possible to identify the specific drivers of land-use dynamics exhibited by soil bacterial communities and to give a complete picture of the bacterial communities in a typical Mediterranean agro-silvo-pastoral system.

Supporting Information

Figure S1 Box-plot analysis showing the frequency distribution of physical-chemical and biological properties of the five Sardinia soils.

Figure S2 Examples of CD-DGGE profiles of the soil bacterial communities associated to the different land uses in May and November. A) From the left to right: hayland pasture rotation (PA), tilled vineyard (TV), grass covered vineyard (CV) in May; B) managed meadow (MM), cork-oak forest (CO), hayland pasture rotation (PA) in May; C) grass covered vineyard (CV) in May, grass covered vineyard (CV) in November, tilled vineyard (TV) in May; D) cork-oak forest (CO), hayland pasture rotation (PA), managed meadow (MM) in November.

Figure S3 Examples of T-RFLP profiles obtained after digestion with *Msp*I (A) and *Rsa*I (B) restriction enzymes of amplified of 16S rRNA gene sequences.

Figure S4 Cluster analysis of T-RFLP patterns generated by *Msp*I and *Rsa*I digestion of 16S rRNA gene sequences. The UPGMA cluster analysis based on Jaccard similarity matrix was calculated for each set of samples using the "hclust" function of the R "stats" package. The scale bar represents the percent of dissimilarity.

Table S1 Pedological profiles and classification of the soils investigated.

Table S2 Determination of total organic carbon soil (C_{org}), extractable carbon (C_{ext}), humified carbon (C_{HA+FA}), non humified carbon (C_{NH}) and humification parameters of the five Sardinian soils.

Table S3 Biochemical parameters measured in the five Sardinian soils.

Table S4 r/k bacterial strategists, total culturable bacteria and EPI index measured in soils under different long-term management practices.

Table S5 Diversity indices of total bacterial communities as inferred from T-RFLP profiles.

Table S6 Phylogenetic affiliations of 203 randomly selected soil bacterial isolates based on comparative analysis of their 16S rRNA gene sequences.

Table S7 The ratio of diversity indices related to the abundance of the different genera detected in cultured isolates between November and May.

Acknowledgments

We are grateful to P.P. Roggero (University of Sassari) and his team (G. Seddaiu, G. Urracci and L. Doro) for their contribution to collect the Sardinian soils and for providing data on experimental area. The experimental site was chosen on the basis of vegetation and soil surveys made in collaboration with S. Madrau, S. Bagella, R. Filigheddu, M.C. Caria, and E. Farris (University of Sassari). We thank R. Pastorelli for her help in soil sampling, S. Tabacchioni and L. Chiarini for bacterial isolation, C. Cantale and M. Sperandei for figure preparation, M.T. Rubino for chemical-biochemical analysis, and S. Cesarini and L. Pirone for their valuable suggestions. The research was carried out in the context of the FISR SOILSINK research project coordinated by R. Francaviglia (CRA-RPS, Rome) (?http://soilsink.entecra.it).

Author Contributions

Conceived and designed the experiments: A. Bevivino RF CD A. Benedetti. Performed the experiments: A. Bevivino PP MSP MCP CD LL. Analyzed the data: A. Bevivino PP GB AF AM CD LL. Contributed reagents/materials/analysis tools: A. Bevivino PP GB AF MSP AM LL MCP RF CD A. Benedetti. Contributed to the writing of the manuscript: A. Bevivino RF GB AF CD LL A. Benedetti.

References

1. Tiedje JM, Asuming-Brempong S, Nusslein K, Marsh TL, Flynn SJ (1999) Opening the black box of soil microbial diversity. Appl Soil Ecol 13: 109–122. doi:10.1016/S0929-1393(99)00026-8.
2. Nannipieri P, Ascher J, Ceccherini MT, Landi L, Pietramellara G, et al. (2003) Microbial diversity and soil functions. Eur J Soil Sci 54: 655–670. doi:10.1046/j.1365-2389.2003.00556.x.
3. Lauber CL, Ramirez KS, Aanderud Z, Lennon J, Fierer N (2013) Temporal variability in soil microbial communities across land-use types. ISME J 7: 1641–1650. doi:10.1038/ismej.2013.50.
4. Myers N, Mittermeier RA, Fonseca GAB, Fonseca GAB, Kent J (2000) Biodiversity hotspots for conservation priorities. Nature 403: 853–858. doi:10.1038/35002501.
5. Puddu G, Falcucci A, Maiorano L (2011) Forest changes over a century in Sardinia: implications for conservation in a Mediterranean hotspot. Agrofor Syst 85: 319–330. doi:10.1007/s10457-011-9443-y.
6. Médail F, Quézel P (1999) Biodiversity Hotspots in the Mediterranean Basin: Setting Global Conservation Priorities. Conserv Biol 13: 1510–1513. doi:10.1046/j.1523-1739.1999.98467.x.
7. Blondel J, Aronson J (1999) Biology and wildlife of the Mediterranean region. Oxford University Press, Oxford.
8. Salis L, Marrosu M, Bagella S, Sitzia M RP (2010) Grassland management, forage production and plant biodiversity in a Mediterranean grazing system. In: Porqueddu C, Ríos S, editors. The contributions of grasslands to the conservation of Mediterranean biodiversity. Zaragoza: CIHEAM/CIBIO/FAO/SEEP, Vol. 185. 181–185.
9. Gonçalves P, Alcobia S, Simões L, Santos-Reis M (2012) Effects of management options on mammal richness in a Mediterranean agro-silvo-pastoral system. Agrofor Syst 85: 383–395. doi:10.1007/s10457-011-9439-7.
10. Bacchetta G, Bagella S, Biondi E, Farris E, Filigheddu R, et al. (2004) A contribution to the knowledge of the order Quercetalia ilicis Br. -Bl. ex Molinier 1934 of Sardinia. Fitosociologia 41: 29–51.
11. Orgiazzi A, Lumini E, Nilsson RH, Girlanda M, Vizzini A, et al. (2012) Unravelling soil fungal communities from different Mediterranean land-use backgrounds. PLoS One 7: e34847. doi:10.1371/journal.pone.0034847.
12. Francaviglia R, Benedetti A, Doro L, Madrau S, Ledda L (2014) Influence of land use on soil quality and stratification ratios under agro-silvo-pastoral Mediterranean management systems. Agric Ecosyst Environ 183: 86–92.
13. Council of the European Communities (1992) Council Directive 92/43/EEC of 21 May 1992 on the conservation of natural habitats and of wild fauna and flora. Off J Eur Communities 35: 7–50.
14. Council of the European Communities (2001) Commission Regulation (EC) No 1808/2001 of 30 August 2001 laying down detailed rules concerning the implementation of Council Regulation (EC) No 338/97 on the protection of species of wild fauna and flora by regulating trade therein. Off J Eur Communities L250: 1–43.
15. Aru A, Baldaccini P, Delogu G, Dessena M, Madrau S, et al. (1990) Carta dei Suoli della Sardegna, scala 1/250.000. Assessorato alla programmazione e all'assestamento del territorio, Centro Regionale Programmazione, Dip. Sc. della Terra, Università di Cagliari, Italy.
16. Bacchetta G, Bagella S, Biondi E, Farris E, Filigheddu R, et al. (2009) Forest vegetation and serial vegetation of Sardinia (with map at the scale 1:350,000). Fitosociologia 46: 3–82.
17. Bagella S, Caria MC (2011) Vegetation series: a tool for the assessment of grassland ecosystem services in Mediterranean large-scale grazing systems. Fitosociologia 48: 47–54.
18. Lumini E, Orgiazzi A, Borriello R, Bonfante P, Bianciotto V (2010) Disclosing arbuscular mycorrhizal fungal biodiversity in soil through a land-use gradient using a pyrosequencing approach. Environ Microbiol 12: 2165–2179. doi:10.1111/j.1462-2920.2009.02099.x.
19. Pastorelli R, Landi S, Trabelsi D, Piccolo R, Mengoni A, et al. (2011) Effects of soil management on structure and activity of denitrifying bacterial communities. Appl Soil Ecol 49: 46–58. doi:10.1016/j.apsoil.2011.07.002.
20. Lagomarsino A, Benedetti A, Marinari S, Pompili L, Moscatelli MC, et al. (2011) Soil organic C variability and microbial functions in a Mediterranean agro-forest ecosystem. Biol Fertil Soils 47: 283–291. doi:10.1007/s00374-010-0530-4.
21. Francaviglia R, Coleman K, Whitmore AP, Doro L, Urracci G, et al. (2012) Changes in soil organic carbon and climate change - Application of the RothC model in agro-silvo-pastoral Mediterranean systems. Agric Syst 112: 48–54.
22. Seddaiu G, Porcu G, Ledda L, Roggero PP, Agnelli A, et al. (2013) Soil organic matter content and composition as influenced by soil management in a semi-arid Mediterranean agro-silvo-pastoral system. Agric Ecosyst Environ 167: 1–11. doi:http://dx.doi.org/10.1016/j.agee.2013.01.002.
23. Lai R, Lagomarsino A, Ledda L, Roggero PP (2014) Variation in soil C and microbial functions across tree canopy projection and open grassland microenvironments. Turkish J Agric For 38: 62–69. doi:10.3906/tar-1303-82.
24. Soil Survey Staff (2006) Keys to soil taxonomy. In: United States Department of Agriculture NRCS, editor. Natural Resoiurces Concervation Service.
25. Doran JW, Elliott ET, Paustian K (1998) Soil microbial activity, nitrogen cycling, and long-term changes in organic carbon pools as related to fallow tillage management. Soil Tillage Res 49: 3–18.
26. O'Brien HE, Parrent JL, Jackson JA, Moncalvo J-M, Vilgalys R (2005) Fungal community analysis by large-scale sequencing of environmental samples. Appl Environ Microbiol 71: 5544–5550.
27. Stotzky G, Goos RD, Timonin MI (1962) Microbial changes occurring in soil as a result of storage. Plant Soil 16: 1–18. doi:10.1007/BF01378154.
28. Springer U KJ (n.d.) Prüfung der Leistungsfähigkeit von einigen wichtigeren Verfahren zur Bestimmung des Kohlenstoffs mittels Chromschwefelsäure sowie Vorschlag einer neuen Schnellmethode. J Plant Nutr Soil Sci 64: 1–26 (in German).
29. Ciavatta C, Govi M, Vittori Antisari L, Sequi P (1990) Characterization of humified compounds by extraction and fractionation on solid polyvinylpyrrolidone. J Chromatogr 509: 141–146.
30. Sequi P, De Nobili M, Leita L, Cercignani G (1986) A new index of humification. Agrochimica 30: 175–179.
31. Vance ED, Brookes PC, Jenkinson DS (1987) An extraction method for measuring soil microbial biomass C. Soil Biol Biochem 19: 703–707. doi:10.1016/0038-0717(87)90052-6.
32. Badalucco L, Grego S, Dell'Orco S, Nannipieri P (1992) Effect of liming on some chemical, biochemical, and microbiological properties of acid soils under spruce (Picea abies L.). Biol Fertil Soils 14: 76–83. doi:10.1007/BF00336254.
33. Riffaldi R, Saviozzi A, Levi-Minzi R (1996) Carbon mineralization kinetics as influenced by soil properties. Biol Fertil Soils 22: 293–298. doi:10.1007/BF00334572.
34. Dilly O, Munch J-C (1998) Ratios between estimates of microbial biomass content and microbial activity in soils. Biol Fertil Soils 27: 374–379. doi:10.1007/s003740050446.
35. Anderson T-H, Domsch KH (1989) Ratios of microbial biomass carbon to total organic carbon in arable soils. Soil Biol Biochem 21: 471–479. doi:10.1016/0038-0717(89)90117-X.
36. Pinzari F, Trinchera A, Benedetti A, Sequi P (1999) Use of biochemical indices in the mediterranean environment: Comparison among soils under different forest vegetation. J Microbiol Methods 36: 21–28.
37. Smalla K, Wieland G, Buchner A, Zock A, Parzy J, et al. (2001) Bulk and Rhizosphere Soil Bacterial Communities Studied by Denaturing Gradient Gel Electrophoresis: Plant-Dependent Enrichment and Seasonal Shifts Revealed. 67: 4742–4751. doi:10.1128/AEM.67.10.4742.

38. De Leij FAAM, Whipps JM, Lynch JM (1994) The use of colony development for the characterization of bacterial communities in soil and on roots. Microb Ecol 27: 81–97.
39. Grifoni A, Bazzicalupo M, Di Serio C, Fancelli S, Fani R (1995) Identification of Azospirillum strains by restriction fragment length polymorphism of the 16S rDNA and of the histidine operon. FEMS Microbiol Lett 127: 85–91. doi:10.1016/0378-1097(95)00042-4.
40. Trabelsi D, Mengoni A, Elarbi Aouani M, Mhamdi R, Bazzicalupo M (2009) Genetic diversity and salt tolerance of bacterial communities from two Tunisian soils. Ann Microbiol 59: 25–32. doi:10.1007/BF03175594.
41. Mengoni A, Tatti E, Decorosi F, Viti C, Bazzicalupo M, et al. (2005) Comparison of 16S rRNA and 16S rDNA T-RFLP approaches to study bacterial communities in soil microcosms treated with chromate as perturbing agent. Microb Ecol 50: 375–384. doi:10.1007/s00248-004-0222-4.
42. Hammer Ø, Harper DAT, Ryan PD (2001) Past: Paleontological Statistics Software Package for Education and Data Analysi. Palaeontol Electron 4: 1–9. doi:10.1016/j.bcp.2008.05.025.
43. Duineveld B, Kowalchuk G, Keijzer A, van Elsas J, van Veen J (2001) Analysis of bacterial communities in the rhizosphere of chrysanthemum via denaturing gradient gel electrophoresis of PCR-amplified 16S rRNA as well as DNA fragments coding for 16S rRNA. Appl Environ Microbiol 67: 172–178.
44. Ausubel FM, Brent R, Kingston RE, Moore DD, Seidman JG, et al. (1987) Unit 2.4 Preparation of genomic DNA from bacteria. Curr Protoc Mol Biol: 2.4.1–2.4.2.
45. Di Cello F, Pepi M, Baldi F, Fani R (1997) Molecular characterization of an n-alkane-degrading bacterial community and identification of a new species, Acinetobacter venetianus. Res Microbiol 148: 237–249.
46. El Fantroussi S, Verschuere L, Verstraete W, Top EM (1999) Effect of phenylurea herbicides on soil microbial communities estimated by analysis of 16S rRNA gene fingerprints and community-level physiological profiles. Appl Environ Microbiol 65: 982–988.
47. Cremonesi L, Firpo S, Ferrari M, Righetti PG, Gelfi C (1997) Double-gradient DGGE for optimized detection of DNA point mutations. Biotechniques 22: 326–330.
48. Shannon C, Weaver W (1963) The mathematical theory of communication. University of Illinois Press, Urbana. doi:10.1145/584091.584093.
49. Simpson E (1949) Measurement of diversity. Nature 163: 688–688. doi:10.1038/163688a0.
50. Whittaker RH (1972) Evolution and measurement of species diversity. Taxon 21: 213. doi:10.2307/1218190.
51. Hall TA (1999) BioEdit: a user-friendly biological sequence alignment editor and analysis program for Windows 95/98/NT. Nucleic Acids Symp 41: 95–98. doi:citeulike-article-id:691774.
52. Bray J, Curtis J (1957) An ordination of the upland forest communities of southern Wisconsin. Ecol Monogr 27: 325–349.
53. Haider K (1992) Problems related to the humification processes in soils of temperate climates. In: Stotzky G, Bollag J-M, editors. Soil Biochemistry. Marcel Dekker, New York. 55–94.
54. Oades JM (1995) An overview of processes affecting the cycling of organic carbon in soils. In: Zepp R, Sonntag C, editors. Role of non living organic matter in the earth's carbon cycle. John Wiley, New York. 293–303.
55. Stevenson FJ, Cole MA (1999) Cycles of soil: carbon, nitrogen, phosphorus, sulfur, micronutrients. 2nd ed. John Wiley, New York.
56. Franzluebbers AJ, Stuedemann JA, Schomberg HH, Wilkinson SR (2000) Soil organic C and N pools under long-term pasture management in the Southern Piedmont USA. Soil Biol Biochem 32: 469–478. doi:10.1016/S0038-0717(99)00176-5.
57. Anderson T, Domsch KH (1993) The metabolic quotient for CO2 (qCO2) as a specific activity parameter to assess the effects of environmental conditions, such as pH, on the microbial biomass of forest soils. Soil Biol Biochem 25: 393–395.
58. Benedetti A, Brookes P, Lynch J (2005) Conclusive remarks. Part I: Approaches to defining, monitoring, evaluating and managing soil quality. In: Bloem J, Hopkins D, Benedetti A, editors. Soil Quality. CABI; Wallingford. 63–70.
59. Conant RT, Paustian K, Elliott ET (2001) Grassland management and conversion into grassland: effects on soil carbon. Ecol Appl 11: 343–355. doi:10.1890/1051-0761(2001)011[0343:GMACIG]2.0.CO;2.
60. Edenborn SL, Sexstone AJ (2007) DGGE fingerprinting of culturable soil bacterial communities complements culture-independent analyses. Soil Biol Biochem 39: 1570–1579.
61. Garland JL (1999) Potential and limitations of BIOLOG for microbial community analysis. In: Bell C, Brylinsky M, Johnson-Green P, editors. Microbial Biosystems: new frontiers. Proceedings of the 8th international symposium on microbial ecology. Atlantic Canada Society for Microbial Ecology. Halifax, Canada. 1–7.
62. Ellis RJ, Morgan P, Weightman AJ, Fry JC (2003) Cultivation-dependent and -independent approaches for determining bacterial diversity in heavy-metal-contaminated soil. Appl Environ Microbiol 69: 3223–3230. doi:10.1128/AEM.69.6.3223.
63. Shade A, Hogan CS, Klimowicz AK, Linske M, McManus PS, et al. (2012) Culturing captures members of the soil rare biosphere. Environ Microbiol 14: 2247–2252. doi:10.1111/j.1462-2920.2012.02817.x.
64. Altschul SF, Madden TL, Schäffer AA, Zhang J, Zhang Z, et al. (1997) Gapped BLAST and PSI-BLAST: a new generation of protein database search programs. Nucleic Acids Res 25: 3389–3402.
65. Shange RS, Ankumah RO, Ibekwe AM, Zabawa R, Dowd SE (2012) Distinct soil bacterial communities revealed under a diversely managed agroecosystem. PLoS One 7: e40338. doi:10.1371/journal.pone.0040338.
66. Weon H-Y, Kim B-Y, Yoo S-H, Kwon S-W, Stackebrandt E, et al. (2008) Chryseobacterium soli sp. nov. and Chryseobacterium jejuense sp. nov., isolated from soil samples from Jeju, Korea. Int J Syst Evol Microbiol 58: 470–473. doi:10.1099/ijs.0.65295-0.
67. Soltani A, Khavazi K, Asadi-Rahmani H, Omidvari M, Abaszadeh Dahaji P (2010) Plant growth promoting characteristics in some Flavobacterium spp. isolated from soils of Iran. J Agri Sci 2: 106–115. Available: http://www.ccsenet.org/journal/index.php/jas/article/view/8404/6221.
68. Fierer N, Bradford MA, Jackson RB (2007) Toward an ecological classification of soil bacteria. Ecology 88: 1354–1364. doi:10.1890/05-1839.
69. Girvan MS, Campbell CD, Killham K, Prosser JI, Glover LA (2005) Bacterial diversity promotes community stability and functional resilience after perturbation. Environ Microbiol 7: 301–313. doi:10.1111/j.1462-2920.2005.00695.x.
70. Nacke H, Thürmer A, Wollherr A, Will C, Hodac L, et al. (2011) Pyrosequencing-based assessment of bacterial community structure along different management types in German forest and grassland soils. PLoS One 6: e17000. doi:10.1371/journal.pone.0017000.
71. Acosta-Martínez V, Dowd S, Sun Y, Allen V (2008) Tag-encoded pyrosequencing analysis of bacterial diversity in a single soil type as affected by management and land use. Soil Biol Biochem 40: 2762–2770. doi:10.1016/j.soilbio.2008.07.022.

14

Development and Applications of a Comprehensive Land Use Classification and Map for the US

David M. Theobald[1,2]*

1 Conservation Science Partners, Inc., Fort Collins, Colorado, United States of America, **2** Department of Fish, Wildlife, and Conservation Biology, Colorado State University, Fort Collins, Colorado, United States of America

Abstract

Land cover maps reasonably depict areas that are strongly converted by human activities, but typically are unable to resolve low-density but widespread development patterns. Data products specifically designed to resolve land uses complement land cover datasets and likely improve our ability to understand the extent and complexity of human modification. Methods for developing a comprehensive land use classification system are described, and a map of land use for the conterminous United States is presented to reveal what we are doing on the land. The comprehensive, detailed and high-resolution dataset was developed through spatial analysis of nearly two-dozen publicly-available, national spatial datasets – predominately based on census housing, employment, and infrastructure, as well as land cover from satellite imagery. This effort resulted in 79 land use classes that fit within five main land use groups: built-up, production, recreation, conservation, and water. Key findings from this study are that built-up areas occupy 13.6% of mainland US, but that the majority of this occurs as low-density exurban/rural residential (9.1% of the US), while more intensive built-up land uses occupy 4.5%. For every acre of urban and suburban residential land, there are 0.13 commercial, 0.07 industrial, 0.48 institutional, and 0.29 acres of interstates/highways. This database can be used to address a variety of natural resource applications, and I provide three examples here: an entropy index of the diversity of land uses for smart-growth planning, a power-law scaling of metropolitan area population to developed footprint, and identifying potential conflict areas by delineating the urban interface.

Editor: Adina Maya Merenlender, University of California, Berkeley, United States of America

Funding: This work was partially funded by the NASA Applied Sciences Program (10-BIOCLIM10-0034), the US Forest Service Rocky Mountain Research Center, and the US EPA National Center for Environmental Assessment. The funders had no role in study design, data collection and analysis, decision to publish, or preparation of the manuscript.

Competing Interests: The author has declared that no competing interests exist.

* E-mail: davet@csp-inc.org

Introduction

The majority of the global biosphere is occupied by human-modified landscapes of agricultural and urban land uses, with less than 20% semi-natural and about 25% "wild" [1]. Because of the extent of human domination, it is globally important to understand the extent and effects of human use of ecosystems [2]. Various aspects have been linked directly to land uses, ranging from the consequences on biodiversity [3], water security [4], and human health [5]. Yet for regional to national extents land use is typically measured and mapped at a coarse spatial resolution (e.g., state or county unit) and for only broad categories of use (e.g., urban vs. agriculture). Some land cover maps depict "developed" or built-up land cover types, but these typically miss low-density development that is either diffuse but has direct and permanent human activities or obscured by canopy cover [6]. This under-representation has been estimated to be roughly twice the extent of built-up land cover in the US [7]. Consequently, data products specifically designed to resolve different land use types and patterns would complement land cover datasets and likely improve our ability to understand the extent and complexity of human modification.

In the US a few efforts have developed land use/cover datasets that typically provide more detail than global approaches [8]. These include the National Land Cover Dataset (NLCD [9]; the USGS Land Cover Trends project [10]; the US Department of Economic Research Service (ERS) that summarizes major land uses [11]; and the Natural Resource Conservation Service's National Resource Inventory (NRI) [12]. Generally, these efforts have focused on characterizing land use/cover in rural areas and have not sought to distinguish land use types in more urban and developed areas (but see [13]). The NRI and ERS provide helpful land use data but at county to state levels and for only very coarse classes (i.e. urban) on private lands. The NLCD provides high spatial resolution (~0.2 ha) but predominately is based on cover, not use [9,10], and is based on the Anderson classification system [14]. Two notable recent efforts have produced an approach to develop a prototype for a national land use product targeted for urban land uses and derived principally from NLCD for US urban areas [15] and combined land use/cover approach for Canadian cities [16]. In short, we lack a detailed, consistent, spatially-explicit dataset that directly maps the patterns of *land use* in the United States.

In this paper I describe the generation and applications of this new, detailed, and high-resolution land use dataset. In doing so, two dichotomies are important to consider. The first is the distinction between land *use* and land *cover*. Indeed, a major challenge for "land change" science is to develop systems for monitoring and observing land use and land use change [17], and a critical aspect to ensure progress is to clearly distinguish land use from land cover [18,19]. Although the terms land use and land cover are often used interchangeably, land use is quite distinct from cover. Land use and land cover typically are defined using a systems description where human driving forces lead to actions or activities that are characterized by a land use [19]. A land use may in turn manipulate land cover to such an extent that it results in conversion (e.g., cropland to residential) or modification (e.g., fertilization of cropland). *Land cover* is defined as the observed bio-physical cover on the earth's surface, whereas *land use* is characterized by the arrangements, activities and inputs of people to produce, change or maintain a land cover. For example, grassland is a cover type, but rangeland is a use.

The second dichotomy is the difference between spatial data on land use vs. the system used to classify the land uses. Surprisingly, few efforts have focused specifically on developing land *use* classification systems, but there are two notable exceptions. US urban planners developed the Land Based Classification Standards [20] to map detailed, parcel-level land uses appropriate mostly for urban settings, and it distinguishes activity, function, structure, site, and ownership characteristics. The Australian Land Use Map [21] has been developed over the past 15 years as a hierarchical structured classification to map the primary land use, typically from interpretation of aerial photography. At the highest level there are six classes: conservation, production from natural environments, dryland agriculture, irrigated agriculture, built-up uses, and water.

My goal in this paper is to describe a comprehensive, detailed, consistent, and high-resolution dataset that characterizes human activities on the landscape — that incorporates and maps all sectors of use: agricultural, stewardship/conservation, and urban. To do this, I: a) created a hierarchical land use classification system; b) constructed a detailed land use map built from nationally-available datasets through application of a set of logical rules; c) compared the general patterns to standard statistical summaries (i.e. NRI) and verified with a sample-based, detailed land use dataset; d) and summarized the national and regional patterns and configurations of various land uses. The general intent of this work is to provide both a consistent and comprehensive land use classification system as well as a new, spatial dataset that provides fundamental information on the primary human activities occurring in the US. This dataset provides a significant improvement in our ability to understand land use patterns, which will have broad utility for natural resource applications.

Materials and Methods

Classification system

Ideally, a classification system should be applicable over broad extents, contain hierarchical classes that would allow aggregation and a comprehensive range of classes so that all areas fit within a class; result in repeatable and consistent results through time and by interpreter; and have reasonably-balanced levels of accuracy amongst classes [14]. Adhering to these criteria and building on other classification systems, I synthesized a variety of classification systems [9,12,13,14,20] to construct a three-level hierarchical system that is comprehensive, consistent, and exclusive. The first level is based on the major use groups of built-up, production, recreation, and conservation. This broadly captures the four main ways that humans interact with their environment: by constructing buildings and living and working at a location (built-up), producing or removing natural resources from an area such as cropland agriculture, grazing, or mining (production), visiting an area for the purposes of enjoyment, but not building significant structures or removing resources (recreation and tourism), or by setting aside places primarily to protect natural systems and wildlife habitat (conservation).

The sub-group and class levels within each of these four major groups further differentiate the primary land use or activity occurring in an area (Table 1), resulting in 69 different classes (plus an additional 11 water-based classes). It is important to note that interacting with the definition of classes (i.e. intensity) is the spatial scale or minimum mapping unit that land use is represented at – in this case the classification is attempting to be consistent with most 30 m resolution datasets (e.g., NLCD) – which is roughly 1 ha (nominally 9 cells). Using the four-group structure, I developed sub-group and classes building on Anderson level 2 codes as well as the typology defined by the North American Industrial Classification System (NAICS). Built-up uses are divided into residential, commercial, industrial, institutional, and transportation. Urban residential uses are defined at 1.6 dwelling units per acre (dua; 3.95 units per ha), based on the US Census definition of urban population of 1000 people per square mi [6], while the urban high category is greater than 10 dua based on typical densities at which public transportation is viable [23]. Suburban areas have residential densities below urban threshold but greater than 1 unit per 1 ha (0.4 dua), which is a typical density at which centralized services such as municipal sewer and water supply are provided versus de-centralized septic and water supply (e.g., ground water well) systems. Production classes include agricultural, mining, timber, energy, animal, and aquaculture and were designed to be roughly consistent with major categories of the Census of Agriculture, the NRI, and NAICS. Recreation classes distinguish uses that typify urban parks such as playgrounds, ball fields, golf courses, and courts, as opposed to less intense uses where natural land cover dominates. "Natural parks" are places that are managed primarily for recreation and occur in relatively natural land cover settings with many natural processes intact. These are typically accessible to the public, and can be further separated into motorized and non-motorized forms of recreation. A developed facility class is reserved for ski resorts, campgrounds, and the like. Facilities such as stadiums and amusement parks are included under the built-up group because they usually occur inside a building and are relatively passive activities. Conservation classes represent areas that are primarily for maintain "nature". "Natural areas" are places where the primary use is to maintain natural vegetation and processes, such as nature reserves, research natural areas, and areas of critical environmental concern, as well as municipal watersheds. These areas usually are not accessible to the public. "Wildlife habitat" is an area that is primarily managed for hunting, fishing, and "watchable" opportunities, which often focus on big game and charismatic species. Areas designated as wilderness are placed in their own, distinct class. Finally, a miscellaneous class includes private land conservation easements, archaeological and historical sites, and other miscellaneous values. Maps that reflect conservation classes such as cemeteries are derived largely from land ownership and management datasets. Figures 1, 2, 3 illustrate the datasets and rules used to classify these classes.

Note that because the variety of uses around land areas covered by water are quite complex and dynamic, here I simply distinguish naturally-occurring (e.g. lakes) from those that have are human-created (e.g. reservoir, canals). This simple distinction, however is quite useful and important, as previous efforts to examine the effects of human modification using land cover data alone have

Table 1. A multi-purpose, hierarchical land use classification system.

Level 1	Level II	Level III	Code
1 Water	Natural - area	Lake	111
		Swamp/marsh	112
		Playa	113
	Human - area	Reservoir	121
	Natural - linear	River	131
		Wash	132
	Estuary	Estuary & complex channels	141
	Human - linear	Canal/ditch	151
	Wetlands	Wetlands	161
	Ocean	Open ocean	171
		Bay inlet	172
2 Built-up	Residential	Dense urban (>0.1 ac)	211
		Urban (0.1–1)	212
		Suburban (1–2.5 ac)	213
		Exurban (2.5–10 ac)	214
		Rural (10–40 ac)	215
	Commercial	Office	221
		Retail/shopping centers	222
		Entertainment (stadiums, amusement, etc.)	223
		Lodge	224
	Industrial	Factory, plant	231
		Landfill (land fill, waste trt.)	232
		Confined animal feeding	233
		Utilities (power, sewage)	234
	Institutional	Schools (dev)	241
		Schools (undeveloped)	242
		Medical (hospitals, nursing home, etc.)	243
		Government/public	244
		Military/DOD/DOE (dev)	245
		Military/DOD (training)	246
		Fire & police stations	247
		Church/religious	248
		Prison/penitentiary	249
	Transportation	Airports (developed)	251
		Highways, railways	252
		Other transportation	253
		Port, train station	254
		Undeveloped	255
	Miscellaneous	Rural buildings, cemetery	261
3 Production	General	General agricultural	310
	Cropland	Cropland/row crops	311
		Pastureland	312
		Orchards	313
		Sod & switch grass	314
		Aquaculture	315
	Rangeland	Grazed	321
		Stock tank	322
	Mining	Mining strip mines, quarries, gravel pits	330
		Mine shafts	331

Table 1. Cont.

Level 1	Level II	Level III	Code
	Timber	Timber harvest	341
		Timber plantations	342
	Extractive/barren land	Oil/Gas wells	351
		Misc. barren	352
4 Recreation	Undifferentiated park	General park	410
	Developed park	Urban park	411
		Golf course	412
		Motorized	413
		OHV staging area/trailhead	414
		Resort/ski area	415
		Marina	416
		Campground/ranger station	417
		Picnic/trailhead	418
		Boat/fishing access	419
	Natural park	Natural park	421
		Designated recreation area	422
		Designated scenic area	423
Conservation	Public	Wildlife habitat (hunting & fishing)	511
		Conservation area (BLM)	512
		Nature reserve	513
		Wilderness	514
		Areas of Critical Env. Concern, Research Natural Area	515
		Fish & Wildlife Service refuge	516
		Wilderness study area	517
		Archaeology, historic site, scenic area	518
		Wild & Scenic river	519
	Public-limited access	Municipal watershed	521
		Corps of Engineers dam	522
		Marine Protected Area	523
	Private easement	Wildlife conservation	531
		Agricultural conservation	532

been forced to place all lakes and reservoirs wholly into either human modified or natural [24].

Compile available datasets and construct map

I attempted to identify and compile a range of publicly-available spatial datasets that could be used to spatially represent the variety of land use classes defined in the classification system, that included a variety of aspects of housing characteristics, employment, recreational facilities, land cover, and transportation (Table 2). For the majority of these classes this was possible, but there remain a minority of classes that were not explicitly recorded in the mapped dataset because of data gaps. I then generated a series of simple rules to map each land use class, following the basic principles of von Thünen/Ricardo land rent theory [25], based on the economic notion of "highest and best use" that changes predictably with distance to markets and resources. In practice though, this dataset should be considered an initial attempt at estimating land use that should be critically evaluated, challenged, and improved on – as currently no other national extent, full sector, high-resolution land use dataset or classification exists. Note that currently the *primary use* is mapped at a given location, but in future versions the dataset could be extended to recognize that there can secondary or even tertiary uses at a given location (e.g., commercial uses mixed with residential).

Data are available at: http://csp-inc.org/public/NLUD2010_20140326.zip.

Processing of datasets

All datasets were projected to Albers Equal Area Conic and converted to raster datasets (at 30 m). Data on the location of airports (AIRP) were obtained from the US Census Bureau (www.census.gov/geo/www/tiger) as simple polygonal boundaries. Block housing density (BHD) data were obtained from the US Census 2010. "Water blocks" (those identifying a lake/reservoir/river), as well as blocks with 0 housing units were removed. Data on census blocks occupied by transportation (BLKS) were obtained from the US Census 2010, and blocks with 0 housing units and those that intersected interstates and US highways were

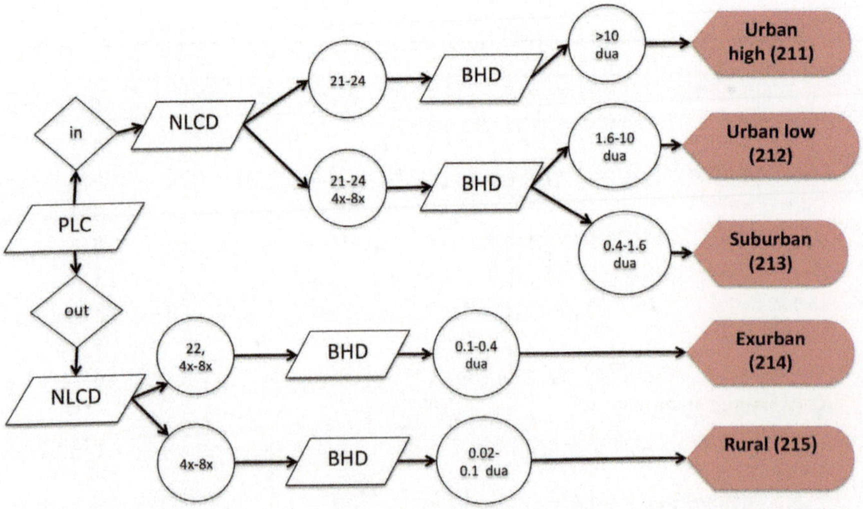

Figure 1. A flow chart depicting the process to map the residential land uses. Locations mapped as high-density urban residential were mapped in the National Land Cover Dataset 2006 (NLCD) as a built-up cover type and had a census block housing density in 2010 (BHD) of at least 10 dwelling units per acre (dua). Low-density urban residential was mapped as within PLC, on land but not water cover types (NLCD water, barren, or wetlands), and had a housing density between 1.6 and 10 dua. Suburban residential was mapped similar to low-density urban but at a density between 0.4 and 1.6 dua. Low-density residential areas were required to be outside of cities/towns (and were defined using Census places; PLC), and exurban residential had a housing density between 0.1 and 0.4 dua, while rural residential had a housing density between 0.025 and 0.1 dua.

identified. These long, linear blocks demarcate transportation, typically interstate corridors and on/off-ramps.

Data on agricultural cropland were obtained from the Cropland Data Layer (CDL) from the National Agricultural Statistical Service (http://nassgeodata.gmu.edu/CropScape/). I grouped the CDL classes into three general agricultural classes. Cropland (row crops) was assigned to various crop types including grains and vegetables such as corn, cotton, rice, etc. Pasture land was assigned

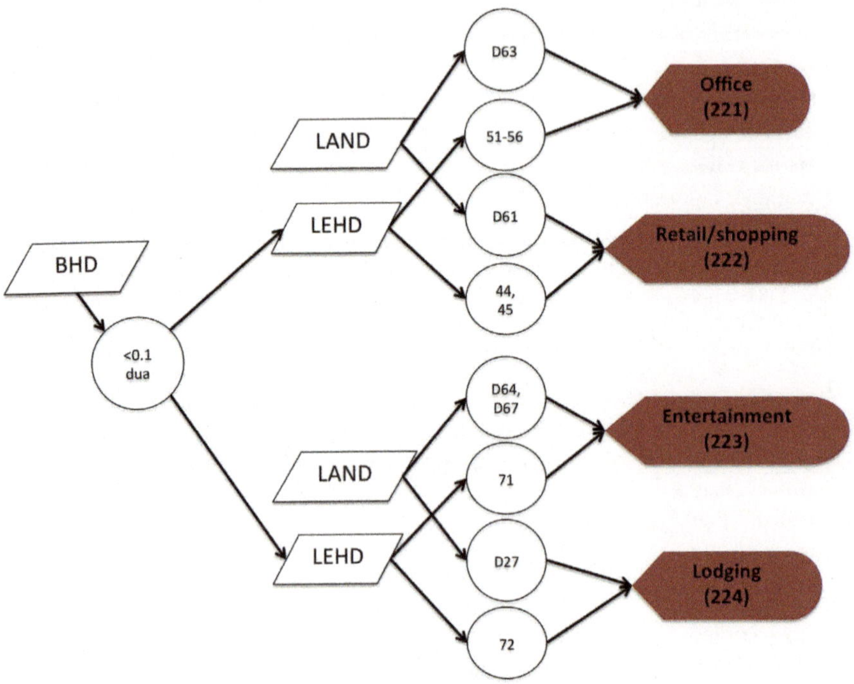

Figure 2. A flow chart depicting the process to map commercial land uses. Locations mapped as office commercial use were defined as an office building or office park by a census landmark polygon (LAND) or by North American Industry Classification System 2012 (NAICS) 2-digit code 51–56, which includes information, finance and insurance, real estate, professional, scientific and technical services, management of companies, and administrative and support services. Because commercial and residential uses are often mixed (e.g. particularly vertically, with commercial on the ground floor and residential above), and because the Longitudinal Employer-Household Dynamics (LEHD) data are mapped at the Census block-level, the residential housing density also had to be low (<0.1 dua). Consequently, areas that are mapped as urban residential land use could also contain some commercial land uses as well.

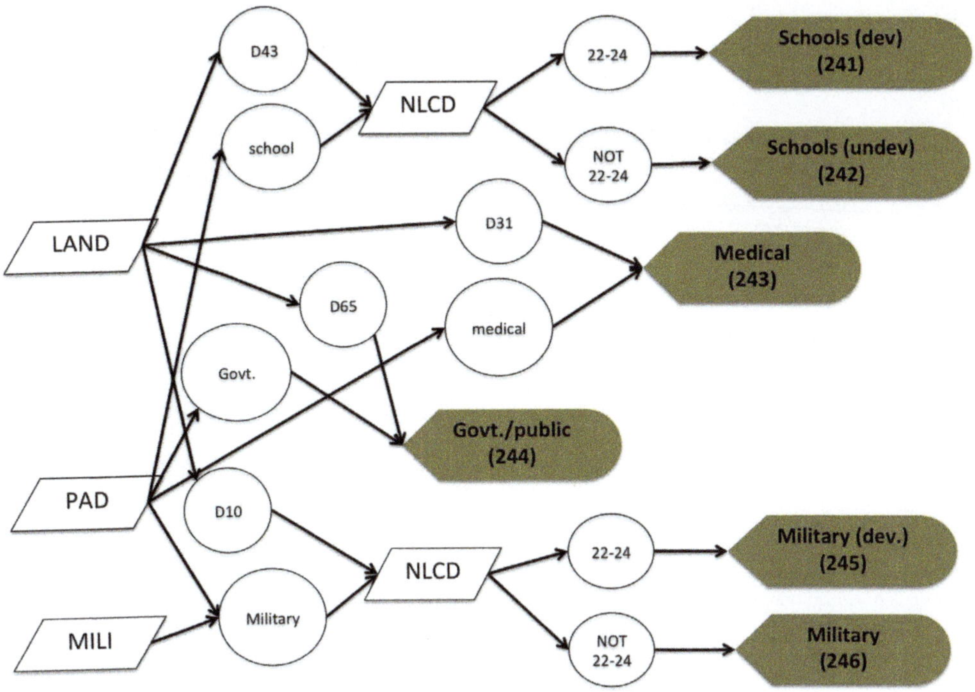

Figure 3. A flow chart depicting the process to map institutional land uses. Educational uses are mapped through Census landmark polygons (LAND) with values of D43 (Educational institution, including academy, school, college, and university) or if designated as a school in the Protected Areas Database US v1.3 (PAD). Educational institutions are further classified as developed (including buildings and playgrounds) using the National Land Cover Dataset 2006 (NLCD) with built-up values 22–24, or undeveloped (that includes grass fields and other vegetation types). Medical uses and government/public uses are mapped similarly, using LAND values = D31 (hospitals) or PAD designation as a hospital or other medical use; or D65 (government complex) or government designation from PAD. Military uses are depicted using LAND with values of D10 or as mapped in PAD or in the military polygons from the National Transportation Atlas Database 2010 (MILI). Developed portions of military lands, which can range from airports to storage facilities to residential uses, are differentiated using NLCD built-up values 22–24 (or if not, then are mapped as military undeveloped).

to alfalfa, pasture/grass, and hay types. An orchard class was assigned to tree and shrub-based production such as cherries, peaches, apples, oranges, etc. Finally, I assigned sod and switch grass to a miscellaneous agricultural land use class.

Data on various land mark features such as shopping malls, schools, etc. were obtained from the US Census 2010 (LAND). These features were re-coded in the following way based on the TIGER Census Feature Class Codes: D10 = military, D31 = hospitals, D37 = correctional facility, D43 = schools, D61 = commercial retail (malls, shopping centers), D63 = office building or office park, D65 = government center; D67 = entertainment/stadium, D81 = golf courses, and D82 = cemetery. Additional data on military lands (MILI) was obtained from the National Transportation Atlas (2010; www.bts.gov/publications/national_transportation_atlas_database). Employment density and type data were obtained from the US Census Longitudinal Employer-Household Dynamics (LEHD; www.lehdmap.did.census.gov). These data provide the number of employees in each block, and using the 2-digit NAICS code I grouped these types into agricultural (11), commercial (22, 23, 44–45, 51–56, 71, 72, 81), industrial (21–23, 31–33, 42, 48–49), government (92), educational (61), health (62), and transportation (48–49). Recreational facilities on US Forest Service lands (INFRA) were mapped using data were obtained from the FS Topo Recreation Facilities (www.fsgeodata.fs.fed.us/vector). These points were buffered by 100 m.

Data on the location of active mines and processing plants (MINE) were obtained from the USGS Mineral Resources Data System (www.mrdata.usgs.gov/mineplant). Only records that were attributed as a current producer or plant were included. These locations were then buffered up to 400 m, but for only those locations mapped as Barren land (rock/sand/clay) in NLCD 2006. Water classes were derived from the US Geological Survey National Hydrography Dataset (high resolution, 1:24,000; nhd.usgs.gov). I converted the areal-based features from NHD (i.e. lakes/reservoirs, wide-rivers, washes, and marsh/wetlands) into 30 m raster cells (center-method). Note that the water class from the NLCD 2006 was removed, so water classes in NLUD are only derived from the NHD data – providing a much more detailed, consistent, and precise representation of water features. This includes removing the water pixels from the original NLCD that are offshore from the coastline feature class from NHD. To distinguish natural water features (e.g., lakes and rivers) from human-made (e.g., reservoirs), I identified the NHD water body features that contained "reservoir" in the GNIS name or if it was the nearest feature (up to 200 m) from a dam identified in the National Inventory of Dams 2009 (NID; www.geo.usace.army.mil).

Land cover data were obtained from the 2006 NLCD (www.mrlc.org). These data were used to differentiate developed from undeveloped areas on institutional land using NLCD values of 21–24 on military or institutional (educational) land use types, then these locations were changed to schools (developed) or military (developed) NLUD classes. Wetland areas are identified partially from NLCD as well as from the NHD. Note that the NLCD data were filtered to remove a number of artifacts, including thin lines of developed areas (primarily along rural roads caused by "burning in" pixels from a highway data layer) that were removed

Table 2. Datasets used to construct the national land use database.

Group	Sub-Group	Datasets
Built-up	Residential	BHD = US Census 2010 block housing density (1:100k)
		LEHD = US Census Longitudinal Employer-Household Dynamics (block level, www.lehdmap.did.census.gov)
		NLCD = National Land Cover Dataset 2006 (30 m; www.mlrc.org;)*
		PLC = US Census 2010 place polygons** (www.census.gov/geo/www/tiger)
	Commercial	
	Commercial	LAND = Census area landmarks (1:100k), LEHD
	Industrial	LAND, LEHD
	Institutional	MILI = Military lands from National Transportation Atlas 2010 (1:100k) (www.bts.gov/publications/national_transportation_atlas_database)
		PAD = Protected Areas Database-US (1:100k; http://gapanalysis.usgs.gov/padus/, v1.3)
		AIRP = US Census TIGER 2010 (1:100k)
	Transportation	BLKS = US Census 2010 blocks with 0 housing units, long and narrow
		TRDS = US Census TIGER 2010 roads, interstates and state highways
	Miscellaneous	PAD, general resource production activities
Production	General	NLCD, Cropland Data Layer (CDL; http://nassgeodata.gmu.edu/CropScape/)
	Cropland	NLCD
	Grazing	NLCD
	Timber	NLCD
	Mining, oil & gas	MINE = Active mines and plants in the US (USGS, 2005; www.mrdata.usgs.gov/mineplant)
Recreation	General	NLCD, PAD
	Urban park	LAND, PARK = US Census parks (1:100k)
	Natural park	LAND, PARK
	Developed faciility	INFRA = US Forest Service FS Topo Recreation Facilities (1:24k; www.fsgeodata.fs.fed.us/vector)
		SKI = Ski resort locations from National Weather Service (1:100k; www.nohrsc.noaa.gov/gisdatasets)
Conservation	Natural area	PAD, areas of critical environmental concern, research natural areas, wildlife refuges, etc.
	Wildlife habitat	PAD, wildlife habitat and state wildlife areas
	Wilderness	PAD, Wilderness areas
	Miscellaneous	PAD, other
Water	water	NHD = National Hydrographic Dataset (1:24k, areas, water bodies, nhd.usgs.gov)
		NID = National Inventory of Dams 2009 (1:24k, geo.usace.army.mil)

*With highways and roads filtered/removed
**Removed county, unknown and Census Designated places.

if they were less than 90 m wide and were within 90 m of an interstate, highway, or county highway (from US Census 2010 TIGER roads). If a pixel was removed, it was filled in with the class of the nearest "natural" cover type using Anderson code level I (e.g., grass, shrub, forest, wetland).

Data on land ownership and management were obtained from the US Geological Survey's Protected Areas Database-US (PAD) v1.3 (http://gapanalysis.usgs.gov/padus). I assigned land use classes based on unique values (n = 1119) from the field "Primary Location Description" and by interpreting information from additional fields such as management type, GAP status code, and name of protected area (a look up table is available on request). I also used morphological analysis to filter small and thin "sliver" polygons, and well as filled in "gaps" between polygons that were caused primarily because of the state-by-state nature of the PAD-US dataset. Data on mostly urban parks (PARK) were obtained from US Census landmark data as well as from the PAD dataset. Data on urban or city areas were obtained from the US Census 2010, using the "places" geography. These polygons were used to exclude certain land uses, such as campgrounds and exurban residential development. Locations of ski resorts (SKI) was obtained from the National Weather Service (www.nohrsc.noaa.gov/gisdatasets) and were buffered by 200 m.

Evaluation

It is difficult to rigorously evaluate the results of this dataset because of the differences of it with other datasets in terms of class definitions, the scale of data (county to state vs. very fine), time differences (2007 vs. 2010), and coverage or geographic extent (e.g., NRI is only for privately-held lands). I do, however, provide a general comparison of the results to the NLCD as well as to the NRI. In addition, I validated the NLUD against a detailed land use dataset that was generated by interpreting land uses from recent (2005–2010, mostly 2007–8) high-resolution (<1 m) aerial photography sampled at ~6,000 random locations across mainland US. For each sample location or "chip" (roughly 600 m×600 m), a trained photo interpreter mapped polygons of each land use type following an established protocol [26]. I

Table 3. Area and proportion of the US for sub-group level of the National Land Use dataset 2010.

Group	Sub-group	Area (km^2)	Proportion of US
Built-up	Residential – Urban	148,502	1.83%
	Residential – rural	732,797	9.05%
	Commercial	20,540	0.25%
	Industrial	11,044	0.14%
	Institutional	71,529	0.88%
	Institutional undeveloped	24,403	0.30%
	Transportation	43,698	0.54%
	Misc. open space	49,592	0.61%
Production	Cropland	1,637,198	20.21%
	Miscellaneous	16,160	0.20%
	Grazing	2,559,409	31.60%
	Mining	938	0.01%
	Timber	20,436	0.25%
Recreation	General	756,049	9.33%
	Urban park	11,202	0.14%
	Golf course	6,813	0.08%
	Natural park	39,683	0.49%
Conservation	Wildlife habitat	190,358	2.35%
	Reserves & conservation	818,840	10.11%
	Wilderness	243,077	10.11%
	Watershed	19,897	3.00%
	Private easements	33,325	0.25%
Water	Natural	250,027	0.41%
	Human	61,094	3.09%
	Wetlands, swamps, playas	301,444	0.75%
	Oceanic	31,813	3.72%

intersected the NLUD data with the chips, such that each of the 1.9 million cells (30 m) that intersected a chip polygon was assigned the underlying land use to assess against the NLUD value.

Results

The final land use classification system consisted of 5 groups, 26 sub-groups, and 79 classes (Table 3; Figure 4). I found that data were insufficient to map 11 classes (e.g., energy, animal hunt & trapping, fish ponds, algal beds, confined animal feeding operations, etc.), so that the first version of the dataset contained 58 classes. Built-up areas occupy 13.6% of mainland US, but that the majority of this occurs as low-density exurban/rural residential (9.1% of the US), while more intensive built-up land uses occupy 4.5%. For every acre of urban and suburban residential land, there are 0.13 commercial, 0.07 industrial, 0.48 institutional, and 0.29 acres of interstates/highways. Areas of resource production occupy the majority of the US at 52%, with 20.0% in cropland. Recreational activities occupy 10%, including 0.14% in urban parks and 0.08% in golf courses. Conservation lands occupy 16%, with 3% of the mainland US in designated wilderness areas. Water (although not a land use *per se*) occupies 6% with natural lakes and rivers (including large portions of the Great Lakes) occupying 1.2%, human constructed facilities of reservoirs and canals occupy 0.78%, wetlands/swamps occupy 3.5%, and playas/washes occupy 0.3%.

There are important geographic variations in the land use patterns (Figure 5). Not surprisingly, built-up areas are more extensive in the eastern US (36.5% northeast and 32.5% southeast) and minimal in the west (4.3% mountain and 5.1% in North-central). Production areas are most extensive in the North-central region (81.7% especially croplands) and the South central region (76.1% especially grazing) and minimal in the Northeast (35.1%). Recreational uses are most extensive in the Pacific (20.0%) and minimal in the South-central region (2.1%). Conservation lands are most extensive in the Pacific and Mountain regions (28.2% and 35.6%), and minimal in the North-central and South-central (2.4% and 2.5%).

The results of the accuracy assessment of NLUD against the validation land use dataset were promising (Table 4), although I was unable to validate the recreation and conservation groups because the validation dataset did not adequately distinguish these land use types. The overall accuracy and the group (top) level was 84.2% for built-up and 99.3% for agricultural. At the sub-group level, Residential was correctly classified 74.4% (primarily confused with commercial). Commercial was correctly classified 25.6%, being largely confused with Institutional sub-group. All other sub-groups (Industrial, Transportation, and Miscellaneous) were >91% accurate.

This new map of land use provides a comprehensive, detailed, and spatially-explicit characterization of the primary human activities that are carried out on US lands (Figure 6). In addition to a partial validation, it is useful to compare the NLUD to two existing land use datasets. The first is a comparison of NLUD built-up uses to the NLCD 2006 (Table 5). Although NLCD is one of the three primary input datasets used to create NLUD, there are some important differences. I found that residential urban was predominately identified in the urban cover classes (81.0%), but that 12.9% of urban residential cells were (mis-) identified as forest cover. For exurban and rural residential land uses less than 1 unit per 2.5 acres (0.025–0.4 dua), NLCD identified only 8.9% in the urban cover classes, with over 57.2% identified as forested land cover, and 33.8% in agricultural, grass, and shrubland cover classes. This concurs with previous findings that NLCD misses significant areas of the US that are in low-density residential land use [6,7].

A second comparison is to the NRI dataset. The NRI found that in 2007 about 5.74% of the US was classified as developed land uses (that includes built-up lands permanently removed from rural land use, including urban and rural transportation corridors), which compares roughly to an estimated 6.26% from the NLUD using the built-up group and urban park & golf course classes (minus the low-density exurban class of less than 0.1 dua. Roughly adjusting the NRI estimate to be concordant in time with the 2010 NLUD, the NRI estimate would increase to about 5.87% (the US population increased from 2007 to 2010 by 1.0236%, assuming a linear relationship between population increase and developed land area). In the land use dataset, the built-up group is not limited to private land, and a very small amount of built-up (mostly classed as institutional or industrial) occurs on public lands, which might account for some of the gap. Comparing intensive agricultural land use, the NRI estimated 24.54% in cropland and pastureland uses, as compared to 20.9% in the land use dataset. A smaller area might likely occur because the land use dataset agricultural categories are derived primarily from satellite imagery, and consequently the narrow strips of adjacent lands (fallow, corners of center pivot sections, etc.) that are likely in agricultural use might be classified as grassland, shrubland, or perhaps barren in

Figure 4. The National Land Use dataset for 2010, depicting 59 land use classes for the conterminous US, at a resolution of 30 m (0.09 hectares).

the NLCD. The definition of the other main NRI classes (rangeland and forestland) are simply too different from the land use dataset to provide a reasonable comparison, primarily because activities that occur on these types, such as grazing, can occur on both rangelands and forested land, as well as on both public and private land.

Example applications

In addition to investigating patterns and summaries at local, regional or state-levels, these data can be used to examine many different types of questions, particularly those that aim to quantify the condition of metropolitan and urban areas (Figure 7). For example, the smart growth literature has identified a key measure of lower energy use and higher "walkability", called the land use

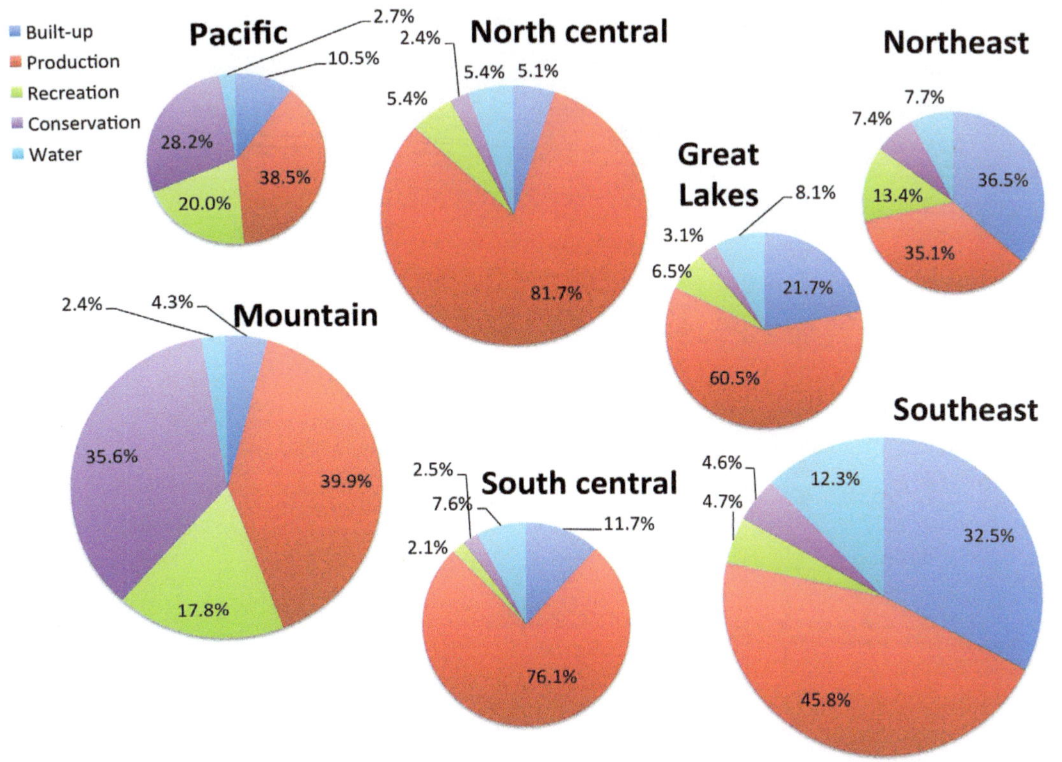

Figure 5. Regional variation of the proportion of major land uses in the conterminous US. The relative extent of 7 regions is depicted by the size of the pie-charts: Pacific (CA, OR, WA); Mountain (AZ, CO, ID, MT, NM, NV, UT, WY); North central (IA, KS, MN, MO, ND, NE, SD); South central (AR, LA, OK, TX), Great Lakes (IL, IN, MI, OH, WI), Northeast (PA and northeastern states), Southeast (DE, KY, WV and southern states). Land in productive uses dominated the major land uses for all regions but the Northeast, which was narrowly dominated by built-up uses. The proportion of land in recreation and conservation land uses exceeded 10% for the Pacific, Mountain, and Northeast regions. Note: "water" is not a land use per se, and includes both water and wetland cover types.

mix, which is commonly calculated at project-level extents using parcel-level data [22]. I measured this using an entropy index of the diversity of land uses within 1 mile:

$$E = -\left\{\left[\sum_{i=1}^{n}(P_i * \ln P_i)\right] \Big/ \ln(n)\right\}$$

where P_i is the proportion of use type i and $n = 4$. Land use classes were combined to reflect types i = live (residential), work (commercial, industrial, & institutional), play (recreation), and shop (retail commercial). This index shows interesting patterns of land use mix (Figure 7) delineating more mixed-use areas from more homogenous parts of a city, or comparing urban areas (e.g., Fort Collins, Colorado and Ashville, North Caroline had a mean land use mix value of 0.498 and 0.229, respectively. Locations with higher land use diversity have been found to be strongly related to "walkability" of a neighborhood as well as lower vehicle miles travelled [23].

A second application is to more precisely estimate the power-law scaling of urban areas [27]. Discussions about urban sprawl have been confounded by poor aggregate indicators such as population density, which do not account for overall population size. I found that the population of US metropolitan areas scaled to the developed footprint: $y = 972.75\ X^{0.974}$, $R^2 = 0.68$, where X is defined as any built-up land use except urban parks/golf courses (Figure 8). Refining earlier work [28], this equation can be used to allow comparison amongst metro areas to allow fairer and more consistent examination of potential efficiencies of resource use (or conversely, "sprawl"). By taking the ratio of the current population to the scale-adjusted population, a relative indication of how efficiently (per capita) land is being used, compared to all other metro areas. For example, of metro areas with 2010 populations of Ashville, NC had an index of 0.46, suggesting that in 2010 its pattern was less efficient than average, whereas Fort Collins, CO was a bit better than average (1.27) and Santa Barbara, CA was well above average (2.76).

A third example is identifying potential conflict areas with natural resources such as the agricultural-urban interface, which has become increasingly contentious and has engendered numerous right to farm acts across the US. Using NLUD I found that in 2010 roughly 13% of intense agricultural lands (cropland and pastureland, 222,610 km^2) were within ½ mile of the developed footprint (of at least 160 acres in area, a reasonable proxy for minimum farm size). Similarly, about 10.5% of the forested ecosystems were within the wildland-urban interface (within ~½ mile of developed lands). I also found that 13% of the (mainland) US coastline was developed.

Discussion

Although the dataset constructed here nominally represents conditions in 2010, the datasets compiled actually represent a range of time frames. The primary datasets of housing density, employment, and transportation do represent 2010 conditions, but land cover is from 2006. The other datasets represent 2010

Table 4. Results of the accuracy assessment testing the National Land Use dataset against a sample-based, detailed land use dataset.

Group	Sub-group	Water		Built-up						Agricultural		
		Natural	Human	Residential	Commercial	Industrial	Institutional	Transportation	Misc	Cropland	Grazing	Timber
Water	Natural	18,551	-									
	Human		13,078			2						
Built-up	Residential		69	55,958	272		2			59	2	
	Commercial			9,644	1,100							
	Industrial		6	655	416	528		44				37
	Institutional				2,465		510					
	Transportation			5,384				475				
	Misc								458			
Agri-cultural	Cropland			27	11	2				441,237		32,736
	Grazing										442	
	Mining		391	1,121			1					
Recreation			18	2,410	28		7					
Unknown			596	14	4					70		3,108
		Water		Built-up						Agricultural		
Water		Natural	Human	Residential	Commercial	Industrial	Institutional	Transportation	Misc	Cropland	Grazing	Timber
	Natural	100.00%										
Built-up	Human		96.40%			0.40%						
	Residential		0.50%	74.40%	6.30%		0.40%				0.50%	
	Commercial			12.80%	25.60%							
	Industrial			0.90%	9.70%	99.20%		8.50%				0.10%
	Institutional				57.40%		98.10%					
	Transportation			7.20%				91.50%				
Agri-cultural	Misc								100.00%			
	Cropland			0.00%	0.30%	0.40%				100.00%		99.90%
	Grazing										99.50%	
	Mining		2.90%	1.50%	0.00%		0.20%					
Recreation			0.10%	3.20%	0.70%		1.30%					
Unknown			4.40%	0.00%	0.10%					0.00%		9.50%

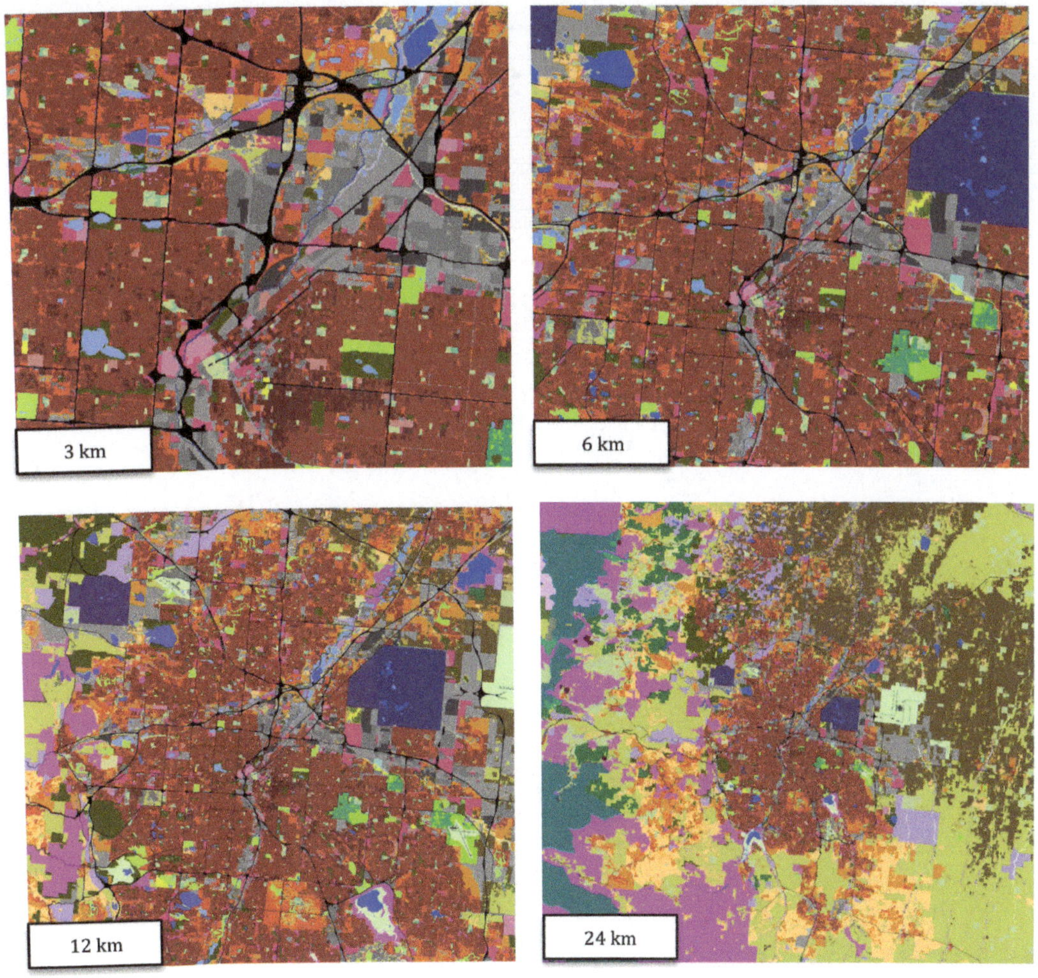

Figure 6. The National Land Use dataset for 2010 centered on Denver, Colorado (I-25 and I-70), showing NLUD at 4 different scales. Detailed industrial (light grey), commercial (dark grey), and transportation (black) land uses can be readily scene in the detailed map (3 km scale), while regional land use patterns surrounding the Denver metro area, including "open space" lands composed of recreational, conservation, and production lands, are shown in the coarse map (24 km scale). The full legend is provided in Figure 4.

conditions (mostly used in the buil-up group — landmarks, institutional, airports, military, etc.), while the recreation and conservation groups represent conditions that range from early 2000s to 2010. This requires a bit of interpretation, particularly for the conservation classes. Another opportunity is to update the land use dataset when the 2011 version of NLCD 2011 becomes available (anticipated in early 2014), or potentially generate a

Table 5. A comparison of the National Land Use database (2010) against the National Land Cover Dataset (2006) for built-up uses.

	Water	Urban				Barren	Forest	Shrub	Grass	Ag.	Wetland
NLUD Built-up	**1**	**21**	**22**	**23**	**24**	**3**	**4**	**5**	**7**	**8**	**9**
Res. Urban	0.00%	27.40%	37.30%	13.90%	2.40%	0.00%	12.90%	2.10%	2.40%	1.50%	0.00%
Res. Rural	0.00%	6.80%	1.80%	0.30%	0.10%	0.00%	57.20%	8.90%	9.30%	15.60%	0.00%
Commercial	1.30%	10.90%	11.10%	24.40%	19.50%	1.40%	9.20%	5.20%	3.80%	13.00%	0.20%
Industrial	1.30%	11.40%	10.80%	21.40%	16.90%	1.10%	10.00%	4.80%	4.00%	18.20%	0.10%
Institutional	0.90%	3.60%	16.10%	20.20%	9.10%	0.30%	22.50%	4.40%	2.10%	5.00%	15.80%
Inst. Undevel.	0.10%	3.60%	0.30%	0.10%	0.10%	6.60%	12.40%	62.30%	10.20%	2.00%	2.30%
Transportation	0.20%	35.80%	29.20%	12.70%	4.70%	0.20%	3.90%	2.80%	2.60%	7.10%	0.90%
Miscellaneous	1.30%	8.70%	5.90%	10.10%	4.60%	1.50%	18.80%	15.60%	8.00%	24.70%	0.80%

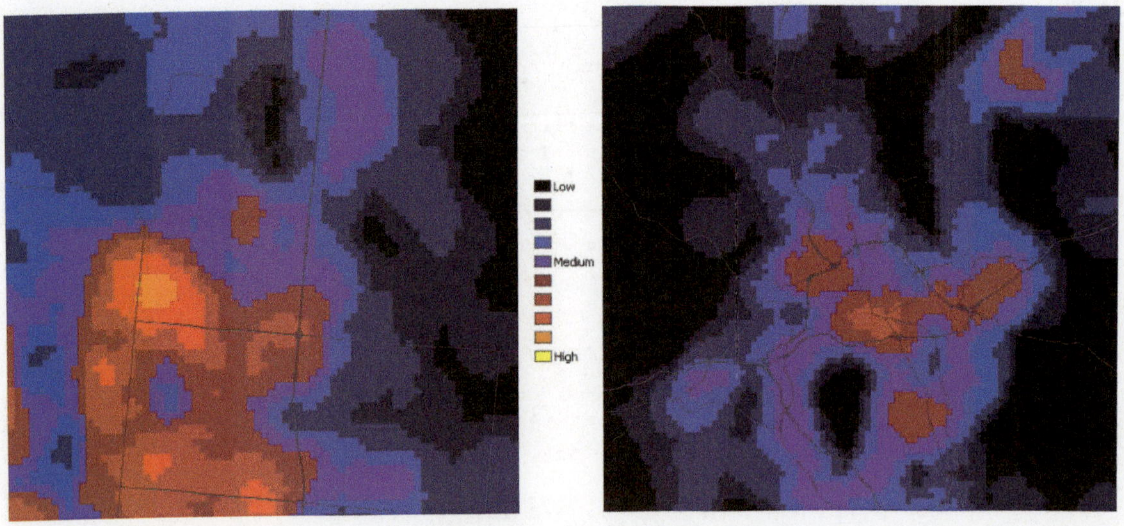

Figure 7. The diversity of land uses shown for Fort Collins, CO (left) and Ashville, NC (right). Higher diversity is depicted in yellow/red, with lower diversity shown in blue and dark blue. Major highways and interstates are labeled for reference.

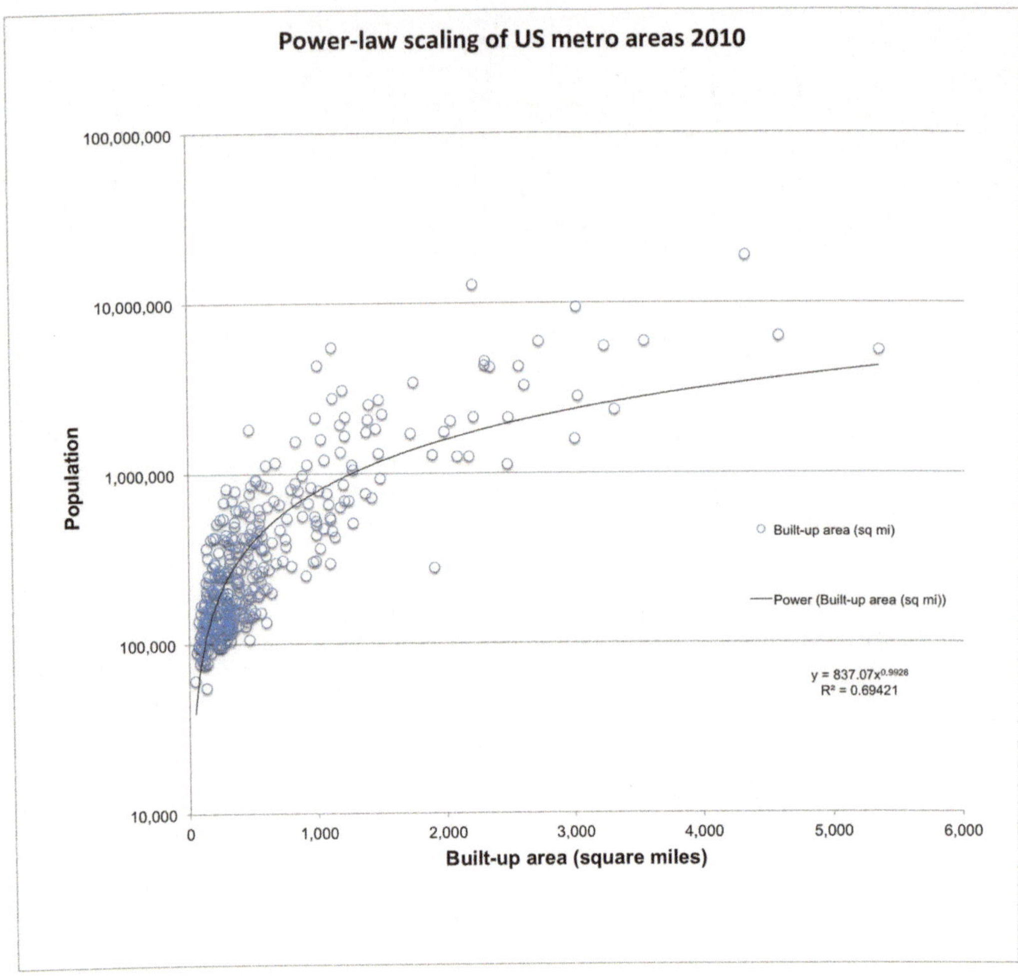

Figure 8. The population of US metropolitan areas follows a power law-scaling to the developed "footprint", which includes built-up land uses except for urban parks and golf courses. Computing the ratio of the raw population to the footprint (scale) adjusted population provides a relative measure of land use efficiency for metropolitan areas.

series of national land use datasets to be consistent with the NLCD series (2001, 2006, 2011).

Both the level of detail in the land use classes, and the high-resolution of the dataset make it particularly vulnerable to potential misclassifications and errors. The general accuracy of the NLUD at sub-group levels is on par with other previous efforts (e.g., NLCD 2006; [29]), but represents a substantial improvement over urban/built-up classes of NLCD (84% vs. 56%) and is roughly similar to agricultural cover types (77 and 88%).

The greatest uncertainty in mapping land use centers on the uses in the grazing/rangeland sector of agriculture, and recreation, and conservation on public lands. This is in part because these are multi-use lands, where multiple uses are occurring simultaneously (e.g., many national forest lands provide grazing, recreation, and timber, and national parks have a dual mission to preserve resources and to provide for the enjoyment of parks). It is also due to the relatively crude surrogate used — ownership/management from the Protected Areas Database — which could be improved by incorporating more detailed information from resource and travel management plans. Similarly, because many of the parks and protected areas are represented solely as their ownership boundaries, the variety of uses within a park typically are not differentiated.

Notably, the grazing land use class has a high degree of uncertainty, as it was mapped simply as privately owned lands dominated by shrub and grass land cover. Many privately-owned, conserved lands could also be incorporated into the NLUD, such as from the recently-created National Conservation Easement Database.

Key findings from this study are that built-up areas occupy 13.6% of mainland US, but that the majority of this occurs as low-density exurban/rural residential (9.1% of the US), while higher-density built-up land uses occupy 4.5%. For every acre of urban and suburban residential land, 0.13 commercial, 0.07 industrial, 0.48 institutional, and 0.29 acres of interstates/highways. Three example applications of the dataset were provided, but there are a wide variety of additional applications and uses of this dataset.

Acknowledgments

I thank B. Monahan from the National Park Service, J. Falcone and P. Claggett with the US Geological Survey, P. Morefield and B. Bierwagen with the US Environmental Protection Agency, M. Nelson from the US Forest Service, and Sarah Reed from Wildlife Conservation Society for comments on earlier drafts of this manuscript, and to the reviewers and editor for their helpful comments and suggestions. I also thank two land use visionaries who inspired this work: Senator Henry Jackson and Bill Travis (Riebsame).

Author Contributions

Conceived and designed the experiments: DT. Performed the experiments: DT. Analyzed the data: DT. Contributed reagents/materials/analysis tools: DT. Wrote the paper: DT.

References

1. Ellis EC, Goldewijk KK, Siebert S, Lightman D, Ramankutty N (2010) Anthropogenic transformation of the biomes, 1700 to 2000. Glob Ecol Biogeogr 19: 589–606.
2. Foley JA, DeFries R, Asner GP, Barford C, Bonan J, et al. (2005) Global consequences of land use. Science doi:10.1126/science.1111772
3. Scholes RJ, Biggs R (2005) A biodiversity intactness index. Nature 434: 45–49.
4. Vorosmarty CJ, McIntyre PB, Gessner MO, Dudgeon D, et al. (2010) Global threats to human water security and river biodiversity. Nature Doi:10.1038/nature09440.
5. Patz JA, Daszak P, Tabor GM, Aguirre AA, et al. (2004) Unhealthy landscapes: policy recommendations on land use change and infectious disease emergence. Environ Health Perspect 112(10:1092–1098.
6. Theobald DM (2001) Land-use dynamics beyond the American urban fringe. Geogr Rev 91(3): 544–564.
7. Irwin EG, Bockstael NE (2007) The evolution of urban sprawl: evidence of spatial heterogeneity and increasing land fragmentation. Proc Natl Acad Sci U S A Doi:10.1073/pnas.0705527105.
8. Di Gregorio A (2005) *Land cover classification system: Classification concepts and user manual.* FAO, Rome. Food and Agriculture Organization. Global Forest Resources Assessment 2005: Progress towards Sustainable Forest Management. United Nations. FAO Forestry Paper 147.
9. Homer C, Huang C, Yang L, Wylie BK, Coan M (2004) Development of a 2001 National Landcover Database for the United States. Photogramm Eng Remote Sensing 70:829–840.
10. Loveland TR, Sohl TL, Stehman SV, Gallant AL, Sayler KL, et al. (2002) A strategy for estimating the rates of recent United States land cover changes. Photogramm Eng Remote Sensing Vol. 68, No. 10, pp. 1091–1099.
11. Lubowski RN, Vesterby M, Bucholtz S, Baez A, Roberts MJ (2006) *Major uses of land in the United States, 2002.* USDA Economic Research Service, Economic Information Bulletin Number 14. Washington, DC.
12. US Department of Agriculture (NRI) (2009) *Summary report: 2007 National Resources Inventory,* Natural Resources Conservation Service, Washington, DC, and Center for Survey Statistics and Methodology, Iowa State University, Ames, Iowa. 123 Pages.
13. Cadenesso ML, Pickett STA, Schwarz K (2007) Spatial heterogeneity in urban ecosystems: reconceptualizing land cover and a framework for classification. Front Ecol Environ 5(2): 80–88.
14. Anderson JR, Hardy EE, Roach JT, Witmer RE (1976) *A land use and land cover classification system for use with remote sensor data.* US Geological Survey Professional Paper 964.
15. Falcone JA, Homer CG (2012) Generation of a US National urban land-use product. Photogramm Eng Remote Sensing 78:1057–1068.
16. Leroux A, Gauthier JP, Lemonsu A, Mailhot J (2009) Automated urban land use and land cover classification for mesoscale atmospheric modeling over Canadian cities. Geomatica 63:393–406.
17. Turner BL II, Lambin EF, Reenberg A (2007) The emergence of land change science for global environmental change and sustainability. Proc Natl Acad Sci U S A 104(52): 20666–20671.
18. Verburg PH, van de Steeg J, Veldkamp A, Willemen L (2009) From land cover change to function dynamics: A major challenge to improve land characterization. *J* Environ Manage 90: 1327–1335.
19. Turner BL II, Moss RH, Skole DL (eds) (1993) *Relating land use and global land-cover change: A proposal for an IGBP-HDP core project.* Report from the IGBP-HDP Working Group on Land-Use/Land-Cover Change. Joint publication of the International Geosphere-Biosphere Programme (Report No. 24) and the Human Dimensions of Global Environmental Change Programme (Report No. 5). Stockholm: Royal Swedish Academy of Sciences.
20. American Planning Association (APA) (2001) Land-based classification standards. URL: http://www.planning.org/lbcs (accessed 31 January 2011).
21. Bureau of Rural Services (BRS) (2006) Guidelines for land use mapping in Australia: principles, procedures and definitions: a technical handbook supporting the Australian Collaborative Land Use Mapping Programme Ed. 3. 84 pgs. (Canberra, ACT).
22. Salafsky N, Salzer D, Stattersfield AJ, Hilton-Taylor C, Neugarten R, et al. (2008) A standard lexicon for biodiversity conservation: unified classifications of threats and actions. Con Biol 22(4):897–911.
23. Ewing R, Cervero R (2010) Travel and the built environment. J Am Plann Assoc 76(3): 265–294.
24. Riitters KH, Wickham JD, Wade TG (2009) An indicator of forest dynamics using a shifting landscape mosaic. Ecol Indic 9: 107–117.
25. Chisholm M (1962) *Rural settlement and land use.* Hutchinson, London.
26. Leinwand IIF, Theobald DM, Mitchell J, Knight RL (2010) Land-use dynamics at the public-private interface: a case study in Colorado. Landsc and Urban Plan 97(3): 182–193.
27. Batty M (2008) The size, scale, and shape of cities. Science Doi:10.1126/science.1151419.
28. Sutton PC (2003) A scale-adjusted measure of "urban sprawl" using nighttime satellite imagery. Remote Sens Environ 86:353–369.
29. Wickham JD, Stehman SV, Gass L, Dewitz J, et al. (2013) Accuracy assessment of NLCD 2006 land cover and impervious surface. Rem Sens of Environ 130: 294–304.

15

Digital Mapping of Soil Organic Carbon Contents and Stocks in Denmark

Kabindra Adhikari[1]*, Alfred E. Hartemink[1], Budiman Minasny[2], Rania Bou Kheir[3†], Mette B. Greve[3], Mogens H. Greve[3]

1 Department of Soil Science, University of Wisconsin—Madison, Madison, Wisconsin, United States of America, 2 Department of Environmental Sciences, The University of Sydney, Sydney, New South Wales, Australia, 3 Department of Agro-ecology, Aarhus University, Tjele, Denmark

Abstract

Estimation of carbon contents and stocks are important for carbon sequestration, greenhouse gas emissions and national carbon balance inventories. For Denmark, we modeled the vertical distribution of soil organic carbon (SOC) and bulk density, and mapped its spatial distribution at five standard soil depth intervals (0−5, 5−15, 15−30, 30−60 and 60−100 cm) using 18 environmental variables as predictors. SOC distribution was influenced by precipitation, land use, soil type, wetland, elevation, wetness index, and multi-resolution index of valley bottom flatness. The highest average SOC content of 20 g kg^{-1} was reported for 0−5 cm soil, whereas there was on average 2.2 g SOC kg^{-1} at 60−100 cm depth. For SOC and bulk density prediction precision decreased with soil depth, and a standard error of 2.8 g kg^{-1} was found at 60−100 cm soil depth. Average SOC stock for 0−30 cm was 72 t ha^{-1} and in the top 1 m there was 120 t SOC ha^{-1}. In total, the soils stored approximately 570 Tg C within the top 1 m. The soils under agriculture had the highest amount of carbon (444 Tg) followed by forest and semi-natural vegetation that contributed 11% of the total SOC stock. More than 60% of the total SOC stock was present in Podzols and Luvisols. Compared to previous estimates, our approach is more reliable as we adopted a robust quantification technique and mapped the spatial distribution of SOC stock and prediction uncertainty. The estimation was validated using common statistical indices and the data and high-resolution maps could be used for future soil carbon assessment and inventories.

Editor: Dafeng Hui, Tennessee State University, United States of America

Funding: The study was funded by the Danish Ministry of Climate and Energy through SINKS Project (2009–2012). The funders had no role in study design, data collection and analysis, decision to publish, or preparation of the manuscript.

Competing Interests: The authors have declared that no competing interests exist.

* Email: kadhikari@wisc.edu

† Deceased.

Introduction

Digital Soil Mapping uses statistical tools to quantify the spatial relationship between soil property values to its environmental covariates [1]. The digital mapping of soil organic carbon (SOC) at fine resolution is a challenging task [2] and the mapping is also a high priority for SOC assessment and monitoring [3]. Spatial models on SOC prediction has a long history (e.g., [4]). A range of techniques have been used to predict and map SOC from landscape to national or continental levels and Minasny et al. [5] provided a comprehensive review. Several researchers applied splines to model the vertical distribution of SOC in the soil profiles and predicted SOC at a landscape scale using data-mining tools and environmental variables as predictors [6–8]. Mishra et al. [9] calculated SOC pool in each soil horizon and applied geographic weighted regression to map SOC pool at a regional scale in mid-west USA. Odgers et al. [10] used splines to derive SOC content from soil map units and predicted SOC at six standard soil depths for the entire USA. Arrouays et al. [11], Bou Kheir et al. [12], Chaplot et al. [13], Doblas-Miranda et al. [14], mapped SOC at national level using different statistical tools ranging from statistical aggregation to advanced regression tools such as regression trees.

The distribution of SOC changes across the landscape and it also varies by depth. In most soils, SOC is higher in the surface horizons and it decreases with depth. Such depth-wise variability is mostly continuous [15–17] except in soils with a strong human impact like some soils in the Netherlands [18]. Although most SOC studies and inventories are confined to 30 cm soil depth [19–21], the amount of SOC stored below 30 cm is of relevant in many ecosystems [22,23]. For accurate quantification of SOC stocks, a depth function needs to be modeled. Several tools exist: spline function [15,16], exponential decay function [24] or soil-type specific or profile depth functions [18,25]. For modeling SOC with depth, the equal-area spline function has been proven to be useful in several studies [7,8,10]. Spline predicted SOC values with depth act as a geo-referenced point data to which environmental variables are joined and prediction models are generated using digital soil mapping techniques.

In Denmark, studies on SOC dynamics and its quantification has started after a national wide soil database was established between the years 1975–1985. Based on simple statistical scaling-up techniques, Krogh et al. [26] calculated a total stock of about 579 Tg and reported that 69% of it was stored in the soils under agriculture. Greve et al. [27] estimated topsoil SOC contents (g 100 g^{-1}) for the whole country but have not assessed the SOC stocks. Bou Kheir et al. [12] predicted the spatial extent of organic soils across Denmark. Similarly, Olesen [28] and Taghizadeh-Toosi et al. [29] estimated the stock but did not map its distribution. Most of these studies have estimated the SOC content and stock but they have not explored the spatial distribution of SOC stock nor quantified the uncertainty of the SOC predictions.

We applied digital soil mapping techniques to quantify the SOC content and stocks for Denmark. The main objectives of this study were: to model the vertical distribution of SOC content and bulk density in soil profiles, to predict and map their spatial distribution using environmental variables, to identify major environmental variables responsible for SOC distribution, to estimate the SOC stock for the soils of Denmark, and to assess the uncertainties in the SOC predictions.

Materials and Methods

Study area

Denmark is situated in Northern Europe covering an area of approximately 43,000 km^2. The temperate climate is characterized by a mild winters with annual mean winter and summer temperatures of about 0°C and 16°C [30]. Precipitation is well distributed throughout the year with an average annual rainfall of 800 mm in the west to 500 mm to the east. The country is relatively flat with a mean elevation above mean sea level is about 31 m. Denmark is divided into 10 physiographic regions—referred in this paper as geo-regions—based on geographical, climatic and soil-formation criteria.

The soils are coarse sandy to clayey to heavy clays as defined in the Danish Soil Classification System [31]. Soil in the western part

Table 1. List of environmental variables used to predict the distribution of soil organic carbon and its stock in Denmark.

Environmental variables	Scorpan factor	Type of variable	Description	Range of values	Scale or resolution	Reference
Soil map	S	Categorical	Map of Soil types based on soil texture (8 classes)	-	1:50,000	[31]
Precipitation	C	Continuous	Average annual rainfall (mm) (1961–1990)	525 to 905	30.4 m	[31]
Geo-regions	C	Categorical	Scanned geographical regions map (10 classes)	-	1:100,000	[31]
Insolation	C	Continuous	Potential incoming solar radiation (2011)	254 to 698	30.4 m	[61]
Mid-slope position	C, N	Continuous	Covers the warmer zones of slopes	0 to 1	30.4 m	[62]
Land use	O	Categorical	CORINE Land cover data adopted in Denmark (31 classes)	-	1:100,000	[45]
Elevation	R	Continuous	Elevation of the land surface derived from LiDAR (m)	0 to 170	30.4 m	
Slope gradient	R	Continuous	Maximum rate of change between the cells and neighbors (degree)	0 to 90	30.4 m	[63]
Slope aspect	R	Continuous	Direction of the steepest slope from the North (degree)	0 to 360	30.4 m	[63]
Flow accumulation	R	Continuous	Number of upslope cells	1 to 73645	30.4 m	
SAGA wetness index	R	Continuous	Wetness Index. $WI = \ln(A_s / \tan \beta)$: where A_s is modified catchment area and β is the slope gradient	7.2 to 19	30.4 m	[64]
Multi-resolution valley bottom flatness	R	Continuous	Possible depositional areas	0 to 11	30.4 m	[65]
Valley depth	R	Continuous	Extent of the valley depth (m)	0 to 90	30.4 m	
Wetlands	S, R	Categorical	Map showing the presence or absence of wetlands	-	1:20,000	[31]
Landscape	R	Categorical	Landform types (10 classes)	-	1:100,000	[31]
Altitude above channel network	R	Continuous	Vertical distance to channel network base level (m)	0 to 56	30.4 m	
Slope length factor	R	Continuous	LS-factor of Universal Soil Loss Equation (m)	0 to 47	30.4 m	[66]
Geology	P	Categorical	Scanned and registered geological map (86 classes)	-	1:100,000	[31]

S-soil types; C-climate, O-organisms; R-relief; P-parent material; N-spatial position.

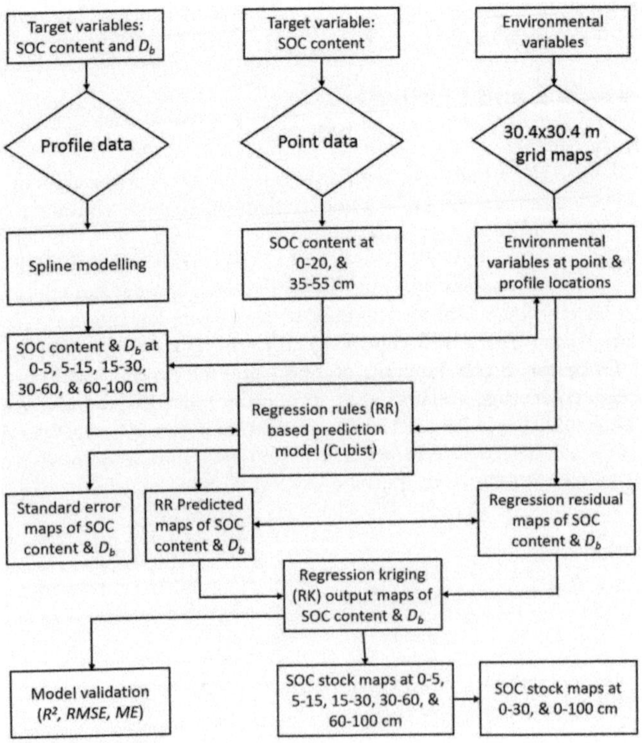

Figure 1. Schematic representation of overall prediction scenario.

of the country are developed in non-glaciated sandy parent material along the glacial flood-plains and Saalian moraine, whereas the soils from the central and eastern region have been developed on relatively young basal till high in finer materials [31]. Most of the soils in the north have been formed in sand mixed with uplifted marine sediments. More than 66% of the soils are classified as Podzols (Spodosols) and Luvisols (Alfisols). Podzols occupy a major portion in the west and Luvisols and Cambisols (Inceptisols, Entisols) in the central and eastern part of the country [32]. Peat soils (Histosols) occur in poorly drained basins throughout the country. About 66% of the total land area is used for agriculture with grain and potato as the main crops. Forest areas include spruce, pine and beach and these cover more than 12%.

Point data

Point (pedon) data on soil organic carbon (SOC) (g kg^{-1}) and bulk density (D_b) (Mg m^{-3}) were derived from two databases: Danish Soil Classification database (DSC) and Danish Soil Profile database (DSP). DSC consists of about 36,000 point observations from the topsoil (0−20 cm) sampled randomly from agricultural fields in the period 1975−1980. From about 6,000 same pedons, soils from the subsoil (35−55 cm) was also sampled. SOC content was determined by dry combustion using a LECO IR-12 furnace. DSP consists of a grid based data (7×7 km spacing) established during the 1990s for improve fertilizer recommendation in Denmark [33]. At each 850 grid intersection, soil samples were collected based on genetic horizons and were analyzed for SOC by dry combustion. For some selected horizons, samples were taken to determine D_b. In addition, soil data from about 1100 profiles were used, and these were collected during the establishment of main gas pipeline system and other research activities across Denmark [34].

In total 40,250 topsoil point samples and 1,994 soil profiles with horizon based SOC data were used in this study. About 1,133 soil profiles included D_b measurements.

Environmental covariates

The environmental covariates data used in this study are terrain parameters from the Digital Elevation Model (DEM) of Denmark derived using Light Detection and Ranging (LiDAR) technology. The LiDAR points were interpolated using Delaunay Triangulation and the output Triangular Irregular Network (TIN) surface was converted to a grid DEM with a 1.6×1.6 m spacing. This study resampled the original 1.6 m DEM to 30.4 m using simple aggregation considering the mean value. This 30.4 m grid size was also used in the previous studies in Denmark (e.g., [32,35]). Before aggregating to a coarser grid, the DEM was corrected by removing the pits and peaks of about 50 cm dimensions in order to ensure a regular flow on the surface. Once the DEM was processed, a number of terrain parameters (e.g., slope aspect, slope gradient, elevation, SAGA wetness index, multi-resolution index of valley bottom flatness (MrVBF), altitude above channel network, slope-length factor, over-land flow distance, and valley depth) were derived. Tools and algorithms incorporated in SAGA GIS [36] and Arc GIS V10.2 [37] were used to process the DEM and to derive these parameters.

Other covariates used were six choropleth maps which were compiled at different cartographic scales including: soil map, landscape types, geo-regions, geology, land use, and wetlands – see Table 1 and for more detail Adhikari et al. [35].

Key environmental variables affecting the spatial distribution of SOC and D_b in Denmark were identified. The relative usage of the environmental variables used during SOC and D_b prediction was calculated for each depth and their importance was expressed in percentage. This was done with *Cubist* software which determines the relative importance of variables based on their usage in the prediction rules. The prediction method adopted in this study is summarized in Figure 1.

Table 2. Average soil bulk density (Mg m^{-3}) for different soil organic carbon levels (g 100 g^{-1}) within the central wetlands [Source: [44]].

SOC content	**Soil depth (cm)**		
	0−30	**30−60**	**60−90**
<6	1.15	0.56	0.76
6−12	0.77	0.61	0.44
>12	0.39	0.25	0.19

Table 3. Descriptive statistics of soil organic carbon content (g kg^{-1}) and bulk density (D_b) (Mg m^{-3}) data used in this study.

Parameters	Spline predicted data										Point observation	
	Soil depth (cm)											
	0–5		5–15		15–30		30–60		60–100		0–20	35–55
	SOC	D_b	SOC	D_b	SOC	D_b	SOC	D_b	SOC	D_b	SOC	
Minimum	0.07	0.47	0.09	0.47	0.02	0.47	0.02	0.47	0.01	0.47	0.10	0.10
Maximum	562.31	1.84	562.1	1.84	562.22	1.99	564.01	2.01	570.01	1.96	562.21	559.22
Interquartile Range	15.92	0.21	14.11	0.20	10.22	0.19	6.12	0.17	2.33	0.16	9.16	7.61
Mean	35.22	1.44	30.71	1.44	23.81	1.46	15.61	1.52	9.91	1.59	19.71	15.01
Std. error of mean	1.45	0.00	1.18	0.00	1.11	0.00	1.14	0.00	1.03	0.00	0.07	0.48
Std. deviation	64.81	0.17	52.9	0.17	49.74	0.15	51.2	0.15	46.41	0.15	15.38	44.85
Coef. variation	184	12.02	175.91	11.71	208.9	10.71	328.31	9.81	465.42	9.51	78.12	298.61
Skewness	4.72	−1.01	5.61	−0.96	6.31	−1.03	7.11	−1.32	7.91	−2.02	15.22	7.91

Spline predicted data represents soil profile data from the Danish Soil Profile Database, whereas point observation represents data from the Danish Soil Classification.

Modeling SOC and D_b distribution

The vertical distribution of SOC and D_b in the soil profiles was modeled with mass preserving equal-area quadratic splines [15] in R [38]. The mathematical derivation of the spline has been described by Malone et al. [8]. As the fitting quality of splines to profile attribute data depends on a smoothing parameter—lambda (λ), we tested seven λ values (0.00001, 0.0001, 0.001, 0.01, 0.1, 1, 10) for SOC and D_b data from all the profiles and selected a λ value that showed the best fit for all the profiles using the root mean square. With increasing λ value, the fit becomes more rough. During the fit, we pegged the spline by introducing a 1 cm thick slice with a same SOC value at the topmost layer to prevent unnecessary extrapolation on the surface horizon.

Once the depth function of SOC and D_b were modeled, a weighted-average value of these properties were derived for five soil depths (0−5, 5−15, 15−30, 30−60, and 60−100 cm) based on the GlobalSoilMap specifications [39]. To these new values for the standard depths from all soil profiles, environmental variables were intersected and used for statistical analysis and spatial prediction.

Mapping to the spatial domain

Spatial prediction of SOC content and D_b at five depths was based on Regression kriging (RK) [40]. RK assumed that the spatial prediction function consists of a deterministic model formed by a regression, and the residuals of the regression (unexplained variation) are spatially correlated. The general principle of RK includes (1) regression, and (2) simple kriging of the residuals from the regression, where outputs from these two steps are added to obtain the final prediction. For the regression step, we adopted Regression-rules (RR) derived using *Cubist* software [41]. This tool generates a set of condition−based rules where each rule comes with a multiple regression prediction function that only operates once the conditions specified by the rule are met [42].

To build the SOC and D_b prediction model in *Cubist*, the data set was split randomly into two sets: 75% for calibration and 25% for model validation. Before the data split, SOC content from the topsoil observations (0−20 cm) were joined to the spline predicted SOC content from 0−5 and 5−15 cm soil depths. Similarly, SOC content from subsoil observations (35−55 cm) were attached to the spline predicted SOC at 30−60 cm soil depth. This way, a larger number of point SOC observations were incorporated to the splined SOC data from the 7×7 km grid profiles. The *Cubist* tool was run for log transformed SOC [log SOC g kg^{-1}] and D_b data from each depth interval and the output was converted to a regular grid map using a program written in FORTRAN. For each calibration location, the difference between measured and RR predicted value was calculated and its spatial distribution over the study area was generated using local variogram and point kriging in VESPER program [43]. This continuous residual surface was added to the corresponding RR output to get a final prediction of SOC and D_b for all five depths. Together with the prediction, a map showing the uncertainty of the prediction was generated. Both SOC prediction and uncertainty maps at each depth were then back-transformed to SOC in g kg^{-1}.

Bulk density in the peat areas

Peat lands are mostly present along the central part of the wetlands across Denmark. Bulk density in those areas were adjusted according to Greve et al. [44]. For the three surface layers (i.e., 0−5, 5−15, and 15−30 cm soil depths), D_b from 0−30 cm was used, and for the 30−60 and 60−100 cm, D_b from 30−60 cm, and 60−90 cm were used (Table 2).

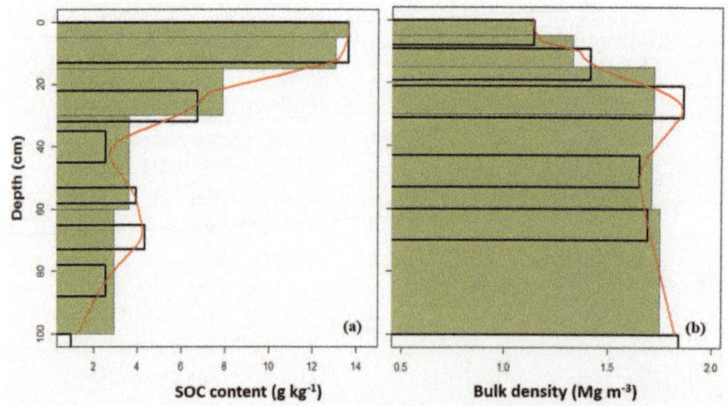

Figure 2. Example of a fitted spline for soil organic carbon content (a), and for bulk density (b). Horizontal bars represent measured soil organic carbon and bulk density at different soil horizons, continuous line through horizons represents a fitted spline, and horizontal olive-green bars give an weighted-average values of these properties at five standard soil depth intervals (i.e., 0−5, 5−15, 15−30, 30−60 and 60−100 cm).

SOC stocks

SOC stock for each soil depth was calculated according to Eq. (1) using SOC content and D_b data. SOC stocks from the five layers were summed to obtain a SOC stock to 1 m soil depth. D_b in Eq. (1) was corrected for gravel content.

$$SOC_{stock} = [SOC_{content} \times D_b \times D]/10 \quad (1)$$

Where SOC_{stock} is the SOC stock (t ha^{-1}), $SOC_{content}$ the SOC content (g kg^{-1}), D_b the soil bulk density (Mg m^{-3}) and D the given soil layer thickness (cm).

Model validation

Model performance in predicting SOC content and D_b was evaluated on 25% of the point data. The following three indices that were calculated:

$$R^2 = \frac{\sum_{i=1}^{n} (pred_i - \overline{obs})^2}{\sum_{i=1}^{n} (obs_i - \overline{obs})^2} \quad (2)$$

$$ME = \frac{1}{n}\sum_{i=1}^{n}(obs_i - pred_i) \quad (3)$$

$$RMSE = \sqrt{\frac{1}{n}\sum_{i=1}^{n}(obs_i - pred_i)^2} \quad (4)$$

Where obs and $pred$ are observed and predicted SOC and D_b values from n number of observations at i^{th} locations, ME is mean error, and $RMSE$ is the root mean square error.

Predicted SOC stock was determined for all five depth intervals and then aggregated to 0−30 cm and 0−100 cm soil depth. These two C stock maps were stratified based on soil and land use types. The soil class map of Denmark based on the FAO legend [32] and a land use map (CORINE data) were used. The CORINE database for Denmark has 31 classes [45] but in this study the legend was reduced to five major land uses types: artificial surface (urban areas, industrial areas, road network and ports etc.), agricultural areas, forest and semi-natural areas, wetlands, and others (e.g., coastal lagoon, estuaries etc.) for ease of comparison to other studies (e.g., [11,22,26]). We also stratified our stock maps based on Danish geo-regions. We first derived the area of each soil, land use and geo-regions class based on the number of predicted pixels within those classes and then calculated an average and total SOC stock for each class at 0−30 and 0−100 cm soil depth. The 0−30 cm depth represents the plough depth from agriculture areas and estimation of carbon for this depth is of interest to farm management. The top 1 m soil depth mostly represents a rooting depth of many field crops and may act as an important soil depth section for carbon balance and accounting studies.

Results

Summary of SOC and bulk density

SOC content was highly variable and ranged from 0 to 562 g kg^{-1} for the topsoil (0−20 cm) and from 0 to 569 g kg^{-1} in the subsoil (Table 3). Mean SOC decreased with soil depth and SOC at 60−100 cm was about four times lower than the SOC in the 0−5 cm layer. With depth, the coefficient of variation (CV) of the SOC content increased. The CV at 0−5 cm was 184% and that for the 60−100 cm was about 466%. The SOC data was positively skewed at all soil depths with a maximum skewness coefficient at 60−100 cm. The equal-area splines modeled the depth-wise distribution and generated a continuous SOC profile to 1 m depth. The best λ value to fit all soil profiles for both SOC and D_b data was 0.1. Also average D_b was found to be increased with soil depth. Up to 30 cm depth the D_b was on average 1.44 Mg m^{-3}, whereas it increased to 1.52 Mg m^{-3} below 60 cm depth. Bulk density appeared to be less variable with depth (Table 3).

Figure 2 shows a measured and spline predicted SOC and D_b for a coarse sandy soil under agriculture area from the Saalian moraine soilscape in western Denmark (West Jutland). Measured SOC from different horizons decreased with depth except at 53−73 cm where it increased due to podsolization. Although a spline should pass through the mid-point of each measured horizon for this soil profile, in this case the spline slightly extrapolated the SOC value at 35−55 cm due to the selected λ at 0.1.

Prediction rules for SOC and D_b

Depending on the soil depth, 17 to 54 condition-based regression rules were generated while predicting SOC and D_b.

Table 4. Relative usage (%) of the environmental variables to predict soil organic carbon at different soil depths in Denmark.

Environmental variables	Soil depth (cm)									
	0–5		5–15		15–30		30–60		60–100	
	CR+	PF+	CR	PF	CR	PF	CR	PF	CR	PF
Wetlands	72	-	55	-	10	-	-	-	-	-
Multi-resolution valley bottom flatness index	64	67	9	81	-	52	-	33	-	5
Geo-regions	62	-	65	-	17	-	73	-	10	-
Soil map	55	-	38	-	28	-	60	-	5	-
Precipitation	54	74	98	98	4	76	66	62	20	25
Landscape	53	-	26	-	15	-	48	-	5	-
Land use	45	-	60	-	5	-	-	-	-	-
Altitude above channel network	37	85	11	87	82	28	7	4	-	3
Elevation	36	88	31	88	50	15	32	21	-	5
Geology	27	-	23	-	30	-	62	-	100	-
SAGA wetness index	22	94	67	94	-	70	5	56	-	2
Valley depth	9	41	16	47	-	26	-	59	-	15
Slope gradient	2	78	8	93	-	54	-	63	-	10
Mid-slope position	2	50	2	52	-	54	4	15	-	2
Flow accumulation	-	30	8	35	-	26	-	-	-	-
Slope length factor	-	85	-	89	-	31	1	52	-	-
Insolation	-	34	-	16	-	-	-	17	5	-
Slope aspect	-	22	-	2	-	15	2	26	-	-

+CR-Variable usage in setting the rule conditions; PF-Variable usage in the linear prediction function.

Table 5. Predicted soil organic carbon content (g kg^{-1}) at five soil depths for each FAO-UNESCO soil groups.

FAO soil groups	Soil depth (cm) 0–5		5–15		15–30		30–60		60–100	
	Mean	Stdev.	Mean	Stdev.	Mean	Stdev.	Mean	Stdev.	Mean	Stdev.
Alisols	20.8	10.4	19.7	10.9	15.4	19.0	9.8	20.3	2.1	0.7
Arenosols	12.5	10.2	11.9	8.8	11.8	12.4	7.8	15.9	1.9	1.1
Cambisols	17.9	8.2	17.0	6.4	12.2	7.5	7.3	8.8	2.2	0.6
Luvisols	18.0	7.1	16.4	5.7	15.7	8.3	6.8	7.2	2.2	0.7
Podzols	21.9	10.9	21.4	14.1	16.6	25.3	9.1	17.4	2.1	1.2
Fluvisols	24.1	12.0	22.7	9.0	16.7	15.3	12.0	27.5	2.6	0.9
Gleysols	22.8	15.2	22.3	15.0	21.4	26.7	11.9	29.5	2.4	0.8
Podzoluvisols	20.8	6.3	20.8	6.6	14.7	10.2	8.7	8.3	2.0	0.7
Histosols	38.9	22.8	37.8	27.1	52.6	52.5	37.5	71.5	2.5	0.7
Unmapped areas	13.1	6.5	16.5	5.9	12.7	11.5	8.0	13.9	2.0	0.6

This paper only included one of the 54 rules produced during SOC prediction at 0–5 cm soil depth, as an example.

Rule 3: [400 cases, mean 3.1, range 2.2 to 4.5, est err 0.21]

if

Georegions in (1, 2, 3, 6)

Landscape in (5, 6, 7)

MrVBF>6.6

Soil map in (3, 4, 6, 7)

then

log SOC g kg^{-1} = 1.527+0.124*MrVBF+0.0067*Elevation+0.084*SAGA wetness index–0.0011*precipitation+0.06*slope gradient.

This rule used elevation, SAGA wetness index, MrVBF, precipitation and slope gradient to predict SOC in the areas where the MrVBF index is higher than 6.6 and consisted of fine sandy soils from a Moraine landscapes from the geo-regions (e.g. Himmerland). This rule was only valid for 400 locations where the mean SOC was 3.1 [log SOC g kg^{-1}] and the prediction error was about 0.2 [log SOC g kg^{-1}].

Identifying predictors

Variables were identified for SOC and D_b prediction based on their relative usages in the model (Table 4). In all models, precipitation appeared to be the most dominant variable followed by altitude above channel network and SAGA wetness index to predict SOC. As an example, to predict SOC at 5–15 cm, precipitation had a usage of 98% for both rule setting and developing a linear prediction model followed by SAGA wetness index which had a relative usage of 67% and 94%. For this model, insolation and slope aspect had the lowest contribution. Geology became robust with increasing soil depth, whereas land use was important for SOC prediction of the surface layers. Geology, soil map, wetlands, land use, precipitation, MrVBF, SAGA wetness index, elevation, slope gradient, slope-length factor, and altitude above channel network were among the predictors that had a relative importance of >60%. Similarly, land use, soil map, geology, slope gradient, SAGA wetness index, MrVBF, and elevation appeared to be the most important variables for predicting D_b at all soil depths.

Predicted maps

Predicted maps of SOC content (Figure 3) and D_b (map not shown) at five soil depths were produced at a resolution of 30.4×30.4 m. The highest mean SOC content was in the 0–5 cm layer (mean 20 g kg^{-1}; sd 11 g kg^{-1}). Predicted SOC content decreased with soil depth and at 60–100 cm, it was on average 2.2 g kg^{-1}. The soils of western Denmark have relatively higher SOC content than the rest of the country. The northern part has a moderate amount of SOC with two large raised bogs with high SOC contents. Along the coastline, especially in the west, soils with lower SOC concentration were mapped.

The prediction errors were higher towards the west and along the coastline. The prediction error increased with soil depth. For the 0–5 cm layer, the mean error was 1.1 g kg^{-1} and it increased to 1.8 g kg^{-1} at 60–100 cm soil depth.

Predicted SOC content for the FAO–UNESCO soil groups is shown in Table 5. Average SOC content ranged between 11.8 to 52.6 g kg^{-1} at 0–30 cm and between 1.9 to 37.5 g kg^{-1} at 30–100 cm. The highest SOC was observed in Histosols and the lowest in Arenosols at most soil depths.

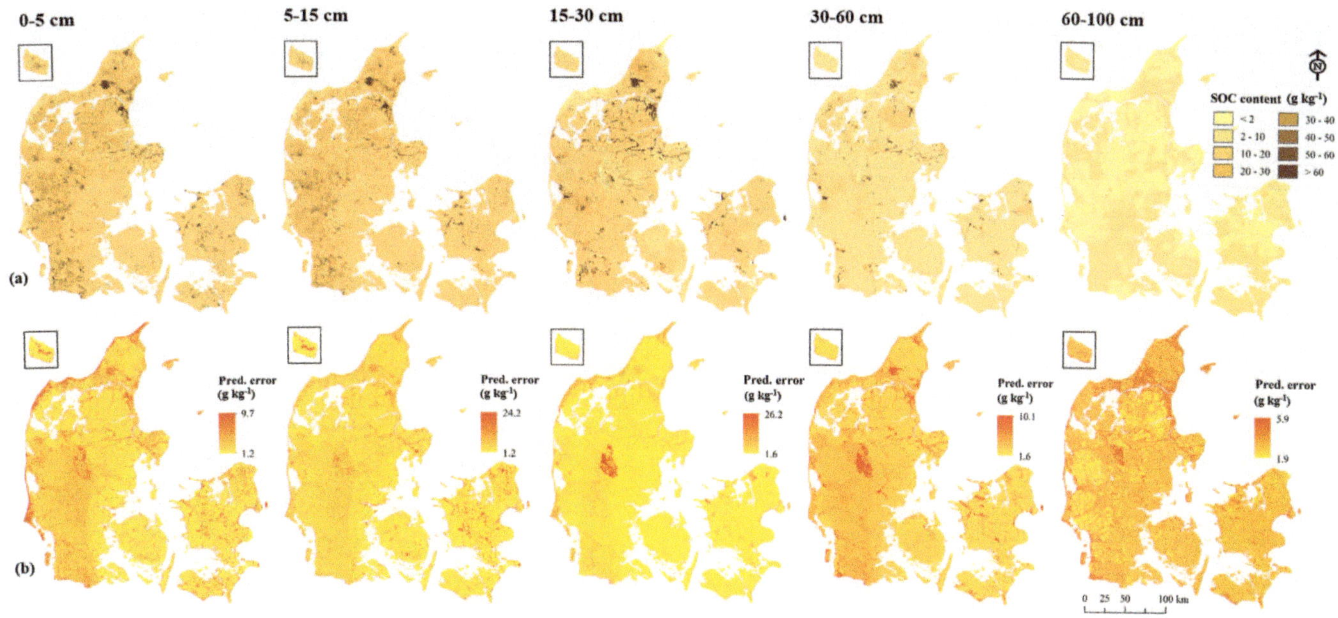

Figure 3. Predicted soil organic carbon content (a), and standard error maps (b) at five soil depths in Denmark.

Model validation

SOC prediction models performance is summarized in Table 6. The best prediction was found at 5−15 cm soil depth for both training and test data sets. The model performance at 60−100 cm was relatively poor compared to the rest of the soil depths. Negative *ME* values suggested that almost all the prediction models were negatively biased suggesting some under prediction of the mapping of SOC distribution.

SOC stocks

SOC stock maps were made for two soil depths (Figure 4). For 0−30 cm, average SOC stocks were about 72 t ha^{-1} and for the top 1 m depth, it was about 120 t SOC ha^{-1}. Most of the western and northern parts of the country have more than 80 t SOC ha^{-1} in the top 30 cm whereas the average stock in the eastern part of the country was less than 80 t SOC ha^{-1}. Total SOC stock was calculated for each geo-region (Figure 5) and Himmerland and West Jutland had an average stock of about 135 t SOC ha^{-1} followed by North Jutland and Thy both having a mean stock of > 120 t SOC ha^{-1}. The soils of West Jutland and East Denmark contain almost 50% of the total SOC stock in Denmark.

Luvisols and Podzols contain about 60% of the total SOC stock (Table 7). Other soil groups that contain significant amounts of SOC were Cambisols (6%), Gleysols (9%), and Arenosols (9%). Although Histosols had SOC stock of 176 t ha^{-1}, its total content was 20 Tg SOC. For all soil groups, more than 58% of the total stock was in the top 30 cm. Unmapped areas in the soil map representing major Danish cities that may also contain a substantial amount of carbon [46].

Of the total stock of 570 Tg SOC about 59% is in the upper 30 cm. Soils under agriculture have an average stock of 121 t ha^{-1} and contain about 444 Tg which is almost 78% of the total estimated stock (Table 8). Another large fraction of SOC stock is found in the soils of the forest and semi-natural areas, and these had a stock of 39 Tg in the top 30 cm and about 67 Tg up to 1 m soil depth. Wetland areas contain large amounts of SOC, and average SOC stock within 1 m soil depth was about 152 t ha^{-1} which is nearly 2% of the total stock. Almost 90% of the total SOC stock within the top 1 m soil depth is found in the soils under agriculture, forest and natural areas.

Discussion

Here we have predicted the distribution of SOC contents and stocks across Denmark including an assessment of the prediction error. We have also estimated the SOC contents and stocks for different land uses and soil types. This discussion focuses on the

Table 6. Model performance to predict soil organic carbon content [log SOC g kg^{-1}] based on Training and Validation datasets.

Soil depth (cm)	Training data			Validation data		
	R^2	*RMSE*	*ME*	R^2	*RMSE*	*ME*
0−5	0.61	0.22	−0.008	0.41	0.24	−0.08
5−15	0.63	0.22	−0.006	0.42	0.24	−0.02
15−30	0.51	0.62	−0.03	0.43	0.66	−0.22
30−60	0.50	0.53	−0.05	0.29	0.56	0.02
60−100	0.28	0.47	−0.06	0.23	0.48	0.12

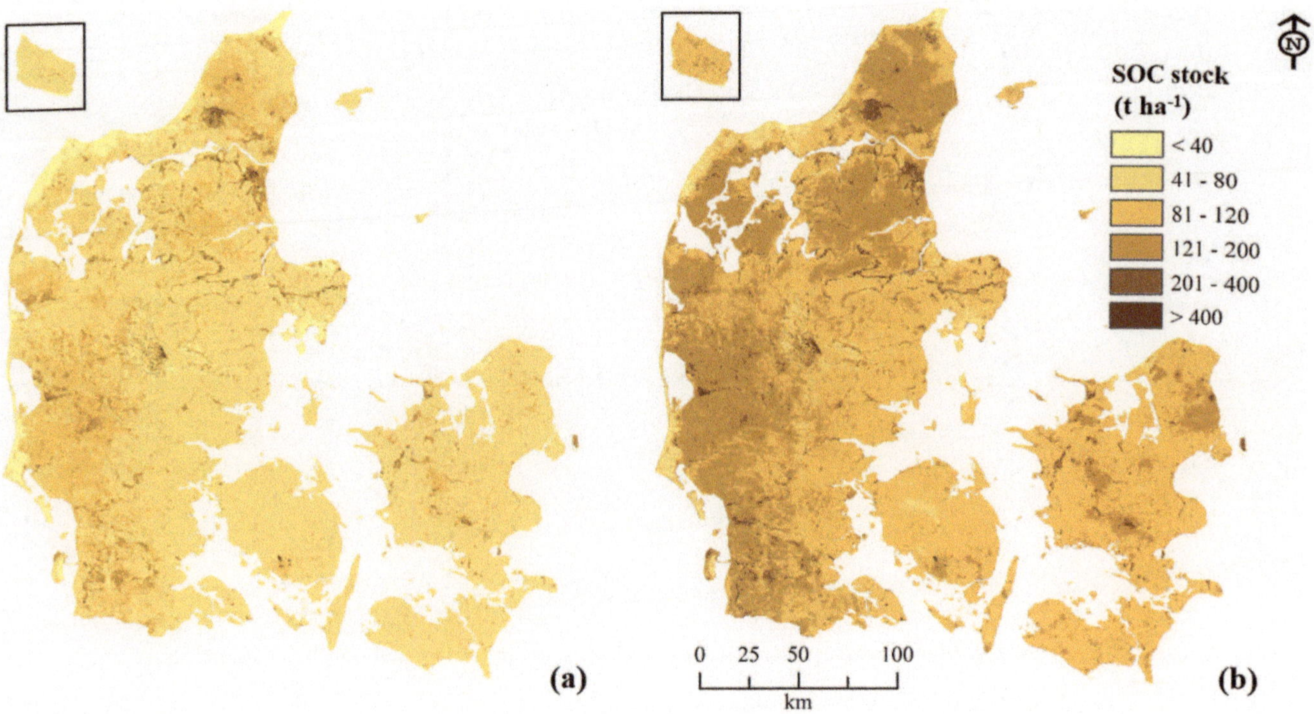

Figure 4. Predicted soil organic carbon stock maps at 0–30 cm (a), and 0–100 cm (b) soil depths for Denmark.

prediction model, the importance of the variables, the uncertainty, and the SOC contents and stocks.

Prediction model

The equal-area spline fit the discrete horizon data but it also harmonized the profile by disaggregating the horizon bulk data and generated a continuous function of SOC distribution. Several other researchers have advocated the usefulness of such splines in depth-wise mapping of SOC in different parts of the world applied from local to regional extents (e.g., [7,8,10]). Pegging of spline by introducing an artificial horizon on the surface benefitted our splines that restricted biased extrapolation of SOC on the surface horizon.

The spatial prediction method (i.e. rule-based regression using the *Cubist* software) was capable of exploring the relationship of SOC to its environment predictors. The prediction rules were conditioned to a given environmental settings such that each rule is valid only to that specific boundary within which SOC distribution presumably less heterogeneous compared to the areas outside where other conditions and rules prevail. For example, SOC content in forests or clayey soils might be different than the SOC from agriculture or sandy soils hence different prediction models operated in these two specific areas. Such a beneficial and advanced function has been used by Lacoste et al. [7] who found regression-rules combined with the spline depth function useful for producing a detailed pseudo-3D map of SOC content in heterogeneous agricultural landscape in France. Minasny and McBratney [42] found that regression rules provided a greater accuracy compared to partial least squares while predicting total carbon content. Several other studies have applied this tool in digital soil mapping (e.g., [35,47,48]).

Variable importance. The environmental variables used to predict SOC content showed a varying level of importance in the model. There was a large influence of precipitation, land use, soil type and some terrain parameters such as elevation, slope gradient, SAGA wetness index, and MrVBF on the spatial distribution of SOC content. The influence of topographic parameters on SOC distribution has been documented in other studies [6,8,9,49–51]. Similarly, land use, precipitation, soil types, wetlands were found important while mapping SOC distribution [11,14,18,52–55]. The influence of terrain parameters on SOC composition and distribution can be linked to its behavior on soil re-distribution through erosion and deposition, in the maintenance of vegetation cover and rooting depths, and in soil drainage that affects SOC decomposition as well as vegetation. In Denmark, the influence of elevation, soil types, geology, and slope gradient was also documented by Bou Kheir et al. [12] when predicting SOC in wet cultivated lands.

The categorical variables such as soil types, geo-regions, and land use were used in defining the rule conditions and continuous variables (terrain parameters) in regression functions. Some terrain parameters like elevation, SAGA wetness index, MrVBF, altitude above channel network etc. were utilized in setting rule conditions. This combined approach in defining the conditions partitioned the study area into several possible strata where SOC distribution was supposed to be more homogeneous and in each stratum several terrain parameters were again used to make sure that the within-stratum SOC variability was well captured by the model. This has made the model robust in predicting SOC content.

Uncertainty. Based on validation indices, the prediction models showed a higher performance (i.e., higher R^2 and a lower *RMSE*) in calibration data (75%) than in validation data (25%). The uncertainty of the SOC prediction increased with depth. The R^2 value ranged between 0.23–0.63 - the higher values for the surface layers. It suggests that our prediction was able to capture up to 63% of total SOC variability. The range of R^2 values was comparable to similar SOC mapping studies where internal validation was applied [24,50,56,57]. Values higher than 0.7 are unusual and values <0.5 are quite common in soil attribute

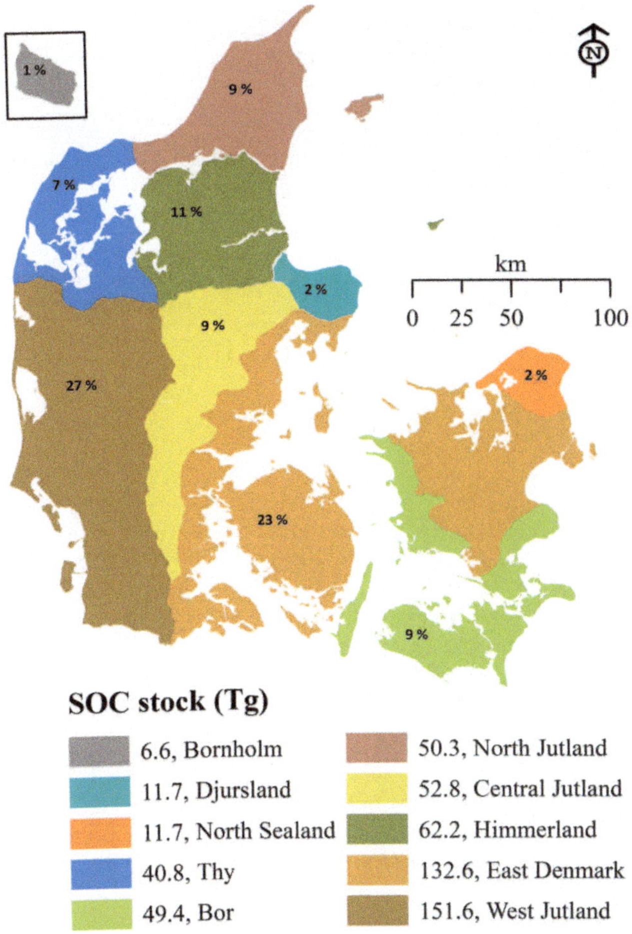

Figure 5. Soil organic carbon stock (1 m depth) for the geo-regions in Denmark. Percentage values represent the fraction of the total soil organic carbon content stock (570 Tg).

predictions [58]. The difference in R^2 value between the two datasets could probably be attributed to the different data density used in prediction. Minasny et al. [5] reviewed several previous SOC mapping studies, and reported that with increasing data density the R^2 of prediction was larger. Similarly, SOC maps from the upper soil depths were associated with a low prediction error compared to the maps with depth. This could be linked to the terrain parameters used as predictors because most of these parameters explain soil surface phenomena and the uncertainty increases with depth [24]. Moreover, higher data density from the surface layers (e.g., 0−5 and 5−5 cm) might have a positive influence on prediction performance.

SOC content and stock. SOC data from two main sources were used in this study: high-density (about 1 observation per km^2) topsoil and subsoil samples, and from soil profiles at each 7-km grid-intersections together with the data from profiles along the pipe lines across Denmark. The grid and point data covered the entire country.

The SOC data were all from dry combustion analysis and from the period 1975−85 so we have modeled and predicted here the SOC contents and stocks for that period. Assuming a steady-state condition at that period, i.e. the carbon levels represent the equilibrium with its physical environment and landuse. Likely, SOC contents have changed since that time, thus the map can be used as a baseline to indicate spatio-temporal changes.

A higher SOC content was found in surface soils, and the lowest average SOC content of 3 g kg^{-1} was found at 60−100 cm soil depth. The western part of the country mostly consists of glacial floodplains and old Saalian moraine landscapes where soils are predominantly sandy. To increase the soil fertility in those areas, farmers has been adding for decades large amounts of manure in the form of pig slurry. This may have led to increased SOC content and gave the highest SOC stock in those areas compared to other geo-regions in Denmark as also reported in Taghizadeh-Toosi et al. [29].

Our results are in line with the previous SOC estimations in Denmark. According to Krogh et al. [26], total stock in Denmark to 1 m soil depth ranged between 563−598 Tg with 579 Tg as the average which was comparable to our prediction of 570 Tg. Similarly, SOC stock from the topsoil (0−28 cm) was about 230 Tg while our prediction showed about 266 Tg. Small differences

Table 7. Soil organic carbon stock in the top 1 m soil depth according to FAO−UNESCO soil groups.

FAO Soil groups	Area (km^2)	Average SOC stock (t ha^{-1})		Total stock		Relative stock
				(Tg)		(%)
		0−30 cm	0−100 cm	0−30 cm	0−100 cm	0−30 cm:0−100 cm
Alisols	921.9	71.3	118.3	7.3 (2%)	12.1 (2%)	60
Arenosols	3,585.9	60.3	105.0	29.5 (9%)	51.3 (9%)	57
Cambisols	2,910.2	64.0	109.9	20.8 (6%)	35.4 (6%)	58
Luvisols	14,499.4	62.3	107.6	100.1 (29%)	172.9 (30%)	58
Podzols	13,745.0	79.6	129.8	115.4 (34%)	189.0 (33%)	61
Fluvisols	879.6	80.2	144.5	7.7 (2%)	13.7 (2%)	56
Gleysols	3,310.0	85.3	140.5	30.3 (9%)	49.7 (9%)	61
Podzoluvisols	698.3	75.7	126.0	5.8 (2%)	9.7 (2%)	60
Histosols	1,039.6	120.8	176.1	14.0 (4%)	21.1 (4%)	69
Unmapped areas	1,320.45	63.1	109.8	9.0 (3%)	15.7 (3%)	57
Sum	42,910.6			340 (100%)	570 (100%)	

Table 8. Predicted soil organic carbon stock in different land use types derived for two different soil depths.

Major Land use types	Area (km^2)	Soil depth (cm)	
		0−30	0−100
		SOC Stock in Tg	
Artificial surface (Urban, Industry, Roads, etc.)	3169.0	22.3 (6%)	38.7 (7%)
Agricultural areas	32942.3	266.4 (78%)	443.9 (78%)
Forest and semi-natural areas	5547.1	38.9 (13%)	66.7 (11%)
Wetlands	860.0	8.3 (2%)	13.5 (3%)
Other (Coastal lagoons, Estuaries, etc.)	391.9	4.1 (1%)	7.2 (1%)
Sum	42910.3	340 (100%)	570 (100%)

in stocks estimates between the two studies were noticed. For example, SOC stock in agricultural areas as predicted by Krogh et al. [26] was 404 Tg, whereas the estimate in the current study is 444 Tg. This could be due to a difference in areas estimated for agricultural lands in two studies. Our prediction used CORINE data that suggested an agricultural area of nearly 32,942 km^2, whereas this area was 28,883 km^2 in the previous study based on AIS (Area Information System). In a separate study, Olesen [28] reported a total stock of 604 Tg from 0−60 cm soil depth calculated for an area of 34,000 km^2 based on AIS. Unlike in our study where bulk density from the peat lands were adjusted, Olesen applied a standard bulk density of 1 Mg m^{-3} for all peats or organic soils and that might have increased the stock leading to over estimation.

Previous estimates of SOC stocks had not quantified the spatial distribution of SOC stock nor validated their prediction. Our approach seemed to be more reliable and the data generated could be useful for future SOC content and stock assessments.

A large amount of SOC was present in the soils under agriculture. Our estimation of 121 t ha^{-1} is in a comparable to the SOC stock of 111 t ha^{-1} estimated for arable lands in Scotland [59] but it was slightly higher than the arable stock in Southeast Germany [60]. Similarly, our predicted average SOC stock of 121 t ha^{-1} from the agricultural areas was lower than the findings of Taghizadeh-Toosi et al. [29]. The later study used a soil-type dependent standard D_b established for 7×7 km grid and perhaps overestimated the SOC stock.

The estimation of average SOC stock in different soil groups within 0−30 cm depth was comparable to Arrouays et al. [11], but the total stock was approximately nine times less than reported for France (which is almost 13 times larger than Denmark). Contrary to our study, the soils under forest soils contained more SOC than the arable soils in France. Similar results were also reported by Chaplot et al. [13] from Laos. A slightly lower average stock in the soils under forest in our study might appear due to the inclusion of non-forested areas (semi-natural areas) in the same class and exclusion of litter layer that possibly lowered its value. Decade long intensive soil management practices such as addition of large amount of manure to the agricultural soils might have increased SOC contents and consequently the SOC stock from those areas [29]. Likewise, our estimated average stock for 1 m soil depth from different soil groups, for example, Cambisols (110 t ha^{-1}), and Gleysols (140 t ha^{-1}) were in a agreement with the global estimate of Batjes [22] using the same soil depth. We also observed >50% stock stored within the top 30 cm soil depth for almost all soil groups which also corresponds to the finding of Batjes [22].

Although the SOC stock in Denmark might have changed substantially over the past few decades [29], our estimation based on the available data has provided a baseline SOC. Together with the estimation of uncertainty, the maps are more reliable and could be useful in environmental research in Denmark. It could support national carbon accounting and carbon balance studies and also act as a back ground information for monitoring temporal SOC changes in Denmark.

Conclusions

This study predicted the spatial distribution of SOC content (g kg^{-1}) at five soil depths intervals (0−5, 5−15, 15−30, 30−60, and 60−100 cm) and quantified its stock (t ha^{-1}) to 1 m soil depth for Denmark. DEM derivatives and soil maps were used as predictors where condition-based regression rules were applied to quantify the spatial relationship between measured SOC and D_b with the predictors. The following can be concluded from this study:

- Equal area spline modeled the continuous depth function of SOC and D_b data from discrete soil horizons in soil profiles from Denmark.
- The most important variables that influenced SOC distribution across Denmark were precipitation, wetlands, land use, soil types, elevation, and saga wetness index.
- Model performance was better for surface soil layers and almost all prediction models suffered from underpredictions.
- The total estimated SOC stock at 0−30 cm soil depth was about 340 Tg and that for 0−100 cm was 570 Tg.
- Almost 60% of the total SOC stock was found in Luvisols and Podzols.
- About 90% of SOC was held in soils under agriculture, forest and semi-natural vegetation. For the soils under agriculture, 60% of the SOC was found in the top 30 cm.
- West Jutland and east Denmark contained almost 50% of the total SOC stock.
- This article is an example for a national level SOC mapping based on GlobalSoilMap procedures and the methods applied can be tested and used in other part of the world.

Acknowledgments

The authors thank all soil surveyors in Denmark for compiling a national Danish soil database used in this study. We are grateful to the GlobalSoilMap consortium for providing the specifications that were used in this paper to predict soil carbon. We also thank the academic editor and two reviewers for their comments on this paper. During the course of writing, we were extremely distressed as our friend and co-author Prof. Rania Bou Kheir suddenly passed away. We dedicate this paper to her.

Author Contributions

Conceived and designed the experiments: KA MHG. Performed the experiments: KA MHG. Analyzed the data: KA AEH BM MHG. Contributed reagents/materials/analysis tools: KA AEH BM RBK MBG MHG. Contributed to the writing of the manuscript: KA.

References

1. McBratney AB, Mendonça-Santos ML, Minasny B (2003) On digital soil mapping. Geoderma 117: 3−52.
2. Hartemink AE, McSweeney K (2014) Soil Carbon. Dordrecht: Springer.
3. Hartemink AE, Lal R, Gerzabek MH, Jama B, McBratney AB, et al. (2014) Soil carbon research and global environmental challenges. PeerJ PrePrints 2: e366v361.
4. Jenny H, Salem AE, Wallis JR (1968) Interplay of soil organic matter and soil fertility with state factors and soil properties. "Study Week on Organic Matter and Soil Fertility", Pontificiae Academiae Scientiarvm Scripta Varia 32: 5−36.
5. Minasny B, McBratney AB, Malone BP, Wheeler I (2013) Digital mapping of soil carbon. Advances in Agronomy 118: 1−47.
6. Dorji T, Odeh IOA, Field DJ, Baillie IC (2014) Digital soil mapping of soil organic carbon stocks under different land use and land cover types in montane ecosystems, Eastern Himalayas. Forest Ecology and Management 318: 91−102.
7. Lacoste M, Minasny B, McBratney AB, Michot D, Viaud V, et al. (2014) High resolution 3D mapping of soil organic carbon in a heterogeneous agricultural landscape. Geoderma 213: 296−311.
8. Malone BP, McBratney AB, Minasny B, Laslett GM (2009) Mapping continuous depth functions of soil carbon storage and available water capacity. Geoderma 154: 138−152.
9. Mishra U, Lal R, Liu D, van Meirvenne M (2010) Predicting the spatial variation of the soil organic carbon pool at a regional scale. Soil Science Society of America Journal 74: 906−914.
10. Odgers NP, Libohova Z, Thompson JA (2012) Equal-area spline functions applied to a legacy soil database to create weighted-means maps of soil organic carbon at a continental scale. Geoderma 189–190: 153−163.
11. Arrouays D, Deslais W, Badeau V (2001) The carbon content of topsoil and its geographical distribution in France. Soil use and Management 17: 7−11.
12. Bou Kheir R, Greve MH, Bøcher PK, Greve MB, Larsen R, et al. (2010) Predictive mapping of soil organic carbon in wet cultivated lands using classification-tree based models: The case study of Denmark. Journal of Environmental Management 91: 1150−1160.
13. Chaplot V, Bouahom B, Valentin C (2010) Soil organic carbon stocks in Laos: spatial variations and controlling factors. Global Change Biology 16: 1380−1393.
14. Doblas-Miranda E, Rovira P, Brotons L, Martínez-Vilalta J, Retana J, et al. (2013) Soil carbon stocks and their variability across the forests, shrublands and grasslands of peninsular Spain. Biogeosciences 10: 8353−8361.
15. Bishop TFA, McBratney AB, Laslett GM (1999) Modelling soil attribute depth functions with equal-area quadratic smoothing splines. Geoderma 91: 27−45.
16. Poncehernandez R, Marriott FHC, Beckett PHT (1986) An improved method for reconstructing a soil-profile from analyses of a small number of samples. Journal of Soil Science 37: 455−467.
17. Hartemink AE, Minasny B (2014) Towards digital soil morphometrics. Geoderma 230−231: 305−317.
18. Kempen B, Brus DJ, Stoorvogel JJ (2011) Three-dimensional mapping of soil organic matter content using soil type-specific depth functions. Geoderma 162: 107−123.
19. IPCC (2003) Good practice guidance for land use, land-use change and forestry. Institute for Global Environmental Strategies.
20. Janssens IA, Freibauer A, Schlamadinger B, Ceulemans R, Ciais P, et al. (2005) The carbon budget of terrestrial ecosystems at country-scale–a European case study. Biogeosciences 2: 15−26.
21. Smith J, Smith P, Wattenbach M, Zaehle S, Hiederer R, et al. (2005) Projected changes in mineral soil carbon of European croplands and grasslands, 1990–2080. Global Change Biology 11: 2141−2152.
22. Batjes NH (1996) Total carbon and nitrogen in the soils of the world. European Journal of Soil Science 47: 151−163.
23. Jobbágy EG, Jackson RB (2000) The vertical distribution of soil organic carbon and its relation to climate and vegetation. Ecological applications 10: 423−436.
24. Minasny B, McBratney AB, Mendonça-Santos ML, Odeh IOA, Guyon B (2006) Prediction and digital mapping of soil carbon storage in the Lower Namoi Valley. Australian Journal of Soil Research 44: 233−244.
25. Mishra U, Lal R, Slater B, Calhoun F, Liu DS, et al. (2009) Predicting soil organic carbon stock using profile depth distribution functions and ordinary kriging. Soil Science Society of America Journal 73: 614−621.
26. Krogh L, Noergaard A, Hermansen M, Greve MH, Balstroem T, et al. (2003) Preliminary estimates of contemporary soil organic carbon stocks in Denmark using multiple datasets and four scaling-up methods. Agriculture, Ecosystems & Environment 96: 19−28.
27. Greve MH, Greve MB, Bøcher PK, Balstrom T, Madsen HB, et al. (2007) Generating a Danish raster-based topsoil property map combining choropleth maps and point information. Geografisk Tidsskrift-Danish Journal of Geography 107: 1−12.
28. Olesen JE (1991) Forelobig beregning af CO_2 emission fra danske landbrugsjord. AJMET Notat no 24 Afdeling for Jordbrugsmeteorologi. Tjele, Denmark: Research Center Foulum, Danish Institute of Agricultural Sciences.
29. Taghizadeh-Toosi A, Olesen JE, Kristensen K, Elsgaard L, Østergaard HS, et al. (2014) Changes in carbon stocks of Danish agricultural mineral soils during 1986−2009. European Journal of Soil Science.
30. Danmarks Meteorologiske Institut (1998) Danmarks Klima 1997. Copenhagen: Danmarks Meteorologiske Institut.
31. Madsen HB, Nørr AH, Holst KA (1992) The Danish soil classification: Atlas over Denmark Copenhagen, Denmark: The Royal Danish Geographical Society.
32. Adhikari K, Minasny B, Greve MB, Greve MH (2014) Constructing a soil class map of Denmark based on the FAO legend using digital techniques. Geoderma 214–215: 101−113.
33. Østergaard HS (1990) Kvadratnettet for nitratundersøgelser i Denmark 1986−89. Skejby, Århus: Landbrugets Rådgivningscenter, Landskontoret for Planteavl.
34. Madsen HB, Jensen NH (1985) The establishment of pedological soil databases in Denmark. Geografisk Tidsskrift-Danish Journal of Geography 85: 1−8.
35. Adhikari K, Kheir RB, Greve MB, Bøcher PK, Malone BP, et al. (2013) High-resolution 3-D mapping of soil texture in Denmark. Soil Science Society of America Journal 77: 860−876.
36. SAGA GIS System for Automated Geoscientific Analyses. Available: http://www.saga-gis.org.
37. ESRI (2012) ArcGIS Desktop: Release 10.1. Redlands, CA: Environmental Systems Research Institute.
38. R Development Core Team (2008) R: A language and environment for statistical computing. Vienna, Austria: R Foundation for Statistical Computing.
39. Arrouays D, Grundy MG, Hartemink AE, Hempel JW, Heuvelink GBM, et al. (2014) GlobalSoilMap: Toward a fine-resolution global grid of soil properties. Advances in Agronomy 125: 93−134.
40. Odeh IOA, McBratney AB, Chittleborough DJ (1995) Further results on prediction of soil properties from terrain attributes - Heterotopic cokriging and Regression-kriging. Geoderma 67: 215−226.
41. Quinlan JR (1993) C4. 5: Programs for machine learning. San Mateo, CA, USA: Morgan Kaufmann.
42. Minasny B, McBratney AB (2008) Regression rules as a tool for predicting soil properties from infrared reflectance spectroscopy. Chemometrics and intelligent laboratory systems 94: 72−79.
43. Minasny B, McBratney AB, Whelan BM (2005) VESPER version 1.62. Australian Centre for Precision Agriculture, McMillan Building A05, The University of Sydney. NSW.
44. Greve MH, Christiansen OF, Greve MB, Bou Kheir R (2014) Change in peat coverage in Danish cultivated soils during the past 35 Years. Soil Science (Accepted).
45. Stjernholm M, Kjeldgaard A (2004) CORINE landcover update in Denmark-Final report. Denmark: National Environment Research Institute (NERI).
46. Vasenev VI, Stoorvogel JJ, Vasenev II, Valentini R (2014) How to map soil organic carbon stocks in highly urbanized regions? Geoderma 226–227: 103−115.
47. Lacoste M, Lemercier B, Walter C (2011) Regional mapping of soil parent material by machine learning based on point data. Geomorphology 133: 90−99.
48. Bui EN, Henderson BL, Viergever K (2006) Knowledge discovery from models of soil properties developed through data mining. Ecological Modelling 191: 431−446.
49. Razakamanarivo RH, Grinand C, Razafindrakoto MA, Bernoux M, Albrecht A (2011) Mapping organic carbon stocks in eucalyptus plantations of the central highlands of Madagascar: A multiple regression approach. Geoderma 162: 335−346.
50. Mueller TG, Pierce FJ (2003) Soil carbon maps: Enhancing spatial estimates with simple terrain attributes at multiple scales. Soil Science Society of America Journal 67: 258−267.
51. Mendonça-Santos ML, Dart RO, Santos HG, Coelho MR, Berbara RLL, et al. (2010) Digital soil mapping of topsoil organic carbon content of Rio de Janeiro State, Brazil. In: J. L Boettinger, D. W Howell, A. C Moore, A. E Hartemink, S Kienast-Brown, editors. Digital Soil Mapping. Springer Netherlands. 255−266.
52. Ross CW, Grunwald S, Myers DB (2013) Spatiotemporal modeling of soil organic carbon stocks across a subtropical region. Science of The Total Environment 461–462: 149−157.
53. Martin MP, Wattenbach M, Smith P, Meersmans J, Jolivet C, et al. (2011) Spatial distribution of soil organic carbon stocks in France. Biogeosciences 8: 1053−1065.
54. Wiesmeier M, Barthold F, Blank B, Kögel-Knabner I (2011) Digital mapping of soil organic matter stocks using Random Forest modeling in a semi-arid steppe ecosystem. Plant Soil 340: 7−24.

55. Meersmans J, van Wesemael B, Goidts E, van Molle M, De Baets S, et al. (2011) Spatial analysis of soil organic carbon evolution in Belgian croplands and grasslands, 1960–2006. Global Change Biology 17: 466−479.
56. Bui EN, Henderson BL, Viergever K (2009) Using knowledge discovery with data mining from the Australian Soil Resource Information System database to inform soil carbon mapping in Australia. Global biogeochemical cycles 23: 1−15.
57. Zhao Y-C, Shi X-Z (2010) Spatial prediction and uncertainty assessment of soil organic carbon in Hebei province, China. Digital Soil Mapping. Springer. 227−239.
58. Beckett P, Webster R (1971) Soil variability: a review. Soils and Fertilizers 34: 1−15.
59. Chapman SJ, Bell JS, Campbell CD, Hudson G, Lilly A, et al. (2013) Comparison of soil carbon stocks in Scottish soils between 1978 and 2009. European Journal of Soil Science 64: 455−465.
60. Wiesmeier M, Spörlein P, Geuβ U, Hangen E, Haug S, et al. (2012) Soil organic carbon stocks in southeast Germany (Bavaria) as affected by land use, soil type and sampling depth. Global Change Biology 18: 2233−2245.
61. Böhner J, Antonić O (2009) Land-surface parameters specific to topo-climatology. In: T Hengl and H. I Reuter, editors. Geomorphometry- Concepts, Software, Applications. New York: Elsevier. 195−226.
62. Bendix J (2004) Geländeklimatologie – Gebrüder Borntraeger. Berlin, Stuttgart.
63. Zevenbergen LW, Thorne CR (1987) Quantitative analysis of land surface topography. Earth Surface Processes and Landforms 12: 47−56.
64. Moore ID, Gessler PE, Nielsen GA, Peterson GA (1993) Soil attribute prediction using terrain analysis. Soil Science Society of America Journal 57: 443−452.
65. Gallant JC, Dowling TI (2003) A multiresolution index of valley bottom flatness for mapping depositional areas. Water Resources Research 39: 1347−1359.
66. Desmet PJJ, Govers G (1996) A GIS procedure for automatically calculating the USLE LS factor on topographically complex landscape units. Journal of Soil and Water Conservation 51: 427−433.

16

Communities of Endophytic Sebacinales Associated with Roots of Herbaceous Plants in Agricultural and Grassland Ecosystems Are Dominated by *Serendipita herbamans* sp. nov.

Kai Riess, Franz Oberwinkler, Robert Bauer, Sigisfredo Garnica*

Plant Evolutionary Ecology, Institute of Evolution and Ecology, University of Tübingen, Tübingen, Germany

Abstract

Endophytic fungi are known to be commonly associated with herbaceous plants, however, there are few studies focusing on their occurrence and distribution in plant roots from ecosystems with different land uses. To explore the phylogenetic diversity and community structure of Sebacinales endophytes from agricultural and grassland habitats under different land uses, we analysed the roots of herbaceous plants using strain isolation, polymerase chain reaction (PCR), transmission electron microscopy (TEM) and co-cultivation experiments. A new sebacinoid strain named *Serendipita herbamans* belonging to Sebacinales group B was isolated from the roots of *Bistorta vivipara*, which is characterized by colourless monilioid cells (chlamydospores) that become yellow with age. This species was very common and widely distributed in association with a broad spectrum of herbaceous plant families in diverse habitats, independent of land use type. Ultrastructurally, the presence of *S. herbamans* was detected in the cortical cells of *Plantago media, Potentilla anserina* and *Triticum aestivum*. In addition, 13 few frequent molecular operational taxonomic units (MOTUs) or species were found across agricultural and grassland habitats, which did not exhibit a distinctive phylogenetic structure. Laboratory-based assays indicate that *S. herbamans* has the ability to colonize fine roots and stimulate plant growth. Although endophytic Sebacinales are widely distributed across agricultural and grassland habitats, TEM and nested PCR analyses reinforce the observation that these microorganisms are present in low quantity in plant roots, with no evidence of host specificity.

Editor: Gabriele Berg, Graz University of Technology (TU Graz), Austria

Funding: This research was funded by the German Foundation Research (OB 24/30-1) and the Open Access Publishing Fund of Tuebingen University. Additional funding was received by a Doctoral Government Scholarship to KR. The funders had no role in study design, data collection and analysis, decision to publish, or preparation of the manuscript.

Competing Interests: All authors have contributed substantially, have read and approved the manuscript. All prevailing local, national and international regulations and conventions, and normal scientific ethical practices, have been respected. The authors have declared that no competing interests exit.

* E-mail: sigisfredo.garnica@uni-tuebingen.de

Introduction

Endophytic fungi are phylogenetically diverse and occur nearly ubiquitously in land plants across a variety of ecosystems. At some stage of their complex life cycle, they inhabit plant tissues without causing any obvious disease symptoms [1]. Plant benefits associated with the infection of endophytic fungi include enhanced performance and fitness, stress tolerance and disease resistance [2]–[6]. In recent years, there has been an increasing level of awareness of the diversity and function of fungal endophytes colonizing aboveground plant organs, whereas investigations focusing on the communities inhabiting roots are extremely limited. These few investigations have demonstrated that the diversity and community structure of belowground fungal endophytes are greatly affected by several factors including seasonal changes [7]–[8], the presence of arbuscular mycorrhizal fungi [7], the application of fungicides [9], and the genotype and defence compounds of their hosts [7], [10]. Some but not all fungal endophytes show distinctive host preferences [7] and in contrast to mycorrhizal fungi, many endophytes can be easily cultured which allows subsequent manipulation leading to better a understanding of their ecological functions and interactions.

The first description of a Sebacinales endophyte (*Piriformospora indica*) was carried out by [11] and the details of the mode of root colonization were accurately documented by [12]. Subsequently, more than 150 plant species, including various plants of economic interest like wheat [13], maize [14], tomato [15] and soya bean [16], and several other studies have demostrated the positive effects of endophytic Sebacinales in co-cultivation experiments [17]. Most of these experiments were performed with *P. indica* isolated from a desert soil in India and strains of the *Serendipita vermifera* complex isolated from the roots of orchids in Australia, whereas one study has been conducted with the European species *P. williamsii* isolated from clover roots [12]. In this study, in order to achieve as complete a picture as possible of the endophytic fungal diversity within the roots of herbaceous plants we used a combination of isolation and polymerase chain reaction (PCR) techniques as suggested by [18].

Although sebacinoid communities are known to be widespread in the field, their composition and function in ecosystems have scarcely been explored. Garnica *et al.* [19] suggested that the diversity, community structure and species composition of endophytic Sebacinales associated with plant communities along an altitudinal gradient can be driven by land use and extremes in soil conditions. Furthermore, the results of this study suggested that Sebacinales are widespread but of low abundance in plant roots, and that some endophytes can colonize multiple host plants. A crucial first step in understanding how these endophyte communities are assembled in plant roots is to characterize their natural composition and occurrence, and subsequently determine whether or not these communities are affected by differences in environmental disturbance or not.

In this paper, our goal was to characterize the patterns of phylogenetic diversity, species distribution and community composition of endophytic Sebacinales from herbaceous plants growing in agricultural and grassland habitats under different land uses. We used a combined approach of strain isolation, microscopy, PCR and co-cultivation experiments to address the following questions: (i) How diverse are endophytic Sebacinales in the roots of herbaceous plants in agricultural and grassland habitats? (ii) To what degree are the diversity and composition of sebacinoid endophyte communities influenced by land use/plant community? (iii) What are the phenotypical effects of a newly isolated sebacinoid strain on selected herbaceous plants? We hypothesized that habitats with similar land use/plant community would show similar endophytic community diversity and composition. In addition, as the newly isolated Sebacinales strain occurs in association with a broad spectrum of herbaceous plants in nature, we used *Arabidopsis thaliana* and *Poa annua* to study the phenotypic response using inoculated and endophyte-free plants.

Material and Methods

Ethics statement

Plant species used in this study are not protected and therefore no specific permits were requested for sampling. For sampling in the Natural Reserve "Filsenberg" near Mössingen (Germany), we received permission from the government of Tübingen, Unit Nature Protection and Landscape Conservation (Germany). Individual owners of the agricultural sites gave us permission to collect plant samples for this study.

Study sites

Herbaceous plants were sampled from various sites surrounding Tübingen, Baden-Württemberg, Germany (Table S1). To more fully determine the phylogenetic diversity and community composition of Sebacinales plant root endophytes, we collected plant material from habitats under different land uses: agricultural habitats (intensive use), adjacent grasslands immediately bordering the agricultural habitats (medium use) and typical grassland habitats (low use). Details concerning land use type and the vegetation composition of the sampled habitats are given in Data S1. Within each habitat, we sampled economically important and co-occurring herbaceous plants (see Table S1). In addition to the individual plants collected in this study, we screened roots from plants where we frequently detected the presence of Sebacinales via PCR [19] as well economically important plants that were not present in the sampled sites. For further diversity and community composition analysis, these were included within the corresponding categories as described above.

Sampling of plant specimens

In total, we sampled 243 plant specimens from agricultural habitats, 296 from adjacent grasslands and 440 from typical grassland habitats. The entire root system of plant samples and a portion of soil were removed carefully with a shovel and then the roots were soaked in tap water to loosen the soil. Each root sample was rinsed extensively in tap water and then at least three times in distilled water. Using a stereomicroscope, foreign and dead roots were removed with sterile forceps. Subsequently, fine roots were randomly selected and cut with the help of a sterile scalpel for strain isolation and culture, microscopic analysis as well as molecular analysis.

Isolation and cultivation of sebacinoid endophytes from roots

Root surface sterilization was performed by rinsing the plant material with 30% hydrogen peroxide for 1 min, followed by 70% ethanol for 5 min and finally sterile double-distilled water (2 min). To increase the chance of isolating Sebacinales strains in culture, several media with and without antibiotics/fungicides were used. Root fragments of approximately 0.5 cm length were placed on Petri dishes containing 1.5% water agar, 1.5% Czapek Dox agar, modified Melin Norkrans agar [20], 1.5% malt extract agar or 1.5% potato dextrose agar supplemented with 0.002% streptomycin and 0.005% tetracycline dissolved in 99% ethanol added after autoclaving or with the fungicide Thiabendazole (4 mg/L; Sigma-Aldrich, Steinheim, Germany) [21] or benomyl (2 mg/L; Riedel-de Haen, Seelze, Germany) and Dichloran (2 mg/L; Riedel-de Haen) diluted in 95% ethanol [22]. Root fragments were carefully submerged in the agar with the help of sterile forceps. All the Petri dishes were incubated at 18°C in the dark to promote the growth of endophytes and were periodically checked for any fungus growth. From the non-sporulating fungus colonies that were microscopically characterized by thin, clampless, colourless hyphae, portions of agar and mycelia were transferred to 0.7% malt yeast pepton (MYP) agar. Subsequently, fungal cultures were identified via PCR with specific fungal primers (see below).

Transmission electron microscopy analyses

To detect morphologically the presence of sebacinoid endophytes in the roots of host plants, we used transmission electron microscopy (TEM). For this evaluation, we screened a total of 150 root samples yielding sebacinoid sequences by TEM examinations following the methodology described in [23].

DNA extraction, PCR, cloning, sequencing and sequence editing

Total genomic DNA was extracted from fine roots and fungal cultures using the InnuPREP Plant DNA Kit (Analytik Jena, Jena, Germany) or the DNAeasy Plant Mini Kit (Qiagen, Hilden, Germany) following the manufacturer's instructions. Two to three fine roots approximately 1 cm long were selected from each root system, or circa 0.5 cm^2 of culture medium containing mycelia was placed in Eppendorf tubes. Samples were previously deep-frozen in liquid nitrogen and then ground several times with a sterile plastic pestle.

For the detection of the sebacinoid fungi in root samples, the internal transcribed spacer (ITS1 and ITS2), including 5.8S and the D1/D2 regions of the nuc-rDNA, were amplified using the primer combinations NSSeb1/NLSeb2R [24] with Phusion High-Fidelity DNA Polymerase (Finnzymes Oy, Vantaa, Finland), followed by a nested PCR with ITS1F [25] /NL4 [26] and a Mango*Taq* DNA Polymerase (Bioline, Luckenwalde, Germany) and PCR profiles as described in [19]. For the identification of the fungal cultures, the ITS and the D1/D2 regions were amplified using the primer pair ITS1F and NL4 and Mango*Taq* Polymerase. All PCR products were checked using agarose gel electrophoresis with ethidium bromide staining. PCR products showing multiple bands and those that could not be directly sequenced were cloned using the Topo TA Cloning Kit for Sequencing (Invitrogen, Life Technologies GmbH, Darmstadt, Germany). Fungal colonies were used directly as a template for a PCR with Mango*Taq* Polymerase and the primer pair M13F/M13R (Invitrogen).

The amplified DNA fragments were cleaned using a 1: 20 diluted ExoSAP-IT reagent (USB Corporation, Cleveland, OH, USA). PCR products were cycle sequenced in both directions with the 1: 6 diluted BigDye Terminator v3.1 Cycle Sequencing Kit (Applied Biosystems, Foster City, CA, USA) on an ABI Prism 3130*xl* Genetic Analyzer (Applied Biosystems). Sequence chromatograms were assembled and manually edited using Sequencher 4.10.1 (Gene Codes Corporation, Ann Arbor, MI, USA). All newly generated Sebacinales DNA sequences have been deposited in GenBank and the plants yielding sebacinoid sequences were deposited at the Herbarium Tubingense (TUB) (see Table S1 and Table S3).

Sequence identity, alignments and phylogenetic analysis

For tentative identification, the ITS sequences of root associated fungi were submitted to BLAST searches against the sebacinoid sequences deposited in the NCBI GenBank database (www.ncbi.nlm.nih.gov). This dataset included a total of 198 ITS+5.8S+D1/D2 LSU sequences, one sequence of the newly isolated Sebacinales strain (*Serendipita herbamans*), 156 newly generated sequences in this study and 41 sequences in total from the studies [19] and [24]. All sequences lacking or having partial ITS1 and/or ITS2 regions and sequences with >3% genetic divergence from our new isolated Sebacinales strain were removed from the final dataset. In addition, to determine the host plant range of *S. herbamans*, we compiled a second dataset with environmental Sebacinales for which large subunit (LSU) sequences are available in GenBank. This dataset included sequences spanning the D1/D2 regions of rDNA with a similarity of 1% [27].

To infer the phylogenetic placement of the newly isolated strain within the order Sebacinales, we assembled two datasets: (i) 35 Sebacinales nuc-rDNA sequences spanning the ITS, 5.8S and D1/D2 regions, and (ii) 48 Sebacinales nuc-rDNA sequences comprising D1/D2 regions from available strains and basidiomata. Because *Serendipita* is not monophyletic and the species delimitations within this genus are still unclear, we have analysed 16 different cultures of the *S. vermifera* complex and also one sequence was selected from each divergent phylogenetic lineage detected in the *Sebacina epigaea* and *S. incrustans* complexes [28].

Sequence datasets were aligned with MAFFT 6.884b using the E-INS-i option [29] and POA 2 under the *-progressive* algorithm [30]. The tool trimAl 1.4 using the algorithm *-compareset* was used to detect the most consistent alignment [31]. Phylogenetic analyses were computed from POA alignments using a maximum likelihood analysis with combined rapid bootstrapping under the GTRCAT model from 1000 runs with RAxML 7.0.4 [32]–[33]. The phylogenetic trees with the best scores were rooted and illustrated using FigTree 1.3 [34].

An uncorrected pairwise distance (*p*-distance) matrix was computed in PAUP* 4.0b10 [35]. Complete ITS sequences with at least 97% similarity were defined as a molecular operational taxonomic unit (MOTU)/species using OPTSIL 1.2 [36].

Haplotype structure and community composition analyses

To estimate the haplotype phases from the ITS sequences of *Serendipita herbamans*, we used PHASE 2.1 [37]–[38] as implemented in DnaSP 5.10.1 [39] under a Markov Chain Monte Carlo algorithm using 1000 iterations with a hybrid model. The haplotype network was constructed using the median joining method [40] as implemented in Network 4.6.1.2 (www.fluxus-engineering.com) with a 'MP' option to identify and eliminate unnecessary median vectors and links [41]. Habitat provenance was coded for each haplotype.

Furthermore, the community structure of Sebacinales endophytes was analysed using Phylocom 4.2 [42]–[43]. To calculate the net relatedness index (NRI) and nearest taxon index (NTI), and to test whether or not endophytic communities have a distinctive phylogenetic structure related to land use, we pooled the samples within the core land use categories as described above. In addition, we compared agricultural habitats + adjacent grasslands vs. typical grasslands, and agricultural habitats vs. adjacent grasslands + typical grasslands. Several null models as implemented in Phylocom were tested. To assign patterns of community structure, we considered NRI and NTI values as significant if values >2 were clustered and <–2 indicated overdispersion, and if both P-values were <0.05 [44].

Co-cultivation assays and effects on plant phenotypes

In our study, we used *Arabidopsis thaliana* and *Poa annua* because of their small size and because both are easily grown on agar contained in Petri dishes, and because the presence of root-associated Sebacinales endophytes have been detected for both species in the field [24]. To obtain fungus-free *A. thaliana* (wild type Columbia Col-0) and *P. annua* (wild type; ProSementis GmbH, Kusterdingen, Germany) seedlings, seeds were washed with double-distilled sterile water, then placed for 1 min in 70% ethanol and 5 min in 1% NaOCl, and finally soaked in double-

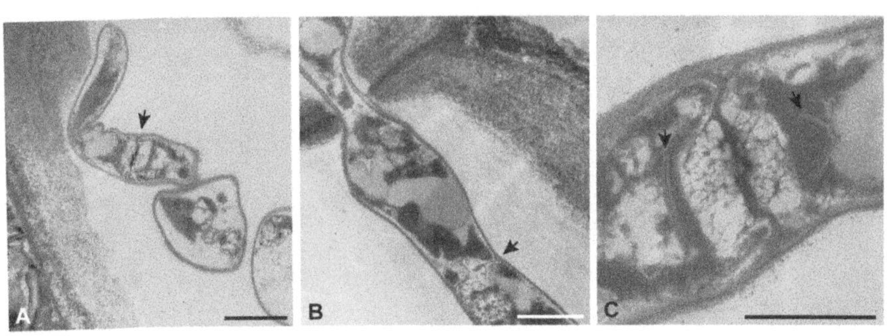

Figure 1. Transmission electron micrographs showing the presence of sebacinoid endophytes in the root cells of *Plantago media*. (A, B) Presence of hyphae with a septum (arrows) in cortical cells. **(C)** Septal porus with imperforate parenthesomes (arrows). Scale bar A, B = 1 μm; C = 0.5 μm.

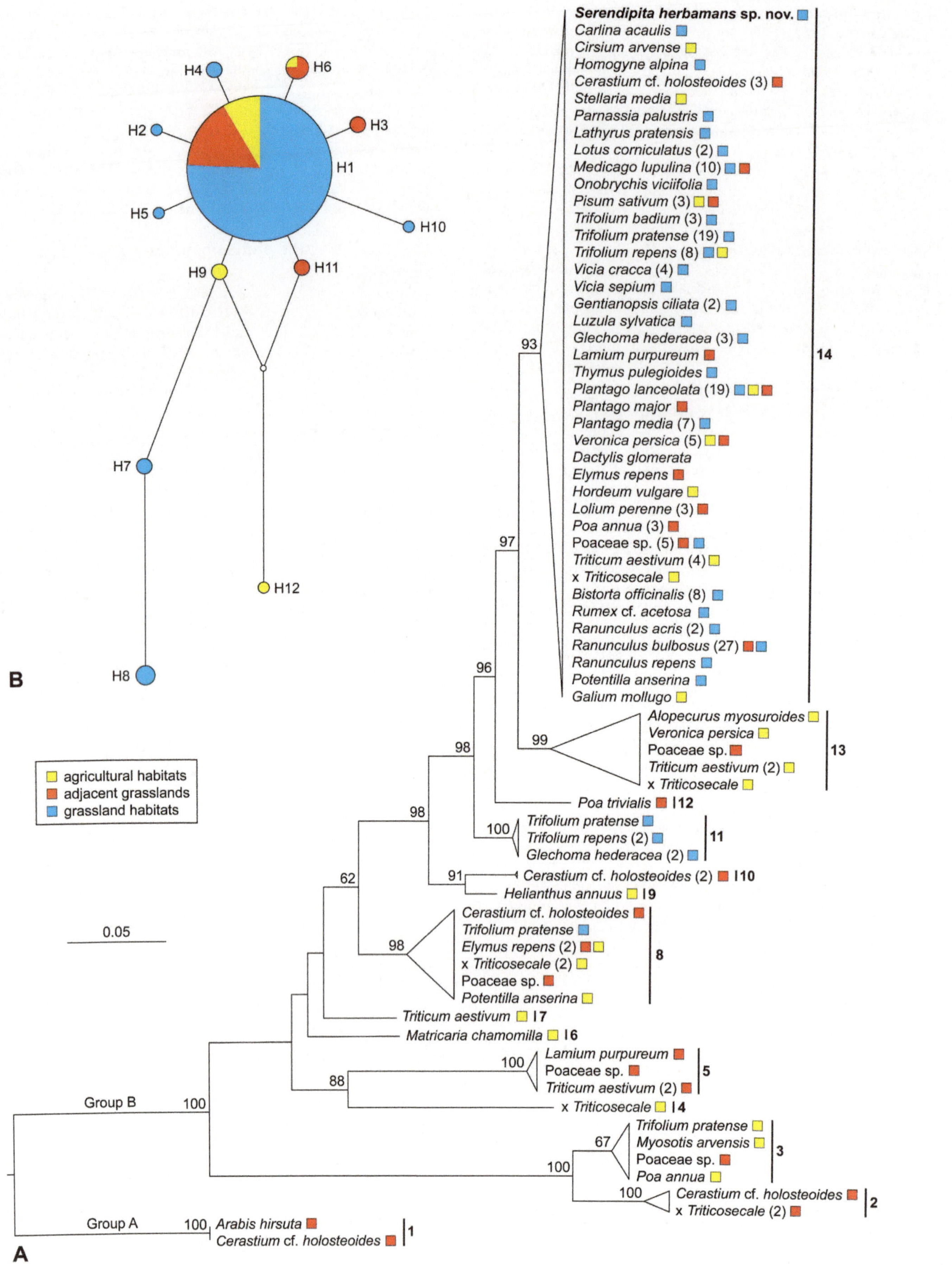
Serendipita herbamans sp. nov.
Carlina acaulis
Cirsium arvense
Homogyne alpina
Cerastium cf. holosteoides (3)
Stellaria media
Parnassia palustris
Lathyrus pratensis
Lotus corniculatus (2)
Medicago lupulina (10)
Onobrychis viciifolia
Pisum sativum (3)
Trifolium badium (3)
Trifolium pratense (19)
Trifolium repens (8)
Vicia cracca (4)
Vicia sepium
Gentianopsis ciliata (2)
Luzula sylvatica
Glechoma hederacea (3)
Lamium purpureum
Thymus pulegioides
Plantago lanceolata (19)
Plantago major
Plantago media (7)
Veronica persica (5)
Dactylis glomerata
Elymus repens
Hordeum vulgare
Lolium perenne (3)
Poa annua (3)
Poaceae sp. (5)
Triticum aestivum (4)
x Triticosecale
Bistorta officinalis (8)
Rumex cf. acetosa
Ranunculus acris (2)
Ranunculus bulbosus (27)
Ranunculus repens
Potentilla anserina
Galium mollugo
14
93
97
96
98
98
62
Alopecurus myosuroides
Veronica persica
Poaceae sp.
Triticum aestivum (2)
x Triticosecale
99
13
Poa trivialis 12
100
Trifolium pratense
Trifolium repens (2)
Glechoma hederacea (2)
11
91
Cerastium cf. holosteoides (2) 10
Helianthus annuus 9
Cerastium cf. holosteoides
Trifolium pratense
Elymus repens (2)
x Triticosecale (2)
Poaceae sp.
Potentilla anserina
98
8
Triticum aestivum 7
Matricaria chamomilla 6
100
Lamium purpureum
Poaceae sp.
Triticum aestivum (2)
5
88
x Triticosecale 4
Group B
100
67
Trifolium pratense
Myosotis arvensis
Poaceae sp.
Poa annua
3
100
100
Cerastium cf. holosteoides
x Triticosecale (2)
2
Group A
100
Arabis hirsuta
Cerastium cf. holosteoides
1
A
0.05
agricultural habitats
adjacent grasslands
grassland habitats
H1
H2
H3
H4
H5
H6
H7
H8
H9
H10
H11
H12
B

Figure 2. Phylogenetic diversity and relationships of endophytic Sebacinales associated with herbaceous plants in agricultural and grassland habitats. (**A**) Tree topology assessed using a maximum likelihood approach under a GTRCAT model of DNA evolution from 198 sequences spanning the ITS1+5.8S+D1/D2 LSU regions (alignment length = 1300 bp). Numbers above the branches are bootstrap values obtained from 1000 replicates (values ≥60 are shown). Numbers in parentheses indicate the number of sequences found for the same host plant. Bold numbers (1–14) represent MOTUs/species using a 3% cut-off. Coloured boxes indicate land use types (if these were available). Note: MOTU 14 (*Serendipita herbamans*) corresponds to clade 4 in [24] and MOTU 54 in [19]. (**B**) Median-joining network for MOTU 14 (*S. herbamans*) inferred from ITS sequences excluding indels [53–54] and recombining blocks [55–56]. Circle sizes are proportional to haplotype frequency and connecting lines are proportional to mutation events between haplotypes.

distilled sterile water for 10 min. The seeds were subsequently placed in ½ Murashige and Skoog (MS) medium [45]. Seed dormancy was broken using a MS medium with the addition of 1% agar, 1% saccharose and 0.01% casamino acids (pH 5.8), followed by incubation period of 48 h at 4°C in the dark. The germination of the seeds was performed at a temperature of 16°C with 8-hour photoperiod.

Prior to the interaction assays, the strains *Piriformospora indica* DSM 11827 and *Serendipita herbamans* DSM 27534 were cultivated on MYP agar for 4 weeks at 18°C in the dark. For each treatment, a total of 40 seedlings of *A. thaliana* and *P. annua* were placed with a sterile tweezers on a MS medium, and an agar block (0.5 cm^2) containing mycelia of *P. indica* or *S. herbamans* was added near the principal root. Seedlings lacking fungus were used as controls. Interaction experiments were carried out at 16°C with 8-hour photoperiod. The presence of fungal hyphae was analysed using cotton blue in lactic acid under the light microscope. Measurements of the root surface were performed after 24 days for *A. thaliana* and 34 days for *P. annua* using ImageJ 1.45s [46]. Petri dishes containing inoculated and fungus-free seedlings were scanned and aboveground plant parts were cut out from the image. By setting a scale and defining agar, mycelium and Petri dishes as background noise, the root surface for each plant was measured automatically from a planar picture. The fresh and dry weight (dried for 2 days at 60°C) of the roots and shoots were measured and the data were pictured in Origin 9 (ADDITIVE GmbH, Friedrichsdorf, Germany).

Nomenclature

The electronic version of this article in Portable Document Format (PDF) in a work with an ISSN or ISBN will represent a published work according to the International Code of Nomenclature for algae, fungi, and plants, and hence the new names contained in the electronic publication of a PLoS ONE article are effectively published under that Code from the electronic edition alone, so there is no longer any need to provide printed copies.

In addition, the new species name contained in this work was submitted to MycoBank, from where they will be made available to the Global Names Index. The MycoBank number can be resolved and the associated information viewed through any standard web browser by appending the MycoBank number contained in this publication to the prefix www.mycobank.org/MB. The online version of this work is archived and available from the following digital repositories: PubMed Central, LOCKSS.

Results and Discussion

Strain isolation and morphological detection of roots associated Sebacinales

One endophytic Sebacinales strain (MOTU 14, *Serendipita herbamans*) was successfully isolated in water agar medium from non-ectomycorrhizal roots of *Bistorta vivipara* (Polygonaceae). From the 150 root samples yielded sequences of sebacinoid endophytes analysed, was only detected morphologically the presence of Sebacinales in *Plantago media* TUB 019555 (Plantaginaceae, Fig. 1). Previously, the presence of the newly isolated strain was also detected using light and transmission electron microscopy in field samples of *Potentilla anserina* (Rosaceae; [19]) and *Triticum aestivum* (Poaceae; [24]). The hyphae had dolipore septa with imperforate parenthesomes and were present in the outer cortical cells of the fine roots (Fig. 1). Apparently, the cortical cells colonized by the sebacinoid endophytes were not completely vital. In both *P. media* and *P. anserina*, only single hyphae were found; therefore, we cannot make a comparison with the colonization patterns observed in *T. aestivum*. Relevant aspects of the morphological structures, occurrence, phylogenetic relationships and *in vitro* effects on selected herbaceous plants are given below.

Occurrence, phylogenetic diversity and relationships of Sebacinales root endophytes

Using a nested PCR approach, we were able to generate 156 new rDNA sequences comprising the ITS+5.8S+D1/D2 regions of Sebacinales root endophytes from agricultural and grassland sites in 36 herbaceous species belonging to 11 plant families (Fig. 2A, Table S1). In agreement with [19], [24] and [47], we detected endophytes from plants belonging to Asteraceae, Boraginaceae, Brassicaceae, Caryophyllaceae, Fabaceae, Lamiaceae, Plantaginaceae, Poaceae, Ranunculaceae, Rosaceae and Rubiaceae. Overall, we detected Sebacinales sequences in only 15.9% of the plant roots analysed. The widely sampled families Fabaceae, Plantaginaceae and Ranunculaceae had an above average presence of Sebacinales, whereas Poaceae had a low

Table 1. Phylogenetic diversity and structure of endophytic Sebacinales communities associated with herbaceous plants in agricultural and grassland habitats.

Plot	N	S	E	H	D	PD	NRI	NTI	Pattern
Grassland habitats	89	3	0.25	0.28	0.13	0.60	7.95*	−0.66	No trend
Agricultural habitats	29	8	0.79	1.64	0.76	0.28	−2.55	2.31*	No trend
Adjacent grasslands	38	9	0.69	1.51	0.65	0.39	−3.00	0.90	Random

N = Number of sequences, S = MOTUs/species richness, E = Evenness, H = Shannon's diversity index, D = Simpson's diversity index, PD = Faith's index of phylogenetic diversity, NRI = Net Relatedness Index (* = $P<0.05$), NTI = Nearest Taxon Index (*, $P<0.05$). Diversity indexes were calculated as described in [19].

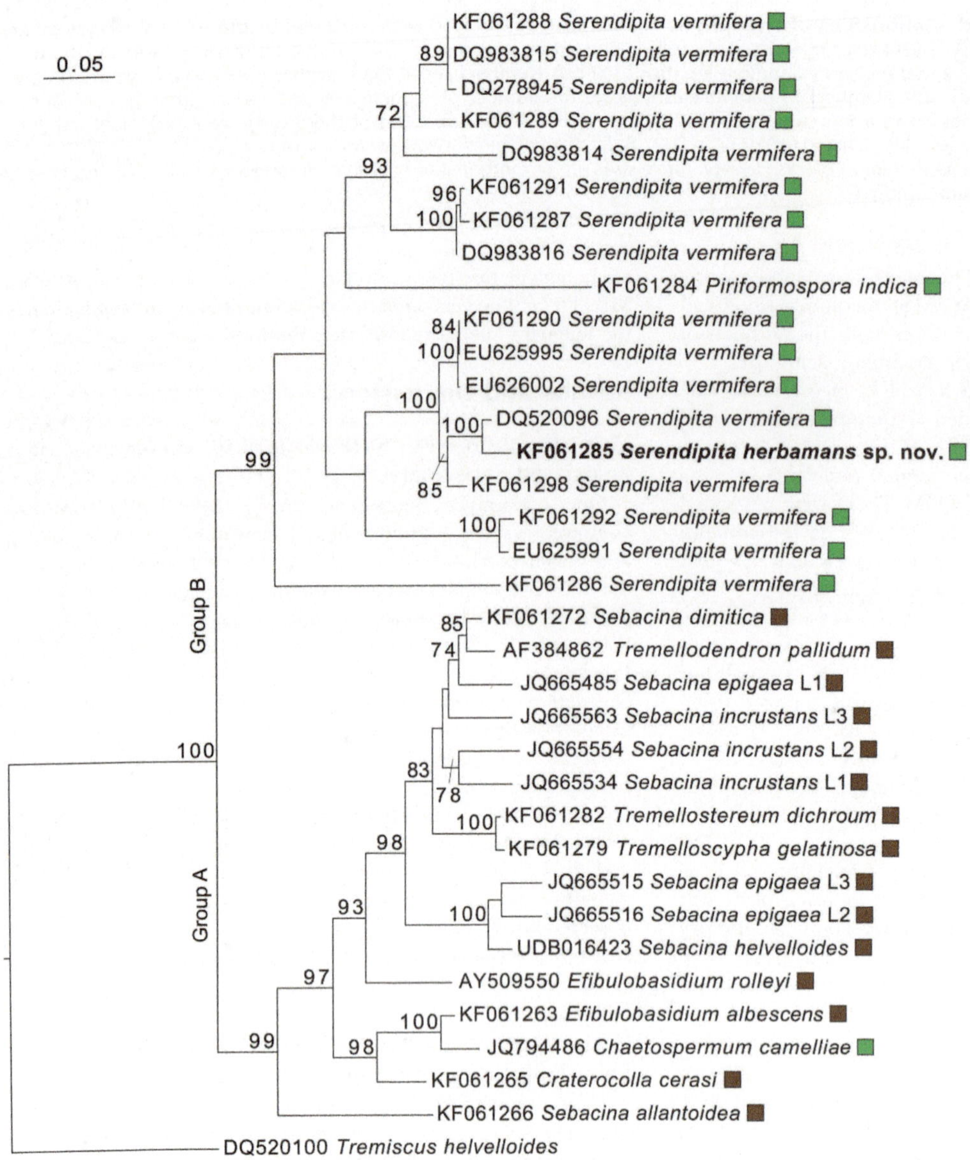

Figure 3. Phylogenetic placement of *Serendipita herbamans* within Sebacinales. Maximum likelihood analysis inferred from 35 ITS+5.8S+D1/D2 rDNA sequences of basidiomata (brown boxes) and culture (green boxes) of the members of the order Sebacinales (alignment length = 1395 bp). *Tremiscus helvelloides* (Auriculariales) was used as an outgroup. Bootstrap supports (≥60%) are shown for each node. For *Serendipita vermifera* MAFF and Warcup's strain numbers [50], see Table S3. L1, L2 and L3 indicate major lineages within *Sebacina epigaea* and *S. incrustans* complexes according to [28].

average presence of the fungi (see Table S2). This is much lower than the 28.9% presence as detected in [47] or 46.5% as reported by [19]. However, these detection rates are difficult to compare because in both studies surveyed, there were more heterogeneous habitats in comparison to agricultural ones. Furthermore, Selosse *et al.* [47] sampled a geographically larger area (Europe and the Caribbean) and Garnica *et al.* [19] also analysed sebacinoids forming mycorrhizae. These results suggest that roots of herbaceous plants in grassland and agricultural ecosystems appear to be sparsely colonized by Sebacinales, which could be partly explained by the influence of intensive land use. Only MOTU 1 clustered in Sebacinales group A, whereas the remaining MOTUs (2–14) are distributed within Sebacinales group B [48]. This is in concordance with previous studies [19], [24] and [47], where most endophytic Sebacinales were included in group B and only a few sequences clustered in group A.

As reported [19], various MOTUs in this study represented singletons: MOTU 4 (x *Triticosecale*), MOTU 6 (*Matricaria chamomilla*), MOTU 7 (*Triticum aestivum*), MOTU 9 (*Helianthus annus*) and MOTU 12 (*Poa trivialis*). The remaining MOTUs were detected in the roots of many different plant species and therefore showed no evidence of host specificity (Fig. 2A). MOTU 14 is described here as a new species, *Serendipita herbamans* (see below), which was the most frequent (76%) endophyte, detected in the roots of 29 herbaceous plant species from eight angiosperm families (Table. S1). We also detected this MOTU in the roots of *Avena sativa* and *Bromus hordeaceus*, but the sequences included only the ITS2 region and therefore, they were removed from the final alignment. Furthermore, based on 1% sequence similarity of the

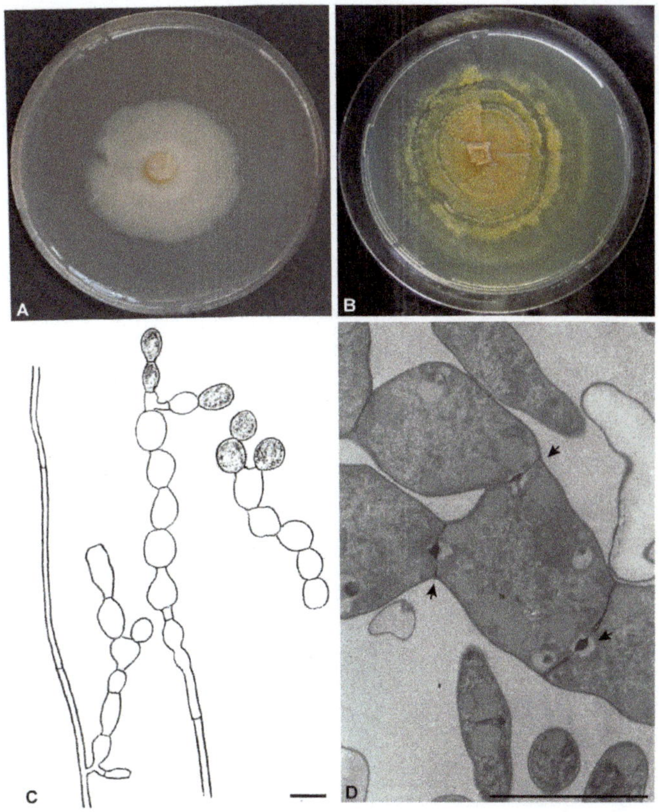

Figure 4. Morphology of *Serendipita herbamans*. (**A**) Colony after two weeks of inoculation on MYP agar. (**B**) Colony after 6 months of inoculation on MYP agar. (**C**) Cylindrical and monilioid hyphae without clamps. Older monilioid hyphae become slightly yellow coloured. (**D**) Transmission electron micrographs showing monilioid hyphae with a septal porus at the constrictions (arrows). Scale bar C, D = 5 µm.

D1/D2 regions of the rDNA, 24 species from 10 plant families represent putative hosts for *S. herbamans* (data not shown). This new species has been previously detected via PCR by [19] (MOTU 54) and [24] (clade 4).

In most cases, molecular detection of sebacinoid endophytes requires nested PCRs since morphologically these fungi have been found only in the root cells of three different plant species. This reinforces the observation that this fungal group seems to be relatively widely distributed but of low quantity in plant roots in the field [24], [47], [49].

Distribution, haplotype and community structure of Sebacinales root endophytes

A total of four MOTUs (4, 6, 7 and 9) were restricted to agricultural habitats and five MOTUs (1, 2, 5, 10 and 12) were found exclusively in adjacent grasslands, whereas one MOTU (11) was detected only in grassland habitats. Considering the low frequency of these MOTUs, we cannot be sure whether or not there is a correlation with land use and/or plant community because the intrinsic nature of these components make it difficult to analyse them separately. On the contrary, MOTUs 8 and 14 (*Serendipita herbamans*) were found to be independent of land use (see Fig. 2A and Table S1). The network analysis for *S. herbamans* resulted in one network with 12 haplotypes: H1 was the most frequent haplotype, and four haplotypes (H2, H5, H10, H12) represented single samples (Fig. 2B). The haplotype H1 was scattered throughout different habitats, whereas the remaining haplotypes were restricted either to adjacent grasslands (H3, H11) or typical grassland habitats (H2, H4, H5, H7, H8, H10). The haplotypes H9 and H12 were restricted to agricultural habitats.

Sebacinoid endophyte communities from agricultural and grassland habitats showed no trend of phylogenetic clustering, whereas adjacent grasslands had a random structure (Tab. 1). Similar endophytic community trends were observed if agricultural habitats and adjacent grasslands or all grassland habitats were pooled together (data not shown). At first glance, this is in contrast to [19], who found significant clustering in root-associated Sebacinales from meadows under different land use intensities. However, if we considered that MOTU 14 (*Serendipita herbamans*) was dominant across all investigated sites (comprising almost 80% of our sequences), it is not surprising that we have not revealed a distinctive phylogenetic structure for the analysed endophytic communities. A similar community structure was observed if the dominant MOTU 14 was excluded from the Phylocom analysis (data not shown). In general, MOTUs/species were more diverse in typical agricultural habitats and adjacent grasslands than in grassland habitats (Table 1). Because our study resulted in a low detection rate (15.9%) of endophytic Sebacinales associated with plant roots, we were not able to compare diversity and community assemblage at a finer scale (e.g. organic vs. no-tillage vs. conventional farming, see Data S1).

Phylogenetic placement of *Serendipita herbamans* within Sebacinales

The newly isolated Sebacinales strain (*Serendipita herbamans*) clusters in Sebacinales group B and forms a well-supported clade with some members of the genus *Serendipita* (Fig. 3). The phylogeny based on 34 LSU sequences confirmed the ITS position of *S. herbamans* within the Sebacinales (data not shown). The closest related species to *S. herbamans* is *S. vermifera* strain MAFF 305828, which was isolated from the Australian orchid *Eriochilus cucullatus* [50]. *S. herbamans* differs by 29 bp (4.0%) from *S. vermifera* MAFF 305828 in the ITS region. Morphologically, *S. herbamans* has a closer similarity with the anamorphic *Piriformospora indica*, but they differ in the size of chlamydospores [11] (see Fig. 4). Phylogenetically, *S. herbamans* is distantly separated from *P. indica* with a sequence divergence of 27.7% in the ITS region.

Interaction assays between Sebacinales strains and herbaceous plants

Co-cultivation of *Serendipita herbamans* with *Arabidopsis thaliana* and *Poa annua* showed that hyphae were abundant on the root surface. After 14 days, the fungus penetrated single root hair cells but only a small portion of the cortical cells in the elongation zone was fully colonized. Hyphae were hardly branched and coiled. The fungus was only detected in the outer cortical cells, but not in the endodermis or the central cylinder. This observation agrees with *in vitro* interaction experiments between *Piriformospora indica* [12], [14], [51] and *S. vermifera* [52] and various host plants. The field colonization pattern of *S. herbamans* in *Triticum aestivum* based on a single snapshot [24] showed a high level of agreement with that in *A. thaliana* and *P. annua* under greenhouse conditions.

Our experimental interaction assays with *S. herbamans* and *P. indica* demonstrated that these have positive effects on the growth of the roots of *A. thaliana* (see Fig. 5). Seedlings inoculated with the fungi showed stronger root branching compared to the fungal-free plants. The root fresh and dry weight and the root area/surface of *A. thaliana* were significantly higher when grown with Sebacinales than in seedlings lacking of fungi (Fig. 5). On the other hand, no

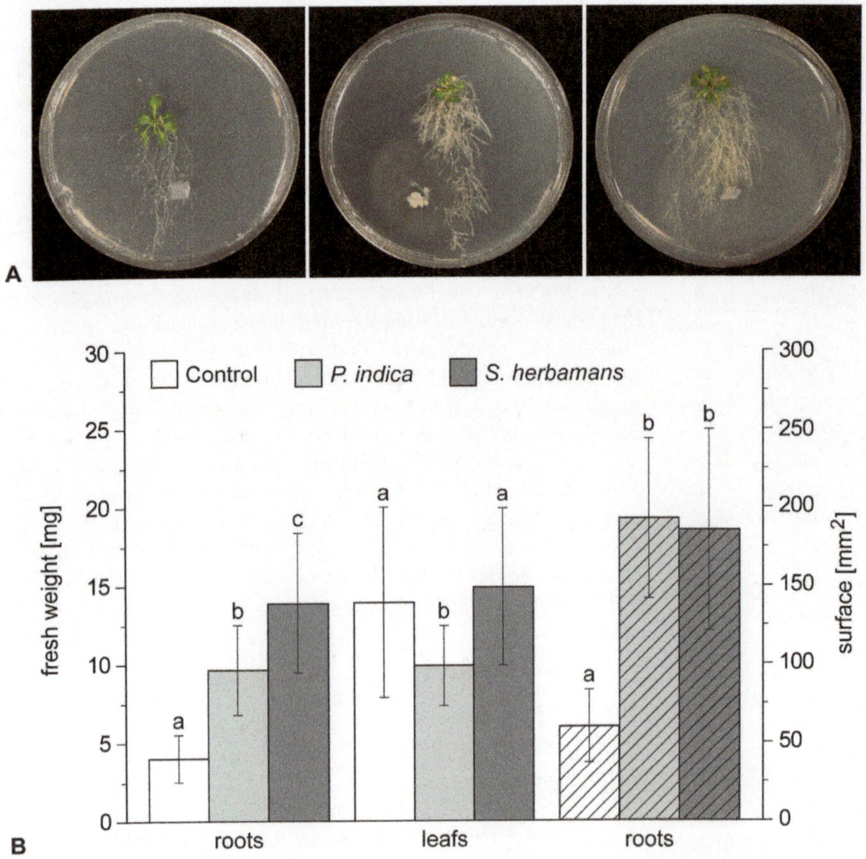

Figure 5. Interaction assays between Sebacinales strains and *Arabidopsis thaliana* under controlled conditions. (**A**) Control (left), inoculated with *Piriformospora indica* (middle) and inoculated with *Serendipita herbamans* (right) seedling phenotypes after 24 days at 16°C. (**B**) Total and root fresh weight and root surface of seedlings after 24 days at 16°C. Error bars give the standard deviation. Different letters indicate significant differences at the confidence level of $P<0.001$ for fresh weight and root surface.

significant differences in leaf fresh and dry weight were observed between inoculated and control seedlings. In addition, *S. herbamans* and *P. indica* were co-cultivated with *P. annua*, confirming the positive effect on the root growth as described for *A. thaliana* (data not shown). Our analyses indicate that *S. herbamans*, like *P. indica* and *S. vermifera* [12], has potential to promote plant growth. Further experiments are planned to address whether *S. herbamans* may also increase plant leaf, seed and/or fruit biomass. The wide occurrence of *S. herbamans*, particularly with economically important plant species from various sites in Europe, opens up huge further potential use of *S. herbamans* as beneficial symbiont in the field.

Serendipita herbamans K. Riess, Oberw., Garnica, sp. nov.

(Figs. 1–4)

[MycoBank MB 807011]

Etymology

Named for its preference to occur in the roots of several herbaceous plant species as revealed via PCR and sequencing.

Diagnosis

The anamorphic *Serendipita herbamans* forms monilioid hyphae (chlamydospores) (Fig. 4), which have not been reported in the genus until now. Based on the complete ITS region (KF061285), it is phylogenetically closely related to *S. vermifera* strain MAFF 305828 (Fig. 3), but differs from it by 29 bp (4.0%) and *S. vermifera* develops teleomorph stage after a few weeks in culture [50].

Holotype

Isolated from vital, surface sterilized fine roots of *Bistorta vivipara* (Polygonaceae) growing in a poor meadow lacking ectomycorrhizal trees within a radius of at least 50 m, collected on July 30, 2010 in Bad Hindelang (47°31′ N, 10°24′ E), Bavarian Alps, Germany. The strain has been deposited in Deutsche Sammlung für Mikroorganismen und Zellkulturen (DSMZ), Braunschweig, Germany (DSM 27534).

Description

Colonies growth rapid on MYP agar, reaching a radius of 13.0±1.2 mm in 14 days at 18°C without light. The fungus has an appressed growth habit and is weakly zonate, and mostly submerged at the edges. It is slightly glossy and cream to ochre-coloured, becoming orange brown with age (Fig. 4A, B). Hyphae were 1–1.5(−2) µm in diameter, cylindrical, thin-walled, clampless and colourless. Young hyphae are cylindrical; with age, these often become septate with constrictions at the septa. This monilioid cells of 3–5 µm×4–5 µm are smooth and colourless, becoming yellow with age (Fig. 4C, D). Monilioid cells (chlamydospores) are more frequent on media with low C and N contents. Basidia and basidiospores were not observed. Hyphae containing typical dolipore septa with imperforate parenthesomes were observed in root cells of *Plantago media* (Fig. 1), *Triticum aestivum* [24] and *Potentilla anserina* [19].

Geographic distribution

This species has been detected by means of PCR in the roots of 55 species of herbaceous plants belonging to 18 angiosperm families from various agricultural and grassland sites in Europe (Germany, Italy, France) and North America (USA) (Fig. 2A).

Conclusions

A new endophyte species, *Serendipita herbamans*, is common and widely distributed in plant roots across agricultural and grassland ecosystems. This endophyte grows easily on artificial media and has the potential to promote the growth of herbaceous plants under laboratory conditions. There was no distinctive correlation of diversity and composition of Sebacinales communities with land use in the sampled habitats. Finally, our study indicates that endophytic Sebacinales are widely distributed in the roots of herbaceous plants in agricultural and grassland ecosystems, but they are of low quantity and show no host specificity.

Supporting Information

Table S1 Plant species yielding the sebacinoid endophytes detected in this study. The respective land use category, type of farming/grassland, host plant species, host plant family, sample date, sample location, collector(s), Herbarium Tubingense (TUB) number, GenBank accession number, Sebacinales molecular operational taxonomic unit (MOTU)/species, haplotype designation, sequence source and morphological detection are given. Plant samples TUB 019485–019567 were processed by D. Kuhn in the context of her Diploma thesis.

Table S2 Total number of herbaceous plant families sampled to detect the presence of sebacinoid fungi in their roots.

Table S3 Sebacinales building basidiomata or strains isolated in culture are classified into Sebacinales groups A and B. The respective GenBank accession number, Herbarium Tubingense (TUB) number/strain designation, collector(s) with collection/isolation number, basidiomata substrate/isolation source and country of collection are given. New sequences in this study use the acronymous KF. Asterisks indicate that only an ITS sequence was available.

Data S1 Description of habitats, including land use and vegetation composition.

Acknowledgments

We thank D. Kuhn for helping with the sequencing of Sebacinales endophytes, C. Karasch-Wittmann and S. Silberhorn for assistance with the laboratory work, C. Krause for providing seeds of *Arabidopsis thaliana*, J. Bauer for help with the statistical analysis and J. Ammirati for the valuable comments on the manuscript. We are grateful to V. Bandala, L. Ryvarden and P. Welt for providing fungal material and to the owners of the agricultural sites for their kind permission to take plant samples for this study. Sampling of plant material in the Natural Reserve Filsenberg (Germany) was permitted by the Regierungspräsidium of Tübingen.

Author Contributions

Conceived and designed the experiments: KR SG. Performed the experiments: KR SG. Analyzed the data: KR SG. Contributed reagents/materials/analysis tools: RB SG FO. Wrote the paper: KR SG.

References

1. Rodriguez RJ, White JF, Arnold AE, Redman RS (2009) Fungal endophytes: diversity and functional roles. New Phytol 182: 314–330.
2. Yue C, Miller CJ, White JEJ, Richardson M (2000) Isolation and characterization of fungal inhibitors from *Epichloä festucae*. J Agric Food Chem 48: 4687–4692.
3. Rodriguez RJ, Henson J, van Volkenburgh E, Hoy M, Wright L, et al. (2008) Stress tolerance in plants via habitat-adapted symbiosis. The ISME Journal 2: 404–416.
4. Achatz B, von Rüden S, Andrade D, Neumann E, Pons-Kühnemann, et al. (2010) Root colonization by *Piriformospora indica* enhances grain yield in barley under diverse nutrient regimes by accelerating plant development. Plant Soil 333: 59–70.
5. Redman RS, Kim Y-O, Woodward CJDA, Greer C, Espino L, et al. (2011) Increased fitness of rice plants to abiotic stress via habitat adapted symbiosis: a strategy for mitigating impacts of climate change. PLoS ONE 6(7): e14823. doi:10.1371/journal.pone.0014823.
6. Saari S, Faeth SH (2012) Hybridization of *Neotyphodium* endophytes enhances competitive ability oft he host grass. New Phytol 195: 231–236.
7. Wearn JA, Sutton BC, Morley NJ, Gange AC (2012) Species and organ specificity of fungal endophytes in herbaceous grassland plants. J Ecol 100: 1085–1092.
8. Garcia E, Alonso A, Platas G, Sacristán S (2013) The endophytic mycobiota of *Arabidopsis thaliana*. Fungal Divers 60: 71–89.
9. Pancher M, Ceol M, Corneo PE, Longa CMO, Yousaf S, et al. (2012) Fungal endophytic communities in grapevines (*Vitis vinifera* L.) respond to crop management. Appl Environ Microbiol 79: 4308–4317.
10. Saunders M, Kohn LM (2009) Evidence for alteration of fungal endophyte community assembly by host defense compounds. New Phytol 182: 229–238.
11. Verma S, Varma A, Rexer K-H, Hassel A, Kost G, et al. (1998) *Piriformospora indica*, gen. et sp. nov., a new root-colonizing fungus. Mycologia 90: 896–903.
12. Deshmukh S, Hückelhoven R, Schäfer P, Imani J, Sharma M, et al. (2006) The root endophytic fungus *Piriformospora indica* requires host cell death for proliferation during mutualistic symbiosis with barley. Proc Natl Acad Sci U S A 103: 18450–18457.
13. Serfling A, Wirsel SG, Lind V, Deising HB (2007) Performance of the biocontrol fungus *Piriformospora indica* on wheat under greenhouse and field conditions. Phytopathology 97: 523–531.
14. Varma A, Verma S, Sudha, Sahay N, Butehorn B, et al. (1999) *Piriformospora indica*, a cultivable plant-growth-promoting root endophyte. Appl Environ Microbiol 65: 2741–2744.
15. Fakhro A, Andrade-Linares DR, van Bargen S, Bandte M, Büttner C, et al. (2010) Impact of *Piriformospora indica* on tomato growth and on interaction with fungal and viral pathogens. Mycorrhiza 20: 191–200.
16. Kharkwal AC, Prasad R, Kharkwal H, Das A, Bhatnagar K, et al. (2007) Co-cultivation with Sebacinales. In: Varma A, Oelmüller R, editors. Advanced Techniques in Soil Microbiology. Berlin: Springer. pp. 247–270.
17. Oberwinkler F, Riess K, Bauer R, Selosse M-A, Weiβ M, et al. (2013) Enigmatic Sebacinales. Mycol Prog 12: 1–27.
18. Arnold AE, Henk DA, Eells RL, Lutzoni F, Vilgalys R (2007) Diversity and phylogenetic affinities of foliar fungal endophytes in loblolly pine inferred by culturing and environmental PCR. Mycologia 99: 185–206.
19. Garnica S, Riess K, Bauer R, Oberwinkler F, Weiβ M (2013) Phylogenetic diversity and structure of sebacinoid fungi associated with plant communities along an altitudinal gradient. FEMS Microbiol Ecol 83: 265–278.
20. Kottke I, Guttenberger M, Hampp R, Oberwinkler F (1987) An *in vitro* method for establishing mycorrhizae on coniferous tree seedlings. Trees 1: 191–194.
21. Luz C, Netto MCB, Rocha LFN (2007) *In vitro* susceptibility to fungicides by invertebrate-pathogenic and saprobic fungi. Mycopathologia 164: 39–47.
22. Worrall JJ (1991) Media for selective isolation of Hymenomycetes. Mycologia 83: 296–302.
23. Bauer R, Begerow D, Sampaio JP, Weiβ M, Oberwinkler F (2006) The simple-septate basidiomycetes: a synopsis. Mycol Prog 5: 41–66.
24. Weiβ M, Sýkorová Z, Garnica S, Riess K, Martos F, et al. (2011) Sebacinales everywhere: previously overlooked ubiquitous fungal endophytes. PLoS ONE 6: e16793. doi:10.1371/journal.pone.0016793.
25. Gardes M, Bruns TD (1993) ITS primers with enhanced specificity for basidiomycetes: application to the identification of mycorrhizae and rusts. Mol Ecol 1993: 113–118.
26. White TJ, Bruns T, Lee S, Taylor JW (1990) Amplification and direct sequencing of fungal ribosomal RNA genes for phylogenetics. In: Innis MA, Gelfand DH, Sninsky JJ, White TJ, editors. PCR Protocols: a Guide to Methods and Applications. New York: Academic Press. pp. 315–322.
27. Setaro SD, Garnica S, Herrera PI, Suárez JP, Göker M (2012) A clustering optimization strategy to estimate species richness of Sebacinales in the tropical

Andes based on molecular sequences from distinct DNA regions. Biodivers Conserv 9: 2269–2285.
28. Riess K, Oberwinkler F, Bauer R, Garnica S (2013) High genetic diversity at the regional scale and possible speciation in *Sebacina epigaea* and *S. incrustans*. BMC Evol Biol 13: 102. doi:10.1186/1471-2148-13-102.
29. Katoh K, Kuma K, Toh H, Miyata T (2005) MAFFT version 5: improvement in accuracy of multiple sequence alignment. Nucleic Acids Res 33: 511–518.
30. Lee C, Grasso C, Sharlow M (2002) Multiple sequence alignment using partial order graphs. Bioinformatics 18: 452–464.
31. Capella-Gutiérrez S, Silla-Martínez JM, Gabaldón T (2009) trimAl: a tool for automated alignment trimming in large-scale phylogenetic analyses. Bioinformatics 25: 1972–1973.
32. Stamatakis A (2006) RAxML-VI-HPC: Maximum likelihood-based phylogenetic analyses with thousands of taxa and mixed models. Bioinformatics 22: 2688–2690.
33. Stamatakis A, Hoover P, Rougemont J (2008) A rapid bootstrap algorithm for the RAxML web servers. Syst Biol 57: 758–771.
34. Göker M, García-Blázquez G, Voglmayr H, Tellería MT, Martín MP (2009) Molecular taxonomy of phytopathogenic fungi: a case study in *Peronospora*. PLoS One 4(7): e6319. doi:10.1371/journal.pone.0006319.
35. Rambaut A (2009) FigTree drawing tool. Version 1.3.1. Available online at: http://tree.bio.ed.ac.uk/software/figtree [accessed on 7 May 2012].
36. Swofford DL (2002) PAUP*. Phylogenetic Analysis Using Parsimony (*and other methods). Version 4.0b10. Sunderland, Massachusetts: Sinauer Associates.
37. Stephens M, Smith NJ, Donnelly P (2001) A new statistical method for haplotype reconstruction from population data. Am J Hum Gen 68: 978–989.
38. Stephens M, Donnelly P (2003) A comparison of Bayesian methods for haplotype reconstruction from population genotype data. Am J Hum Gen 73: 1162–1169.
39. Librado P, Rozas J (2009) DnaSP v5: a software for comprensive analysis of DNA polymorphism data. Bioinformatics 25: 1451–1452.
40. Bandelt HJ, Forster P, Röhl A (1999) Median-joining networks for inferring intraspecific phylogenies. Mol Biol Evol 16: 37–48.
41. Polzin T, Daneshmand SV (2003) On Steiner trees and minimum spanning trees in hypergraphs. Oper Res Lett 31: 12–20.
42. Webb CO, Ackerly DD, Kembel AW (2008) Phylocom: software for the analysis of phylogenetic community structure and trait evolution. Bioinformatics 24: 2098–2100.
43. Webb CO, Ackerly DD, Kembel AW (2009) Phylocom: software for the analysis of phylogenetic community structure and character evolution (with Phylomatic and Evolve). Version 4.1. Available online at: http://www.phylodiversity.net/phylocom [accessed on 6 November 2013].
44. Gotelli NJ, Rohde K (2002) Co-occurrence of ectoparasites of marine fishes: a null model analysis. Ecol Lett 5: 86–94.
45. Murashige T, Skoog F (1962) A revised medium for rapid growth and bio-assays with tobacco tissue cultures. Physiol Plant 15: 473–497.
46. Abramoff MD, Magalhães PJ, Ram SJ (2004) Image processing with ImageJ. Biophotonics International 11: 36–42.
47. Selosse MA, Dubois MP, Alvarez N (2009) Do Sebacinales commonly associate with plant roots as endophytes? Mycol Res 113: 1062–1069.
48. Weiβ M, Selosse M-A, Rexer K-H, Urban A, Oberwinkler F (2004) Sebacinales: a hitherto overlooked cosm of heterobasidiomycetes with a broad mycorrhizal potential. Mycol Res 108: 1003–1010.
49. Setaro SD, Kron KA (2011) Neotropical and North American Vaccinioideae (Ericaceae) share their mycorrhizal Sebacinales – an indication for concerted migration? PLoS Curr 3: RRN1227. doi:10.1371/currents.RRN1227.
50. Warcup JH (1988) Mycorrhizal associations of isolates of *Sebacina vermifera*. New Phytol 110: 227–231.
51. Peškan-Berghöfer T, Shahollari B, Giong PH, Hehl S, Markert C, et al. (2004) Association of *Piriformospora indica* with *Arabidopsis thaliana* roots represents a novel system to study beneficial plant–microbe interactions and involves early plant protein modifications in the endoplasmic reticulum and at the plasma membrane. Physiol Plant 122: 1399–3054.
52. Ghimire SR, Charlton ND, Craven KD (2009) The mycorrhizal fungus, *Sebacina vermifera*, enhances seed germination and biomass production in switchgrass (*Panicum virgatum* L). Bioenergy Res 2: 51–58.
53. Price EW, Carbone I (2005) SNAP: workbench management tool for evolutionary population genetic analysis. Bioinformatics 21: 402–404.
54. Aylor DL, Price EW, Carbone I (2006) SNAP: Combine and Map modules for multilocus population genetic analysis. Bioinformatics 22: 1399–1401.
55. Hudson RR, Kaplan NL (1985) Statistical properties of the number of recombination events in the history of a sample of DNA sequences. Genetics 111: 147–164.
56. Myers SR, Griffiths RC (2003) Bounds on the minimum number of recombination events in a sample history. Genetics 163: 375–394.

17

Climate and Land Use Controls on Soil Organic Carbon in the Loess Plateau Region of China

Yaai Dang[1,2,3]¶, Wei Ren[2]¶, Bo Tao[2], Guangsheng Chen[2], Chaoqun Lu[2], Jia Yang[2], Shufen Pan[2], Guodong Wang[3], Shiqing Li[1], Hanqin Tian[2]*

1 State Key Laboratory of Soil Erosion and Dryland Farming on the Loess Plateau, Institute of Soil and Water Conservation, Northwest A&F University, Yangling, Shaanxi, China, **2** International Center for Climate and Global Change Research, School of Forestry & Wildlife Sciences, Auburn University, Auburn, Alabama, United States of America, **3** College of Science, Northwest A&F University, Yangling, Shaanxi, China

Abstract

The Loess Plateau of China has the highest soil erosion rate in the world where billion tons of soil is annually washed into Yellow River. In recent decades this region has experienced significant climate change and policy-driven land conversion. However, it has not yet been well investigated how these changes in climate and land use have affected soil organic carbon (SOC) storage on the Loess Plateau. By using the Dynamic Land Ecosystem Model (DLEM), we quantified the effects of climate and land use on SOC storage on the Loess Plateau in the context of multiple environmental factors during the period of 1961–2005. Our results show that SOC storage increased by 0.27 Pg C on the Loess Plateau as a result of multiple environmental factors during the study period. About 55% (0.14 Pg C) of the SOC increase was caused by land conversion from cropland to grassland/forest owing to the government efforts to reduce soil erosion and improve the ecological conditions in the region. Historical climate change reduced SOC by 0.05 Pg C (approximately 19% of the total change) primarily due to a significant climate warming and a slight reduction in precipitation. Our results imply that the implementation of "Grain for Green" policy may effectively enhance regional soil carbon storage and hence starve off further soil erosion on the Loess Plateau.

Editor: Xiujun Wang, University of Maryland, United States of America

Funding: This study was supported by NASA Land Cover and Land Use Change Program (NNX08AL73G), NASA Interdisciplinary Science Program (NNG04GM39C), Chinese Universities Scientific Fund (z10921007), US National Science Foundation Grants (AGS-1243220, CNS-1059376), State Key Laboratory of Soil Erosion and Dryland Farming on the Loess Plateau Foundation (K318009902-1410), and Shaanxi Administration of Foreign Expert Affairs Science and Technology Activities Fundation (201327). Study design was supported by NASA Land Cover and Land Use Change Program (NNX08AL73G), NASA Interdisciplinary Science Program (NNG04GM39C), and Chinese Universities Scientific Fund (z10921007). Data collection and analysis was supported by NASA Land Cover and Land Use Change Program (NNX08AL73G), NASA Interdisciplinary Science Program (NNG04GM39C), and State Key Laboratory of Soil Erosion and Dryland Farming on the Loess Plateau Foundation (K318009902-1410). Preparation of the manuscript was supported by NASA Land Cover and Land Use Change Program (NNX08AL73G), US National Science Foundation Grants (AGS-1243220, CNS-1059376), and State Key Laboratory of Soil Erosion and Dryland Farming on the Loess Plateau Foundation (K318009902-1410).

Competing Interests: The authors have declared that no competing interests exist.

* E-mail: tianhan@auburn.edu

¶ These authors contributed equally to this work.

Introduction

Soil organic carbon (SOC), the major component of soil organic matter, plays a key role in the terrestrial carbon cycle and thus has drawn great attention from scientific community. It is a dynamic component of terrestrial systems, affecting carbon exchange between terrestrial ecosystem and the atmosphere [1,2]. SOC storage is nearly three times as large as carbon storage in vegetation and twice as large as global atmospheric carbon storage [3]. Soil has higher potential to sequester more carbon (such as converting the type of land use) in the future [2,4], therefore, increasing soil carbon storage is one of the most economical and effective ways to alleviate the greenhouse effect, which has become a hot scientific and political issue during the past decades.

Changes in climate and land use, caused by both natural and anthropogenic processes, have greatly influenced the terrestrial carbon balance during the past decades [5,6,7,8]. It was reported that about one fourth of anthropogenic CO_2 emissions were due to land cover and land use change (LCLUC), especially deforestation [9]. Long-term experimental studies have confirmed that SOC is highly sensitive to land conversion from natural ecosystems, such as forest or grassland, to agricultural land, resulting in substantial SOC loss [6,10]. In addition, LCLUC may also cause carbon depletion by influencing soil respiration [11]. It was estimated that global carbon release from SOC mineralization owing to agricultural activities was approximately 0.80 Pg C/year (1 Pg = 10^{15} g) [12]. Globally, land use change resulted in a carbon release of (1.6±0.8) Pg C per year to the atmosphere during the period of 1990s [13]. However, the effects of conversions from cropland to grassland/forest on the SOC storage have not been fully understood and there still remains large uncertainty.

The Loess Plateau of China (Figure 1), located in the geographic center of China (33°43′N 100°54′E to 41°16′N 114°33′E), covers a total area of 628,000 km^2, which is about 6.5% of China's total land area. The Loess Plateau is characterized by highly erodible soils, steep slopes, being subjected to heavy rain, and low vegetation coverage due to excess exploitation of land resource and improper land use [4,14]. During the past decades, serious soil erosions caused by natural and anthropogenic disturbances (e.g., climate change, natural disasters, LCLUC etc.) occurred in the area of the Loess Plateau. Previous reports also indicated that the warming and drying climate in this region has significantly aggravated soil erosion [15]. As a result, the large amount of fine surface soil eroded from the loess area is transported into the Yellow River and acts as the main source of sediment of this river, which runs through the Loess Plateau and is considered to be the most turbid river in the world. Due to these disturbances, soil carbon storage on the Loess Plateau is much lower compared to other regions in China [16]. In general, adjusting the land use pattern so as to restore the degraded ecosystems and to modify the local rural income structure is regarded as the main measures to control soil and water erosion and to improve farmers' living conditions on the Loess Plateau. Since the 1950s, a series of conservation policies have been implemented in this region, such as extensive tree planting since the 1970s, integrated soil erosion controls on the watershed scale in the 1980s and the 1990s [17,18], and the government-funded project "Grain for Green" in 1999, aiming at transforming the low-yield slope cropland into grassland/forest. The implementation of these policies improved vegetation coverage, altered land use patterns, and changed the SOC storage. Although many field experiments have been performed to explore the impacts of both drying and warming climate and LCLUC on soil carbon storage on the Loess Plateau [19,20], little attention has been paid to the regional impacts of these factors and their interactions.

Over recent decades, many field observations and control experiments have been conducted to explore the effects of climate and land use change on the SOC in this region and make it possible to study the regional effects of climate change and LCLUC on SOC. In addition, many approaches, including eddy covariance flux tower, inventory, remote sensing techniques, forward and inversion models, have been used to examine the regional carbon budget on the Loess Plateau [21,22,23,24,25]. Among them, process-based ecosystem modeling is one of the most effective approaches to estimate regional SOC storage and fluxes in different terrestrial ecosystems driven by multiple global changes factors [7,11,24,26,27]. To address the effects of changes in climate, land use, and other environmental factors on SOC storage in this region, the Dynamic Land Ecosystem Model (DLEM), a highly integrated process-based model [28], was applied to evaluate the spatial and temporal patterns of SOC storage on the Loess Plateau during 1961–2005. The objectives of this study are: 1) to investigate the temporal and spatial patterns of SOC storage on the Loess Plateau; and 2) to identify the relative contribution of climate and land use changes to the SOC storage changes.

Methods

Model Description

The DLEM is a process-based terrestrial ecosystem model, which aims at simulating the impacts of natural and anthropogenic disturbances on the structure and functions of terrestrial ecosystems over the spatial and temporal contexts. The DLEM has been widely used to simulate the effects of climate variability and change, elevated atmospheric CO_2, tropospheric ozone pollution, land use change, and increasing nitrogen deposition, etc. on terrestrial carbon storage and fluxes in China and other regions across the globe [15,29,30,31,32,33].

In this study, DLEM simulates two kinds of LCLUC: land conversion from natural ecosystem to cropland, and cropland abandonment. In the DLEM model, the balance of soil organic matter depends on the transformation of litter (LIT) to soil organic matter, the fractions of conversion from gross primary production (GPP) to dissolved organic carbon (DOC), the returned organic matter from production decay (PRD) (e.g., manure), the growth of microbe, the methane production from dissolved organic carbon, and the carbon loss from soil organic matter (SOM) decomposition.

$$\frac{dC_{som}}{dt} = k_{tr}LITC_{loss} + k_{gppdoc}GPP + k_{prd}PRD_{docom} - k_{rh}SOMC_{docom} - k_{Lucc}C_{som} - DOC_{loss,\,methane}$$

where k_{tr} is the transfer rate of decomposed LIT to SOM; k_{gppdoc} is the fraction of GPP converted to soil DOC; k_{prd} is the returned rate of decomposed (or consumed) PRD to SOM pools as manure; k_{rh} is the fraction of decomposed SOM that is converted to CO_2 through heterotrophic respiration; k_{Lucc} is coefficient for quick carbon loss from SOM due to land use conversion; $DOC_{loss,methane}$ is DOC consumed for the growth of production of methane. More

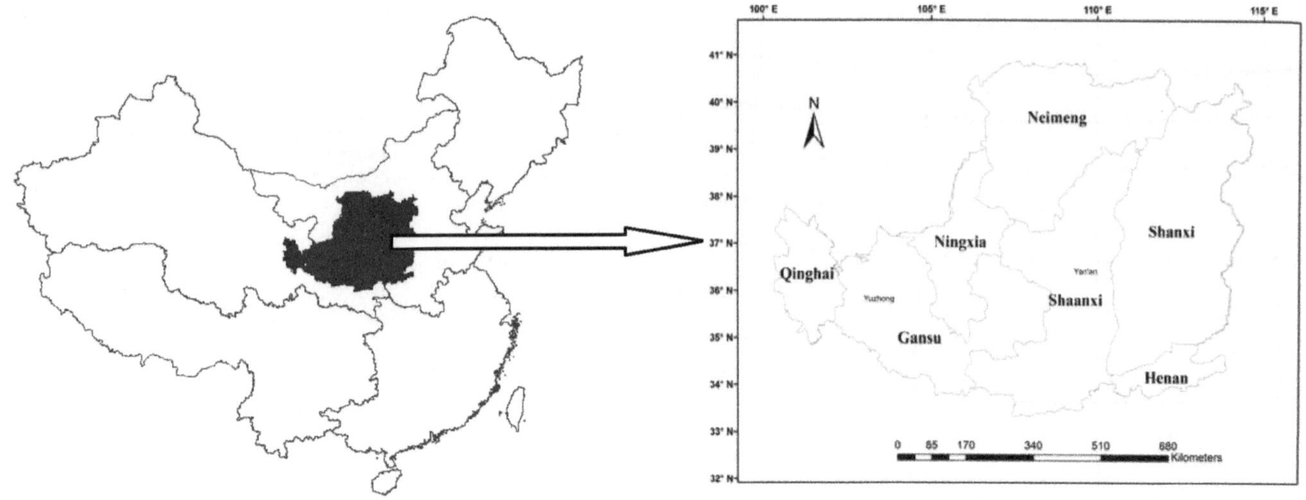

Figure 1. Location of the Loess Plateau, China.

detailed processes were described in our previous papers [15,29,30].

Input Data Description

The major input data in the DLEM include: (1) daily climatic data (i.e. maximum, minimum, average temperature, precipitation, relative humidity, and radiation) and atmospheric chemistry data (i.e. tropospheric O_3, atmospheric CO_2 and nitrogen deposition); (2) soil properties (including soil type, bulk density, depth, pH, soil texture) which are derived from the 1:1 million soil map based on the Second National Soil Survey of China [34,35,36]; (3) contemporary vegetation map for 2000 which was developed from Landsat Enhanced Thematic Mapper (ETM) imagery [37]; (4) long-term land use history which was developed on the basis of three recent (1990, 1995 and 2000) land cover maps and historical census datasets [38,39]. All the input datasets were developed at the spatial resolution of 10 km×0 km. Detailed information about other input data were described in our previous studies [15,29,30,40].

Climate change. Average air temperature on the Loess Plateau increased at a rate of 0.030°C/year from 1961 to 2005 (Figure 2a), higher than those reported for the entire China (about 0.029°C/year) and the global average level (about 0.010°C/year) [41] in the same period. The most rapid increase in temperature occurred during the 1990s. Air temperature increased from the north to the south of the Loess Plateau (Figure 2b), with the greatest increase in the northern Loess Plateau (e.g., the north of Shanxi and the northwest of Inner Mongolia).

The Loess Plateau can be mainly divided into three climate zones according to the precipitation: the northern Loess Plateau with precipitation below 400 mm, the central Loess Plateau with precipitation between 400 and 500 mm, and the southern Loess Plateau with precipitation above 500 mm [42,43]. The precipitation less than 550 mm/year occurred across most areas of the Loess Plateau. Over the past 45 years, the mean annual precipitation was approximately 423 mm, with the lowest of 288 mm in 1997 and the highest of 661 mm in 1964 (Figure 2c). A slightly decreasing trend at a rate of 1.27 mm/year in precipitation was found on the Loess Plateau from 1961 to 2005. This temporal trend of precipitation was consistent with a previous study based on the meteorological observations according to 99 stations for the period 1956–2005 on the Loess Plateau [44]. Figure 2d further indicated that decreases in precipitation occurred in the most areas of the central and southern Loess Plateau from 1961–1990 to 1991–2005. The regions with the most obvious drying trend were located in the central and southern Loess Plateau, especially in the north of Shaanxi and the center of Shanxi Province with a reduction of more than 80 mm in the recent 15 years (1991–2005) compared to the 1961–1990 average.

LCLUC. Expansion of cropland and pasture was driven by social-economic factors on the Loess Plateau during the past decades, though to some extent, the soil erosion and frequent drought events limited the massive expansion of cropland. Since the 1950s, many measures have been implemented to alleviate and control serious soil erosion on the Loess Plateau. For example, "Grain for Green" project, which was launched in 1999, recommended that cropland with slopes greater than 15° should be converted back to natural vegetation. Driven by such governmental policies, the Loess Plateau experienced a remarkable change in land use, characterizing by a large area of

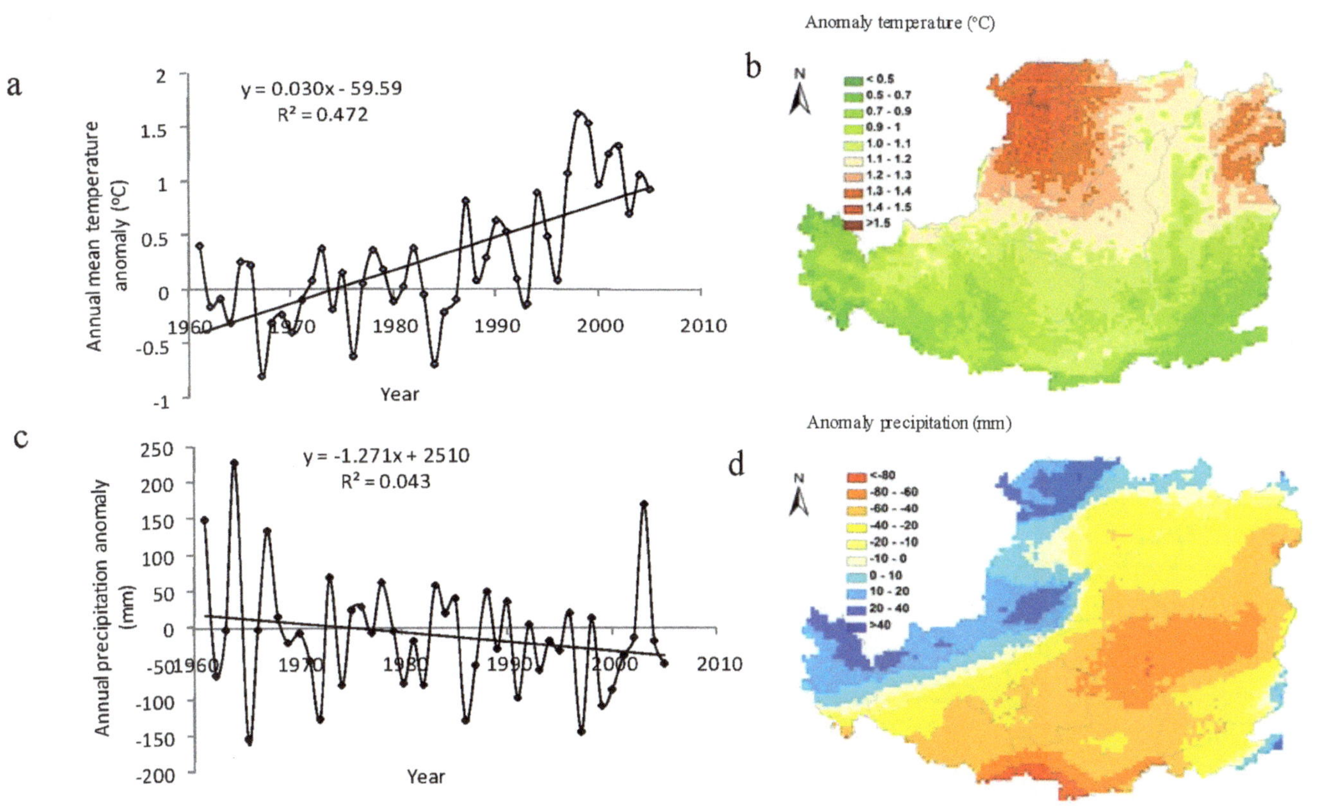

Figure 2. Anomaly of annual mean temperature (a), and precipitation (c) on the Loess Plateau from 1961 to 2005; and spatial distribution of temperature anomaly (b) and precipitation anomaly (d) during 1991–2005 (relative to 1961–1990 average);

conversion from cropland to grassland/forest. In order to well understand the LCLUC history on the Loess Plateau, we analyzed the spatial and temporal variations from 1961 to 2005 (Figure 3–4). Figure 3 showed that cropland area on the Loess Plateau decreased slowly until 1999, and followed by a sharp decrease. However, the grassland area showed an opposite changing trend in the same periods. Since 1961, the area of cropland decreased by 19.61%, while grassland and forest increased by 7.31% and 6.75%, respectively.

Land use patterns exhibited large spatial variations on the Loess Plateau (Figure 4). Grassland was the dominant vegetation type in the northern Loess Plateau. Cropland, grassland and shrubland were the main vegetation type in the middle Loess Plateau. In contrast, cropland and forest occupied large area of the southern Loess Plateau. Compared to 1960, the coverage of cropland decreased mainly due to land conversion from cropland to grassland, especially in the middle and southern part of Loess Plateau. During 1999–2005, cropland largely shrank in some areas of the middle Loess Plateau, especially in the northern Shaanxi and middle of Shanxi Province. These results were consistent with previous studies [18,45].

Model Parameterization and Evaluation

DLEM has been well calibrated and intensively validated against the site-level observed carbon fluxes and pool sizes from Chinese Ecosystem Research Network (CERN) and other previous studies [29,38,46]. In this study, we further compared our simulated results with field observation data and survey data from the Second National Soil Survey (1979–1983) to evaluate the model performance in simulating the SOC storage on the Loess Plateau. The simulated SOC storage as influenced by multiple environmental factors significantly correlated with the observed data ($R^2 = 0.738$, $p<0.01$, Figure 5), indicating that DLEM could capture the spatial and temporal patterns of SOC storage on the Loess Plateau. Several factors may contribute to the difference between model results and field observations. First, the input datasets were developed at a spatial resolution of 10 km×10 km for driving DLEM simulations. Each grid was assumed to have the uniform climate, land use type, and vegetation cover in the model. As a well-known climate-sensitive zone and a fragile ecological belt, some subtle changes in climate, land use or other factors on the Loess Plateau might cause large difference in the SOC storage even within one grid cell. This difference in the same grid will be reflected in the field experiments but neglected in the model

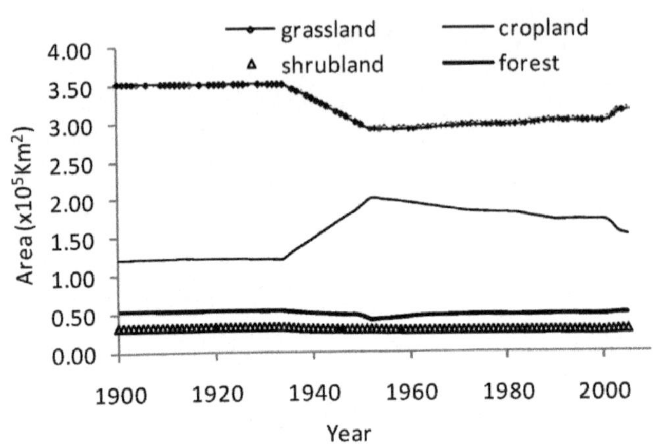

Figure 3. Area of major land use cover on the Loess Plateau during 1900–2005.

simulations. Second, the shortage of field observation data may weaken the capability of model to realistically capture the magnitude and patterns in SOC changes, which has long been identified as one of the biases in the large-scale model development. In addition, model simplification and neglecting microbial biomass might be another potential reason [13,30].

Simulation Experimental Design and Model Run

In this study, four main simulation experiments were designed to analyze the effects of climate change alone, LCLUC alone, the interaction of climate and LCLUC, and the combined effects of all environmental factors on the SOC storage on the Loess Plateau (Table 1). The experiment I was designed to provide a 'best-estimate' of spatial and temporal patterns of SOC driven by major environmental factors changes including climate variability, LCLUC, elevated CO_2, N fertilizer, N deposition, and O_3 pollution, etc. In experiment II and III, we simulated the contributions of climate variability alone and the LCLUC alone, respectively. In the experiment IV, we tried to understand the interactive effect between climate change and LCLUC on the SOC storage. The model simulations began with an equilibrium run to obtain the baseline carbon pools for each grid. A spin-up of about 100 years was applied if the climate change was included in the simulation experiments. Finally, the model was run in transient mode driven by the daily climate data and other time-variant or invariant input data.

Results and Analysis

Temporal Changes in SOC on the Loess Plateau

Model simulation indicated that the SOC storage over the entire Loess Plateau displayed substantial temporal and spatial variations in the context of multiple environmental factors changes (Figure 6a, b). As a whole, the combination of all these environmental factors considered in this study (e.g. climate, LCLUC and others environment factors) caused a net increase of about 0.27 Pg C in SOC storage from 1961 to 2005. The SOC storage kept relatively stable before the 1970s, and then gradually increased in the following two decades, and rapidly increased since 2000 (Figure 6a). Our further analysis found that the decrease of cropland area was relatively slower from the 1960s to 1999, and became rapid after then. Most of the abandoned cropland were replaced by grassland (Figure 3). Our results implied that temporal pattern in SOC change was partly related to the land use change on the Loess Plateau.

Spatial Variation in SOC on the Loess Plateau

We found that the spatial patterns of SOC storage were primarily controlled by the precipitation distribution. Spatially, the SOC storage increased gradually from the north to the south along an increasing precipitation gradient on the Loess Plateau. Large SOC increases were found in the southern and central Loess Plateau (south of Shaanxi and Shanxi Province in particular); and some other areas show a slight increase of SOC storage in the past decades. Due to changes in multiple environmental factors, the SOC storage increased throughout the majority of the Loess Plateau over the past 45 years (Figure 6b). Some areas showed a significant increase in SOC storage of more than 200 g C/m^2. However, a significant decrease occurred in the middle of central Loess Plateau (especially in the Northern Shaanxi Province), where experienced the most obvious warming and drying tendency in the past decades (Figure 2 b,d), releasing more than 100 g C/m^2 during the study period. We further found that the SOC increase

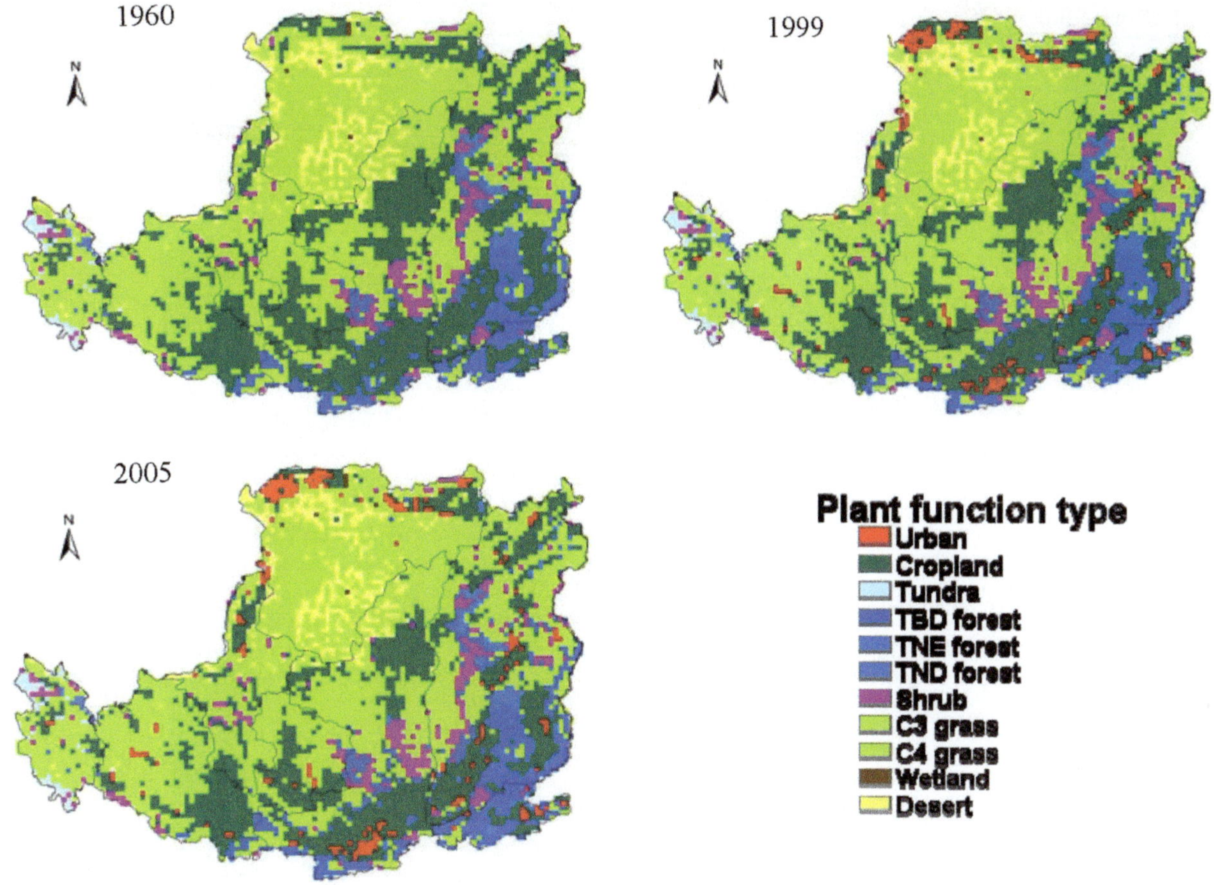

Figure 4. Spatial variations in LCLUC on the Loess Plateau in different year.

in the northern Loess Plateau was lower than that in the southern Loess Plateau during the past decades.

Individual Factorial Contributions to Changes in SOC Storage

To well understand the influence of climate on the SOC storage, we simulated the SOC storage on the Loess Plateau under the influence of climate change alone. DLEM simulated results

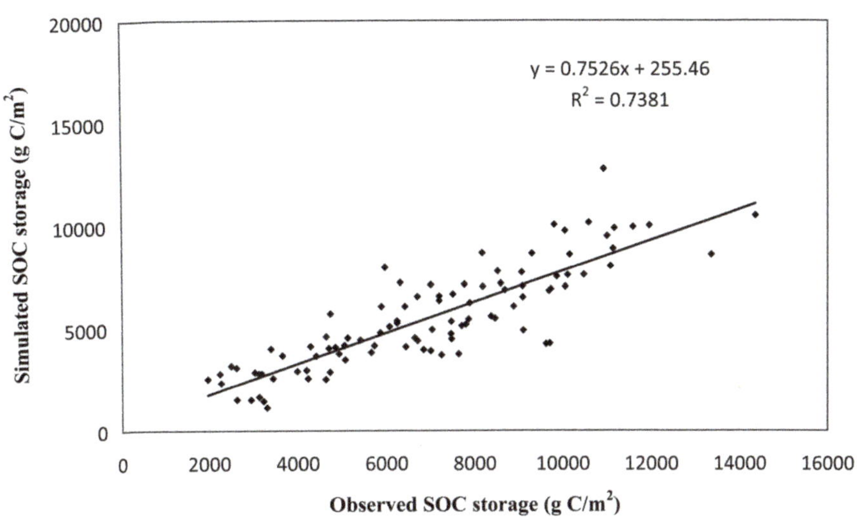

Figure 5. Correlation between simulated and observed SOC storage based on 91 soil samples.

Table 1. Simulation experiments.

Simulation Experiment	Environmental factors		
	Climate	LCLUC	Others
I. All	1960–2005	1960–2005	1960–2005
II. Climate only	1960–2005	1960	1960
III. LCLUC only	1960	1960–2005	1960
IV. Climate-LCLUC	1960–2005	1960–2005	1960

Notes: Climate-LCLUC means the combination effects of climate and LCLUC. Others include the effects of atmospheric CO_2, ozone pollution (AOT 40 index), nitrogen deposition, nitrogen fertilizer application, etc. All means the simulation experiment which includes all above environmental factors.

showed that SOC storage decreased by about 0.05 Pg C with substantial inter-annual fluctuations throughout the Loess Plateau from 1961 to 2005 (Figure 6c). SOC storage showed a slowly decreasing trend before the late 1990s, followed by a sharp decrease until the early 21st century. Since 1961, SOC storage decreased in most areas of the Loess Plateau under the influences of the climate, with a maximum carbon release of 200 g C/m^2 in some areas of the southern and the central Loess Plateau (Figure 6d). Figure 6d also indicated that SOC storage decreased in the southeastern humid monsoon climatic regions, and to a lesser extent in the continental dry climatic regions in the northern Loess Plateau during 1961–2005.

Considering the single effect of LCLUC, our results showed that the SOC storage increased by 0.14 Pg C during 1961–2005 (Figure 6e). We found that the most rapid increase in the SOC storage occurred after the late 1970s (Figure 6e). DLEM simulation results also showed that obvious increases of SOC storage were found in most areas of the central and the southern Loess Plateau (Figure 6f).

Relative Contributions of Climate, LCLUC, and their Interactions

Our simulated results indicated that with multiple environmental changes, SOC continuously increased from 1961 to 2005 on the Loess Plateau, resulting in a net increase in SOC storage by 0.27 Pg C (5.97 Tg C/year) (Figure 7). Among these factors, LCLUC was obviously the major factor affecting the magnitude of SOC change on the Loess Plateau, leading to a significant increase (0.14 Pg C) in SOC storage, accounting for 55% of the net increase in SOC in the past 45 years. However, in the same period, warming and drying climate greatly reduced SOC storage by 0.05 Pg C, approximately 19% of the total SOC storage change. The interaction between climate and LCLUC contributed to the net SOC increases by 3%.

Discussion

SOC Storage Change on the Loess Plateau

In this study, the combination of all environmental factors (climate, LCLUC and others environment factors) caused a net increase of about 0.27 Pg C in SOC storage, indicating that soil acted as a weak carbon sink on the Loess Plateau during the past 45 year. The temporal pattern of SOC storage was largely influenced by land use change and relevant land use policies.

Except the central Loess Plateau, the SOC storage increased throughout the entire region, with a smaller sink in the northern Loess Plateau and a larger sink in the southern Loess Plateau during 1961–2005, which is consistent with previous reports [43,47,48]. The less increase of SOC storage in the northern Loess Plateau might be due to more sandy soil, lower soil carbon input from plant biomass, and larger soil carbon loss from erosion. This is consistent with previous reports, indicating the soil in the northern Loess Plateau contained more sand which could accumulate less carbon than the soil with more clay [49,50,51,52]. On the other hand, the area of forest and shrubland, characterized by higher aboveground biomass and productivity than cropland and grassland [43,53], decreased from the south to north over the Loess Plateau (Figure 4). This also contributed to realtively less SOC increase in the northern Loess Plateau. In addition, the extensive soil erosions in the northern Loess Plateau was suggested to further explain the lower contents of SOC storage [2,17,54].

Comparisons with Field Observations

Our estimation of SOC storage and its change on the Loess Plateau were comparable to other studies. Based on the 0–100 cm SOC data collected from 382 sampling sites across the entire Loess Plateau in 2008, Liu et al. [55] indicated that mean SOC storage was 7.70 kg C/m^2. Using data gathered by the Second National Soil Survey of China (1979–1983), Xu et al. [56] estimated that SOC storage was 1.07 Pg in 0–20 cm soils layers on the Loess Plateau region (with a land area of 429,800 km^2). A soil survey conducted throughout the Loess Plateau region indicated that the SOC storage amounted to 1.23 Pg C in the 0–20 cm soil layers with land area of 592,900 km^2 during 1985–1988, [55]. Converting SOC storage from different soil depth to the same soil depth according to the summary of the vertical distribution of soil organic in the 0–100 cm soil [31], SOC storage were 6.22 kg C/m^2 and 5.19 kg C/m^2 in Xu et al.'s [57] and Liu et al.'s [55] reports (be mentioned but not published) respectively. In the same soil depth, DLEM-estimated average SOC storage was 6.45 kg C/m^2 in 2005, which falls in the range of previous studies.

Tian et al. [7] estimated an average SOC sink of 94 Tg C/year in the terrestrial ecosystem of China as influenced by multiple global change factors during 1961–2005. Our results showed that soil on the Loess Plateau acted as a weak sink of 5.97 Tg C/year under the combined experiment during the same period. Considering that the Loess Plateau accounts for approximately 6.5% of the country's land surface, implying that the SOC storage on the Loess Plateau was close to the country-level average. At national level, Wang et al. [35] suggested that SOC storage in China decreased by about 5.69 g C/m^2 per year during the 1960s–1980s, while our study showed that the SOC storage on the Loess Plateau significantly increased by 4.30 g C/m^2 per year during the same period. This indicated that successful implementation of land conservation measures was beneficial and effective to reduce soil erosion and improve soil properties, thus enhance the soil carbon sequestration over the Loess Plateau.

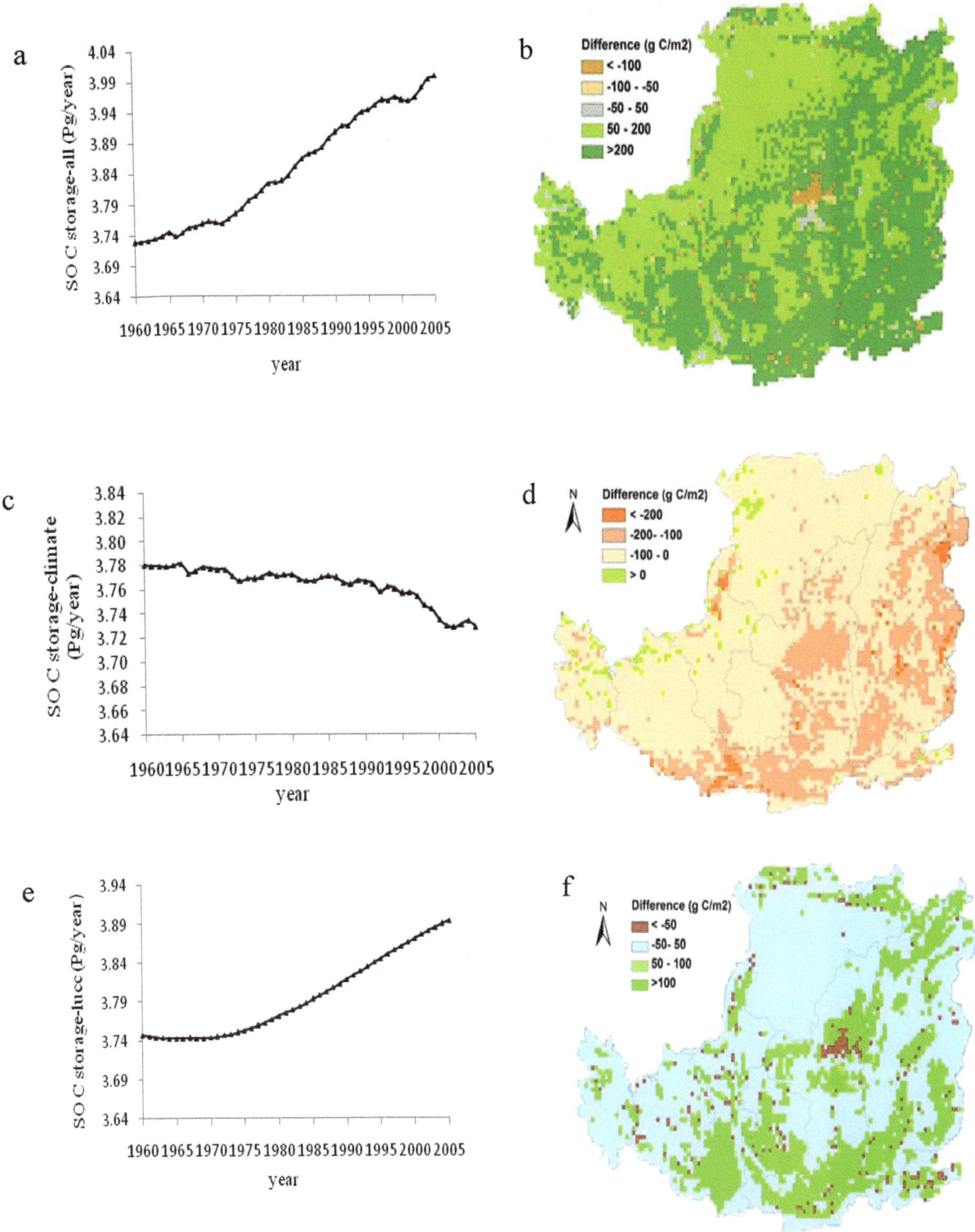

Figure 6. Spatiotemporal variations of SOC storage under experiments I, II, and III on the Loess Plateau from 1960–2005. Note: The "difference" means change in SOC storage between 2005 and 1960.

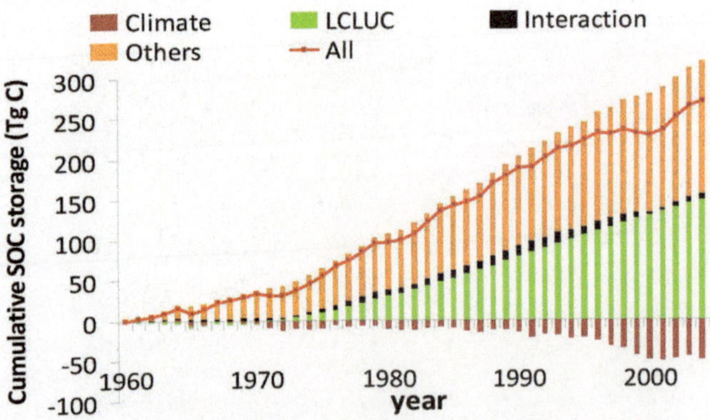

Figure 7. Contributions of multi-factor global changes to accumulative SOC change during 1961–2005. Note: "interaction" referrers to the interactive effects between climate and LCLUC.

Climate Controls on the Spatiaotemporal Patterns of SOC Storage

Climatic factors, especially precipitation and temperature, play an important role in long-term variations of SOC due to their effects on the quantity and quality of organic residue inputs and on the rates of soil organic matter and litter decomposition [1,12,58]. DLEM-simulated results showed that SOC storage decreased by about 0.05 Pg C with a significant inter-annual fluctuations throughout the Loess Plateau from 1961 to 2005 as influenced by climate change alone (Figure 6c). As shown in Figure 2, precipitation decreased at a rate of 1.27 (mm/year) during 1961–2005, which was a notable factor that influenced the change of SOC on the Loess Plateau, especially after 1990. Meanwhile, increased air temperature (0.030°C/year) might accelerate the evapotranspiration and potentially aggravate the water deficiency, thus cause the formation of drying soil layer and suppress the growth of vegetation. Due to the warming and drying climate, dried soil layer was widely distributed in the hilly and gully areas of the Loess Plateau. The development of drying soil layer has been regarded as a key cause for the decrease of SOC storage in some areas on the Loess Plateau in the past decades [59].

Climate also had a substantial effect on the spatial distribution of SOC storage. Our results suggested that SOC storage decreased in most areas of the Loess Plateau under the influences of the climate change since 1961 (Figure 6d). This result could be explained partly by the distribution of temperature and precipitation, which is well-known to have a positive relationship with SOC decomposition [60]. During 1961–2005, both temperature and precipitation were higher in the central and southern Loess Plateau, compared to the northern Loess Plateau. Therefore, the higher loss of soil carbon from decomposition might be an important cause for lower SOC accumulation in the central and southern Loess Plateau. This result was supported by previous studies [43,50,57].

Land Use Controls on the Spatiaotemporal Patterns of SOC Storage

During the past decades, the Loess Plateau has experienced a complex change in land use pattern (Figure 3). Considering the LCLUC alone, DLEM-simulated results showed that the SOC storage increased by 0.14 Pg C during 1961–2005 (Figure 6e), which was significantly larger than the influence of climate change alone. However, in the first few years, the SOC changed slowly, even had a decreasing tendency in the 1960s, and then increased gradually since the 1980s.

Since the 1950s, various soil and water conservation measures including afforestation, cropland abandonment, and terrace construction etc., have been implemented in this region [17,18]. All of these measures directly or indirectly affected the vegetation cover and further influenced the SOC storage on the Loess Plateau. Previous studies demonstrated that soil can lose up to 20–40 percent of organic carbon into the atmosphere when perennial vegetation land was converted into cultivation land [61,62]. On the contrary, conversion from cropland to perennial vegetation land or shrubland was found to accumulate SOC by increasing carbon derived from new vegetation and decreasing carbon loss from decomposition and erosion [11,43,51,63]. Liu et al. [47] reported that SOC in shrubland, which was converted from

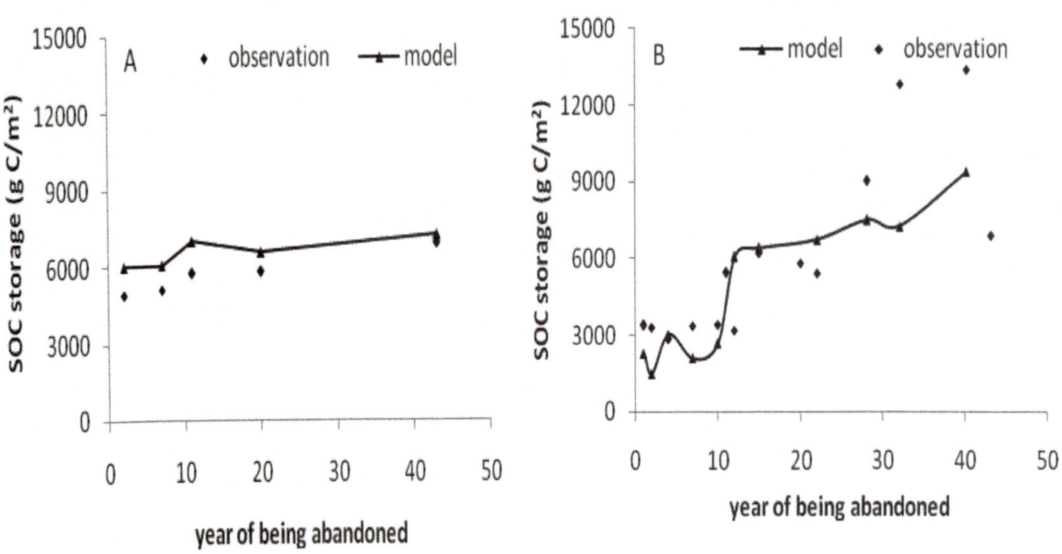

Figure 8. Comparison between simulated (DLEM) and observed SOC storage. Notes: A: cropland abandoned inYuzhong (Gansu Province), the observation data come from Jinping Jiang et al. [67]; B: cropland abandoned in Yan'an (Shaanxi Province), the observation data come from Junmin Wang et al. [2].

cropland in 1985, were 27.7%–34.8% higher than that of the cropland in 2010 on the Loess Plateau. Fu et al. [64] suggested that cropland abandonment significantly increased the density and stock of SOC in 0–100 cm soil profiles on the Loess Plateau. Feng et al. [48] found a total of 96.1 Tg of additional carbon had been sequestered on the Loess Plateau since China's "Grain for Green" program during 2000–2008, by using remote sensing techniques and ecosystem modelling. Our results are comparable with those previous findings.

LCLUC might induce an immediate change in vegetation coverage but a lagged effect on the change of SOC storage [65]. In this study, the SOC changed slowly, and even had a decreasing trend in the 1960s. We also found that the most rapid increase in the SOC storage occurred after the late 1970s, which might be partly due to extensive and conducted tree planting projects for alleviating soil erosion. The large-scale cropland abandonment in response to the recently implemented "Grain for Green" policy also contributed to the increase of SOC storage on the Loess Plateau. The plantation from "Grain for Green" Project would keep a large proportion of carbon in wood which need long time to return to soil. During the initial period of land conversion from cropland to forests, leaf biomass of trees were very low, so the litters on the ground decreased and resulted in a slight change (even decrease) in SOC. During the period of implementing "Grain for Green" project, litters on the ground accumulated gradually with the growth of planted trees, resulting in increasing SOC storage after a certain time period. Compared with simulated results in the context of multiple environmental factors, a similar temporal pattern for the SOC change was found when considering LCLUC alone from 1961 to 2005 (Figure 6a,e). This further implied that LCLUC was the dominant factor controlling temporal variations of SOC storage.

DLEM simulated results also showed obvious increases of SOC storage in most areas of the central and southern Loess Plateau (Figure 6f). Further analysis found that changes in SOC storage were smaller in the LCLUC alone simulation experiment than that with the combination of all environmental change (Figure 6b). However, SOC storage changed slightly in most areas of the northern Loess Plateau from 1961 to 2005, which were also suggested by other research results (e.g., [18,43]).

Interactive Effects of LCLUC and Climate on SOC Storage

Based on DLEM simulations, Tian et al. [7] found that LCLUC accounted for 17% of the net carbon increase, but climate change reduced it by 4% in China in the past decades. However, we found that LCLUC increased SOC storage by 0.14 Pg C, accounting for 55% of the net increase in SOC in the past 45 years over the Loess Plateau. In the same period, warming and drying climate greatly reduced SOC storage by 0.05 Pg C, approximately 19% of the total SOC storage change. These results implied that the SOC was more sensitive to LCLUC and climate change on the Loess Plateau comparing to other regions in China. In the past decades, LCLUC, particularly in the southern Loess Plateau, enhanced the vegetation coverage and reduced the anthropogenic disturbance, which further enhanced the SOC storage.

The interaction between climate and LCLUC contributed to the net SOC increases by 3%. Our results also showed that the interactive effects among environmental factors can't be neglected in attributing the changes of SOC storage in response to environmental factors on the Loess Plateau. Although the interaction among environmental factors has been recognized long before [66], most of the field experiments still overlook it. This study further demonstrated that the modeling approach may serve as one complementary tool for the field experiments in addressing interactive effects among multiple environmental factors.

Effects of Cropland Abandonment on SOC Storage

Cropland abandonment and natural vegetation recovery were important implementations to mitigate soil loss on the Loess Plateau. Previous studies demonstrated that former land use types, soil property, climate change, and soil management practices were crucial to changes of SOC storage in the establishment of perennial vegetation type [49,50]. Other factors, such as the abandonment age, have also been found to play a significant role in SOC accumulation and should not be ignored [4,50].

In this study, we chose two sites (Yuzhong City and Yan'an City) (Marked in Figure 1) located on the Loess Plateau to explore SOC storage pattern at different abandonment stages (or restoration age) and to further evaluate the DLEM simulated results against the observations (Figure 8). It showed that DLEM results could well capture the distribution characteristics of the SOC storage in different abandonment stages. Both model results and field observations demonstrated that the restoration age played a key role in SOC accumulation. Jiang et al. [67] indicated that SOC storage changed little in early abandonment stage in a semiarid hilly area of Yuzhong City, followed by an obvious increase after 9–12 years, and then increased stably (Figure 8A). Wang et al. [2] found the similar temporal pattern through studying the change of SOC storage during different successional stages of rehabilitated grassland in Yan'an City. Compared to cropland, the rehabilitated grassland had a lower SOC storage at the early stage (1–12 years). However, SOC storage was higher than that in cropland after 15 years and then increased steadily (Figure 8B). Generally, long-term abandonment (>10 years) and the following colonization of natural vegetation could lead to substantial increase of SOC storage. The difference of the SOC storage between two sites over the early abandonment stage might be attributed to difference in environmental factors such as climate and soil property. Yan'an City is located in a semi-arid and warm temperate zone with less precipitation and more obvious drying and warming tendency, compared with Yuzhong City during the past years. This also partly explained the lower SOC storage in cropland in this climate zone than others over the Loess Plateau.

Our results indicated that cropland abandonment could increase SOC storage, improve soil quality and promote ecosystem restoration, especially in the warm and dry climate zone. However, these effects will emerge after a relative longer time period (e.g., >10 years) under local environmental conditions.

Uncertainties

This study examined temporal and spatial patterns of SOC storage and attributed these patterns to multiple environmental factors on the Loess Plateau during 1961–2005. There were several uncertainties which need to be addressed in our future work. First, impacts of some ecological processes, such as soil erosion, were not separated from other environmental factors in this study. The soil erosion area could be as high as 45.4×10^4 km^2 (72.3% of the land area) over the Loess Plateau. Soil erosion has long been identified as one of key factors controlling the change of SOC storage [47]. Further efforts should be put on interactions among soil erosion and other environmental factors. Second, the input datasets were developed at the spatial resolution of 10 km×10 km which was the finest dataset for the Loess Plateau. However, it is still difficult to capture subtle change of SOC storage due to high heterogeneity in land surface processes. In the long run, finer gidded datasets would be greatly helpful for further quantifying temporal and spatial changes in SOC storage on the Loess Plateau. In addition, the uncertainties from other input data, model structure, and parameterization need to be further specified in the furture efforts.

Conclusions

This study examined effects of climate and land use changes on SOC storage on the Loess Plateau of China in the context of multiple global changes by using an integrated ecosystem model. The results showed that temperature on the Loess Plateau has significantly increased, while precipitation slightly decreased during 1961–2005. Meanwhile, this region experienced a remarkable change in land cover and land use, characterized by conversions from cropland to grassland/forest owing to the government policies to alleviate soil and water losses during the past decades. The overall change in SOC storage due to multiple environmental factors was estimated to be a net increase of SOC storage by 0.27 Pg C during 1961–2005, indicating that soil on the Loess Plateau acted as a carbon sink in this period. Among multiple factors, LCLUC led to a significant increase in SOC storage of 0.14 Pg C, accounting for 55% of the net increase in SOC. In contrast, climate change reduced SOC storage by 0.05 Pg C (approximately 19% of the total SOC change). The interaction of climate and LCLUC accounted for 3% of the net increase in SOC. Our results were consistent with field observation data and both of them suggested that SOC storage could be enhanced significantly by the conversion of cropland to grassland along with the increasing abandonment age on the Loess Plateau. However, the magnitude could be influenced by local environmental conditions.

This study provides the first attempt to quantify relative effects of multiple environmental factors (climate and LCLUC in particular) on regional SOC storage on the Loess Plateau over the past decades. The results drawn from this study provide insight for land management as well as policy-making to enhance carbon sequestration and alleviate the serious soil erosion conditions on the Loess Plateau. To reduce uncertainties in estimating effects of climate and land use changes on SOC storage, it is needed to put further efforts in developing more reliable and fine-resolution input data and improve model representation of some other processes relevant to SOC, such as soil erosion.

Acknowledgments

This study was supported by NASA Land Cover and Land Use Change Program (NNX08AL73G), NASA Interdisciplinary Science Program (NNG04GM39C), Chinese Universities Scientific Fund (z10921007), US National Science Foundation Grants (AGS-1243220, CNS-1059376), State Key Laboratory of Soil Erosion and Dryland Farming on the Loess Plateau Foundation (K318009902-1410), and Shaanxi Administration of Foreign Expert Affairs Science and Technology Activities Fundation (201327). Thank Mingliang Liu, Xiaofeng Xu, and Jiyuan Liu for their contributions in the development of spatial data and the DLEM model.

Author Contributions

Conceived and designed the experiments: HT YD WR. Performed the experiments: YD WR. Analyzed the data: YD WR BT GC. Contributed reagents/materials/analysis tools: GC BT JY SP. Wrote the paper: YD WR HT BT GC CL JY SP GW SL.

References

1. Lal R (2004) Soil carbon sequestration impacts on global climate change and food security. Science 304: 1623–1627.
2. Wang Y, Fu B, Lü Y, Chen L (2011) Effects of vegetation restoration on soil organic carbon sequestration at multiple scales in semi-arid Loess Plateau, China. Catena 85: 58–66.
3. Post WM, Emanuel WR, Zinke PJ, Stangenberger AG (1982) Soil carbon pools and world life zones.
4. Zhang K, Dang H, Tan S, Cheng X, Zhang Q (2010) Change in soil organic carbon following the 'Grain-for-Green' programme in China. Land Degradation & Development 21: 13–23.
5. Eglin T, Ciais P, Piao S, Barre P, Bellassen V, et al. (2010) Historical and future perspectives of global soil carbon response to climate and land-use changes. Tellus B 62: 700–718.
6. Martin D, Lal T, Sachdev C, Sharma J (2010) Soil organic carbon storage changes with climate change, landform and land use conditions in Garhwal hills of the Indian Himalayan mountains. Agriculture, Ecosystems & Environment 138: 64–73.
7. Tian H, Melillo J, Lu C, Kicklighter D, Liu M, et al. (2011) China's terrestrial carbon balance: Contributions from multiple global change factors. Global Biogeochemical Cycles 25.
8. Wang X, Feng Z, Ouyang Z (2001) The impact of human disturbance on vegetative carbon storage in forest ecosystems in China. Forest ecology and management 148: 117–123.
9. Barnett TP, Adam JC, Lettenmaier DP (2005) Potential impacts of a warming climate on water availability in snow-dominated regions. Nature 438: 303–309.
10. Paul EA (1997) Soil organic matter in temperate agroecosystems: long term experiments in North America: CRC PressI Llc.
11. Post WM, Kwon KC (2000) Soil carbon sequestration and land-use change: processes and potential. Global change biology 6: 317–327.
12. Lal R, Bruce J (1999) The potential of world cropland soils to sequester C and mitigate the greenhouse effect. Environmental Science & Policy 2: 177–185.
13. Schimel J (2001) 1.13-Biogeochemical Models: Implicit versus Explicit Microbiology. Global Biogeochemical Cycles in the Climate System: 177–183.
14. Liu S, Bliss N, Sundquist E, Huntington TG (2003) Modeling carbon dynamics in vegetation and soil under the impact of soil erosion and deposition. Global Biogeochemical Cycles 17.
15. Zhang C, Tian H, Pan S, Liu M, Lockaby G, et al. (2008) Effects of forest regrowth and urbanization on ecosystem carbon storage in a rural–urban gradient in the southeastern United States. Ecosystems 11: 1211–1222.
16. Li Z, Liu W-z, Zhang X-c, Zheng F-l (2009) Impacts of land use change and climate variability on hydrology in an agricultural catchment on the Loess Plateau of China. Journal of hydrology 377: 35–42.
17. Chen L, Gong J, Fu B, Huang Z, Huang Y, et al. (2007) Effect of land use conversion on soil organic carbon sequestration in the loess hilly area, loess plateau of China. Ecological Research 22: 641–648.
18. Xin Z, Xu J, Zheng W (2008) Spatiotemporal variations of vegetation cover on the Chinese Loess Plateau (1981–2006): impacts of climate changes and human activities. Science in China Series D: Earth Sciences 51: 67–78.
19. Fu B-J, Wang Y-F, Lu Y-H, He C-S, Chen L-D, et al. (2009) The effects of land-use combinations on soil erosion: a case study in the Loess Plateau of China. Progress in Physical Geography 33: 793–804.
20. Wang G, Huang J, Guo W, Zuo J, Wang J, et al. (2010) Observation analysis of land-atmosphere interactions over the Loess Plateau of northwest China. Journal of Geophysical Research: Atmospheres (1984–2012) 115.
21. Zhao M, Heinsch FA, Nemani RR, Running SW (2005) Improvements of the MODIS terrestrial gross and net primary production global data set. Remote Sensing of Environment 95: 164–176.
22. Fang J, Guo Z, Piao S, Chen A (2007) Terrestrial vegetation carbon sinks in China, 1981–2000. Science in China Series D: Earth Sciences 50: 1341–1350.
23. Peters W, Jacobson AR, Sweeney C, Andrews AE, Conway TJ, et al. (2007) An atmospheric perspective on North American carbon dioxide exchange: CarbonTracker. Proceedings of the National Academy of Sciences 104: 18925–18930.
24. Piao S, Ciais P, Friedlingstein P, de Noblet-Ducoudré N, Cadule P, et al. (2009) Spatiotemporal patterns of terrestrial carbon cycle during the 20th century. Global Biogeochemical Cycles 23.
25. Li K, Wang S, Cao M (2004) Vegetation and soil carbon storage in China. SCIENCE IN CHINA SERIES D EARTH SCIENCES-ENGLISH EDITION- 47: 49–57.
26. King J, Bradley R, Harrison R, Carter A (2004) Carbon sequestration and saving potential associated with changes to the management of agricultural soils in England. Soil Use and Management 20: 394–402.
27. Yan H, Cao M, Liu J, Tao B (2007) Potential and sustainability for carbon sequestration with improved soil management in agricultural soils of China. Agriculture, Ecosystems & Environment 121: 325–335.
28. Tian H, Chen G, Liu M, Zhang C, Sun G, et al. (2010) Model estimates of net primary productivity, evapotranspiration, and water use efficiency in the terrestrial ecosystems of the southern United States during 1895–2007. Forest ecology and management 259: 1311–1327.
29. Ren W, Tian H, Liu M, Zhang C, Chen G, et al. (2007) Effects of tropospheric ozone pollution on net primary productivity and carbon storage in terrestrial ecosystems of China. Journal of Geophysical Research: Atmospheres (1984–2012) 112.

30. Xu X, Tian H, Zhang C, Liu M, Ren W, et al. (2010) Attribution of spatial and temporal variations in terrestrial methane flux over North America. Biogeosciences Discussions 7: 5383–5428.
31. Tian H, Xu X, Lu C, Liu M, Ren W, et al. (2011) Net exchanges of CO2, CH4, and N2O between China's terrestrial ecosystems and the atmosphere and their contributions to global climate warming. Journal of Geophysical Research: Biogeosciences (2005–2012) 116.
32. Ren W, Tian H, Tao B, Huang Y, Pan S (2012) China's crop productivity and soil carbon storage as influenced by multifactor global change. Global Change Biology 18: 2945–2957.
33. Tian H, Chen G, Zhang C, Liu M, Sun G, et al. (2012) Century-scale responses of ecosystem carbon storage and flux to multiple environmental changes in the southern United States. Ecosystems 15: 674–694.
34. Shi H, Shao M (2000) Soil and water loss from the Loess Plateau in China. Journal of Arid Environments 45: 9–20.
35. Wang S, Tian H, Liu J, Pan S (2003) Pattern and change of soil organic carbon storage in China: 1960s–1980s. Tellus B 55: 416–427.
36. Zhang XP, Zhang L, McVicar TR, Van Niel TG, Li LT, et al. (2008) Modelling the impact of afforestation on average annual streamflow in the Loess Plateau, China. Hydrological Processes 22: 1996–2004.
37. Liu J, Tian H, Liu M, Zhuang D, Melillo JM, et al. (2005) China's changing landscape during the 1990s: Large-scale land transformations estimated with satellite data. Geophysical Research Letters 32.
38. Liu M, Tian H (2010) China's land cover and land use change from 1700 to 2005: Estimations from high-resolution satellite data and historical archives. Global Biogeochemical Cycles 24.
39. Tian H, Chen G, Zhang C, Melillo JM, Hall CA (2010) Pattern and variation of C: N: P ratios in China's soils: a synthesis of observational data. Biogeochemistry 98: 139–151.
40. Chen H, Tian H, Liu M, Melillo J, Pan S, et al. (2006) Effect of land-cover change on terrestrial carbon dynamics in the southern United States. Journal of environmental quality 35: 1533–1547.
41. Trenberth K, Jones P, Ambenje P, Bojariu R, Easterling D, et al. (2007) Observations: Surface and Atmospheric Climate Change, chap. 3 of Climate Change 2007: The Physical Science Basis. Contribution of Working Group I to the Fourth Assessment Report of the Intergovernmental Panel on Climate Change [Solomon, S., Qin, D., Manning, M., Marquis, M., Averyt, KB, Tignor, M., Miller, HL and Chen, Z.(eds.)]., 235–336. Cambridge University Press, Cambridge, UK and New York, NY, USA.
42. Li R, Yang W, Li B (2008) Research and future prospects for the Loess Plateau of China. SciencePress, Beijing (in Chinese).
43. Chang R, Fu B, Liu G, Liu S (2011) Soil carbon sequestration potential for "Grain for Green" project in Loess Plateau, China. Environmental management 48: 1158–1172.
44. Xin Z, Xu J, Ma Y (2009) Spatio-temporal variation of erosive precipitation in Loess Plateau during past 50 years. Scientia Geographica Sinica 29: 89–104.
45. Song Y, Ma M-g (2007) Study on vegetation cover change in Northwest China based on SPOT VEGETATION data. Journal of Desert Research 27: 89–93.
46. Ren W, Tian H, Xu X, Liu M, Lu C, et al. (2011) Spatial and temporal patterns of CO2 and CH4 fluxes in China's croplands in response to multifactor environmental changes. Tellus B 63: 222–240.
47. Liu X, Li F-M, Liu D-Q, Sun G-J (2010) Soil organic carbon, carbon fractions and nutrients as affected by land use in semi-arid region of Loess Plateau of China. Pedosphere 20: 146–152.
48. Feng X, Fu B, Lu N, Zeng Y, Wu B (2013) How ecological restoration alters ecosystem services: an analysis of carbon sequestration in China's Loess Plateau. Scientific reports 3.
49. Guo Z, Ruddiman WF, Hao Q, Wu H, Qiao Y, et al. (2002) Onset of Asian desertification by 22 Myr ago inferred from loess deposits in China. Nature 416: 159–163.
50. Paul K, Polglase P, Nyakuengama J, Khanna P (2002) Change in soil carbon following afforestation. Forest ecology and management 168: 241–257.
51. LAGANIÈRE J, ANGERS DA, PARÉ D (2010) Carbon accumulation in agricultural soils after afforestation: a meta-analysis. Global Change Biology 16: 439–453.
52. Cheng L, Zhao X (2011) Soil mineralized nutrients changes and soil conservation benefit evaluation on 'green project grain' in ecologically fragile areas in the south of Yulin city, Loess Plateau. African Journal of Biotechnology 10: 2230–2237.
53. Xiao Y (1990) Comparative studies on biomass and productivity of Pinus tabulaeformis plantations in different climatic zones in Shaanxi Province [China]. Acta Phytoecologica et Geobotanica Sinica 14.
54. Fu X, Shao M, Wei X, Horton R (2010) Soil organic carbon and total nitrogen as affected by vegetation types in Northern Loess Plateau of China. Geoderma 155: 31–35.
55. Liu Z, Shao Ma, Wang Y (2011) Effect of environmental factors on regional soil organic carbon stocks across the Loess Plateau region, China. Agriculture, Ecosystems & Environment 142: 184–194.
56. Xu X, Zhang K, Peng W (2003) Spatial distribution and estimating of soil organic carbon on Loess Plateau. J Soil Water Conserv 17: 13–15.
57. Xu X, Liu W, Kiely G (2011) Modeling the change in soil organic carbon of grassland in response to climate change: effects of measured versus modelled carbon pools for initializing the Rothamsted Carbon model. Agriculture, Ecosystems & Environment 140: 372–381.
58. Schlesinger WH (1990) Evidence from chronosequence studies for a low carbon-storage potential of soils. Nature 348: 232–234.
59. Han F, Hu W, Zheng J, Du F, Zhang X (2010) Estimating soil organic carbon storage and distribution in a catchment of Loess Plateau, China. Geoderma 154: 261–266.
60. Lehmann J, Skjemstad J, Sohi S, Carter J, Barson M, et al. (2008) Australian climate–carbon cycle feedback reduced by soil black carbon. Nature Geoscience 1: 832–835.
61. Houghton R, Hackler J, Lawrence K (1999) The US carbon budget: contributions from land-use change. Science 285: 574–578.
62. van der Werf GR, Morton DC, DeFries RS, Olivier JG, Kasibhatla PS, et al. (2009) CO2 emissions from forest loss. Nature Geoscience 2: 737–738.
63. Richter DD, Markewitz D, Trumbore SE, Wells CG (1999) Rapid accumulation and turnover of soil carbon in a re-establishing forest. Nature 400: 56–58.
64. Fu B-J, Zhang Q-J, Chen L-D, Zhao W-W, Gulinck H, et al. (2006) Temporal change in land use and its relationship to slope degree and soil type in a small catchment on the Loess Plateau of China. Catena 65: 41–48.
65. Kuzyakov Y, Gavrichkova O (2010) Review: Time lag between photosynthesis and carbon dioxide efflux from soil: a review of mechanisms and controls. Global Change Biology 16: 3386–3406.
66. Dermody O (2006) Mucking through multifactor experiments; design and analysis of multifactor studies in global change research. New Phytologist 172: 598–600.
67. Jiang J-P, Xiong Y-C, Jiang H-M, Ye D-Y, Song Y-J, et al. (2009) Soil microbial activity during secondary vegetation succession in semiarid abandoned lands of Loess Plateau. Pedosphere 19: 735–747.

18

Local Scale Comparisons of Biodiversity as a Test for Global Protected Area Ecological Performance

Bernard W. T. Coetzee[1,2]*, Kevin J. Gaston[3], Steven L. Chown[2]

1 Centre for Invasion Biology, Department of Botany and Zoology, Stellenbosch University, Stellenbosch, Western Cape, South Africa, **2** School of Biological Sciences, Monash University, Melbourne, Victoria, Australia, **3** Environment and Sustainability Institute, University of Exeter, Penryn, Cornwall, United Kingdom

Abstract

Terrestrial protected areas (PAs) are cornerstones of global biodiversity conservation. Their efficacy in terms of maintaining biodiversity is, however, much debated. Studies to date have been unable to provide a general answer as to PA conservation efficacy because of their typically restricted geographic and/or taxonomic focus, or qualitative approaches focusing on proxies for biodiversity, such as deforestation. Given the rarity of historical data to enable comparisons of biodiversity before/after PA establishment, many smaller scale studies over the past 30 years have directly compared biodiversity inside PAs to that of surrounding areas, which provides one measure of PA ecological performance. Here we use a meta-analysis of such studies (N = 86) to test if PAs contain higher biodiversity values than surrounding areas, and so assess their contribution to determining PA efficacy. We find that PAs generally have higher abundances of individual species, higher assemblage abundances, and higher species richness values compared with alternative land uses. Local scale studies in combination thus show that PAs retain more biodiversity than alternative land use areas. Nonetheless, much variation is present in the effect sizes, which underscores the context-specificity of PA efficacy.

Editor: Robert Guralnick, University of Colorado, United States of America

Funding: This work was funded by the National Research Foundation of South Africa, Stellenbosch University, and by a Harry Crossley grant to BWTC. The funders had no role in study design, data collection and analysis, decision to publish, or preparation of the manuscript.

Competing Interests: The authors have declared that no competing interests exist.

* Email: bwtcoetzee@gmail.com

Introduction

Nearly 12% of the world's terrestrial surface is now classified as some form of protected area (PA) [1]. Indeed, the designation and maintenance of PAs are considered key global strategies to address the growing extinction crisis [1]. The unprotected world has been so transformed by human activity that it can now be characterized more readily by a set of human biomes than by the classic biogeographic regions [2]. Therefore, affording an area protection, a long-standing and current centrepiece of conservation strategy [3], appears to be an effective means of conserving its biodiversity features. Conservation scientists have rightly been concerned, however, that simple assumptions of positive ecological performance may be misleading [4–9]. Studies have recognized that effective PA management is key to biodiversity protection [7], and demonstrated that PA designation achieves good conservation return on investment at a relatively low cost [10]. Evidence exists, however, that in many cases PA systems are inefficiently planned to maximize benefits to biodiversity often owing to their spatial location [4,6], and worrying declines in biodiversity even within PAs in particular regions have been identified [5,8].

Much interest has focused on determining PA effectiveness in terms of preventing landscape cover changes (e.g. [11–13]), but these assessments serve only as a proxy for PA performance, as the measures used cannot necessarily capture the implications of land use change for biodiversity features. Where the latter are investigated, outcomes are typically available for specific areas, such as the tropics [8], or particular taxa, such as birds [14] or mammals [5]. Given that negative pressures on biodiversity and evidence for population declines are global in extent [15,16], the overall significance of terrestrial PAs for maintaining biodiversity values thus remains unclear. Protected area policy demonstrating their efficacy should ideally be evidence-based, that is, informed by rigorously established objective scientific evidence, as should be the case for conservation policy generally [17]. However, such evidence is not as well developed as it should be [18], despite urgent calls for so doing both in the scientific [18,19] and policy [3] arenas.

Protected area ecological performance would best be assessed by determining for every established PA what the overall biodiversity status is compared with what would have happened in the absence of protection. Plainly such a comparison cannot readily be achieved. One experimentally tractable alternative is the assessment of biodiversity before and after land cover change, but such studies are extremely rare (although see [20] for a notable exception). The scarcity of data to enable comparisons before/after PA establishment almost invariably necessitates comparisons of each PA with some other area that is unprotected, but similar in all but this designation. As a consequence, assessments of PA performance are typically restricted to small spatial scales and particular taxa (e.g. [14,21]), but there is no clear indication of the generality of their often-contrasting outcomes [18,21,22]. The biodiversity response when comparing a PA with some other area that is unprotected can vary widely, with different studies finding both higher and/or lower biodiversity values across areas [8,20–22]. Results from such studies suggest that PA ecological performance is context-specific and can be influenced by several local factors [8,18,22]. As a consequence, the generality of PA efficacy in maintaining biodiversity across regions remains unclear.

Here we use local scale studies comparing biodiversity between PAs and surrounding alternative land use areas to test if PAs contain higher biodiversity values than surrounding areas, using a meta-analysis. Specifically, our aim is to assess the ecological performance of terrestrial PAs, compared with areas in close proximity that are not protected, thus outside PAs, using three key biodiversity attributes: the abundances of individual species (hereafter 'species abundances'), assemblage abundances (summed across species) and assemblage species richness. These are key measures of biodiversity [23]. We also use an information theoretic approach with candidate explanatory variables to explore reasons for the variation in effect size. We then consider sources of bias in interpreting results, and highlight the benefits and shortcomings of our approach in determining PA efficacy.

Methods

Literature search

Our search and data extraction protocol follows best practice guidelines in conducting meta-analysis (see Appendix S1). We used keyword searches in Web of Science, Scopus, and Google Scholar for relevant papers published from 1975–2011, and their references, and included those reporting pairwise comparisons of biodiversity measurements either inside and outside protected areas (PAs) or between areas within PAs (details follow). The initial search string was: (bird* OR mammal* OR reptile* OR amphibia* OR arthropod* OR insect* OR fish* OR plant* OR vegetation*) AND ("protected area" OR "protected areas" OR "national park" OR "national parks" OR "reserve" OR "reserves" OR "game reserve" OR "game reserves") AND (effective* OR inside* OR performance* OR assessment* OR evaluation* OR estimate* OR comparison* OR contrast*) AND (outside* OR adjacent* OR neighbour* OR adjoining* OR bordering* OR near*). We also used unstructured and opportunistic literature searches with sections of the initial string (particularly to identify studies on different taxonomic groups), expert knowledge of available data sets, and results of a recent multi-database systematic review on protected area efficacy [9]. Grey literature (informally published written material [such as reports, theses and books]) was targeted but exceedingly rare, as found by others [9].

Data capture

From suitable papers we retained studies that measured biodiversity responses inside PAs, and outside PAs, in a replicated study design. These data represent three response variables: (i) species abundances (the abundances of individual species), (ii) abundances per assemblage, (iii) species richness per assemblage, following [24]. Species abundance represents indices of abundance; for example counts, density, capture frequency, occupancy estimates and biomass, for a single species both inside and outside PAs, but only in cases where taxonomy was resolved to the species level (species included are listed in the Dataset S1; N = 243). Assemblage abundance represents indices of abundance; for example counts, density, capture frequency, occupancy estimates and biomass, from cases where abundance was reported across species assemblages or could be calculated across sampling sites. We included these data as estimated by the original authors, but ensured that authors followed a replicated study design, and identical calculations of such indices both inside and outside PAs, and so we consider this response variable as an additional indicator of abundance across species groups (following [24]). Species richness included, for example, observed/estimated/rarefied richness, species density and genera/family richness, from cases where richness was reported or could be calculated across sampling sites. The use of these three variables as estimates of biodiversity was constrained by the approaches adopted by the studies we examined. Although biodiversity can be measured in a variety of ways, these three measures are commonly used as effective measures of biodiversity [23].

Most data we used included mean, standard deviation and sample sizes both inside and outside PAs. Where standard deviation was not reported (20%; 300/1484), we calculated it from either standard error or confidence intervals using imputation methods [25]. Studies with no suitable variance measures were omitted. Where the species could not be correctly identified to species level, data were omitted from the species abundance analysis, but included in the assemblage abundance analyses in cases where these values could be summed across groups of unidentified species. Invasive and domesticated species were omitted as they are not considered here to be of conservation interest. Observations on presence or absence of species were omitted as effect size cannot be calculated. Where data were reported across mixed habitats and/or vegetation types, we only took those from matched pairs, i.e. in areas where the same habitat type was reported. To avoid inadequate treatment of fragmentation effects, data from habitat fragments, typically forest fragments, were excluded following [24]. We did not find studies comparing biodiversity features before and after PA establishment. Few studies considered census-area effects [26] directly, but as individual studies had similar (often identical) sampling designs inside and outside the PAs and in consequence comparable sampling areas, we consider the potential confounding effects of sampling area negligible. Data reporting diversity indices were omitted to avoid pseudoreplication [24], as they are secondary (or derived) measures of species richness and/or abundance, and data on demographics or community structure were omitted because the direction of the expected response was not straightforward to interpret [24,27]. WebPlotDigitizer v2.4 [28] was used to capture data from figures, which is considered a robust technique [29].

We found 861 pairwise observations inside and outside PAs from 86 studies distributed amongst 32 countries and 57 PAs that met our criteria (Figure 1; all studies used in the meta-analysis are in Dataset S1; Figure S1 in File S1). During the search, we also discovered comparisons within PAs only, that typically included a pristine baseline site (as judged by the authors) in the PA and an anthropogenically disturbed area also inside the PA (disturbances such as logging, clearing or hunting pressure). To determine if PA designation may offset negative anthropogenic influences, such as land transformation, we compared this portion of the dataset to the comparisons made inside and outside PAs. If the effect sizes of comparisons between areas within PAs are lower than for comparisons made inside and outside PAs, we can infer that PAs offset negative anthropogenic influences, such as land transformation, to a greater degree than no PA designation. We identified an additional 623 such pairwise comparisons from 41 studies between sites within PAs only.

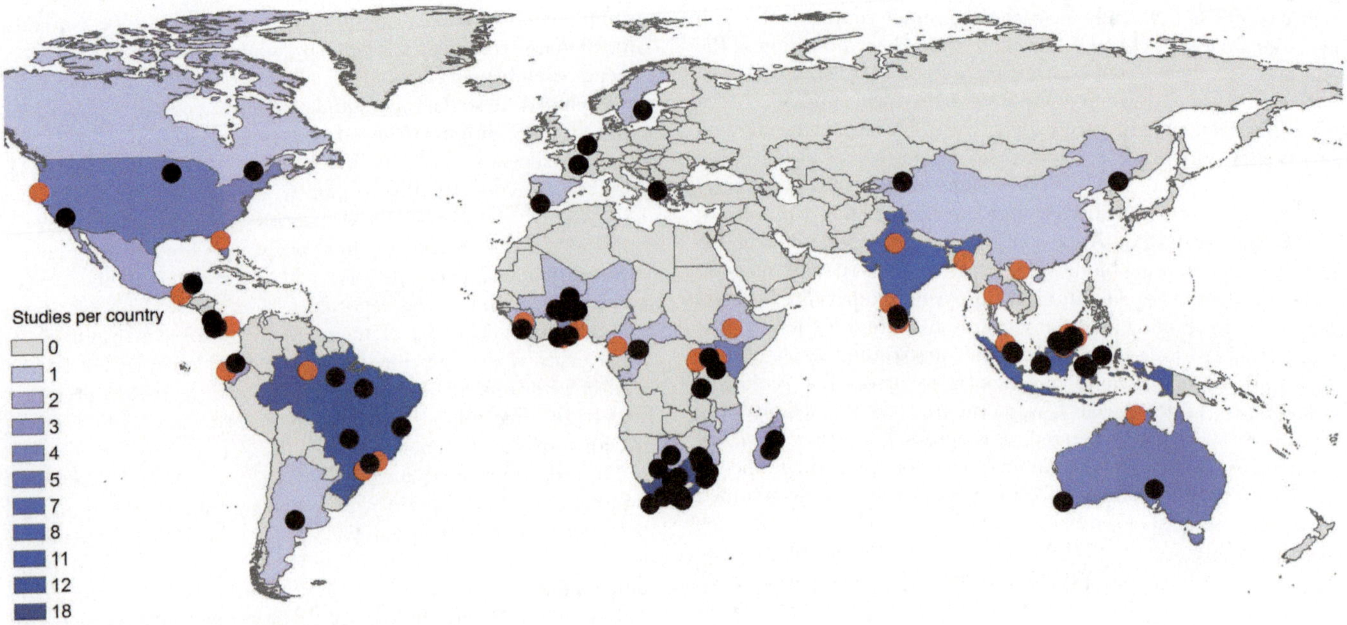

Figure 1. Map of the study sites by the centroid coordinates of protected areas for inside-outside pairwise comparisons (black dots; n = 71) and inside only comparisons (red dots; n = 32). Both categories include data where studies reported across clusters of protected areas.

Data analysis

To estimate effect size, we calculated the Hedges g metric for pairwise comparisons. This is the weighted average of the mean standardized difference (based on pooled variance measures). The metric is the most commonly used, and preferred, to compare pairs of means where variance is available, and unlike others is insensitive to unequal sampling variances in paired groups [27,30]. It is defined as

$$g = \frac{x_{inside} - x_{outside}}{SD_{pooled}}$$

where

$$SD_{pooled} = \sqrt{\left(\frac{(n_{inside} - 1)SD^2_{inside} + (n_{outside} - 1)SD^2_{outside}}{n_{inside} + n_{outside} - 2}\right)}$$

Since Hedges g is a biased estimator of population effect size [27,30], we used the commonly applied conversion factor J to compute the bias-corrected Hedge's g^* metric [27], or $g^* = gJ$, where

$$J = 1 - \frac{3}{4(n_{inside} + n_{outside} - 2) - 1}$$

We then calculated effect sizes using a random-effects model that weights individual comparisons by the inverse of within-study variance plus between study variance, following [24,27,30], with a maximum likelihood variance estimator. Random effects models encompass the variance both between-studies and within-studies, and as such, they are the most appropriate for the majority of ecological meta-analyses because they account for variation in study-specific effects [27]. All analyses were conducted in the "metafor" package [31] in R [32].

We set the direction of the sign of the effect size as positive when the biodiversity value for PAs was greater than outside PAs, implying that PAs contain higher biodiversity values [33]. We selected a single effect size measure that could incorporate variance, which was standardized across response variables. We calculated effect sizes for pairwise comparisons across data inside and outside PAs (N = 861), and also for comparisons inside PAs only (N = 623). We then calculated effect sizes for the three response variables. For inside and outside comparisons only we further calculated effect sizes for (i) five major taxonomic groups (mammals, birds, herptiles [reptiles and amphibians combined], arthropods, plants), (ii) continents (excluding Antarctica with no data), and (iii) International Union for Conservation of Nature's Protected Area Management Categories (a globally recognised PA categorisation system [34]). The categories are primarily based on their management objectives, in which categories 1–4 reflect stricter goals for biodiversity conservation [1 being the strictest], and 5–6 generally allow extractive use via exploitation of biodiversity features [34], see full definitions in Appendix S2). Finally, (v) we calculated effect sizes for the status of species on the IUCN Red List (a global inventory of the threat status of species according to predetermined criteria [35]).

We calculated two commonly used metrics to characterize heterogeneity between pair-wise comparisons as employed in the "metafor" package [31] in R [32]: the Q-statistic and the I^2 value. Total heterogeneity in effect size can be tested with a Q-statistic where a significant value indicates that the estimated effect size is more heterogeneous than expected by chance [27]. The total heterogeneity of the study, Q_{TOTAL} or Q_T can be calculated as

$$Q_T = \sum_{k=1}^{K} W_k(\hat{\theta}_k - \hat{\mu})^2$$

The I^2 value is presented as the total percentage of heterogeneity that can be attributed to between-study variance [27]. The metric quantifies the heterogeneity by comparing the calculated Q_T to its expected value under the assumption of homogeneity [27], as

$$I^2 = Max\left[100 \times \frac{Q_T - (K-1)}{Q_T}, 0\right]$$

We conducted three commonly used tests for detecting bias in meta-analysis in the "metafor" package [31] in R [32]: funnel plot, cumulative meta-analysis and Orwin's fail safe N [27,30,31]. A funnel plot graphs effect size against standard error, and assumes that studies with the largest sample sizes will have lower standard error, and so will be near the average effect size, while studies with smaller sample sizes will be spread on both sides of the average effect size. Variation from this assumption can indicate bias, although the source of such bias may be unclear from a funnel plot [27]. However, positive asymmetry is typically taken to mean publication bias, in that those studies with positive effects are submitted and/or accepted for publication with a greater frequency then those with negative effects [27]. A cumulative meta-analysis sorts all pair-wise comparisons by precision, thus starting with the studies with the largest standard error, after which the comparison with the next largest standard error is added and the effect size is recalculated, and so continues iteratively [27]. The resulting graph enables inspection of the development of the observed effect size with the addition of more precise data. Orwin's fails safe N is a metric of the number of studies averaging null results that would have to be added to the observed outcomes to reduce the average effect size to half the observed effect size. All tests for publication bias in this study showed that it was negligible [27] (Table S1; Figure S2 & S3 in File S1).

To address the potential spatial pseudoreplication in the dataset that could arise from multiple responses reported within studies [22], PAs, countries or species, we recalculated effect sizes after sampling one pairwise comparison only per study, PA, country or species, respectively. This resampling was repeated 10 000 times for each of these four parameters and the estimated mean and 95% confidence intervals compared with the overall effect size, which was conducted in R [32].

We used an information theoretic approach [36] to assess the influence of a candidate set of models and variables to explain the variation in effect size, where data were available for all variables. Models tested the influence of (i) pre-planned subgroups in the meta-analysis (variables: response variable, taxonomic group, PA IUCN Category), (ii) design, location and structural attributes of the PAs (variables: continent, latitude, longitude, PA area in km^2, and PA establishment date, using [34]) and (iii) influence of socio-economic conditions of the countries in which PAs are located (variables: World Governance Index [37], Gross Domestic Product (GDP), Country Population size and the Gini coefficient of income inequality [38]). We followed an exhaustive search approach, which entails fitting all possible model formulations, with a Generalized Linear Model (GLM; [39]). We assumed a Gaussian distribution with a log link function, which was identified as the appropriate family and link function by visual inspection of quantile-quantile plots, using the "glmulti" package [40], in R [32]. We selected the highest ranked model based on the lowest Akaike Information Criterion [39,40] value. Furthermore, to address possible pseudoreplication, one pairwise comparison per study was selected at random and the respective GLM model refitted as above. We selected the highest ranked model based on the lowest Akaike Information Criterion value [39,40], and repeated this procedure 1000 times, to calculate the proportionally highest ranked model for each candidate dataset.

At least some of the variation in effect size may also be accounted for by the scale over which studies were conducted. To test the influence of distance between PA boundaries, and the maximum distance between pair-wise comparisons, a best GLM model by exhaustive fit was performed as above [40]. These independent variables were (i) the maximum distance to protected area boundary within studies, and (ii) the maximum distance between pair wise comparisons within studies, meaning, within each study, the maximum distance between sampling points assigned to all points in that study. Since the data on the distance between comparisons were only available for a reduced subset of pair-wise comparisons (N = 569), we performed a separate GLM as above with only the distance variables. We also estimated Pearson's product moment correlation coefficient between Hedges g* metric values for pair-wise comparisons and the two distance variables in R [32].

Results

The mean effect size using the random effects model, which provides an indication of the general trend across all 861 comparisons, was 0.444 (95% confidence intervals 0.324–0.564; Table 1). Substantial variation was present in the direction and size of effects in response variables for different pairwise comparisons (I^2>87.9%; Q-statistic significant at <0.01; Table 1). However, when fitting the random effects model, PAs had higher species abundances (Figure 2A; N = 330), assemblage abundances (Figure 2A; N = 297) and assemblage species richness (Figure 2A; N = 234) than land use areas outside PAs (see Table 1).

Effect sizes for PAs with no IUCN category designation were lower than those with a designation, but remained positive and overlapped with the overall effect size and so we included them here (Table 1). Studies that reported across clusters of PAs rather than individual PAs remained positive and were thus included in the overall assessment of effect size (Table 1). When resampling effect sizes to account for pseudoreplication, they remained positive and overlapped with the overall effect size for both inside-outside PAs and inside PA only comparisons, but were less positive for species responses (Table S2). The variance of these resampled effect sizes also increased, but we note that effect size precision increases with additional data (Figure S3 in File S1). Thus, the present results overall can be considered robust to pseudoreplication.

Although variable, the mean effect sizes confirm that on average PAs contain significantly higher numbers of species and more individuals for mammals, birds, herptiles and arthropods, but the effect is non-significant for plants (Figure 2B; Table 1). Small mammals showed a smaller effect size for species abundance (< 1 kg; N = 25; 0.042; CI: −0.236–0.320) than did large mammals (>1 kg; N = 114; 0.372; CI: 0.131–0.613). These results suggest that while most species benefit from PA establishment, a suite of them, typically plants, fare better outside PAs in typically anthropogenically transformed habitat.

Protected area efficacy by continent generally showed positive effect sizes, apart from South America (strongly negative and only just non-significant) and Australia (also non-significant; Figure 2C). We note that sample sizes for Europe and Australia are low.

Improved biodiversity outcomes with an increase in IUCN management category would seem an obvious *a priori* outcome, but such a simple relationship was not clear (Figure 2D). IUCN

Table 1. Effect sizes (ES), lower bound (lbCI) and upper bound (ubCI) confidence intervals and sample sizes (N), Tau square (I^2), Q statistic (Q) with its p-value (Qp) for at different subgroup designations.

Description	Subgroup	ES	lbCI	ubCI	N	I^2	Q	Qp
Overall	Inside Outside PAs	0.444	0.324	0.564	861	94.9	8620.1	<0.01
Overall	Comparisons inside PAs only	0.172	0.083	0.261	623	46.8	1439.9	<0.01
Overall	PAs with IUCN designation	0.621	0.488	0.755	496	92.9	3622.3	<0.01
Overall	PAs with no IUCN designation	0.161	−0.050	0.372	365	95.9	4910.5	<0.01
Overall	Clusters of PAs	0.560	0.400	0.719	189	86.1	980.9	<0.01
Overall	Unique PAs identified	0.413	0.264	0.562	672	95.8	7638.9	<0.01
Variable	Species abundance	0.517	0.382	0.652	330	87.9	1869.6	<0.01
Variable	Assemblage abundance	0.349	0.083	0.615	297	97.6	4997.7	<0.01
Variable	Assemblage species richness	0.457	0.238	0.676	234	93.2	1666.9	<0.01
Variable	Species abundance (inside)	−0.086	−0.235	0.063	295	50.2	721.4	<0.01
Variable	Assemblage abundance (inside)	0.165	0.044	0.286	152	9.9	270.8	<0.01
Variable	Assemblage species richness (inside)	0.529	0.364	0.694	176	51.7	418.5	<0.01
Taxon	Mammals	0.179	0.010	0.344	216	88.2	1129.6	<0.01
Taxon	Birds	0.657	0.410	0.910	332	97.2	5415.6	<0.01
Taxon	Herptiles	0.487	0.159	0.816	55	89.6	425.4	<0.01
Taxon	Arthropods	0.654	0.421	0.882	152	64.1	406.1	<0.01
Taxon	Plants	0.043	−0.214	0.301	106	95.4	1167.8	<0.01
Continent	Africa	0.450	0.319	0.581	422	92.9	3320.0	<0.01
Continent	Asia	0.986	0.769	1.203	185	71.2	571.4	<0.01
Continent	South America	−0.441	−0.883	0.002	152	98.2	3950.6	<0.01
Continent	North America	0.587	0.340	0.835	78	84.0	387.2	<0.01
Continent	Australasia	0.056	−0.160	0.273	18	4.2	16.9	<0.01
Continent	Europe	2.544	0.362	4.726	6	97.1	133.3	<0.01
IUCN Cat.	IUCN 1	0.316	0.038	0.594	35	0.0	32.1	0.560
IUCN Cat.	IUCN 2	0.754	0.581	0.926	367	94.8	3204.3	<0.01
IUCN Cat.	IUCN 3	0.646	−0.473	1.764	1	NA	0.0	0.999
IUCN Cat.	IUCN 4	−0.099	−0.222	0.025	50	42.1	104.5	<0.01
IUCN Cat.	IUCN 5	1.371	0.838	1.904	27	83.8	112.6	<0.01
IUCN Cat.	IUCN 6	−0.044	−0.382	0.295	16	0.0	6.1	0.978

Table 1. Cont.

Description	Subgroup	ES	lbCI	ubCI	N	I^2	Q	Qp
IUCN Cat.	No IUCN Designation	0.161	−0.050	0.372	365	95.9	4910.5	<0.01
Red List	Not Evaluated	0.596	0.234	0.957	62	91.0	402.8	<0.01
Red List	Data deficient	−0.479	−1.170	0.212	6	85.2	32.4	<0.01
Red List	Least Concern	0.460	0.291	0.628	168	83.4	744.8	<0.01
Red List	Near Threatened	1.027	0.262	1.791	27	95.2	323.8	<0.01
Red List	Vulnerable	0.814	0.333	1.294	28	89.2	195.4	<0.01
Red List	Endangered	0.557	0.266	0.848	22	60.9	50.1	<0.01
Red List	Critically Endangered	−0.119	−0.548	0.311	6	66.3	14.5	0.013
	Small mammals	0.042	−0.236	0.320	25	3.4	21.0	0.637
	Large mammals	0.372	0.131	0.613	114	93.1	796.6	<0.01

PA = Protected area. Cat = Category.

Category 2 PAs, and to a lesser extent Category 1 PAs, had a high positive effect size. However, although the sample sizes are low, so did IUCN Category 5 PAs that allow much extractive use within their borders. PAs with no IUCN designation, and those of categories 4 and 6 had no significant effect and few data are available on Category 3 PAs.

Species listed as Least Concern, Near Threatened, Vulnerable or Endangered by the IUCN generally had greater abundances inside than outside PAs (Figure 3; Table 1). A small sample size of only six observations from one species (*Gorilla gorilla*) that is Critically Endangered was non-significant. However, at least for species abundance, few data on those species of greatest conservation concern, as measured by their Red List status, are available (Figure 3). Indeed, published studies comparing the abundances of highly threatened species both inside and outside PAs are rare, and many such species do not occur in PAs [18,41].

The overall effect size from pairwise comparisons inside PAs only (0.172; 95% confidence intervals: 0.083–0.261) was lower than that of the inside and outside comparisons only (their 95% confidence intervals did not overlap). When fitting the random effects model, pristine areas in PAs only, did not have significantly higher species abundances (−0.086; CI: −0.235–0.063; n = 295), but did have significantly higher assemblage abundances (0.165; CI: 0.044–0.286; n = 152) and significantly higher assemblage species richness (0.529; CI: 0.364–0.694; n = 176) than anthropogenically disturbed areas.

The variation explained by the fitted explanatory models was low, with the meta-analytical and socio-economic models each accounting for about 5% and 7%, respectively, of the variation in effect size (Table S3). By contrast, the PA-model accounted for 25% of the variation (Table S3). For all candidate explanatory variables there were multiple competing best-fit models (Table S4). Distance among comparison sites explained only *c.* 1% of the variation in effect size for studies included in our meta-analysis (Table S5). Indeed, despite being significant the relationship between effect size and the greatest distance between comparison sites was weak (Pearson's $r = 0.146$; $p<0.001$) as was the relationship with distance to PA boundary ($r = 0.085$; $p<0.05$).

Discussion

An initial assessment of global PA efficacy should be to determine if differences exist in biodiversity between PAs and unprotected land in a direction demonstrating higher biodiversity values in the former. Most of the studies included here did indeed find higher species abundances, assemblage abundance, and species richness inside PAs compared to areas outside them. This pattern holds across taxonomic groups (although non-significant for plants) and continents (although non-significant for South America and Australia).

What mechanisms underlie the non-significant effects for plants and for South America and Australia are not entirely obvious. Clearly, habitat change has major effects on plants, with many studies documenting replacement of particular species and changes in habitat structure [2,24,42,43]. Likewise there has been a growing focus on the scarcity of large old trees [44]. The plant studies analysed here included 53 pairwise comparisons of species richness, 41 of assemblage abundance and 12 of species abundances. All of the 12 species abundance studies were concerned with trees. This balance of investigations to date may account for the lack of an effect, and may rather reflect the globally emerging fingerprint for biodiversity loss under global change, where community turnover, but not necessarily species richness change is observed [43]. In the case of Australia, the small

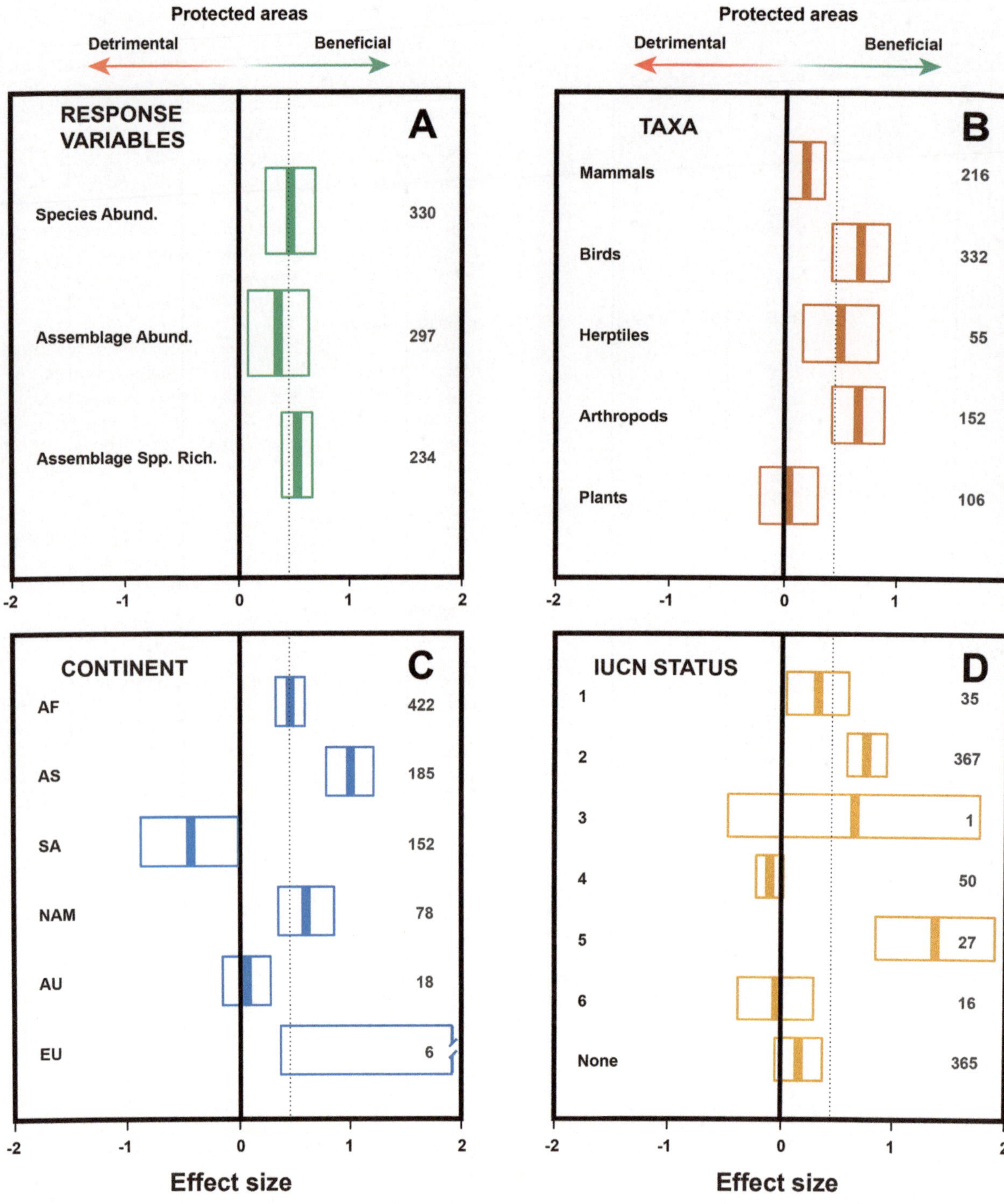

Figure 2. Effect sizes and 95% confidence intervals of response variables (A), taxa (B), continent (C) and the Protected Areas IUCN category (D). Positive boxplot values indicate a net positive impact of protected areas (PAs) on biodiversity. Sample sizes are in grey, the vertical black lines show a zero effect size, while the dashed lines show the overall effect size of 0.444. The effect size for the truncated bar (Europe; Panel C) with large variance due to low sample sizes is 2.54 CI: 0.36–4.73. Abund = Abundance; Spp. Rich. = Species richness; AF = Africa; AS = Asia; SA = South America; NAM = North America; AU = Australasia; EU = Europe. IUCN categories are detailed in Appendix S2.

sample size may be driving the absence of an effect size. Alternatively, most data from Australia (84%; 16/19) came from only one study that documented the recovery of small mammal populations after their isolation from invasive predators [45], and so it is not clear to what extent this outcome reflects general patterns on the continent. For South America the situation is also difficult to explain. The taxon with the most data for South America was the mammals (53%; 81/152), many of them large (> 1 kg; 54%; N = 44), and their increase outside PAs contributes to the non-significant outcome. One explanation may be hunting pressure. Carrilo et al. [46] found that increasing hunting pressure inside a PA diminished large mammal populations in Costa Rica.

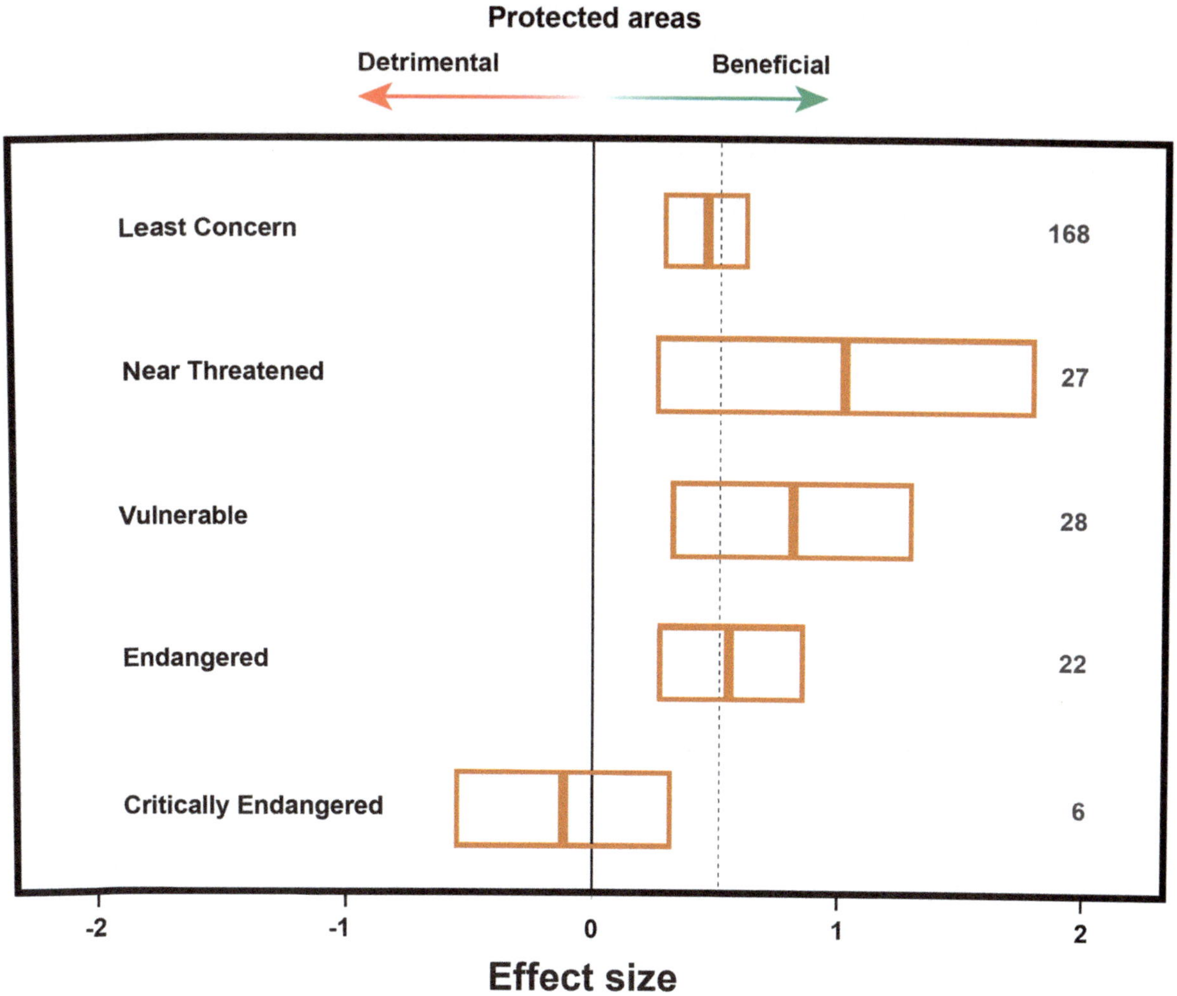

Figure 3. Effect sizes and 95% confidence intervals for species abundance responses by their IUCN Red List status. Positive boxplot values indicate a net positive impact of protected areas on species abundance. Note that multiple responses may be reported for one species (unique species = 168, total cases = 251) and taxonomic uncertainties, Not Evaluated and Data Deficient species are excluded (n = 79). The vertical black line shows a zero effect size while the dashed line indicates the overall effect size for species abundance responses (0.517). Taxa included mammals, birds, herptiles and plants.

In contrast to our general finding, Negroes et al. [47] found that private forest reserves in Brazil were responsible for conserving medium to large-sized vertebrates, more so than PAs. While primary forest is globally irreplaceable for conserving biodiversity [24], land use areas under low extraction, or regenerating forests, seem to contribute to a degree to an integrated landscape level conservation strategy, which may be particularly true in South America [48]. Furthermore, in a comparable meta-analysis, Gibson et al. [24] found that primary forests in South America retain more biodiversity than transformed forests, but since they did not focus on PAs exclusively, they had a larger sample size (N = 909). This finding emphasises that greater sample sizes could increase a positive effect size signal between transformed and more natural areas, a finding that our data corroborate (Figure S3 in File S1).

Despite effect sizes generally being positive, PA efficacy clearly varies considerably amongst PAs, species and local contexts, as demonstrated here and in region-specific studies [5,8,17,21]. Determining the mechanisms driving the pattern of higher species abundances, assemblage abundance, and species richness inside compared to outside PAs remains challenging. Using GLMs we sought to explore a range of proximate factors that might explain this variation. We included explanatory variables such as spatial structure, socio-economic conditions, and structural attributes of PAs, which have been shown elsewhere to have effects on biodiversity values [4,6,49–55]. These assessments provided some insight, but did not account for much of the variation in effect sizes found here. Since comparisons inside and outside PAs in our database were mainly made at the local scale, the geographic context explained little of the observed effect size variation. In consequence, the extent of site matching, which may play a role in increasing estimates of PA efficacy outcomes in some cases [56], may be less important in our database comparing biodiversity features themselves, rather than proxies such as deforestation (but see [56]). Socio-economic factors influence conservation outcomes in some regions [50–52], but here also fared relatively poorly at explaining variation. Likewise structural variables such as PA size and location explained little of the variation. Some of these variables are important drivers of biodiversity variation generally, such as area and latitudinal position (mostly via energy availability) [57,58,59]. However, they have much less of an influence *per se* on differences in diversity inside and outside PAs. Such an outcome suggests that PA efficacy itself is invariate, at least with regard to these variables. In other words the largest effect is of the PAs themselves, as we found here. Our meta-analysis of studies which compare pristine and transformed areas within PAs themselves bears out this suggestion. The pristine areas typically have higher biodiversity values than the transformed areas. Moreover, the effect size here is weaker than the effect size found when comparisons are between sites inside and outside PAs. Together with evidence that pristine forests retain biodiversity features to a greater degree than transformed areas [24], these outcomes suggest that PAs must offset negative anthropogenic influences to a greater degree than no PA designation.

Nonetheless, additional variation may be attributable to other factors such as characteristics of the organisms themselves. For mammals at least, we were able to show that larger species in particular are reduced outside PAs, in keeping with other evidence that smaller mammals are typically better able to tolerate conditions outside PAs [8,21,53,60]. Others have also highlighted the contrasting responses in biodiversity documented by assessments among PAs, but comparing different taxonomic groups [8,21]. Interspecific variability in population responses to landscape change is well-known (e.g. [61]), and some taxa obviously fare better in transformed landscapes, even within PAs. Whether this is the case for species that have a high threat status is more difficult to discern given that information for such species is so scarce [41]. Unfortunately, the scope of our data did not enable us to pursue in more detail possible factors underlying effect size variation among continents and other taxa, largely because of the risk of misleading subgroup effects [25,27]. A range of factors could explain the variation we found given substantial life history differences among these taxa (e.g. dispersal distances, life span, migration propensity, tolerance of disturbance [62–64]), and considerable differences in the evolutionary history of species on continents [59,65].

Similarly, the relationship between the designated management status of a PA and effective conservation of its biodiversity features also seems to be unclear. This outcome provides further evidence for calls that the IUCN PA management categories should be reassessed to reflect biodiversity outcomes rather than management objectives (see [66]). Thus, challenges for conservation science include determining which mechanisms drive positive PA efficacy, under which conditions PAs fail species with differential responses to threats, and concomitantly, IUCN threat designations, and how management categories and biodiversity outcomes can best be aligned.

In addition to these proximate mechanisms that might influence effect size, several ultimate mechanisms may also be important. These include (i) the observation that PAs are non-randomly placed, typically biased towards areas of inaccessibility, which in itself would reduce threatening anthropogenic processes [54]; (ii) the persistence of existing differences in abundance and richness between the areas at the time of PA designations, as a result of the choice of location [18]; (iii) that lower levels of threatening processes, such as habitat alteration or exploitation, have prevailed inside PAs than elsewhere; (iv) that active management of PAs has maintained or increased abundance and richness relative to outside PAs [18]; and/or (v) leakage effects, where threatening processes are displaced from PAs to surrounding areas [55].

Given that data on biodiversity condition before and after establishment of PAs is so rare, studies have understandably had to focus on some comparison of PAs to other areas that may embody likely outcomes if the area had not received PA designation. However, the effect size as calculated here can inherently not explicitly consider the counterfactual (the biodiversity outcome that would have occurred if there had been no PA designation, see [18,56,67]). Site matching approaches attempt to address this bias arising when observable biophysical and socioeconomic factors may affect biodiversity in addition to PA designation [56,67]. While these studies may detect a weaker signal for the influence of PA designation itself, they still typically focus on proxies of biodiversity to determine performance outcomes, such as deforestation offsets. Although the irreplaceability of primary forests for biodiversity has been established [24], measures of deforestation cannot characterize changes in species richness and abundance itself. As a consequence, the vast majority of published studies on biodiversity have taken a broader view of the counterfactual, and made what for the investigators seemed *a priori* sensible comparisons between PAs and other areas that they held to embody likely outcomes if the former had not received designation. However, due to the way the original studies were designed, the dataset developed here cannot be analysed with site matching approaches. Thus some concern might remain that the observed effects have been influenced by any one or many of the mechanisms above. Nonetheless, the ultimate mechanisms that drive patterns of higher biodiversity retention within PAs are clearly far from universal, and their geographic distribution and intensity is poorly known [18,67]. Therefore, a comprehensive assessment of PA effectiveness can benefit from assessments that consider the net outcomes of both observed effects (as is the case here) in addition to those approaches that can better quantify assessments of effects of bias (i.e. site matching approaches [56] or experimental designs [20]).

In conclusion, despite the ultimate mechanisms underlying our findings not being firmly established, and much variation in effect sizes across regions and among taxa remaining unexplained, a signal is clear: PAs have positive biodiversity values compared with alternative land uses. In consequence, our results, together with emerging qualitative evidence [9], suggest that in general PA establishment itself may confer a net benefit to biodiversity. Thus, our approach provides a quantitative demonstration of the value of PAs as an effective strategy for conserving biodiversity. In other words, studies undertaken at a local scale to date clearly indicate that an ecological foundation exists from which the economic, political and social benefits of PAs are being realized [68]. This outcome provides evidence in support of the value of the Aichi Biodiversity Target 11 of the Convention on Biological Diversity's (CBD) new Strategic Plan for Biodiversity [69], the nationally agreed goals to fulfil signatory countries commitments under the CBD [69]. Target 11 calls for the expansion and effective management of PAs. The outcomes of our analyses show why this Target is worth achieving.

Supporting Information

File S1 Contains the following files: **Table S1.** Orwin's fail safe N is 1238 to reach an overall effect size of 0.222. **Table S2.** Effect sizes determined by resampling one pairwise comparison per unit of study, per species, per country and per protected area (PA), to assess the potential spatial pseudoreplication in our dataset arising from multiple responses. **Table S3.** Best GLM models by exhaustive fit for the Meta Analysis model, Protected Areas (PA) model and Socio-Economic model. **Table S4.** Proportion of five highest ranked models for Meta Analysis model, PA model and the Socio-Economic model. **Table S5.** Best GLM model by exhaustive fit for two variables, the maximum distance to protected area boundary within studies, and the maximum distance between pair wise comparisons within studies. **Figure S1.** PRISMA flow diagram, depicting the flow of information through different phases of the search process conducted. **Figure S2.** Funnel plot of effect size standard error plotted against effect size for all inside-outside pairwise comparisons. **Figure S3.** Cumulative meta-analysis of the dataset sorted by precision, with effect sizes and 95% confidence intervals (n = 861). **Appendix S1.** PRISMA (Preferred Reporting Items for Systematic Reviews and Meta-Analyses) checklist. **Appendix S2.** Detailed descriptions of IUCN protected area management categories.

Dataset S1 Complete dataset and references included in the meta-analysis.

Acknowledgments

Comments by Melodie A. McGeoch, Richard A. Fuller and anonymous reviewers improved the manuscript. This paper is dedicated to Cor J.S. Coetzee.

Author Contributions

Conceived and designed the experiments: BWTC SLC. Performed the experiments: BWTC. Analyzed the data: BWTC. Contributed reagents/materials/analysis tools: KJG. Contributed to the writing of the manuscript: BWTC KJG SLC.

References

1. Jenkins CN, Joppa L (2009) Expansion of the global terrestrial protected area system. Biol Conserv 142: 2166–2174.
2. Ellis EC, Ramankutty N (2008) Putting people in the map: anthropogenic biomes of the world. Front Ecol Environ 6: 439–447.
3. Bertzky B, Corrigan C, Kemsey J, Kenney S, Ravilious C, et al. (2012) Protected Planet Report 2012: Tracking progress towards global targets for protected areas. UNEP-WCMC, Cambridge. Available: http://www.unep-wcmc.org/ppr2012_903.html. Accessed February 2014.
4. Rodrigues ASL, Andelman SJ, Bakarr MI, Boitani L, Brooks TM, et al. (2004) Effectiveness of the global protected area network in representing species diversity. Nature 428: 640–643.
5. Craigie ID, Baillie JEM, Balmford A, Carbone C, Collen B, et al. (2010) Large mammal population declines in Africa's protected areas. Biol Conserv 143: 2221–2228.
6. Fuller RA, McDonald-Madden E, Wilson KA, Carwardine J, Grantham HS, et al. (2010) Replacing underperforming protected areas achieves better conservation outcomes. Nature 466: 365–367.
7. Leverington F, Hockings M, Paveses H, Costa K, Courrau J (2010) Management effectiveness evaluation in protected areas - a global study. The University of Queensland, Australia. Available: http://www.wdpa.org/me/PDF/global_study_2nd_edition.pdf. Accessed February 2014.
8. Laurance WF, Useche DC, Rendeiro J, Kalka M, Bradshaw CJA, et al. (2012) Averting biodiversity collapse in tropical forest protected areas. Nature 489: 290–294.
9. Geldmann J, Barnes M, Coad L, Craigie ID, Hockings M, et al. (2013) Effectiveness of terrestrial protected areas in reducing habitat loss and population declines. Biol Conserv 61: 230–238.
10. Balmford A, Bruner A, Cooper P, Costanza R, Farber S, et al. (2002) Economic reasons for conserving wild nature. Science 297: 950–953.
11. Defries R, Hansen A, Newton AC, Hansen MC (2005) Increasing isolation of protected areas in tropical forests over the past twenty years. Ecol Appl 15: 19–26.
12. Joppa LN, Loarie SR, Pimm SL (2008) On the protection of "protected areas." Proc Natl Acad Sci USA 105: 6673–6678.
13. Nagendra H (2008) Do parks work? Impact of protected areas on land cover clearing. Ambio 37: 330–337.
14. Greve M, Chown SL, van Rensburg BJ, Dallimer M, Gaston KJ (2011) The ecological effectiveness of protected areas: a case study for South African birds. Anim Cons 14: 295–305.
15. Butchart SHM, Walpole M, Collen B, van Strien A, Scharlemann JPW, et al. (2010) Global biodiversity: indicators of recent declines. Science 328: 1164–1168.
16. Chown SL, Lee JE, Hughes KA, Barnes J, Barrett PJ, et al. (2012) Challenges to the future conservation of the Antarctic. Science 337: 158–159.
17. Sutherland WJ, Pullin AS, Dolman PM, Knight TM (2004) The need for evidence-based conservation. Trends Ecol Evol 19: 4–7.
18. Gaston KJ, Jackson SF, Cantú-Salazar G, Cruz-Piñón G (2008) The ecological performance of protected areas. Annu Rev Ecol Syst 39: 93–113.
19. Ferraro PJ, Pattanayak SK (2006) Money for nothing? A call for empirical evaluation of biodiversity conservation investments. PLoS Biol 4: e105.
20. Laurance WF, Camargo JLC, Luizão RCC, Laurance SG, Pimm SL, et al. (2011) The fate of Amazonian forest fragments: a 32-year investigation. Biol Cons 144: 56–67.
21. Gardner TA, Caro T, Fitzherbert EB, Banda T, Lalbhai P (2007) Conservation value of multiple-use areas in East Africa. Cons Biol 21: 1516–2155.
22. Caro TM, Gardner TA, Stoner C, Fitzherbert E, Davenport TRB (2009) Assessing the effectiveness of protected areas: paradoxes call for pluralism in evaluating conservation performance. Divers Distrib 15: 178–182.
23. Magurran AE, McGill BJ (2011) Biological diversity: Frontiers in measurement and assessment. Oxford University Press, Oxford. 345 p.
24. Gibson L, Lee TM, Koh LP, Brook BW, Gardner TA (2011) Primary forests are irreplaceable for sustaining tropical biodiversity. Nature 478: 378–381.
25. Higgins JPT, Green S (2011) Cochrane handbook for systematic reviews of interventions. Version 5.0.2 [updated September 2009]. Available: http://www.cochrane.org/sites/default/files/uploads/Handbook4.2.6Sep2006.pdf. Accessed February 2014.
26. Gaston KJ, Blackburn T, Gregory RD (1999) Does variation in census area confound density comparisons? J Appl Ecol 36: 191–204.
27. Koricheva J, Gurevitch J, Mengersen K (2013) Handbook of meta-analysis in ecology and evolution. Princeton University Press, New Jersey, 498 p.
28. Rohatgi A (2014) WebPlotDigitizer v2.4. Available: http://arohatgi.info/WebPlotDigitizer. Accessed February 2014.
29. Schmid CH, Stewart GB, Rothstein HR, Lajeunesse MJ, Gurevitch J (2013) Software for statistical meta-analyis. In: Eds Koricheva J, Gurevitch J, Mengersen K (2013) Handbook of meta-analysis in ecology and evolution. Princeton University Press, New Jersey, 498 p.
30. Borenstein M, Hedges LV, Higgins JPT, Rothstein HR (2009) Introduction to meta-analysis. Wiley-Blackwell, UK. 450 p.
31. Viechtbauer W (2012) Conducting meta-analyses in R with the metafor package. J Stat Soft 3: 1–48.
32. R Foundation (2014) R: A language and environment for statistical computing. Available: http://www.R-project.org. Accessed February 2014.
33. Gaston KJ, Charman K, Jackson SF, Armsworth PR, Bonn A, et al. (2006) The ecological effectiveness of protected areas: the United Kingdom. Biol Conserv 132: 76–87.
34. IUCN UNEP (2014) The world database on protected areas. Available: http://www.protectedplanet.net. Accessed February 2014.
35. IUCN (2014) The IUCN Red List of threatened species. Available: www.iucnredlist.org. Accessed February 2014.
36. Burnham K, Anderson DR (2002) Model selection and multimodel inference: a practical information-theoretic approach. Springer-Verlag, New York. 520 p.
37. World Governance Index (2014) Available: www.govindicators.org. Accessed February 2014.
38. CIA (2012). The World Factbook Available: https://www.cia.gov/library/publications/the-world-factbook/index.html. Accessed February 2014.
39. Bolker BM (2008) Ecological models and data in R. Princeton University Press, New Jersey. 396 p.
40. Calcagno V, Mazancourt C (2010) glmulti: An R package for easy automated model selection with (Generalized) Linear Models. J Stat Soft 34: 1–29.
41. Ricketts TH, Dinerstein E, Boucher T, Brooks TM, Butchart SHM, et al. (2005). Pinpointing and preventing imminent extinctions. Proc Natl Acad Sci USA 102: 18497–18501.
42. Foley JA, Defries R, Asner GP, Barford C, Bonan G, et al. (2005) Global consequences of land use. Science 309: 570–574.
43. Dornelas M, Gotelli NJ, McGill B, Shimadzu H, Moyes F, et al. (2014) Assemblage time series reveal biodiversity change but not systematic loss. Science 344: 296–299.
44. Lindenmayer D, Laurance W, Franklin J (2012) Global decline in large old trees. Science 338: 1305–1306.
45. Moseby KE, Hill BM, Read JL (2009) Arid recovery - a comparison of reptile and small mammal populations inside and outside a large rabbit, cat and fox-proof exclosure in arid South Australia. Austral Ecol 34: 156–169.
46. Carrillo E, Wong G, Cuarón AD (2000) Monitoring mammal populations in Costa Rican protected areas under different hunting restrictions. Conserv Biol 14: 1580–1591.
47. Negrões N, Revilla E, Fonseca C, Soares AMVM, Jácomo ATA, et al. (2011) Private forest reserves can aid in preserving the community of medium and large-sized vertebrates in the Amazon arc of deforestation. Biodivers Conserv 20: 505–518.
48. Barlow J, Gardner TA, Araujo IS, Ávila-Pires TC, Bonaldo AB, et al. (2007) Quantifying the biodiversity value of tropical primary, secondary, and plantation forests. Proc Natl Acad Sci USA 104: 18555–18560.
49. Thomson JR, Weiblan G, Thomson BA, Alfaro S, Legendre P (1996) Untangling multiple factors in spatial distributions: lilies, gophers and rocks. Ecology 77: 1698–1715.
50. Reyers B, van Jaarsveld AS, McGeoch MA, James AN (1998) National biodiversity risk assessment: a composite multivariate and index approach. Biodivers Conserv 7: 945–965.

51. Veech JA (2003) Incorporating socioeconomic factors into the analysis of biodiversity hotspots. Appl Geog 23: 73–88.
52. Balmford A, Moore JL, Brooks TM, Burgess N, Hansen LA, et al. (2001) Conservation conflicts across Africa. Science 291: 2616–2619.
53. Woodroffe R, Ginsberg JA (1998) Edge effects and the extinction of populations inside protected areas. Science 280: 2126–2128.
54. Joppa LN, Pfaff A (2009) High and far: biases in the location of protected areas. PLoS ONE 4: e8273.
55. Ewers RM, Rodrigues ASL (2008) Estimates of reserve effectiveness are confounded by leakage. Trends Ecol Evol 23: 113–116.
56. Andam KS, Ferraro PJ, Pfaff A, Sanchez-Azofeifa GA, Robalino JA (2008) Measuring the effectiveness of protected area networks in reducing deforestation. Proc Natl Acad Sci USA 105: 16089–16094.
57. Rosenzweig ML (1995) Species diversity in space and time. Cambridge University Press, Cambridge.
58. Gaston KJ (2000) Global patterns in biodiversity. Nature 405: 220–227.
59. Jetz W, Fine PVA (2012) Global gradients in vertebrate diversity predicted by historical area-productivity dynamics and contemporary environment. PLoS Biol 10: e1001292.
60. Peres CA (2004) Effects of subsistence hunting on vertebrate community structure in Amazonian forests. Conserv Biol 14: 240–253.
61. Henle K, Davies KF, Kleyer M, Margules CR, Settele J (2004) Predictors of species sensitivity to fragmentation. Biodivers Conserv 13: 207–251.
62. Nicolakakis N, Sol D, Lefebvre L (2003) Behavioural flexibility in birds predicts species richness, but not extinction risk. Anim Behav 65: 445–452.
63. Hays GC, Scott R (2013) Global patterns for upper ceilings on migration distance in sea turtles and comparisons with fish, birds and mammals. Func Ecol 27: 748–756.
64. Di Marco M, Buchanan GM, Szantoi Z, Holmgren M, Grottolo Marasini G, et al. (2014) Drivers of extinction risk in African mammals: the interplay of distribution state, human pressure, conservation response and species biology. Proc R Soc B 369: 20130198.
65. Hawkins BA, McCain CM, Davies TJ, Buckley LB, Anacker BL, et al. (2012) Different evolutionary histories underlie congruent species richness gradients of birds and mammals. J Biogeogr 39: 825–841.
66. Boitani L, Cowling RM, Dublin HT, Mace GM, Parrish J (2008) Change the IUCN protected area categories to reflect biodiversity outcomes. PLoS Biol 6: 436–438.
67. Joppa LN, Pfaff A (2010) Global protected area impacts. Proc R Soc B 278: 1633–1638.
68. Stolton S, Dudley N (2010) Arguments for protected areas: multiple benefits for conservation and use. Routledge, London. 296 p.
69. CBD (2012) Convention on Biological Diversity: Aichi Biodiversity Targets. Available: http://www.cbd.int/sp/targets/. Accessed February 2014.

A Mixed Methods Approach to Exploring the Relationship between Norway Rat (*Rattus norvegicus*) Abundance and Features of the Urban Environment in an Inner-City Neighborhood of Vancouver, Canada

Chelsea G. Himsworth[1,2*], Kirbee L. Parsons[1], Alice Y. T. Feng[1], Thomas Kerr[3,4], Claire M. Jardine[5], David M. Patrick[1]

1 School of Population and Public Health, University of British Columbia, Vancouver, British Columbia, Canada, **2** Animal Health Centre, British Columbia Ministry of Agriculture, Abbotsford, British Columbia, Canada, **3** British Columbia Centre for Excellence in HIV/AIDS, St. Paul's Hospital, Vancouver, British Columbia, Canada, **4** Department of Medicine, University of British Columbia, Vancouver, British Columbia, Canada, **5** Department of Pathobiology, University of Guelph, Guelph, Ontario, Canada

Abstract

Urban rats (*Rattus* spp.) are among the most ubiquitous pest species in the world. Previous research has shown that rat abundance is largely determined by features of the environment; however, the specific urban environmental factors that influence rat population density within cities have yet to be clearly identified. Additionally, there are no well described tools or methodologies for conducting an in-depth evaluation of the relationship between urban rat abundance and the environment. In this study, we developed a systematic environmental observation tool using methods borrowed from the field of systematic social observation. This tool, which employed a combination of quantitative and qualitative methodologies, was then used to identify environmental factors associated with the relative abundance of Norway rats (*Rattus norvegicus*) in an inner-city neighborhood of Vancouver, Canada. Using a multivariate zero-inflated negative binomial model, we found that a variety of factors, including specific land use, building condition, and amount of refuse, were related to rat presence and abundance. Qualitative data largely supported and further clarified observed statistical relationships, but also identified conflicting and unique situations not easily captured through quantitative methods. Overall, the tool helped us to better understand the relationship between features of the urban environment and relative rat abundance within our study area and may useful for studying environmental determinants of zoonotic disease prevalence/distribution among urban rat populations in the future.

Editor: Georges Chapouthier, Université Pierre et Marie Curie, France

Funding: This work was supported by the Canadian Institutes of Health Research [MOP – 119530 and CGV – 104833]. The funders had no role in study design, data collection and analysis, decision to publish, or preparation of the manuscript.

Competing Interests: The authors have declared that no competing interests exist.

* E-mail: Chelsea.Himsworth@gov.bc.ca

Introduction

Norway rats (*Rattus norvegicus*) are among the most widespread rodent species, inhabiting every continent except for Antarctica [1]. The ubiquity of this species is largely attributable to their remarkable ability to quickly and successfully adapt to new environments and resources [1]. Norway rats are best suited to close cohabitation with people, and are considered a true commensal species as few populations are found in a truly sylvatic state [1].

Their propensity to exploit human settlements has caused Norway rats to become a particularly ubiquitous and problematic pest in cities around the world. Within urban settings, Norway rats can damage property, contaminate food, and act as a source of infectious disease for people [1,2]. For these reasons there is considerable interest in identifying and understanding the factors that promote or deter rat infestations in urban centers.

Previous studies have indicated that the environment is probably the single most important determinant of rat density and distribution across the urban landscape [1,3]. More specifically, it is clear that rats require suitable harborage and food sources in order to establish an infestation [1,3]. However, the adaptability of Norway rats means that it is often difficult to identify what specific features of the urban environment combine to create optimal rat habitat.

Researchers have employed a variety of methods in an effort to identify environmental features that promote or deter rat infestations. For example, a study in Buenos Aires, Argentina, compared the relative abundance of Norway rats in several locations demonstrating varying degrees of human modification [4]. They found that Norway rats were more common in shantytowns and parklands compared to industrial-residential neighborhoods and natural preserves [4]. Another study in the same city found that Norway rats were more abundant in shantytowns compared to parklands [5]. In Salzburg, Austria, Norway rats were trapped in a variety of urban locations with distinct environmental features [6]. This study found that Norway rats were most common in habitats with water, vegetation, natural soil, and organic waste, while rats were rarely found in heavily built-up areas and areas frequently utilized by people [6].

The main drawback of the aforementioned approaches is that they are capable of capturing only 'high-level' environmental impacts on rat populations (i.e., differences in rat abundance between different ecosystems). Rat population density and distribution, however, may vary considerably even over very short geographic distances within a single urban ecosystem [7,8], suggesting that minute differences in the micro-environment of a city block could determine its capacity to support rat populations. These micro-environmental data, however, are relatively difficult to collect in a systematic fashion, and is seldom captured by cursory observation or pre-established databases.

A similar problem has been encountered in the social and health sciences, where researchers have attempted to identify links between an individual's health, welfare, behaviour, etc., and the features of the neighborhood in which they reside. In order to overcome this problem, researchers have developed the science of systematic social observation (SSO) [9,10]. The tools employed by SSO require the researcher to closely observe and quantify multiple different aspects of a city block, resulting in the creation of a comprehensive dataset that captures the complexity of the block environment [9,10]. These tools have revolutionized our understanding of impact of the urban environment on the people who reside within it.

Given that the urban environment also influences rat populations, it seems possible that a similar tool could be developed and used to predict rat abundance. There are several studies that have used a similar approach to study rats. However the tools and methodologies employed by these studies had several limitations including the following: capturing a limited number of environmental features [11], being specific to rural [12] or developing areas [13,14], and being difficult to reproduce [13,14].

The goals of this study were: 1) to develop a systematic environmental observational tool that captures and quantifies features of a developed urban microenvironment that could influence rat populations; 2) to develop an approach to using that tool that includes analysis of both quantitative and qualitative data; and 3) to use that tool/approach to identify environmental factors associated with relative rat abundance within an inner-city neighborhood of Vancouver, Canada.

Methods

This study was a part of a larger project (the Vancouver Rat Project, www.vancouverratproject.com) the goal of which is to characterize the health risks that rats pose to people through the transmission of infectious diseases. More specifically, the project aims to understand the ecology of zoonoses in rats and people by understanding the dynamic interplay among rats, people, pathogens, and the urban environment.

Ethics Statement

This study was approved by the University of British Columbia's Animal Care Committee (A11-0087) and adhered to national guidelines set out by the Canadian Council on Animal Care (www.ccac.ca), including those pertaining to animal user training, euthanasia, protocol review, and wildlife (http://www.ccac.ca/en_/standards/guidelines). This study did not involve any endangered or protected species.

Study Area

The study area was composed of 43 contiguous city blocks (0.82 km^2) which included the core of Vancouver's Downtown Eastside (49°15″0′N/123°8″0′W). The Downtown Eastside (DTES) is an impoverished inner-city neighborhood characterized by a large number of single-room occupancy hotels, high rates of homelessness, and a large open air illicit drug scene [15]. It was selected as a priority for study within the city because residents and community groups identified rat infestations as a significant but neglected issue in the area. Also included in the study were several blocks from the more affluent, adjacent Gastown district. All trapping took place on public property only and no specific permissions were required.

Development of an Environmental Observation Tool

The environmental observation tool (Appendix S1) was modeled after the SSO tool developed by Parsons et al. (2010). We utilized scales, items, and constructs from SSO tools [10,16–18] that measured factors with the potential to influence rat populations, as well as from prior studies of rat abundance [11,13,14]. We also developed new items based on a review of the literature regarding rat ecology [1] and our field observations. The final tool included 58 items covering 6 major domains: (1) land use characteristics, (2) property condition, (3) green space characteristics, (4) alley surface characteristics, (5) presence of waste, and (6) alley use (i.e., for loitering and transportation). Additionally, there was a free form section at the end of the tool where observers could make notes about the block under evaluation. For this free form section, particular attention was paid to identifying and describing features that were perceived to be associated with rat presence/abundance, as well as unusual or unique features of the block under study.

In order to facilitate the use of the tool in this study, and in the future, each domain and item was accompanied by detailed instructions regarding its purpose and/or use. Additionally, a supplementary document was produced with photographic examples of different items and ratings (see Appendix S2).

Tool Implementation

Two observers were trained in the use of the tool, and these raters completed their observations together between 8:00 h and 16:00 h, with final ratings for each item being arrived at via consensus. To complete their observations, observers evaluated all aspects of the block including block faces (i.e., the area of the block facing a street), alleyways, and aerial photographs.

Determining Rat Abundance

This study was part of a larger project characterizing the ecology of zoonotic pathogens in rat populations for which whole rat carcasses were required for testing. For this reason a trapping-removal method (vs. a non-lethal method, such as mark-recapture) was required. It should be noted, however, that systematic trapping-removal is considered to be among the most accurate methods for enumerating rats, and has been used frequently in the past to determine rat abundance in urban centers [7,19–21].

Each city block randomly assigned to a 3-week trapping period over the course of 1 year in order to capture seasonal variations in rat abundance within the study as a whole. Of the 43 blocks included in the study, 9 were trapped in the fall, 10 in the winter, 14 in the spring and 12 in the summer.

Within each block, approximately 20 Tomahawk Rigid Traps for rats (Tomahawk Live Trap, Hazlelhurst, USA) per block were set out along each side of the back alley that bisected the block.

Traps were evenly spaced where possible, but had to be placed on outdoor public property in a location where they did not obstruct traffic and could be secured to prevent theft. Traps were pre-baited (filled with bait but fixed open) for one week to acclimatize rats to trapping equipment and bait, followed by two weeks of active trapping. Baits used included peanut butter, bacon fat, flour, and oats. Traps were set at 16:00 h and checked the following morning at 8:00 h.

Trapped rats were anesthetized with isoflurane in a rodent Inhalation Narcosis Chamber (Harvard Apparatus, Holliston, USA) prior to pentobarbital euthanasia via intracardiac injection. Rats were identified to species by external morphology [22].

Dependent Variable

Of the 673 rats trapped over the course of the study, 665 (98.8%) were Norway rats (*Rattus norvegicus*) and 8 (1.2%) were black rats (*Rattus rattus*). Given the relatively small number of black rats trapped, and the fact that the ecology of black rats has been shown to be significantly different than Norway rats [1], black rats were excluded from the analysis.

The unit of analysis was the city block (n = 43). The dependent variable was trap success rate in each block, with trap success rate being equal to the number of rats trapped in a block over the entire trapping period divided by the total trap effort over that period. The total trap effort (# traps set x # nights) was adjusted according to the method described by Nelson and Clark [23]. This method involves subtracting half a unit from the total trap effort for each trap spring for any cause (e.g., trapping of a rat, trapping of a non-target species, tripping of a trap). This adjustment was considered particularly important because of the marked variation in rat abundance among blocks, and because of the frequency with which traps were sprung by non-target species and by members of the public, both of which can significantly bias trap success indices if they are not taken into account [23].

Given that each block under study had roughly the same geometric surface area (1.2 ha) and was trapped in the same manner, and given that rat populations are thought to be relatively isolated to the level of a city block with minimal movement of animals among blocks [1], it is appropriate to use the city block as the unit of analysis and the number of rats trapped offset by trap effort as a proxy for relative population density/rat abundance.

To confirm the absence of geographic autocorrelation of rat abundance among the blocks, a georeferenced map of block outlines within the study area was imported into ArcGIS 10.0 (ESRI, Redlands, USA). Global Moran's I was used to test the null hypothesis that trap success (# rat trapped/trap effort) within a block was randomly distributed (i.e., no spatial autocorrelation) using inverse distance for conceptualization of spatial relationships and a Euclidean distance method.

Predictor Variables

A total of 59 potential predictor variables were included for initial consideration. Each of the 58 items on the environmental observational tool was used to create 1 predictor variable (Table S1 – note that variables were numbered according to the item numbering in the observation tool). The season in which the block was trapped (fall = September – November, winter = December – February, spring = March – May, summer = June – August) was also included as a predictor.

Seven variables (variable nos. 1, 2, 5, 13, 25, 31, and 41) were eliminated initially because they captured the same and/or less information as subsequent variables. For example, variable 13 is a nominal-categorical variable that characterizes the predominant housing type in a block, with 8 categories that each correspond to a class of housing. However, variable nos. 14–19 are ordinal variables that capture both the presence of each class of housing in a block, as well as the proportion of each block occupied by each class of housing.

Next, variables representing different aspects of the same concept were consolidated into a single new composite variable representing the weighted average of the underlying component variables. For example, variable nos. 26–30 all represent building condition (the underlying concept of interest), and pertain to the proportion of each block occupied by buildings in 'extremely poor', 'poor', 'fair', 'good', and 'excellent' condition. These variables were combined into a single composite variable, as were variable nos. 32–36 and 42–44.

Predictor variables where > 95% of observations had the same score (variable nos. 8, 12, 18, and 54) were deemed non-discriminatory and removed. Variable no. 46 (number of rat holes) was also removed because it was deemed to be a sign of rat abundance rather than a cause.

A total of 37 predictor variables were considered for statistical modeling.

Model Building

The goal of the model-building protocol was to identify the most parsimonious set of predictor variables that best explained the outcome, and all predictor variables were given equal consideration going into the model building procedure (i.e., we had no a priori hypotheses regarding the relative importance of specific predictors). Similarly, we had no a priori hypothesis regarding interactions among predictors, and, given the large number of potential predictors (n = 37) relative to the number of observations (n = 43), no interaction terms were included for consideration.

A zero-inflated negative binomial (ZINB) model was used to model the relationship between the predictors and outcome in order to account for excess zeros (no rats were trapped in 11 of 43 blocks) and overdispersion in the outcome variable. This ZINB model combined a binary model with a logit link to model the odds of not trapping any rats (zero-inflation model), and a negative binomial model with a log link to model the number of rats trapped offset by the natural logarithm of the trap effort (count model). The unit of analysis was the city block (n = 43).

The relationship between each predictor and the outcome was first modeled in a bivariate manner, including the predictor in the binary and count portions of the model independently (Table S2). The majority of the predictor variables captured the proportion of the block occupied by an environmental factor of interest, and were therefore modeled in a linear manner. For several ordinal variables (variable nos. 48, 49, 57, and 58) where the spacing between categories was more subjective (e.g., 'none' vs. 'some' vs. 'a little' vs. 'a lot'), bivariate models treating the variable in a linear and categorical manner were generated and compared using the Bayesian Information Criterion (BIC) [24]. For all 4 predictors, the linear form of the variable was superior to the categorical form in both the binary and count portions of the model, therefore these variables were treated as numeric variables. Season was treated as a categorical variable.

In order to narrow down the pool of potential predictors for inclusion in a multivariate model, a program was created in R (R Development Core Team, Vienna, Austria) that performed automated combined forward and backward selection on each part of the model independently using Akaike's Information Criterion (AIC) to select the best potential group of predictors for further consideration [24]. Briefly, in order to select the binary model, the negative binomial model was set at 1 (offset by the natural logarithm of the trap effort). The null binary model (a binary model with no predictors) was used as the starting point, and predictors were added sequentially based on their associated AIC (i.e., the predictor that produced the lowest AIC for the binary model was added first). After each addition, the model was checked to determine if removal of one or more predictors included in the model affected the AIC. If removal of the predictor(s) decreased the AIC for the binary model, then the predictor(s) was eliminated. If removal increased the AIC, then the predictor(s) was retained. This process was repeated until addition or elimination of any subsequent predictors could not further decrease the binary model AIC. The selection procedure described above was then repeated for the count model holding the binary model at 1. All 37 predictor variables, including season, were allowed to compete equally for both components of the model, with the stepwise selection procedure ensuring that only those variables that improved model fit (according to the AIC) were retained. This process should control for confounding among predictors because it will select the variables with the strongest relationship to the outcome in the presence of all other predictor variables.

Variables identified through automated selection for the binary and count components of the ZINB model were then combined in a single model and manual forward and backward selection was performed using the BIC to maximize model fit and parsimony (as the BIC penalizes model complexity more heavily than the AIC) [24]. The final model was that with the lowest BIC.

To investigate collinearity among explanatory variables in the final model, Spearman's rank correlation (ρ) was used to examine bivariate relationships between each of the included predictors [24]. Any highly correlated variables ($\rho > 0.8$) were modeled separately, with only the variable making the most significant contribution to the model (based on the BIC) being retained [24]. Finally, the variables included in the final model were used to build a negative binomial (NB) model and the Vuong test was used to compare the NB and ZINB models with the null hypothesis that the two models were equivalent and the alternative hypothesis that the ZINB model was superior.

Qualitative Analysis

Thematic and narrative qualitative analysis [25] was used in order to: 1) provide insight into the results of the quantitative analysis, 2) capture information not captured by the quantitative aspects of the tool, 3) provide increased resolution for environmental data by allowing the observer to note special features within the block (vs. the quantitative aspects of the tool, which pertained to the block as a whole), and 5) aid in triangulation for the purposes of identifying the environmental characteristics most significantly associated with rat abundance.

Data included in the qualitative analysis included written notes recorded in the environmental observational tool, as well as field notes recorded during a two-day long debrief at the end of the project. This debriefing involved re-visiting each block included in the study area and discussing observer opinions of the block environment and its relationship to rat abundance.

Handwritten notes were transcribed into Microsoft Excel (Microsoft, USA) and labeled with block number and block trap success. Emergent thematic analyses of early observations were discussed among the two observers and served to inform the focus of subsequent observations, as well as ongoing analyses. The coding framework employed a priori codes derived from the quantitative observational tool, as well as emergent codes based upon ongoing observational work. All qualitative data were reviewed, and text segments related to each individual code were categorized/classified (to identify commonalities among blocks and special features within a block). Subsequent coding passes were used to refine and expand code categories and to identify instances of negative evidence.

Results

Quantitative Findings

There was no significant autocorrelation of rat abundance among blocks ($p = 0.14$) based on Global Moran's I, supporting our assertion that the city blocks could be treated as independent units. Bivariate relationships between the outcome and each of the 37 potential predictors are presented in Table S1. The final ZINB model included one variable in the zero-inflation portion of the model and six variables in the negative binomial portion of the model (Table 1). None of the variables included in the final model were highly correlated ($\rho < 0.45$ for all comparisons) and the Vuong test indicated that the ZINB model was superior to the NB model ($P = 0.03$).

The final model indicated that as the proportion of a block occupied by institutional parcels (i.e., those dedicated to not-for-profit services, such as churches, outreach centres, etc.) increased, the odds of not trapping any rats decreased (Table 1). As the proportion of a block occupied by single-family homes increased, the trap success rate decreased. In contrast, as the proportion of the block occupied by housing over commercial and by abandoned parcels (i.e., those with structures that were previously used but are now vacant and not undergoing construction or demolition) increased, the trap success rate increased. As general building condition increased, trap success rate decreased, whereas trap success rate increased as the as the amount of garbage/trash/junk/litter in the alley increased. Finally, as the use of an alley as a transportation corridor increased, trap success rate decreased.

Qualitative Findings

Relationship between rat abundance and land use. Observers often reported trapping rats around abandoned buildings, demolished buildings, and vacant lots. In one block, a large parcel filled with debris from a demolished building was observed to be heavily infested with rats.

Blocks with low trap success tended to be described as being primarily residential (particularly single family homes) or primarily industrial (including industrial sites producing food), while blocks with high trap success tended to be described as being more mixed-use. Observers noted that there was no clear association between the presence of facilities dealing with food (e.g., restaurants, groceries, etc.) and rat presence/abundance.

Relationship between rat abundance and socioeconomic status of a block. Observers reported that rats were much more abundant in the impoverished DTES neighborhood than the adjacent more affluent Gastown neighborhood. Observers also reported that blocks composed primarily of single-family homes appeared to be much more affluent than other residential areas, and that many of the low-rise apartments and housing over commercial was low income housing and single-room occupancy hotels (SROs). Low-income apartments, SROs, and outreach centres for impoverished community members were more commonly reported in blocks with high trap success. Also, within a block, observers often reported trapping rats near SROs.

Relationship between rat abundance and property/building condition. Buildings in disrepair were much more commonly reported in blocks with high trap success vs. blocks with low trap success. Also, within a block, observers often reported trapping rats around buildings in disrepair. Conversely, buildings and properties were much more commonly reported to be in good to excellent condition in blocks with low trap success. However,

Table 1. Relationship between features of the urban environment and relative abundance of Norway rats (*Rattus norvegicus*).

Variable #	Variable description	β	SE	P-Value	exp (β)	95% CI
Count Model (negative binomial with log link)						
10	Proportion of block occupied by **abandoned** parcels	0.70	0.32	0.027	2.02	1.08–3.77
14	Proportion of block occupied by **single family houses**	−0.67	0.25	0.007	0.51	0.31–0.83
19	Proportion of block occupied by **housing over commercial**	0.65	0.21	0.002	1.91	1.27–2.87
c26t30	General **building condition**	−2.01	0.58	0.001	0.13	0.04–0.42
48	Amount of **garbage/trash/junk/litter**	0.83	0.28	0.003	2.29	1.33–3.93
58	Amount of **transport**	−0.99	0.24	< 0.001	0.37	0.23–0.59
Zero-Inflation Model (binomial with logit link)						
5	Proportion of block occupied by institutional parcels	−3.42	1.44	0.012	0.03	0.002–0.55

Zero-inflated negative binomial model for the number of rats trapped in a city block offset by the natural logarithm of the trap effort in that block.

observers reported trapping rats around buildings in good repair and vice versa. Observers noted that all aspects of a block had to be assessed to adequately evaluate repair. In some blocks, the street face appeared to be in good repair while the alley face was in very poor repair.

Relationship between rat abundance and green space. Well maintained green space and food gardens were more often reported in blocks with low trap success. Unkempt green space was reported in blocks with high and low trap success. However, within a block, observers often reported trapping rats near unkempt green space.

Relationship between rat abundance and waste. Observers reported debris, strewn garbage, and overflowing garbage bins much more commonly in blocks with high trap success. Within a block, observers also reported trapping rats in close proximity to areas with accumulated refuse and around dumpsters. Blocks with low trap success were most commonly described as clean with very little garbage. However, observers also noted that some blocks with high trap success were relatively clean. Observers commented that rats were not caught around organic waste bins or compost piles.

Relationship between rat abundance and condition of the alley surface. Blocks with high trap success tended to be described as having pavement in disrepair, cracked pavement, unpaved areas, and rat holes, while blocks with low trap success tended to be described as being well paved and having pavement in good repair. Within an alley, observers reported trapping rats around areas where the pavement was in disrepair and/or deeply cracked, around unpaved areas, and around rat holes, regardless of the condition of the rest of the alley. However, some blocks with high trap success were reported to have good alley surface condition and vice versa.

Sources of harborage for rats. Observers suggested that most rats likely live in buildings or in underground burrows (rat holes were commonly seen in blocks with high trap success). However, observers also reported trapping rats near areas with stored wood pallets, plastic barrels, and other containers. In four blocks, observers reported frequently catching rats near rat corridors (i.e., access points, such as gaps between buildings, which connected the alley face to deeper aspects of the block).

Effect of human activity on rat presence. Heavy use of alleyway by people (particularly people loitering) was reported in blocks with high and low trap success, but was more commonly reported in blocks with high trap success.

Observers noted that the alleyways more heavily used for loitering tended to be dirtier (i.e., have more accumulated refuse).

Discussion

Using quantitative and qualitative analysis, we were able to identify a variety of environmental factors associated with rat abundance. We also found that once observers were comfortable with the use of the tool, its implementation was relatively efficient and straight forward.

As has been found in previous studies using SSO, the combination of quantitative and qualitative analysis was particularly beneficial. Not only did the qualitative data support and elucidate the quantitative results, these data also provided additional insight into environmental variations within a block, and drew our attention to factors not well captured by the quantitative aspect of the tool (e.g., the apparent association between rat abundance and poverty), as well as inconsistencies in the relationship between certain environmental factors and relative rat abundance.

This study was able to identify a number of factors associated with relative rat abundance that have not been studied and/or well characterized previously (e.g., specific land use, residential density, and building disrepair). Several of our findings, however, are in agreement with previous research on the relationship between environmental factors and rat abundance. For example, studies have been able to show that the presence of exposed garbage is a strong predictor/promoter of rat infestations [6,13]. Our study also found that rats were relatively less common in blocks where the land use was predominantly industrial and single-family residential, and this is consistent with the findings of Cavia et al. (2009), who also found that Norway rats were uncommon in industrial and residential neighborhoods.

Other studies have also indicated that factors associated with poverty can promote rat infestations, likely due to a combination of factors including infrastructural disrepair and decreased environmental hygiene, which combine to create ideal urban rat habitat [3,5]. Poverty might also impair an individual's or community's ability to prevent or eliminate infestations (e.g., through neighborhood rehabilitation or employing pest control professionals). In our study, there were several variables that, on first glance, are difficult to intuitively link to rat abundance, including proportion of block occupied by institutional parcels and housing over commercial. However, upon further examination, and in light of the qualitative data, it became clear that much of the housing over commercial was low-income housing, including single room occupancy hotel rooms (SROs). Similarly, institutional parcels in the area are primarily composed of outreach centers for impoverished members of the community. Observers specifically noted that rats were often trapped around SROs and that blocks with high trap success tended to have more SROs, low-income housing, and outreach centers. Observers also noted that far more rats were trapped in the impoverished DTES compared to the adjacent, more affluent Gastown district, and that blocks composed primarily of single-family residences were also more affluent than the rest of the DTES. Factors potentially associated with poverty were associated with trap success independent of variables accounting for infrastructural disrepair and accumulation of waste. This suggests that there are likely one or more environmental variables associated with poverty that were not accounted for in this study.

It is interesting to note that season did not appear to have a significant relationship with rat presence or abundance. Indeed, there has been considerable debate in the literature regarding the impact of season on rats in urban ecosystems [1]. The absence of seasonal variation in urban rat population size might not be surprising given that urban ecosystems, being largely under anthropogenic control, would seem to be less prone to seasonal fluctuation in resource availability (i.e., the presence of suitable food and harborage) compared to more sylvatic settings. However, given that this study only included one year's worth of data, the true impact of season could not be definitively determined [26].

There were a number of findings in our study that contrast with those found in past research on the subject. For example, green space has been reported to be a strong promoter of rat infestations [4,6]. In particular, a study by Traweger et al. (2008) found that rats were most abundant in more natural areas with abundant vegetation, and that trap success was very low in areas with little vegetation, as well as heavily built-up areas, and areas frequented by humans. Our study found that there was no relationship between amount of green space and rat abundance, that rats were very common in built up areas, and that human activity (in the form of loitering) appeared to have a negligible to positive relationship with rat abundance. Increased use of an alley as a transport corridor did appear to have a negative association with the number of rats trapped. However, it is difficult to determine if this because rats avoid areas of increased activity or if alleys used for transport tended to be in better condition or in areas with other features that deter rat infestations. Although Traweger et al. (2008) speculate that natural areas are essential to providing appropriate rat habitat, the results of our study indicate that Norway rats can thrive even in the most urbanized of settings.

Interestingly, the qualitative data also indicated the presence of conflicts within our own study. For example, although strewn garbage seemed to be strongly associated with rat abundance in general, there were some alleys and/or alley segments where rats were caught that appeared relatively clean. Similarly, although many rats were caught in alleys with unpaved areas, cracked pavement, and/or rat holes, some alleys with robust rat populations appeared to have relatively good pavement condition. These conflicts, in combination with the variety of environmental factors that were associated with trap success in our study, suggest that the relationship between rat abundance and the environment is a function of a complex interaction of factors. In other words, there is not one single environmental factor that is a necessary determinant of rat abundance. Rather, a variety of different factors can combine to create suitable rat habitat.

Despite the success of the methodology, this study had several limitations. The relatively low number of observations ($n = 43$) limited the number of variables that could be included in the model. Future studies should seek to increase power (i.e., the number of blocks included) in order to better investigate the full suite of variables that might predict rat abundance. With regard to our trapping methodology, only rats accessing the back alley were accounted for in our measure of rat abundance. Although this 'alley trapping' approach is consistent with previous trapping studies of urban rats [7,8], it is possible that some rats could reside entirely indoors and/or not access the alley. This may particularly be the case for black rats, as it has been suggested that these species prefer to reside in human structures [1]. Indeed, in this study, it could not be determined if the low number of black rats was a result of the trapping methodology, or whether this species is truly rare within this urban area. Ideally, future studies should account for all indoor and outdoor rat populations within a block. However, as this would require intensive trapping on private property, it may not be a feasible proposition. Finally, it should be noted that while many factors have a clear direct relationship with rat abundance (e.g., presence of exposed garbage); other factors (e.g., land use) likely represent some underlying and unmeasured factor(s) that influence rat populations. Although these environmental 'proxies' may be valuable for predicting rat abundance, they are less useful for developing intervention strategies. After all, it is not reasonable to change the land use in a block to prevent rat infestations. It may therefore be useful for future studies to focus on and dissect the relationship between rat abundance and factors like land use and poverty.

Overall, this tool proved to be an efficient and effective way to examine the effect of urban environmental factors on relative rat abundance. In addition to their effects of rat abundance, environmental factors are also known to influence pathogen prevalence in rat populations [2]. In the future, we hope to determine whether this tool could also be used to identify features of the urban environment associated with the prevalence of zoonotic microbes in rats.

Supporting Information

Appendix S1 A systematic environmental observation tool to predict relative rat abundance in urban centres.

Appendix S2 Guidance document for the systematic environmental observation tool.

Table S1 Quantitative predictors generated from the environmental observation tool.

Table S2 Bivariate relationships between features of the urban environment and relative abundance of Norway rats (*Rattus norvegicus*). Zero-inflated negative binomial model for the number of rats trapped in a city block offset by the natural logarithm of the trap effort in that block.

Acknowledgments

The field work was made possible by the assistance of the City of Vancouver (Murray Wightman and Stuart McMillan), the Urban Health Research Initiative, the Vancouver Injection Drug Users Study, and the Vancouver Area Network of Drug Users. We would like to thank Rick White of the University of British Columbia Statistical Consulting and Research Laboratory for his help with the statistical analysis.

Author Contributions

Conceived and designed the experiments: CGH KLP AYTF. Performed the experiments: CGH KLP AYTF. Analyzed the data: CGH TK CMJ DMP. Wrote the paper: CGH KLP AYTF TK CMJ DMP.

References

1. Feng AYT, Himsworth C (2013) The secret life of the city rat: A review of the ecology of urban norway and black rats (*Rattus norvegicus* and *Rattus rattus*). Urban Ecosyst 17: 149–162.
2. Himsworth CG, Parsons KL, Jardine C, Patrick DM (2013) Rats, cities, people, and pathogens: A systematic review and narrative synthesis of literature regarding the epidemiology of rat-associated zoonoses in urban centers. Vector Borne Zoonotic Dis 13: 349–359.
3. Davis DE (1953) The characteristics of rat populations. Q Rev Biol 28: 373–401.
4. Cavia R, Ruben Cueto G, Virginia Suarez O (2009) Changes in rodent communities according to the landscape structure in an urban ecosystem. Landscape Urban Plann 90: 11–19.
5. Vadell MV, Cavia R, Suárez OV (2010) Abundance, age structure and reproductive patterns of *Rattus norvegicus* and *Mus musculus* in two areas of the city of Buenos Aires. Int J Pest Manage 56: 327–336.
6. Traweger D, Travnitzky R, Moser C, Walzer C, Bernatzky G (2006) Habitat preferences and distribution of the brown rat (*Rattus norvegicus* berk.) in the city of Salzburg (Austria): Implications for an urban rat management. J Pest Sci 79: 113–125.
7. Easterbrook JD, Shields T, Klein SL, Glass GE (2005) Norway rat population in Baltimore, Maryland, 2004. Vector Borne Zoonotic Dis 5: 296–299.
8. Glass GE, Gardner-Santana L, Holt RD, Chen J, Shields TM, et al. (2009) Trophic garnishes: Cat-rat interactions in an urban environment. PLoS ONE: e5794.
9. Raudenbush SW, Sampson RJ (1999) Ecometrics: Toward a science of assessing ecological settings with application to the systematic social observation of neighborhoods. Sociol Methodol 29: 1–41.
10. Parsons JA, Singh G, Scott AN, Nisenbaum R, Balasubramaniam P, et al. (2010) Standardized observation of neighbourhood disorder: Does it work in Canada? Int J Health Geogr 9: 6-072X-9-6.
11. Figgs LW (2011) Assessing the association between brown rat (*Rattus norvegicus*) activity and food and harbourage characteristics in a residential community of a medium-sized United States city. Int J Pest Manage 57: 217–222.
12. Yin JX, Geater A, Chongsuvivatwong V, Dong XQ, Du CH, et al. (2008) Predictors for presence and abundance of small mammals in households of villages endemic for commensal rodent plague in Yunnan province, China. BMC Ecology 8: 18.
13. Promkerd P, Khoprasert Y, Virathavone P, Thoummabouth M, Sirisak O, et al. (2008) Factors explaining the abundance of rodents in the city of Luang Prabang, Lao PDR, as revealed by field and household surveys. Integr Zool 3: 11–20.
14. Frantz SC, Comings JP (1976) Evaluation of urban rodent infestation - an approach in Nepal. Proceedings of the 7th Vertebrate Pest Conference. Paper 19.
15. Smith HA (2000) Where worlds collide: Social polarization at the community level in Vancouver's Gastown/Downtown Eastside. Available: https://circle.ubc.ca/handle/2429/10914. Accessed 31 January 2014.
16. Caughy MO, O'Campo PJ, Patterson J (2001) A brief observational measure for urban neighborhoods. Health Place 7: 225–236.
17. Weich S, Burton E, Blancard M, Prince M, Sproston K, et al. (2001) Measuring the build environment: Validity of a site survey instrument for use in urban settings. Health Place 7: 283–292.
18. RAND Corperation (2011) The Los Angeles family and neighborhood survey. Available: http://lasurvey.rand.org/. Accessed 8 September 2013.
19. Emlen J, Stokes A, Davis D (1949) Methods for estimating populations of brown rats in urban habitats. Ecology 30: 430–442.
20. Villafane IE, Cavia R, Vadell MV, Suarez OV, Busch M (2013) Differences in population parameters of *Rattus norvegicus* in urban and rural habitats of central Argentina. Mammalia 77: 187–193.
21. Leslie PH, Davis DHS (1939) An attempt to determine the absolute number of rats on a given area. J Anim Ecol 8: 94–113.
22. Nagorsen DW (2002) An identification manual to the small mammals of British Columbia. Victoria: Province of British Columbia.
23. Nelson L, Clark FW (1973) Correction for sprung traps in catch/effort calculations of trapping results. J Mammal 54: 295–298.
24. Dohoo I, Martin W, Stryhn H (2007) Veterinary epidemiologic research. Charlottetown: AVC Inc.
25. Bricki N, Green J (2007) A guide to using qualitative research methodology. New York: Medicins Sans Frontieres.
26. Krebs C (1999) Current paradigms of rodent population dynamics - What are we missing? In: Singleton GR, Hinds LA, Leirs H, Zhang Z, editors. Ecologically-Based Management of Rodent Pests. Canberra: ACIAR. pp. 33–48.

Uncertainties around the Implementation of a Clearing-Control Policy in a Unique Catchment in Northern Australia: Exploring Equity Issues and Balancing Competing Objectives

Vanessa M. Adams[1,2]*, Robert L. Pressey[2]

1 Northern Australia National Environmental Research Program Hub and Research Institute for the Environment and Livelihoods, Charles Darwin University, Darwin, NT, Australia, 2 Australian Research Council Centre of Excellence for Coral Reef Studies, James Cook University, Townsville, QLD, Australia

Abstract

Land use change is the most significant driver linked to global species extinctions. In Northern Australia, the landscape is still relatively intact with very low levels of clearing. However, a re-energized political discourse around creating a northern food bowl means that currently intact ecosystems in northern Australia could be under imminent threat from increased land clearing and water extraction. These impacts are likely to be concentrated in a few regions with suitable soils and water supplies. The Daly River Catchment in the Northern Territory is an important catchment for both conservation and development. Land use in the Daly catchment has been subject to clearing guidelines that are largely untested in terms of their eventual implications for the spatial configuration of conservation and development. Given the guidelines are not legislated they might also be removed or revised by subsequent Territory Governments, including the recently-elected one. We examine the uncertainties around the spatial implications of full implementation of the Daly clearing guidelines and their potential effects on equity of opportunity across land tenures and land uses. We also examine how removal of the guidelines could affect conservation in the catchment. We conclude that the guidelines are important in supporting development in the catchment while still achieving conservation goals, and we recommend ways of implementing the guidelines to make best use of available land resources for intensified production.

Editor: Paul Adam, University of New South Wales, Australia

Funding: This project was supported by Project 1.1 (Catchment to Coast Conservation Planning) under the North Australia Hub of the National Environmental Research Program. RLP also acknowledges support from the Australian Research Council. The funders had no role in study design, data collection and analysis, decision to publish, or preparation of the manuscript.

Competing Interests: The authors have declared that no competing interests exist.

* E-mail: vanessa.adams@cdu.edu.au

Introduction

The Earth is experiencing a new era, the Anthropocene, in which human actions have become the main driver of global environmental change, with many planetary boundaries being approached or already transgressed, and an already significant and accelerating loss of biodiversity [1]. Species extinction rates are estimated to be 100 to 1000 times their pre-human levels [2]. Human activities are the main drivers of greatly increased species extinctions, with land-use change being the most significant [3]. Despite a global commitment to protect biodiversity through the Convention on Biological Diversity, current indices show that biodiversity continues to decline while human pressures increase [4].

Typical approaches to addressing impacts of clearing of native vegetation on biodiversity are to set protection objectives so that minimum areas with priority for conservation are not cleared or, alternatively, to set limits such that clearing cannot exceed a specified level. In order for either approach to be effective in mitigating proximate threats to biodiversity, conservation planners must first be able to predict changes in the extent and intensity of threatening processes, such as land conversion, so that potential loss can be minimized [5,6]; both conservation objectives and limits on clearing should be informed by potential clearing of particular ecosystems. Future patterns of land use, in terms of extent and rates of change in response to current and emerging driving forces, are not uniform within regions but can be understood with spatially explicit models [7–10]. If planners have spatially explicit data on potential future patterns of land-use change such as areas of high likelihood of clearing of native vegetation, these data can be used to minimize the loss of biodiversity by: 1) adjusting conservation objectives, 2) avoiding more threatened areas where there are spatial options, 3) selecting threat-specific actions, and 4) scheduling conservation actions [5,11]. For example, scheduling acquisition of protected areas can minimize the extent to which conservation objectives are compromised by land clearing while the protected area system is being established [12–14].

Land-use change models identify areas of high likelihood of clearing and can be used to inform policies for protected areas or clearing controls that aim to reduce threats to biodiversity from clearing. However, standard approaches to land-use change modelling require historical data on clearing (typically at least 3 points in time to develop and validate the model) and assume that future clearing patterns will reflect the same driving factors as historical patterns [8]. These assumptions are unlikely to hold true for areas that have experienced low historical rates of clearing or that are experiencing changes in land-use drivers. Alternative methods of evaluating likelihood of clearing and implications for biodiversity conservation will therefore be needed in these cases.

Northern Australia contains nearly a third of the total global area of remaining tropical savanna [15] and is thus significant globally for savanna conservation. Much of Northern Australia remains sparsely populated with a relatively intact environment [15]. However, Northern Australia is often a focus of political discourse in Australia as a potential area for future development and expansion of agriculture [16], so future patterns and rates of clearing are likely to differ from historical trends. In particular, the Daly Catchment in the Northern Territory is an area of interest for future development due to a unique combination of suitable soils, year-round water supplied by large aquifers and the perennial Daly River, and suitable climatic conditions (adequate rainfall during the growing period) for rain-fed crops [17]. Current land use in the Daly River catchment is predominantly pastoral; however, there has been recent interest in clearing for both improved pastures and cropping. In response to concerns over potential clearing impacts on conservation values in the Daly catchment, the Northern Territory Government designed clearing guidelines that set limits on percentages cleared by property, vegetation type, sub-catchment, and the whole catchment (Table 1). The approach taken is unique in that it extends classic clearing-control approaches to include a number of nested hierarchical "caps", thus termed the "cascade rules", to ensure that clearing is distributed evenly across different biophysical features such as sub-catchments and vegetation types [18]. The approach ensures that overall clearing levels are controlled and that clearing does not substantially reduce vegetation types in areas attractive to clearing.

This is an interesting clearing-control approach that could be considered more broadly for protection of native vegetation across northern Australia and internationally. However, the eventual outcome of the restrictions, when all clearing opportunities have been taken, is likely to be sensitive to the order in which properties take up clearing options. The approach therefore needs to be tested more thoroughly to ensure against perverse outcomes. For example, if a sub-catchment reaches the allowed clearing because one or two large properties have used their clearing options, then the remaining properties in the sub-catchment cannot clear.

Understanding the different ways in which clearing guidelines in the Daly catchment could unfold, given scope for extensive further clearing under the cascade rules, has important implications for land management generally and conservation management specifically. The sequence with which properties are cleared is affecting and will further affect equity of opportunity between grazing and horticultural enterprises. This is particularly relevant considering that 26% of the land available for clearing in the catchment is Indigenous land (land either held or managed by Indigenous Australians) without immediate plans for development. Additionally, predicted patterns of clearing flowing from the cascade rules can inform recommendations on the most important areas to prioritize for development to ensure that areas of high production value are developed within clearing limits. Conversely, if areas of high conservation value are identified, the same predictions can be used to prioritize those that are also vulnerable to clearing.

Uncertainties around eventual clearing patterns in the Daly catchment were widened by a change of government in the Northern Territory in September 2012. The new Government is developing new policies for water resources [19] and, given the clearing guidelines are not legislated, the new Government, or subsequent Governments, could revise the cascade rules or replace them. It is therefore critical to understand the potential of existing policies such as the clearing guidelines to contribute to conservation objectives and to explore potential clearing patterns for the catchment in the absence of any guidelines.

Currently, only about 5% of the Daly Catchment has been cleared but a lack of pre-clearing vegetation mapping means we cannot interpret clearing by vegetation structure or assess previous losses by vegetation type. Furthermore, previous clearing is unlikely to reflect future clearing patterns given the change in political focus toward agricultural development and significant changes in policies for land and water resources. Additionally, the cascade clearing guidelines have the potential to constrain future clearing, even in areas suitable for development, so standard methods for land-change modelling are not appropriate. Our study takes an alternative approach to modeling future clearing patterns in the Daly catchment with three main objectives:

- To develop maps of potential clearing for use in regional planning to prioritize areas for competing objectives, namely development and conservation;

Table 1. Summary of cascade rule caps for clearing specified in the clearing guidelines for the Daly catchment [18].

Feature	Percentage clearing cap
Streams – 250 m buffer	Clearing prohibited[a]
Wetlands – 250 m buffer	Clearing prohibited[a]
Daly River – 1000 m buffer	Clearing prohibited[a]
Rainforest – 250 m buffer	Clearing prohibited[a]
Property	70%
Sub-catchment	40%
Vegetation type	30%
Catchment	20%

[a]Buffer zones have generally been supported by the process of assessing clearing applications. However, requests from landholders for exceptions could be approved, with the risk of buffer zones being reduced in unpredictable ways. Neither the extent nor the distribution of ad hoc clearing in buffer zones could be modelled for this study.

- To explore the potential influences of the clearing guidelines on spatial patterns of clearing and equity of opportunity to clear between different land uses and tenures;
- To explore the potential extent and pattern of clearing that could occur if the clearing guidelines were removed, as compared to potential clearing under the guidelines explored in 2.

We analysed scenarios to simulate patterns of clearing under varying assumptions, with and without the cascade rules. This approach allowed us to identify the sensitivity of potential clearing patterns to factors such as land suitability for pastoral and agricultural uses, land tenure, and property size and to make recommendations on implementation of the guidelines to ensure that clearing opportunities are equitable between stakeholder groups across the catchment.

Materials and Methods

Study region

The study region was the whole of the Daly River catchment in the Northern Territory, which is approximately 5.2 million ha, extending from the coastline south-west of Darwin to 250 km inland (Figure 1A). The Daly River catchment has substantial conservation values, including five sites of conservation significance identified by the Northern Territory Government [20], extensive gallery rainforest, and habitats for important wildlife populations, especially of fish, turtles, and waterbirds. The sites of conservation significance within the Daly River catchment have been assessed as either nationally or internationally significant. Although there are no Ramsar-listed wetlands within the catchment, Chatto [21] noted that the Daly River estuary and lower floodplain are likely to qualify for Ramsar listing based on waterbird numbers. More recently, additional studies have identified areas such as the Daly middle reaches and floodplain as high conservation priority [22,23]. Approximately 13% of the catchment is protected by national parks, such as Nitmiluk Gorge, and Indigenous protected areas (Indigenous-owned land enrolled in the Australian national reserve system with the purpose of promoting conservation of biodiversity and cultural resources) such as Fish River (Figure 1A). Of the area potentially available for clearing (87% of the catchment), 10% is government held, 30% is aboriginal land, and 60% is private property (predominantly pastoral). The average size of private properties in the Daly is ~10,500 ha, with properties larger than 5,000 ha representing approximately 13% of landholders but about 90% of the catchment's private land.

The cascade rules

The Daly catchment has been recognized both for its high conservation values as well as its potential for further development. Therefore, although only 5.4% of the catchment has so far been cleared, in 2010 the Northern Territory Government designed clearing guidelines in the form of the cascade rules in response to expected pressure for clearing. Major aims of the approach are to ensure that areas suitable for development are available for clearing while areas of conservation value are adequately protected [18]. The cascade rules involve clearing "caps" specified for features defined at different resolutions (Table 1). When a cap is reached at any level, future clearing is precluded for that feature. The clearing guidelines specify that no more than 20% of the catchment area can be cleared. The cascade rules are supported by vegetation mapping across the catchment (Figure 1B) and there are designated buffer zones around sensitive habitats, such as streams and rivers, in which no clearing is allowed (Figure 1C).

Under the clearing guidelines, all landholders of properties greater than 100 ha must submit a clearing application adhering to the cascade rules. Landholders of properties under 300 ha in size can submit clearing applications without an environmental impact assessment (EIA), while landholders of properties larger than 300 ha must have an EIA before an application will be approved. EIAs can be costly and might therefore deter larger properties from submitting requests to clear. While aboriginal land comprises over one quarter of the catchment, there are no known development plans on this tenure. In comparison, many private landholders have expressed an interest in immediate clearing for intensive land uses.

Because of the nested nature of the caps, the order of clearing and the size and location of cleared properties can preclude other properties from clearing. For example, properties subject to the clearing guidelines range in size from 100–400,000 ha with an average size of 15,000 ha. This means the largest property could singly clear up to 5% of the catchment (and 25% of the available cap) by exercising the right to 70% clearing (assuming no other caps are met within the property). Furthermore, depending on the number, size and tenure of properties within a sub-catchments and vegetation types, properties could be excluded from clearing because neighbouring properties clear up to the sub-catchment or vegetation-type limit.

Aside from potential inequities between tenures and properties, the implications of clearing caps for sub-catchments and vegetation types are difficult to anticipate. Even though all sub-catchment caps are the same (Table 1), sub-catchments vary widely in tenure, including percentages in conservation reserves (Table 2), so the potential for and implications of clearing vary widely between sub-catchments. Similarly, caps for all vegetation types are the same, although vegetation types vary in conservation status and threats beyond the Daly catchment, in conservation status of associated species, and possibly in spatial turnover of species. So the implications of clearing are also likely to vary between vegetation types. Unfortunately, species mapping for the region is limited so it is difficult to properly assess the relative importance of different vegetation types for conservation of individual species. Future improvement in species data would contribute to a more in-depth understanding of whether variable clearing caps would be appropriate and whether clearing should avoid localised areas of particular value to certain species.

Another potential risk from applying the cascade rules is that, depending on the sequence of clearing approvals below the caps in Table 1, clearing of land with low to moderate production value could pre-empt development of areas with high production value.

Data

To conduct our analysis we used the cadastre for the catchment to define property boundaries (Figure 1A). We used vegetation mapping for the Daly river catchment in order to calculate existing percentage cleared and current extent of vegetation types [24]. This vegetation mapping has 98 vegetation types mapped and attributed to 15 broad scale vegetation groups Figure 1B). However, this vegetation mapping does not have an associated pre-clearing product. Therefore, in order to estimate current clearing by vegetation types we used the vegetation groups specified within the mapping and related to the pre-clearing product available for these broad groups [25] to calculate current percentage cleared. We assumed that clearing levels were uniform across vegetation types associated within a broad vegetation group.

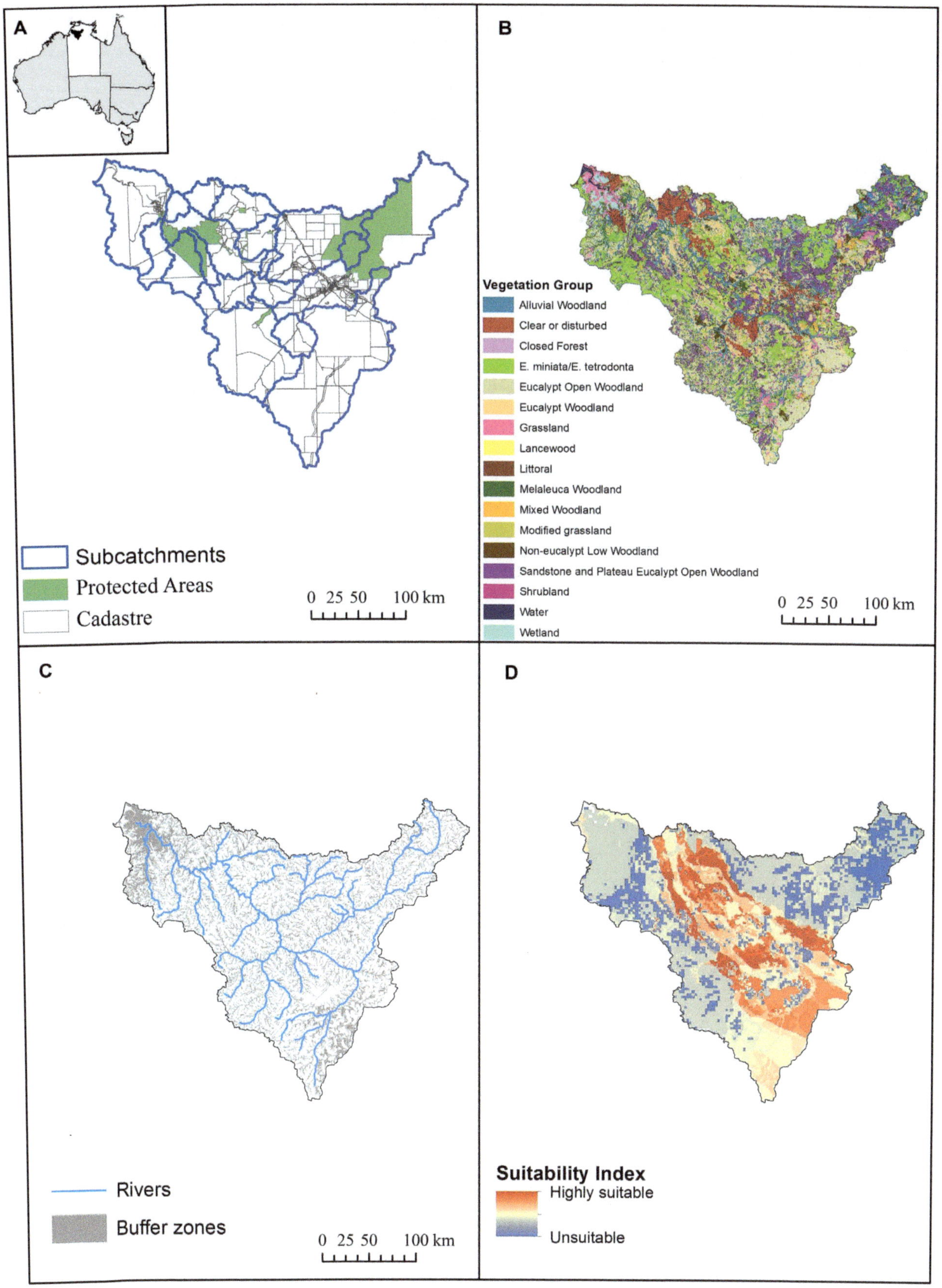

Figure 1. Daly River catchment. Inset in panel A shows Australian states in grey, the Northern Territory in white and the Daly catchment in black. A. Daly River catchment, property boundaries, protected areas (National Parks, such as Nitmiluk Gorge in the northeast, Conservation Areas and Indigenous Protected Areas, such as Fish River in the northwest) and sub-catchments. B. Fifteen vegetation groups based on a hierarchical

classification. The cascade rules relate to 98 vegetation types mapped as subdivisions of these groups. C. Major rivers and buffer zones along streams, rivers, sinkholes and wetlands. D. Suitability index. The index indicates suitability of land for pastoral and agricultural clearing. Suitability ranges from zero (unsuitable, shown in dark blue) to highly suitable (400). Highly suitable land is appropriate for multiple land uses including modified pastures, rainfed and irrigated crops, and perennial and annual horticulture.

We mapped the buffer zones using spatial maps of streams, rivers, wetlands and sinkholes (Figure 1C).

As a guide to possible future patterns of clearing, and to help understand the effects of different clearing sequences on development potential in the catchment, we created a land suitability index. We based our index on land suitability mapping completed by Pascoe-Bell et al. [17] which assessed land suitability using available data on land systems, soil, slope, rainfall, and access to ground water and surface water. Land suitability was mapped as a percentage of land system (0–100%) suitable for four land uses: improved pasture (includes introduced pasture species, usually grasses in combination with legumes), irrigated field crops and perennial horticulture, irrigated annual horticulture, and rainfed field crops and perennial horticulture [17]. For our study, we created an overall clearing suitability index that summed the suitable percentage of each land system across the four land uses (Figure 1D). Our index ranged from 0 – 400, with 0 being completely unsuitable across all land uses (dark blue in Figure 1D) and 400 being completely suitable across all land uses. Although summing values for individual uses gave values larger than 100% for some land systems, we assumed that the range of values larger than 100 indicated relative demand for clearing due to production potential and flexibility of uses.

Simulation of catchment-wide clearing sequences subject to clearing guidelines

The cascade rules were released in 2010 and have only recently been implemented in clearing permits. Very few caps have been reached because only 5.4% of the catchment has been cleared; therefore the guidelines' influence on spatial patterns of clearing is not well understood. We designed four scenarios that allowed us to investigate the influence of different sequences of clearing on the spatial patterns of clearing as well as well as relative amounts of clearing across land tenures. When designing the scenarios we took into account factors such as tenure (aboriginal land versus private properties), size of property, and average suitability of land. The four scenarios were:

- Random – Properties were selected in a random order. We use this as our baseline for comparison with the other scenarios guided by the cascade rules.
- Non-aboriginal, large properties first – Properties were selected from largest to smallest with non-Aboriginal properties clearing first. This scenario reflects the apparent lack of plans to clear aboriginal properties over the short term and the assumption that larger properties would have more financial resources to complete the required EIAs.
- Non-aboriginal, small properties first – Properties were selected from smallest to largest with non-aboriginal properties clearing first. This scenario reflects the apparent lack of plans to clear aboriginal properties over the short term and the assumption that EIAs could be a barrier to larger properties clearing.
- Directed clearing – Properties were selected based on clearing suitability averaged across the property, in descending order. This scenario reflects the case in which clearing is guided to maximize the use of the most suitable land within the caps.

We simulated clearing for each scenario by selecting the property order based on the rules described above and implementing an algorithm for clearing within the cascade rules (Figure 2). Properties with existing clearing have percentage areas

Table 2. Sub-catchment details.

Sub-catchment	Number of properties	Reserve	Aboriginal
Daly River	86	9.05%	50.04%
Chilling Creek	6	1.71%	11.95%
Hayward Creek	4	0.53%	0.00%
Fish River	5	40.08%	71.70%
Bamboo (Moon Boon) Creek	3	53.04%	100.00%
Green Ant Creek	8	0.00%	0.00%
Douglas River	26	2.12%	1.56%
Stray Creek	20	3.46%	4.43%
Bradshaw Creek	7	0.00%	50.31%
Dead Horse Creek	5	0.00%	27.86%
Fergusson River	66	20.38%	54.19%
Flora River	17	1.11%	37.33%
Katherine River	156	32.09%	74.61%
Limestone Creek	4	0.00%	0.00%
King and Dry Rivers	52	2.03%	4.80%
Seventeen Mile Creek	8	100.00%	100.00%

Number of properties, percentage reserved by area and percentage of aboriginal land by area are given for each subcatchment.

cleared ranging from 1–100% with an average of 25%. Given this historical clearing is unlikely to reflect changing land uses and therefore changing patterns of clearing, we used a lower bound of property clearing of 30% and the upper bound of 70% specified by the guidelines. For each selected property we identified a percentage level of clearing based on a uniform distribution between 30% and 70%. We assumed that property owners would first clear land most suitable for production to maximize profits. Within each selected property, areas with the same suitability index were intersected with vegetation types such that the property was divided into units of unique combinations of suitability and vegetation type. For each property, we sequentially simulated clearing of areas in order of decreasing suitability, clearing 100% of each unit with the highest suitability if no caps had been reached and otherwise taking into account caps imposed by previous clearing of vegetation types and sub-catchments until we reached the identified percentage of the property to be cleared, or until all available suitable land (suitability >0) had been cleared if this was less than the identified percentage. We followed this stepwise process for each property until the catchment-wide clearing limit of 20% was reached. Because of the stochastic allocation of clearing percentages to properties, even in scenarios 2–4, we ran each scenario 100 times.

Simulation of catchment-wide clearing sequences with no clearing guidelines

When considering only land suitable for clearing (all areas except those classified as 'unsuitable' or suitability =0 in Figure 1D) outside of buffer zones and currently protected areas, about 2.3 million ha or 44% of the catchment could be cleared. This means that, in the absence of the cascade rules and assuming that future clearing will occur only on suitable land, a total of 49.4% of the catchment might eventually be cleared (44% potential and 5.4% existing).

Given the guidelines are not legislated and could therefore be relaxed or removed, we investigated the extent and potential spatial patterns of unconstrained clearing by running a clearing simulation without any caps related to vegetation types, sub-catchments, or the whole catchment. We used the random ordering applied in Scenario 1 but retained the uniform distribution of property clearing from 30–70% to reflect the likelihood that most properties will retain some vegetation around residences and for grazing of livestock on native pastures (pastures dominated by native plant species) - a major land use currently in the catchment. We also retained the buffer areas because they reflect best-practice clearing recommendations outside of the guidelines for preventing erosion and maintaining water quality. We ran the simulation 100 times. The worst-case scenario of clearing with no guidelines, assuming that landholders are profit-driven and would not invest in clearing unsuitable land with little promise of financial returns, is given by our calculation above of 49.4% of the catchment. Our simulation in the absence of guidelines constrains this potential clearing somewhat by assuming variable clearing percentages across properties.

Simulation Results

For each scenario - four constrained by the cascade rules and one unconstrained - we calculated the average percentage cleared for each vegetation type on each property across the 100 runs. We also calculated the average percentage area cleared by sub-catchment and by vegetation type across the entire catchment. For properties that cleared land, we recorded their number, tenure, average suitability, and average size.

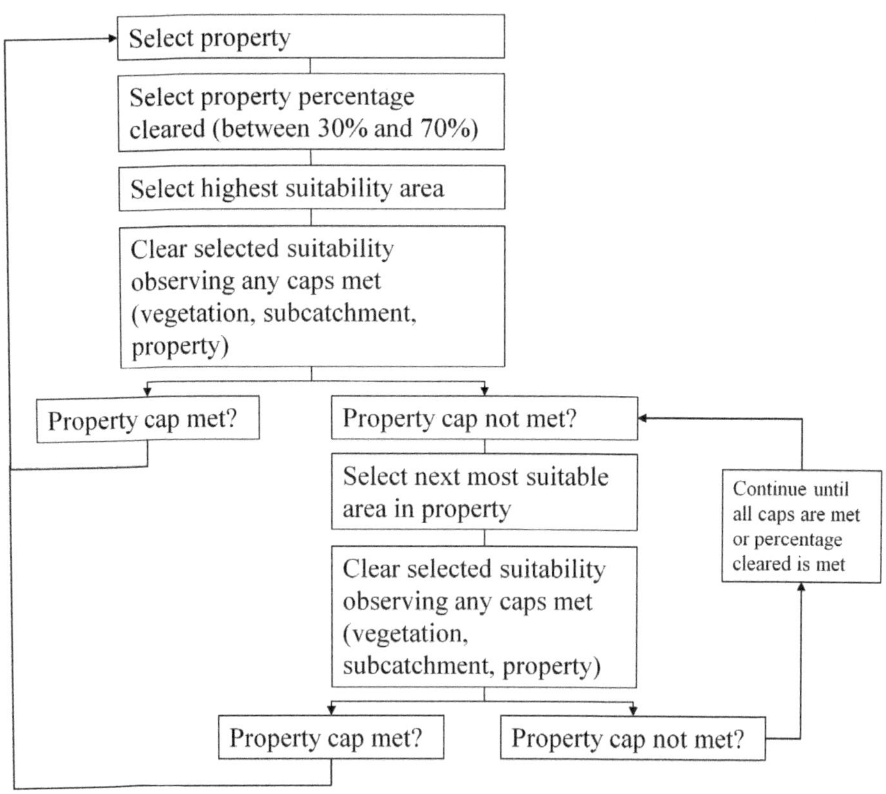

Figure 2. Algorithm for simulating clearing of properties based on cascade rules.

Results

The four scenarios constrained by the clearing guidelines and the single scenario ignoring the clearing guidelines produced markedly different spatial patterns of clearing across the catchment (Figure 3). For example, scenario 1 (random) resulted in moderate levels of clearing across the entire catchment (Figure 3A) while scenario 4 (directed clearing) resulted in high levels of clearing constrained to the areas of highest suitability in the catchment (Figure 3D, and compare Figure 1D). The potential influence of the clearing guidelines on the overall magnitude of clearing can be seen by comparing scenario 1 (random, constrained) with scenario 5 (random, unconstrained). While both scenarios result in widespread clearing, the overall amount of clearing in scenario 5 (>1.7 million ha of new clearing resulting in a total of 38% of the catchment cleared) is more than double that in scenario 1 and in the other constrained scenarios (Table 3) because all properties that are open for clearing take that opportunity.

The average size of property that cleared and the total area cleared varied across scenarios. The assumption that large properties cleared first (Scenario 2) resulted in substantially fewer, much larger properties with clearing (Table 3). This strong difference reflects large variation in property sizes across the catchment. The largest properties in the catchment exceed 100,000 ha. If these large properties all clear large percentages of their holdings, the total available area for clearing across the catchment can be reached with clearing on 15– 27 properties. The sequence with small properties clearing first (scenario 3) led to many more properties having opportunities to clear.

The effect of land tenure in the clearing sequence is evident in scenarios 2 and 3. In both scenarios, opportunities for clearing across the catchment were exhausted on non-aboriginal land before any aboriginal properties had the opportunity to clear (Table 3). In the guided scenario 21% of land cleared was aboriginal, slightly below the overall percentage of aboriginal land (26% of available land). In scenarios that ignored tenure and land quality (1,5), a larger proportion of aboriginal properties were cleared (scenario 1 – 42.5% of total clearing; scenario 5 – 38.6%).

As expected, the directed clearing of scenario 4 selected properties with the highest average suitability (227, Table 3). The random scenarios (1 and 5) produced average property suitabilities somewhat below the catchment-wide average of 178. The lowest average property suitability was 167 for scenario 2 (large properties first), reflecting the larger proportion of land on

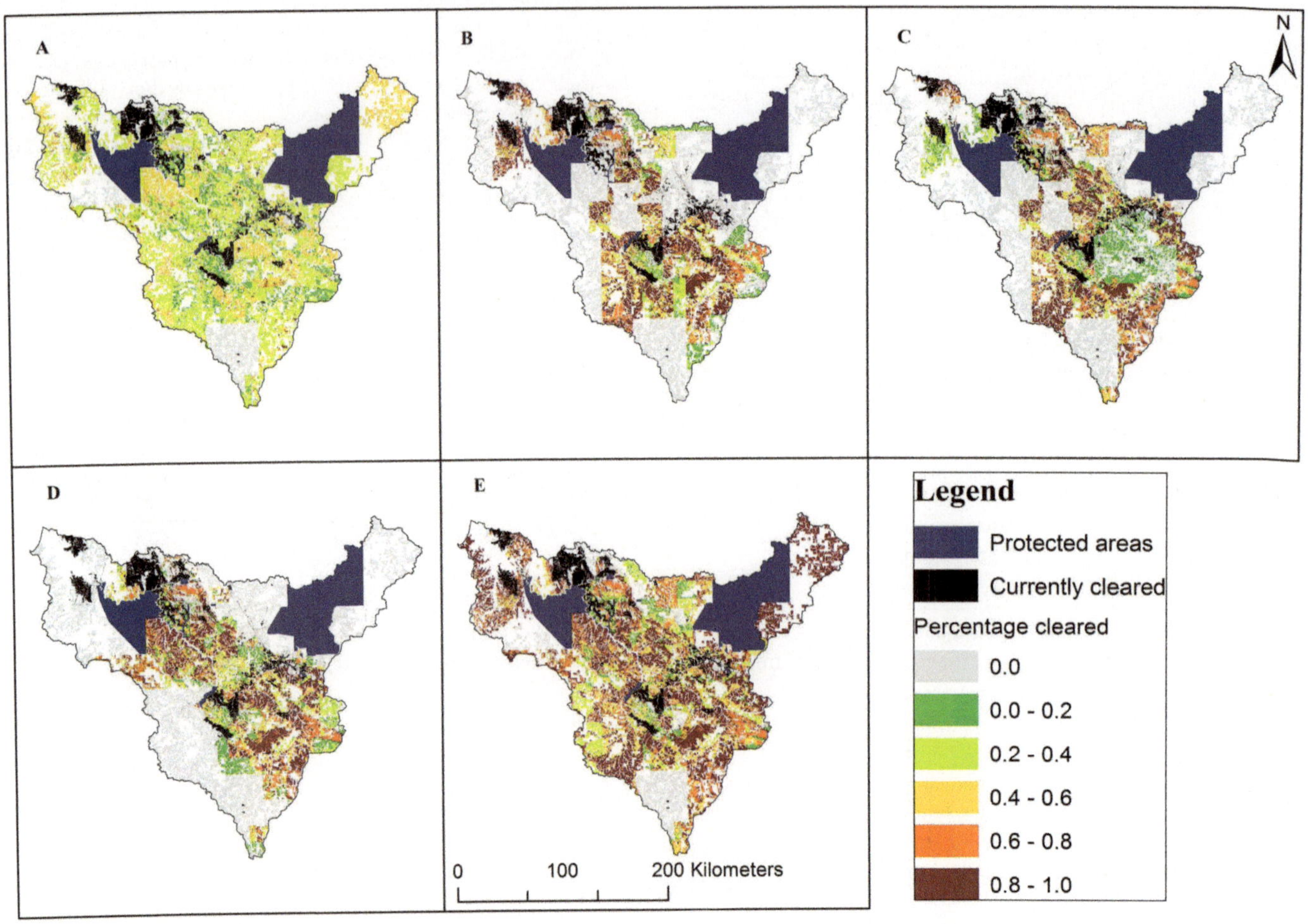

Figure 3. Patterns of clearing from simulations for five scenarios. Percentages cleared are shown by vegetation type within property stratified by land suitability. Currently cleared areas are shown in black and protected areas are shown in blue. Areas available for clearing but not cleared are in pale gray and areas not available for clearing due to being within a buffer zone or unsuitable (index = 0) are in white. A. Scenario 1 – Random, constrained by cascade rules. B. Scenario 2 - Non-aboriginal, large properties first, constrained by cascade rules. C. Scenario 3- Non-aboriginal, small properties first, constrained by cascade rules. D. Scenario 4 – Directed clearing, constrained by cascade rules. E. Scenario 5 – Random, not constrained by clearing guidelines.

Table 3. Summary statistics related to properties, by scenario.

	Scenario 1 – Random	Scenario 2 – Large first	Scenario 3 – Small first	Scenario 4 - Directed	Scenario 5 - No guidelines
Average number of properties with clearing	120	19	228	131	270
Average size (ha) of properties with clearing	18,104	109,930	9,273	15,961	17,140
Average total area cleared (ha)	799,134	774,676	787,104	772,242	1,710,709
Average area cleared - aboriginal properties (ha)	339,531	0	0	164,852	661,155
Average area cleared - non-aboriginal properties (ha)	459,603	774,676	787,104	607,390	1,049,554
Average property suitability (index 0–400)[a]	171	167	180	227	172

[a]Average property suitability is the average suitability across all land on properties selected for clearing. The minimum average property suitability was 3.6 and the maximum was 301. The average suitability across all properties in the Daly catchment was 178.

these properties classified as suitable only for modified pastures and not for other agricultural uses.

Differences in spatial clearing patterns were also evident in the average percentages of sub-catchments cleared (Table 4). There was a large effect of land tenure. For example, Bamboo Creek sub-catchment, totally within aboriginal tenure (Table 2), remains totally uncleared in scenarios 2 and 3 that allocate no clearing to aboriginal land, but is moderately cleared in other scenarios. Differences between scenarios 2 and 3 in terms of percentages of sub-catchments cleared relate to the distribution across sub-catchments of properties of different sizes. Directed clearing (scenario 4) resulted in lower levels of clearing in sub-catchments dominated by low-suitability land, such as Chilling Creek, and higher levels of clearing in highly suitable sub-catchments regardless of tenure, such as Bradshaw Creek and Limestone Creek. As expected, most sub-catchments were more extensively cleared in scenario 5, without clearing guidelines, compared to scenario 1.

The average percentage cleared by vegetation type was similar across scenarios 1–4, and clearing was on average about twice as much by vegetation type without clearing guidelines (scenario 5) (Table 5). For each scenario we compared the percentage area cleared for each vegetation type. Across all vegetation types, there were strong positive correlations between scenarios in terms of percentage clearing across all 98 vegetation types, with ρ ranging from 0.6491 to 0.9165 (Table 6). Scenarios 1 and 5, with random selection of properties, were most similar ($\rho = 0.9165$). This was expected because the only difference between these two scenarios was the limit on total clearing. The next most similar scenarios were 2 and 3 ($\rho = 0.8945$), both constrained to non-aboriginal

Table 4. Average percentage clearing by sub-catchment, current and by scenario.

Sub-catchment	Current	Scenario 1 – Random	Scenario 2 – Large first	Scenario 3 – Small first	Scenario 4 - Directed	Scenario 5 - No guidelines
Daly River	7.91%	21.03%	18.35%	16.03%	21.17%	37.46%
Chilling Creek	1.51%	11.37%	16.28%	4.87%	2.41%	25.55%
Hayward Creek	4.02%	11.84%	19.61%	8.86%	15.88%	21.25%
Fish River	0.00%	5.37%	0.00%	0.00%	10.47%	10.76%
Bamboo (Moon Boon) Creek	0.00%	13.02%	0.00%	0.00%	24.23%	25.75%
Green Ant Creek	60.35%	60.35%	60.35%	60.35%	60.35%	60.35%
Douglas River	18.39%	30.42%	35.95%	39.24%	36.52%	41.30%
Stray Creek	10.15%	25.87%	36.85%	37.64%	36.89%	40.66%
Bradshaw Creek	0.94%	20.43%	24.11%	24.85%	35.64%	39.91%
Dead Horse Creek	2.92%	23.94%	35.56%	36.37%	38.28%	41.83%
Fergusson River	1.44%	20.98%	11.80%	22.75%	12.55%	40.50%
Flora River	3.57%	22.89%	29.73%	29.29%	11.50%	42.56%
Katherine River	4.48%	19.34%	8.12%	10.06%	9.87%	37.93%
Limestone Creek	16.32%	28.23%	37.70%	26.16%	37.16%	39.54%
King and Dry Rivers	0.94%	18.03%	25.35%	24.48%	29.54%	38.73%
Seventeen Mile Creek	0.00%	17.78%	0.00%	0.00%	0.00%	42.50%
Total	5.41%	20.68%	20.22%	20.45%	20.17%	38.10%

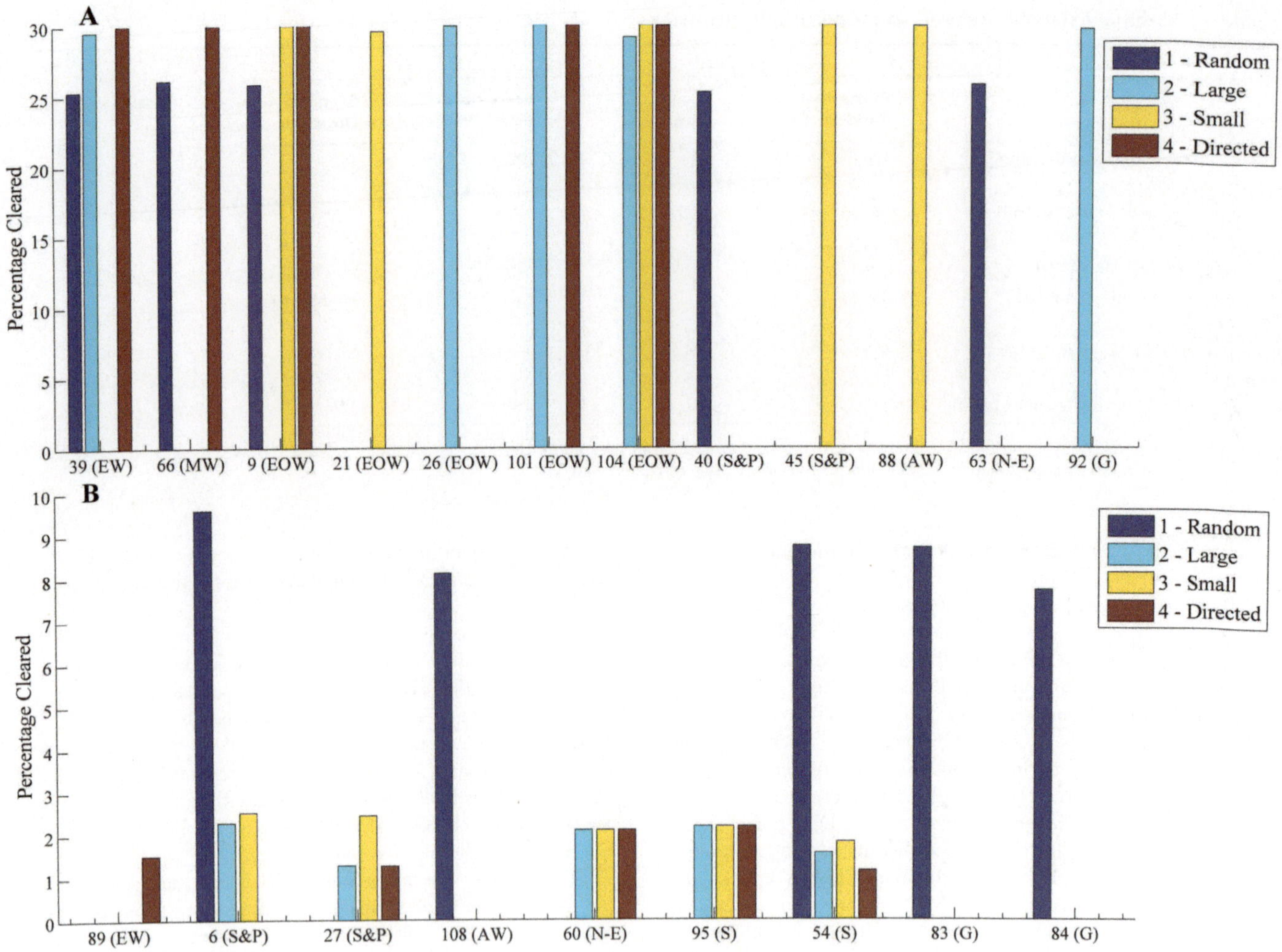

Figure 4. Vegetation types with the highest and lowest levels of clearing in the clearing guideline scenarios. A) Vegetation types with highest clearing in guideline scenarios (1–4). The top five cleared vegetation types in one or more scenarios are shown with percentages cleared by scenario. B) Vegetation types with lowest clearing in guideline scenarios (1–4). The bottom five cleared vegetation types in one or more scenarios are shown with percentages cleared by scenario. Note differences in scale of *y*-axes. Vegetation codes and names of broad vegetation group are displayed [a] [24]. [a] Vegetation groups are shown in Figure 2B. Closed Forest (CF), Eucalypt Woodland (EW), Mixed Woodland (MW), E. miniata/E. tetrodonta (E. min/E. tet), Eucalypt Open Woodland (EOW), Sandstone and Plateau Eucalypt Open Woodland (S&P), Lancewood (L), Melaleuca Woodland (MelW), Alluvial Woodland (AW), Non-eucalypt Low Woodland (N-E), Shrubland (S), Grassland (G), Littoral (Li).

land, indicating that clearing small properties first (scenario 3) results in only a small percentage of catchment-wide clearing, allowing substantial scope for subsequent clearing on larger properties, similar to that in scenario 2. However, there were differences in the levels of clearing within individual vegetation types. Scenarios 1–4 had highly variable impacts on the vegetation types most affected by clearing (Figure 4A), but Scenarios 2–4 targeted similar levels of clearing for the least cleared vegetation types (Figure 4B). Scenario 1 had overall higher levels of clearing across the least-cleared vegetation types due to its more dispersed pattern of clearing.

Discussion

The clearing guidelines constrained total clearing in the catchment to 20% compared to a potential maximum of 49.4%

Table 5. Average percentage cleared and standard deviation across vegetation types (n = 98), by scenario.

	Scenario 1 - Random	**Scenario 2 – Large first**	**Scenario 3 – Small first**	**Scenario 4 - Directed**	**Scenario 5 - No guidelines**
Average percentage cleared by vegetation type	16.81%	14.51%	15.32%	12.68%	32.48%
Standard deviation of percentage cleared by vegetation type	6.02%	8.82%	8.65%	8.67%	14.50%

Table 6. Spearman's Rank Correlation Coefficient (ρ) between pairs of scenarios in terms of percentage vegetation type cleared (n = 98).

	Scenario 1	Scenario 2	Scenario 3	Scenario 4	Scenario 5
Scenario 1	1				
Scenario 2	0.7020	1			
Scenario 3	0.7245	0.8945	1		
Scenario 4	0.7168	0.8355	0.8139	1	
Scenario 5	0.9165	0.6491	0.6451	0.6615	1

All $p<0.001$.

if all available suitable land were cleared and a simulated 38% in our unconstrained scenario, considering realistic percentages of properties cleared. A policy shift to removing the cascade rules, along with other perceived impediments to development in the Daly catchment, could expose some vegetation types and their associated flora and fauna to high levels of clearing. Extensive clearing of the Daly catchment or other parts of northern Australia involves substantial risks for biodiversity. Broad-scale grazing and changed fire regimes associated with intensified land uses have been implicated in declines in small mammals and granivorous birds in northern Australia [26–28]. Furthermore, invasive species originally introduced for improved pastures pose a significant threat to both environmental and economic values [29]. These risks to the Daly's biodiversity apply even with the cascade rules in place. While the guidelines ensure that 80% of the catchment will remain uncleared, and riparian habitat is protected in buffer zones, they still allow significant clearing of some properties and sub-catchments, with poorly understood implications for conservation. In addition, the impacts of land clearing extend beyond the loss of vegetation. For example, land clearing and associated development is often associated with the introduction of invasive species and changed fire regimes whose impacts extend well beyond the cleared sites. Policy makers should carefully monitor and regulate these impacts as clearing proceeds.

The cascade rules have been criticized for lack of flexibility to accommodate development in the catchment. However, our directed clearing scenario demonstrates that the guidelines can allow for clearing of highly suitable land for intensification or expansion of agriculture or pastoralism (scenario 4). Thus, the cascade rules, if implemented carefully, present one policy approach to addressing competing land use objectives of conservation and development. Our analyses also indicate that, even in the absence of the clearing guidelines but with realistic percentage clearing of properties, cleared areas by sub-catchment are typically about 42% or less. So the 40% sub-catchment cap in the cascade rules will not substantially constrain clearing of suitable land or preclude clearing by property managers who wish to develop.

The catchment contains the Daly Basin bioregion in its entirety as well as portions of several other bioregions. While we used the best available vegetation mapping for our analysis, this mapping does not reflect compositional and fine-resolution structural variation in vegetation types within and between bioregions in and around the catchment. Implementing the cascade rules with the existing vegetation mapping could therefore result in loss of biodiversity, including critical habitat for fauna. For example, a directed application of the guidelines (scenario 4) would focus clearing on highly suitable land occurring mainly within the Daly Basin bioregion and minimize clearing in other bioregions intersecting the catchment [30]. A better understanding of species distributions in the catchment is needed to accurately identify areas of high conservation value. If this data were available, clearing scenarios could be used to identify areas of both high conservation value and high vulnerability to clearing, leading the way to informed tradeoffs between conservation and development goals.

Given the nested nature of the clearing guideline caps, the order in which clearing occurs can strongly influence the spatial pattern of clearing across the catchment. We found that a random order of property clearing with the guidelines produced similar spatial patterns to clearing without guidelines, demonstrating that the guidelines do not inherently bias clearing to any portion of the catchment. However, the guidelines have the potential to preclude clearing by aboriginal properties if managers of these properties delay clearing (scenarios 2 and 3). Importantly, the directed scenario focused on high-quality land cleared a representative proportion of aboriginal and non-aboriginal land, allowing for equitable access to property development across land tenures. Scenario 2 showed that, if the largest properties clear first, much of the highly suitable land in the current agricultural and horticultural zones, such as the Douglas Daly, will remain uncleared under the guidelines. The predominant features of cleared land on large properties are modified and improved pastures for grazing (including a current application for 18,000 hectares for grazing on one of the largest properties in the Daly) [31]. If clearing occurs on large properties first, then the 20% of allowed clearing across the catchment would be allocated mainly to cattle grazing at the expense of agriculture and horticulture.

Overall, it is clear that the sequence of clearing under the cascade rules could influence equity of opportunity to develop between aboriginal and non-aboriginal properties, and could determine the relative extents of grazing and horticulture. This result highlights the need for a policy surrounding the implementation of the guidelines to ensure equity and balance between tenures and uses. One implementation policy that could accompany the guidelines is a directed or zoned approach in which areas that are identified as highly suitable for development are earmarked as available for clearing. For example, our directed clearing scenario led to almost proportional clearing across tenure types, so this strategy would address equity issues while insuring that the most suitable land was available for clearing. Another implementation policy could be to mirror the approach taken in water allocations in which a Strategic Indigenous Reserve (SIR) is included in the allocation to ensure a proportional allocation of water is held for aboriginal development. A SIR could be included in the caps such that an equitable portion of the 20% clearing cap is reserved for clearing on aboriginal properties.

The political discourse around a northern Australia 'food bowl' to supply Asia has recently reignited, with consideration of removing the Daly's cascade rules and a recent announcement by the Queensland Government to release new water licenses in the North for crop production [32]. Similarly, Western Australia has announced plans for Stage 2 of the Ord River irrigation scheme and an announcement is expected from the Northern Territory Government about further expansion into Ord Stage 3 [33]. This push for development across relatively intact savanna landscapes [15] comes with risks to biodiversity that are not well understood. The available data, however, indicate continuing declines of many species [28] and offer nothing to support optimism about the coexistence of northern Australian biodiversity with expanding transformation of landscapes [27]. In this context, and even with uncertainties about their conservation implications, the Daly's cascade rules are precautionary while still providing opportunities for development. With adjustments for equity of opportunity and balanced land uses, and with refinements to ensure that the best ecological data are considered in setting limits to clearing in specific parts of the catchment, the Daly clearing guidelines provide a model for other regions across the northern savannas and beyond.

Author Contributions

Conceived and designed the experiments: VMA RLP. Performed the experiments: VMA. Analyzed the data: VMA. Wrote the paper: VMA RLP.

References

1. Rockstrom J, Steffen W, Noone K, Persson A, Chapin FS, et al. (2009) A safe operating space for humanity. Nature 461: 472–475.
2. Pimm SL, Russell GJ, Gittleman JL, Brooks TM (1995) The future of biodiversity. Science 269: 347–350.
3. Sala OE, Stuart Chapin F, Armesto JJ, Berlow E, Bloomfield J, et al. (2000) Global biodiversity scenarios for the year 2100. Science 287: 1770–1774.
4. Butchart SHM, Walpole M, Collen B, van Strien A, Scharlemann JPW, et al. (2010) Global biodiversity: Indicators of recent declines. Science 328: 1164–1168.
5. Pressey RL, Cabeza M, Watts ME, Cowling RM, Wilson KA (2007) Conservation planning in a changing world. Trends in Ecology and Evolution 22: 583–592.
6. Wilson K, Pressey R, Newton A, Burgman M, Possingham H, et al. (2005) Measuring and incorporating vulnerability into conservation planning. Environmental Management 35: 527–543.
7. Pontius Jr RG, Huffaker D, Denman K (2004) Useful techniques of validation for spatially explicit land-change models. Ecological Modelling 179: 445–461.
8. Verburg P, Schot P, Dijst M, Veldkamp A (2004) Land use change modelling: current practice and research priorities. GeoJournal 61: 309–324.
9. Bryan BA, Crossman ND, King D, Meyer WS (2011) Landscape futures analysis: Assessing the impacts of environmental targets under alternative spatial policy options and future scenarios. Environmental Modelling & Software 26: 83–91.
10. Alo CA, Pontius Jr RG (2008) Identifying systematic land-cover transitions using remote sensing and GIS: the fate of forests inside and outside protected areas of Southwestern Ghana. Environment and Planning B: Planning and Design 35: 280–295.
11. Wilson KA, Underwood EC, Morrison SA, Klausmeyer KR, Murdoch WW, et al. (2007) Conserving biodiversity efficiently: What to do, where, and when. Plos Biology 5: 1850–1861.
12. Menon S, Pontius RG, Rose J, Khan ML, Bawa KS (2001) Identifying conservation-priority areas in the tropics: a land-use change modeling approach. Conservation Biology 15: 501–512.
13. Visconti P, Pressey RL, Bode M, Segan DB (2010) Habitat vulnerability in conservation planning—when it matters and how much. Conservation Letters 3: 404–414.
14. Visconti P, Pressey RL, Segan DB, Wintle BA (2010) Conservation planning with dynamic threats: The role of spatial design and priority setting for species' persistence. Biological Conservation 143: 756–767.
15. Woinarski J, Mackey B, Nix H, Traill B (2007) The nature of Northern Asutralia: natural values, ecological processes and future prospects. Canberra: ANU E press.
16. Webster T, Morison J, Abel N, Clark E, Rippin L, et al. (2009) Irrigated agriculture: development opportunities and implications for northern Australia (chapter 10). Northern Australia Land and Water Science Review full report: Coordinated by the Commonwealth Scientific and Industrial Research Organisation (CSIRO).
17. Pascoe-Bell A, Lynch B, Hill J, Green C, Cameron A, et al. (2011) Identification of potential land for long-term sustainable food production. Palmerston, Northern Territory: Department of Natural Resources, Environment, The Arts and Sport (NRETAS).
18. NRETAS (2010) Department of Natural Resources, Environment, The Arts and Sport land clearing guidelines. Palmerston NT: Northern Territory Department of Natural Resources, Environment, The Arts and Sport.
19. Stewart P (2013) Indigenous water reserve policy tap turned off. ABC Rural, Available from http://wwwabcnetau/news/2013-10-09/nt-indigenous-water-reserve-policy-dropped/5012152 (accessed March 2014).
20. NRETAS (2009) Recognising sites of conservation significance for biodiversity values in the Northern Territory. Palmerston: Biodiversity Conservation Unit, Department of Natural Resources, Environment, The Arts and Sport.
21. Chatto R (2003) The distribution and status of shorebirds around the coasts and coastal wetlands of the Northern Territory. Pamerston, NT: Parks and Wildlife Commission of the Northern Territory.
22. Kennard MJ (ed) (2010) Identifying high conservation value aquatic ecosystems in northern Australia. Interim report for the Department of Environment, Water, Heritage and the Arts and the National Water Commission. Darwin, Available at: http://track.gov.au/publications/registry/843: Tropical Rivers and Coastal Knowledge (TRaCK) Commonwealth Environmental Research Facility, Charles Darwin University.
23. Linke S, Kennard MJ, Hermoso V, Olden JD, Stein J, et al. (2012) Merging connectivity rules and large-scale condition assessment improves conservation adequacy in river systems. Journal of Applied Ecology 49: 1036–1045.
24. DLRM (2013) Vegetation mapping of the Daly River Catchment, Northern Territory (Version 3. Updated January 2013). 3 ed. Darwin NT: Northern Territory Government Department of Land Resource Management.
25. NVIS (2010) National vegetation information system. Version 3.1; available at http://www.environment.gov.au (accessed October 2010).
26. Franklin DC, Whitehead PJ, Pardon G, Matthews J, McMahon P, et al. (2005) Geographic patterns and correlates of the decline of granivorous birds in northern Australia. Wildlife Research 32: 399–408.
27. Woinarski J, Green J, Fisher A, Ensbey M, Mackey B (2013) The effectiveness of conservation reserves: Land tenure impacts upon biodiversity across extensive natural landscapes in the tropical savannahs of the Northern Territory, Australia. Land 2: 20–36.
28. Woinarski JCZ, Legge S, Fitzsimons JA, Traill BJ, Burbidge AA, et al. (2011) The disappearing mammal fauna of northern Australia: context, cause, and response. Conservation Letters 4: 192–201.
29. Setterfield SA, Rossiter-Rachor NA, Douglas MM, Wainger L, Petty AM, et al. (2013) Adding fuel to the fire: The impacts of non-native grass invasion on fire management at a regional scale. PLoS ONE DOI: 10.1371/journal.pone.0059144
30. SEWPaC (2012) Interim biogeographic regionalisation for Australia, Version 7. In: Department of Sustainability E, Water, Populations and Communities, editor. Canberra.
31. Arnost M (2013) Tipperary plan raises environmental concerns. ABC Rural available at http://www.abc.net.au/news/2013-05-20/tipperary-clearing/4701026, accessed June 2013.
32. Tapp V, Roocke N (2013) Water licences open new northern cropping region. ABC Rural: available at http://www.abc.net.au/news/2013-05-28/nrn-northern-irrigation/4717350, accessed June 2013.
33. Nel JL, Reyers B, Roux DJ, Cowling RM (2009) Expanding protected areas beyond their terrestrial comfort zone: Identifying spatial options for river conservation. Biological Conservation 142: 1605–1616.

Deforestation and Benthic Indicators: How Much Vegetation Cover Is Needed to Sustain Healthy Andean Streams?

Carlos Iñiguez-Armijos[1,2]*, Adrián Leiva[1], Hans-Georg Frede[2], Henrietta Hampel[3], Lutz Breuer[2]

1 Departamento de Ciencias Naturales, Universidad Técnica Particular de Loja (UTPL), Sección de Ecología, Loja, Ecuador, **2** Institute for Landscape Ecology and Resources Management (ILR), Research Centre for Bio Systems, Land Use and Nutrition (IFZ), Justus-Liebig-University Giessen, Giessen, Germany, **3** Universidad de Cuenca, Quinta Balzaín, Cuenca, Ecuador

Abstract

Deforestation in the tropical Andes is affecting ecological conditions of streams, and determination of how much forest should be retained is a pressing task for conservation, restoration and management strategies. We calculated and analyzed eight benthic metrics (structural, compositional and water quality indices) and a physical-chemical composite index with gradients of vegetation cover to assess the effects of deforestation on macroinvertebrate communities and water quality of 23 streams in southern Ecuadorian Andes. Using a geographical information system (GIS), we quantified vegetation cover at three spatial scales: the entire catchment, the riparian buffer of 30 m width extending the entire stream length, and the local scale defined for a stream reach of 100 m in length and similar buffer width. Macroinvertebrate and water quality metrics had the strongest relationships with vegetation cover at catchment and riparian scales, while vegetation cover did not show any association with the macroinvertebrate metrics at local scale. At catchment scale, the water quality metrics indicate that ecological condition of Andean streams is good when vegetation cover is over 70%. Further, macroinvertebrate community assemblages were more diverse and related in catchments largely covered by native vegetation (>70%). Our results suggest that retaining an important quantity of native vegetation cover within the catchments and a linkage between headwater and riparian forests help to maintain and improve stream biodiversity and water quality in Andean streams affected by deforestation. This research proposes that a strong regulation focused to the management of riparian buffers can be successful when decision making is addressed to conservation/restoration of Andean catchments.

Editor: Tomoya Iwata, University of Yamanashi, Japan

Funding: This research was funded through a scholarship of the Secretaria Nacional de Educación Superior, Ciencia, Tecnología e Innovación of Ecuador (SENACYT), a DFG (Deutsche Forschungsgemeinschaft) grant (FR615/28-1/-2) and by the Universidad Técnica Partucular de Loja (UTPL). The support from Henrietta Hampel was thanks to the PROMETEO Program of SENESCYT. The funders had no role in study design, data collection and analysis, decision to publish, or preparation of the manuscript.

Competing Interests: The authors have declared that no competing interests exist.

* Email: cainiguez@utpl.edu.ec

Introduction

Tropical deforestation is one of the most important topics worldwide associated with biodiversity loss and ecosystem degradation [1–3]. However, the effects of deforestation on ecological condition of tropical streams have been rarely investigated despite evidence that forest loss negatively affects these streams [4–7]. In Ecuador, deforestation for agriculture and/or urbanization is a critical threat for Andean ecosystems [8], reaching annual deforestation rates of 2.7% between 1989 and 2008 in the montane forests of southern Andes (Tapia Armijos, pers. communication). However many of these ecosystems provide essential environmental services including biodiversity conservation, carbon storage and water supply [9]. Several studies focused along the Ecuadorian Andean region have demonstrated that deforestation affects habitat structure and macroinvertebrate community assemblages in streams [10], litter breakdown rates [11] and hydrological processes [12]. Clearly the Ecuadorian montane streams are influenced by the surrounding landscape and several research groups have recently deployed considerable efforts to assess these relationships. Nonetheless, information gaps concerning the influence of land cover/use and spatial scale on stream condition remains, despite they affect water quality and aquatic diversity [13,14]. These data are crucial for management and conservation of developed catchments to insure long term sustainability [15].

In other regions, organizations such as the European Environment Agency (EEA), the United States Environmental Protection Agency (EPA) and national agencies in New Zealand have identified, and are currently considering within their monitoring programs, the impacts of land cover/use changes on stream ecological conditions based on responses of macroinvertebrate communities [16–19]. In addition, they have developed catchment and/or riparian rehabilitation projects in order to enhance stream condition [20]. In South and Central America such extensive monitoring or restoration programs by governmental initiatives are rare, despite their potential to reduce the impact of deforestation on stream condition [13]. In this regard, a Costa

Rican regulation has showed the effectiveness of protecting riparian buffers to reduce the impacts of deforestation on macroinvertebrate communities and sensitive taxa in tropical streams [4].

Several authors [10,21–23] have applied a variety of macroinvertebrate metrics (e.g. Fisher's index; EPT taxa) to assess the impact of land cover/use on the ecological condition of streams in South America focusing on the structure and composition of benthic communities. These studies have confirmed changes to aquatic biodiversity and water quality because of deforestation around the streams. Nevertheless, how does vegetation cover influences stream ecological condition? In streams running through forested catchments the macroinvertebrate communities are more heterogeneous, diversity of intolerant taxa to environmental stress is higher, and abundance and variety of habitats is greater [10,14,21]. Likewise, streams with forested riparian areas have low variation in water temperature and less nutrients and sediment loads along the entire length [4,13]. At the catchment scale, there is also an influence of vegetation cover on controlling hydrology and the entire diversity of the streams [12]. In this regard, estimating the extent of native vegetation cover in a catchment and the importance of riparian management and headwater forest protection to maintain and improve biodiversity and water quality of streams affected by deforestation is a pressing task for management and conservation [13,14,24].

Geographic Information Systems (GIS) allow researchers to explore correlations in spatial patterns with ecological features in the landscape. Allan [13] describes the ability to apply GIS tools with remote sensing data to assess the influence of land cover/use on stream ecological condition. Studies in temperate areas have described the influence of land cover and spatial scale on stream communities [25,26] and on stream water quality [14] by applying GIS based evaluation techniques to derive land cover data and to relate them with physical, chemical and biological data.

In this study, our objective was to investigate the influence of vegetation cover at three spatial scales (local, riparian buffer and catchment) on structure and composition of macroinvertebrate communities and on water quality of headwater streams in the southern Ecuadorian Andes. These streams are located in two contrasting regions of varying intensities of anthropogenic pressure (i.e. deforestation, population growth, and farming). The questions we addressed were: what spatial scale and how much vegetation cover are associated with healthy streams based on macroinvertebrate metrics? In order to address these questions, we calculated several diversity and water quality metrics from macroinvertebrate communities, derived spatial information on land cover from GIS, and related these metrics statistically to three spatial scales and stream parameters. Finally, we addressed management and conservation problems in order to help mitigating impacts of deforestation on stream ecosystems.

Methodology

For this study, the Ministry of Environment of Ecuador issued us a general research permit for every sampling site and to collect water samples and macroinvertebrate specimens. In the case of the Zamora upstream catchment (Zu), the Municipality of Loja granted a specific permission to work at sites Zu1–2, 4–5, 8 because these streams are used for the water supply of the city. To work at the San Francisco catchment (SF), the administration of the San Francisco Research Station and Nature and Culture International allowed us to access sites SF1–7. For all other sites we used the general research permit.

Most of macroinvertebrates collected in this study were identified at genus level and we do not know if any of these genera are endangered or protected, because of the lack of information for Ecuador.

Study area

We studied 23 streams of two headwater catchments of the Zamora River basin, located in the mountain forests in Ecuadorian southern Andes (Figure 1). The Zamora upstream catchment covers an area of 276.2 km^2 and ranges between 2,020 to 3,250 m a.s.l. It is the most populated catchment within the Zamora River basin with approximately 120, 000 inhabitants in the city of Loja. The San Francisco catchment is located in a sparsely populated area of the Zamora River basin with an area of 75.3 km^2 and an elevation range from 1,800 to 3,250 m a.s.l. The natural vegetation in both study catchments is evergreen mountain rain forests, humid shrub, cloud forest and páramo [27,28]. Anthropic activities in both catchments have substituted the original vegetation mainly to pasture, and to a lesser extent to cropland, pine (*Pinus*) and eucalyptus (*Eucalyptus*) plantations [29]. Urban development has only taken place in the Zamora upstream catchment and has spread along its main river. Both catchments have similar climatic conditions with a wet season from May to September and a dry season from October to April. The average annual precipitation and temperature in the Zamora upstream and San Francisco catchments are about 2,200 mm and 15.2°C respectively.

Stream measurements and macroinvertebrates sampling

We sampled 16 streams in the Zamora upstream catchment and 7 streams in the San Francisco catchment. Sampling focused on riffles with an altitudinal variation of 350 m among sites. Every site consisted of a 30–60 m reach placed at the outlet of each stream in order to capture the influences of land use upstream on macroinvertebrate communities. According to the sampling frequency applied in other studies with benthic macroinvertebrates both in Ecuador and in other Neotropical regions [4,10,11,30–34], we collected physical–chemical and macroinvertebrate data during both dry (November 2010; December 2012) and wet seasons (May 2011; April 2012) at the same stream reach. Physical–chemical parameters such as pH, temperature (*T*), specific conductance (SC), dissolved oxygen (DO), were assessed in–stream in each of the sampling periods (Table S1) with a portable YSI 556 Multi Probe System and were averaged from several measurements during the sampling campaigns. Grab samples were taken for measuring further water chemistry at the laboratory of the Universidad Técnica Particular de Loja (UTPL) using standard methods [35]: nitrate (NO_3^-), phosphate (PO_4^{3-}), biochemical oxygen demand during decomposition in a period of 5 days (BOD_5), total coliforms (TC) and turbidity (Table S1).

For sampling of macroinvertebrate communities several methods exist and each one provides different results, nevertheless a combination of sampling methods is preferable [30]. In this study, qualitative macroinvertebrates samples were taken along the stream reach applying two sampling methods: 1) collecting macroinvertebrates directly in their stream habitats from stones and leaves, and 2) using a D–frame net (500 µm mesh) for kick sampling at 3–6 points selected randomly. Collection method 1 was applied during 10 min along the reach to remove all visible macroinvertebrates from stones and collect leaves respectively. Selected stones collected were totally submerged and had an average diameter of 20 cm approximately. Leaves were stored in a plastic bag and rinsed with tap water to remove macroinvertebrates in the laboratory. For collection method 2, macroinverte-

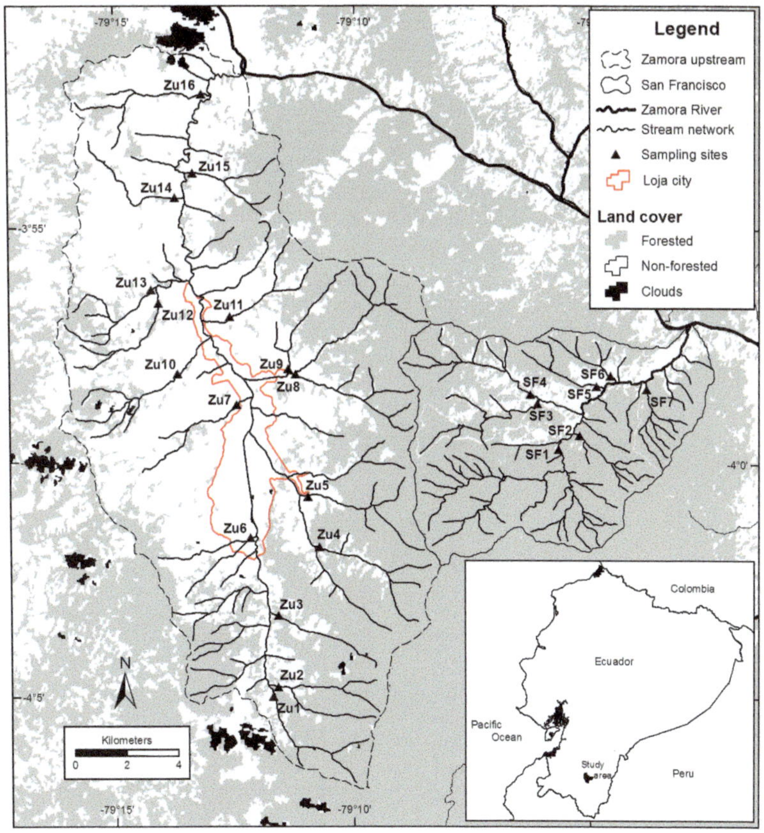

Figure 1. Study area and location of the sampling sites. Codes represent the sampling sites in the Zamora upstream (Zu1 to Zu16) and San Francisco (SF1 to SF7) catchments. Native vegetation cover was classified into forested and non-forested. Only the complete stream network for the study area is shown. Catchment characteristics are further shown in Table S1.

brates were sampled by kicking the substrate while holding the D–frame net downstream at each point. The application of this technique was based on its suitability for sampling in a variety of stream substrates and depths [36,37]. Both collections were combined into a single collection per site in order to avoid differences on macroinvertebrate communities due to seasonality of the Ecuadorian Andes [38,39]. All macroinvertebrates were preserved in 95% ethanol and brought to the laboratory to be sorted and identified to the lowest taxonomical level possible using identification keys for South American macroinvertebrates [40,41].

GIS data and vegetation cover

The Geographic Information System software ArcGIS 9.3 was used to evaluate spatial digital information. Catchments were delineated from a Digital Elevation Model (DEM) with a10 m resolution, available at the Research Unit 816 (RU 816) database (www.tropicalmountainforest.org/) of the Deutsche Forschungsgemeinschaft (DFG). Stream networks were obtained from the national cartography created by the National Mapping Agency (IGM; Instituto Geográfico Militar).

Based on Allan [13], the influence of vegetation cover (VC) on stream condition was evaluated at three spatial scales: 1) catchment scale representing the entire area upstream of the sampling site; 2) riparian scale, using a buffer of 30 m on each bank along the entire length of the stream; and 3) local scale, comprised by the same buffer width along a 100 m reach upstream of the sampling site. Buffer width chosen was based on regulations by local governments (called Municipal Ordinances) in Ecuador. These ordinances reflect several law articles of the Ecuadorian Government in terms of water resources protection, but it does not specify or recommend an amount of riparian vegetation.

A 2010 land use map provided by M.F. Tapia–Armijos (pers. communication) was used to quantify VC within each catchment. This map was generated for southern Ecuador from ASTER images (http://asterweb.jpl.nasa.gov/) with a minimum mapping unit of 1 ha and classified into eight main land use types: páramo, forest, shrub, pasture, crop, plantation, urban and un–vegetated. In Table 1 we show the land use and percentage area covered by each land use type within the 23 studied catchments, with forest and pasture classes as the most dominating ones. Urban land use was absent or very poorly represented. For this study we aggregated this classification to two land cover categories, i.e. forested and non–forested. Forest class included all types of native vegetation and non–forest class all types of disturbed areas (pasture, cropland, plantation, urban, un–vegetated). The forest class was used to determine the VC only at catchment scale, because resolution and accuracy may limit the interpretation of riparian land cover in mountainous areas [13,25,42]. In order to address the riparian and local scales, the VC within the 30 m buffer was digitized and interpreted from 2011 rectified colored orthophotographs (0.30 m resolution). These orthophotographs were obtained and provided by the Ecuadorian project SIG-TIERRAS (Ministerio de Agricultura, Ganadería, Acuacultura y

Table 1. Land use characteristics and area (%) of each land use type of the 23 studied catchments.

Catchment	Páramo	Forest	Shrub	Pasture	Plantation	Crop	Urban	Un-vegetated
Zu1	0.8	69.4	13.4	13.6	0.3	2.3	0.0	0.2
Zu2	9.3	61.1	14.4	12.5	0.5	2.1	0.0	0.0
Zu3	1.4	59.4	10.7	25.0	0.7	2.6	0.0	0.1
Zu4	2.8	46.9	19.3	25.2	0.5	5.3	0.0	0.2
Zu5	4.4	54.9	27.9	10.4	0.9	1.4	0.1	0.1
Zu6	0.3	8.8	17.6	66.0	2.2	2.9	1.5	0.7
Zu7	0.0	4.6	13.2	71.1	0.8	9.0	0.0	1.2
Zu8	2.7	39.0	25.6	30.9	0.7	1.0	0.0	0.2
Zu9	0.5	41.0	12.4	35.9	1.5	8.3	0.2	0.1
Zu10	0.4	8.0	28.0	47.1	2.5	13.4	0.0	0.6
Zu11	0.4	33.7	9.1	42.0	2.7	11.9	0.0	0.2
Zu12	0.2	17.6	4.7	57.3	2.9	16.3	0.0	0.9
Zu13	0.4	16.0	19.2	48.9	1.2	13.5	0.0	0.7
Zu14	0.0	6.5	0.1	73.2	1.9	15.0	0.0	3.3
Zu15	0.0	51.9	9.9	32.9	1.1	4.0	0.0	0.2
Zu16	0.0	15.6	0.2	72.4	2.5	7.0	0.0	2.3
SF1	0.0	82.3	5.3	5.6	0.0	0.0	0.0	6.8
SF2	2.0	93.5	4.0	0.0	0.0	0.0	0.0	0.5
SF3	0.0	78.0	5.0	15.8	0.3	0.0	0.0	0.9
SF4	1.9	87.6	3.7	6.3	0.1	0.0	0.0	0.3
SF5	0.4	81.8	3.0	14.1	0.4	0.0	0.0	0.3
SF6	0.4	76.3	2.8	18.9	1.2	0.0	0.0	0.4
SF7	2.0	92.0	3.9	2.0	0.0	0.0	0.0	0.1

Pesca; Proyecto Sistema Nacional de Información y Gestión de Tierras Rurales e Infraestructura Tecnológica).

Data analysis

We calculated nine indices (also referred as metrics) to describe ecological conditions of the streams. We calculated three structural, two compositional and four water quality indices (Table 2) that allowed us assessing stream conditions in response to increasing VC. The structural indices were richness, Fisher's index, and evenness (Pielou's index). For compositional indices we classified macroinvertebrates by functional feeding groups (FFG) according to [43,44], but we only used percentage of scrappers and shredders because they have an expected decreasing response to perturbation [36], in our case deforestation.

Water quality was assessed with the Water Quality Index (WQI), the EPT index, the Biological Monitoring Working Party index adapted to Colombia (BMWP/Col), and the %5 Dominant taxa. WQI is a single score composed of several measurements of various physical, chemical and bacteriological parameters, where low WQI values show high influence of environmental stress [45]. EPT index shows the taxa richness within the macroinvertebrate groups which are considered sensitive to stressors, and therefore should increase with increasing water quality [46]. BMWP/Col index provides a single value at the family level, representative of the organism's tolerance to stressors, where high tolerance to pollution corresponds to low BMWP scores [40]. %5 Dominant taxa is a measure of the abundance of the five most dominant taxa to the total number of organisms in the sample at each site [36]. When a community is dominated by few taxa it will have high dominance values, thus indicating the community is under the influence of environmental stress.

Two statistical approaches have been used to evaluate relationships between vegetation cover/land use and ecological indices; these are regression [14,42,47] and ordination analyses [4,10,48]. Prior to analyses, macroinvertebrate abundance data (x) were transformed by $\log_{10}(x)$ to reduce the effect of large numbers in single taxa (e.g. *Baetodes*). Rare taxa, i.e. taxa represented by three or less individuals in a sampling site, were excluded in order to avoid the influence of these taxa in the ordination analysis [49]. Because explanatory variables did not have a uniform scale, stream parameters were $\log_{10}(x+1)$ transformed in order to avoid the problem with 0 values and reduce the range of data. All of these variables were analyzed by pairs in scatterplot matrices, and through an exploratory Redundancy Analysis (RDA) to identify the least correlated and most informative variables (PO_4^{3-}, DO, TC and turbidity).

An important issue was to identify whether ecological condition of the streams was attributable to a land use type and/or a stream parameter. We used a stepwise multiple linear regression (SMLR) with calculated macroinvertebrate metrics as dependent variables against land use and PO_4^{3-}, DO, TC and turbidity to determine which of those independent variables affect our response variables. SMLR was applied with the 'stepAIC' function of 'MASS' package in R [50] using backward elimination of variables and Akaike Information Criterion for variables or predictors selection. In addition, we estimated the relative importance of predictors based on their contribution to R^2 using boostrap ('relaimpo' package). All p–values were adjusted with a Bonferroni correction for multiple comparisons during regression analyses. These analyses indicated that forest class (i.e. VC) was consistently more informative to explain macroinvertebrate metrics than other land use types or stream parameter (see Table S2 for SMLR results). Therefore, to examine the linear dependency of each of our response variables with increasing VC, macroinvertebrate metrics (dependent) were regressed against percentages of VC at each spatial scale (independent) using the method of least mean squares ('lm' function of 'stats' package).

We used a RDA ('rda' function of 'vegan' package) to identify macroinvertebrate community assemblages associated with percentages of VC, PO_4^{3-}, DO, TC and turbidity. RDA evaluates multivariate response data and is a powerful tool of constrained ordination in community ecology [51]. RDA works on a matrix Y of response data (e.g. macroinvertebrate data) and a matrix X of explanatory variables (e.g. water chemistry, VC). It identifies groups and shows the vector fitting of variables used in the analysis. While the vector fitting is commonly used for environmental interpretation in ordination diagrams, this implies a linear relationship between ordination scores and the environment

Table 2. Results of regression analysis between calculated macroinvertebrate metrics and percentage of vegetation cover at multiple spatial scales.

Spatial scale	Catchment		Buffer		Local	
Metric	R^2	p	R^2	p	R^2	p
Structural						
Richness	0.30	<0.01	0.34	<0.001	0.10	0.13
Fisher's index	0.50	<0.001	0.38	<0.001	0.01	0.30
Evenness	0.24	<0.01	0.11	0.11	0.00	0.93
Compositional						
% Scrappers	0.02	0.57	0.01	0.72	0.02	0.55
% Shredders	0.42	<0.001	0.46	<0.001	0.12	0.09
Water Quality						
WQI[†]	0.56	<0.001	0.50	<0.001	0.12	0.11
EPT index	0.53	<0.001	0.59	<0.001	0.03	0.39
BMWP/Col	0.44	<0.001	0.46	<0.001	0.17	0.05
%5 Dominant taxa	0.85	<0.001	0.60	<0.001	0.02	0.50

[†]WQI is the unique metric derived from water chemistry data showed in Table S1.

variables and may not always be appropriate if it is desired to understand the ordination of studied sites or species in relation to one or more variables. Thus, we fitted the surface of the best predictor variable, identified with an ANOVA of the RDA results, to detect how macroinvertebrate communities are grouped and spread in relation to this variable. We examined whether macroinvertebrate communities were different between those groups using an Analysis of Similarities (ANOSIM; 'anosim' function of 'vegan' package). ANOSIM is a nonparametric procedure that provides a test of whether there is a significant difference between two or more groups of sampling units.

Results

Stream condition and vegetation cover

Relationships between VC and all metrics are generally stronger at catchment and buffer scale than at local scale where associations are not apparent (Table 2). Significant correlations are found on catchment and buffer scale in 90% and 80% of all cases. Within the structural metrics, only evenness does not show significant relationship with the percentage of VC at buffer scale ($R^2 = 0.11$; $p = 0.11$). In general, correlations decrease with smaller spatial scales, disappearing completely at local scale.

As for compositional metrics, only one FFG shows a strong relationship with VC at the two larger scales ($p<0.001$). The percentage of shredders is more related to VC at catchment and buffer scales (Table 2). Scrappers show weaker regressions at all spatial scales ($R^2<0.02$). At local scale no significant correlation existed at all.

All water quality metrics have stronger relationships at catchment and buffer scale than at local scale ($R^2>0.44$ and $p<0.001$, Table 2). Within water quality metrics, the %5 Dominant taxa depicts the highest correlation across all scales with 0.85 on catchment scale. Regression slopes indicate significant differences in water quality when VC increases at both spatial scales. However, there are some streams with a large area covered by forest at catchment and buffer scale which present low values in water quality indices (Figure 2B–C; F–G). For example, in both spatial scales the EPT index increase with VC ($R^2 = 0.53$; $p<0.001$ and $R^2 = 0.59$; $p<0.001$), but there is one site with approximately 70% of forest cover whose index value is less than 50, indicating fairly poor water quality. In addition, the strongest relationship is inversely correlated between the %5 Dominant taxa and the percentage of VC in the catchment ($R^2 = 0.85$; $p = 0.001$). As a general pattern most of the macroinvertebrate metrics indicate good water quality when the percentage of native VC is higher than 70% within the catchment and the riparian buffer (see Figure 2).

Macroinvertebrate communities and vegetation cover

All structural indices are related with VC at catchment scale rather than at other spatial scales (Table 2). Thus, we decided to further analyze the community assemblage patterns related to catchment forest cover. A correlation analysis run for those variables showed scatterplot matrices and correlation values from 0 to 1.We defined a decision threshold of 0.50 to remove those variables which are correlated with more than one variable. Remaining variables include VC, DO, PO_4^{3-}, TC and turbidity. RDA reveal that 51% of the variation in the data can be explained by these five selected variables and the ANOVA identified that catchment VC is the best predictor for macroinvertebrate composition in the ordination ($p<0.001$). In Figure 3A, most variables are spread along negative axis 1 and negative axis 2, with VC, PO_4^{3-} and turbidity being the most influential factors for axis 1, while DO and TC are more associated with axis 2. Figure 3B depicts the fitted surface of the best predictor variable (i.e. VC) to detect how macroinvertebrate communities are grouped and spread in relation to VC. Thus the ordination plot illustrates two visibly distinguishable groups; one group associated to catchments with a forest cover above 70% which is even more aggregated than the other group related to lower VC. The ANOSIM indicates that both groups of macroinvertebrate assemblages were strongly dissimilar between both ranges of VC ($R = 0.55$; $p = 0.001$). Both RDA plot and Bray–Curtis distance of the ANOSIM suggest that macroinvertebrate assemblages differ in taxonomical composition when there is a decline in the percentage of VC within the catchments.

Overall, macroinvertebrate communities in the 23 study streams present specific patterns. They are composed of a total of 13,512 individuals belonging to 11 orders, 33 families and 53 genera (data available at PANGAEA – Data Publisher for Earth & Environmental Science; doi:10.1594/PANGAEA.831687). Only two taxa (taxa–1 = Empididae; taxa–2 = Pyralidae) were not identified to genera level.

Comparing occurrence and abundance patterns between both groups of catchments (i.e. less than 70% and greater than 70% VC) we detected several singularities. In catchments with a VC between 70 to 100% we collected 4,061 individuals of 11 orders. In contrast 9,451 individuals of nine orders were found in catchments with less VC (Basommatophora and Lepidoptera were absent). We found 33 families, all of them present in catchments highly forested; whilst 10 families were absent in catchments with less than 70% forest cover (Calamoceratidae, Ceratopogonidae, Coenagrionidae, Dolichopodidae, Gyrinidae, Odontoceridae, Physidae, Psephenidae, Ptilodactylidae, and Pyralidae). With regard to the 53 genera found, only *Scirtes* (Coleptera) and *Molophilus* (Diptera) were absent in catchments above 70% forest cover, and 25 genera were absent in catchments with lower forest cover showing the greatest difference between both groups of catchments at this taxonomic level. Here genera absent at <70% catchment VC were: *Physa* (Basommatophora); *Noelmis*, *Phanacerus*, *Dineutus*, *Anchytarsus*, *Psephenops*, *Elodes* (Coleoptera); *Probezzia*, *Aphrosytus*, *Hexatoma* (Diptera); *Andesiops*, *Moribaetis*, *Haplohyphes*, *Tricorythodes*, *Farrodes* (Ephemeroptera); taxa–2 (Lepidoptera); *Argia* (Odonata); *Phylloicus*, *Leptonema*, *Macrostemum*, *Grumichella*, *Nectopsyche*, *Triplectes*, *Marilia*, *Cernotina* (Tricoptera).

A trend can be observed here as well, confirmed by the ANOSIM, as there are differences in macroinvertebrate compositions between both groups of catchments (Table 3). We detect an increase for almost all metrics of the macroinvertebrate communities with an increase of the percentage of VC in the catchment, except for evenness where no changes are observed.

Discussion

Stream condition and vegetation cover

Our results show a stronger relationship between VC and the macroinvertebrate metrics at catchment and buffer scales. This suggests an important linkage between headwater and riparian forests to conserve diversity and water quality of the streams in Ecuadorian Andean catchments. We observe that stream condition begins to improve when native VC within the catchment is over 70% (see Figure 2). Therefore our suggestion is to retain at least 70% native VC within the catchment in order to keep sufficient ecological condition of the streams. Such a percentage of VC can contribute significantly to riparian management for protection or restoration of watersheds where forest has been

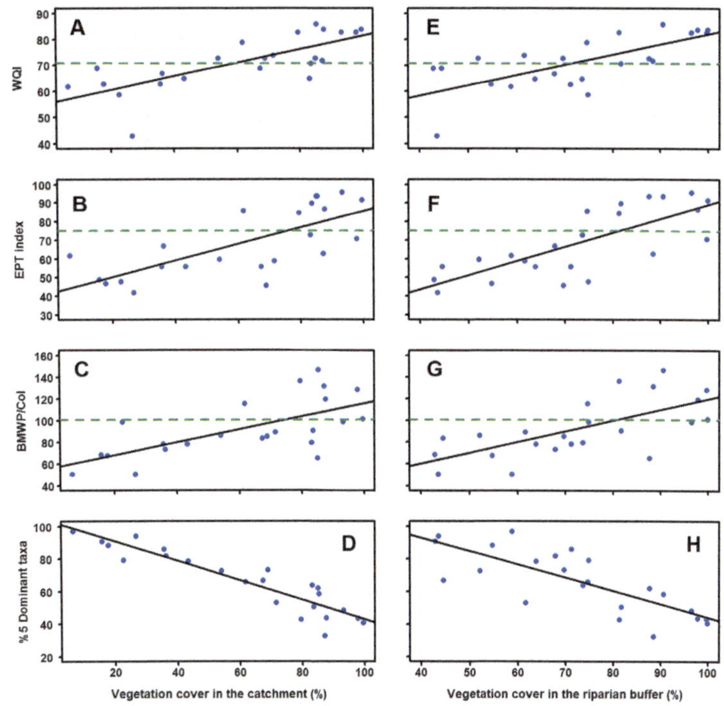

Figure 2. Water quality indices plotted against vegetation cover (%) at catchment and buffer scales. Horizontal green dashed lines represent thresholds of good water quality defined for each metric in the literature. Only the %5 Dominant index does not have ranges of tolerance, but high percentages indicate stress in the communities. For both spatial scales R^2 values are over 0.44 with a *p*–value <0.001, see Table 2 for specific values of each regression.

replaced by grasslands. Further, our findings show that local scale strategies (e.g. conserving small forest patches in riparian areas) do not improve water quality and are unlikely to improve the overall ecological condition of streams. According to [13,47], the maintenance of ecological conditions may require conservation of more than only a few native forest patches, and efforts should

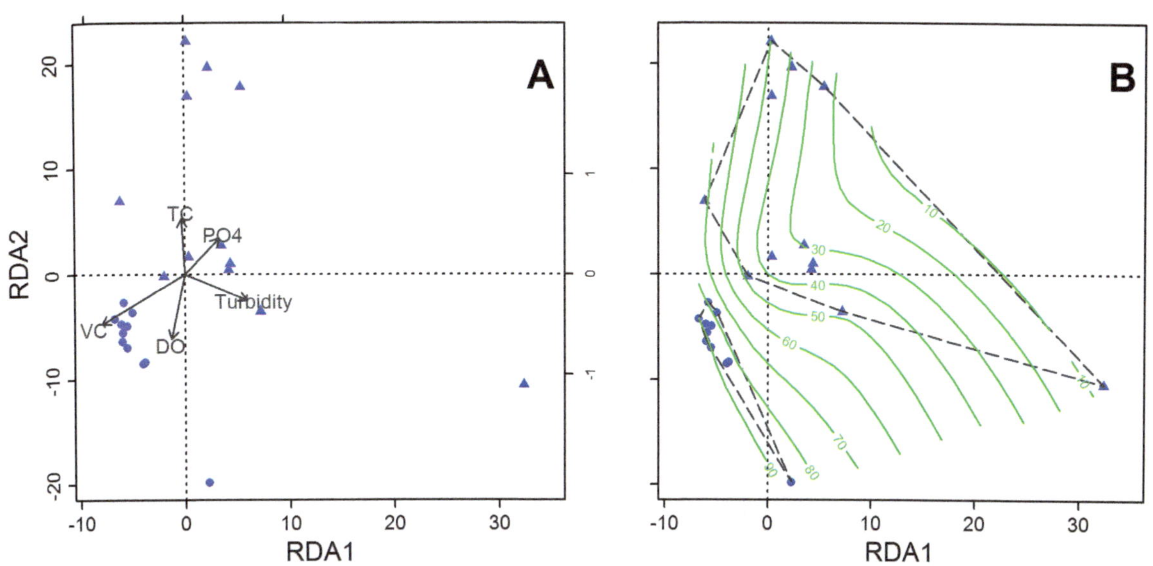

Figure 3. Redundancy Analysis (RDA) of macroinvertebrate community assemblages from catchments with different vegetation covers (VC). This ordination analysis detected two groups of macroinvertebrate assemblages: solid circles represent macroinvertebrate communities in catchments with 70–95% in VC and triangles indicate macroinvertebrate communities in catchments with less than 70% in VC. Each symbol represents the community of a single catchment, i.e. macroinvertebrate community at catchment SF3. A Shows the RDA ordination with vector fitting of five variables selected during analysis. DO = dissolved oxygen, TC = total coliforms, PO_4 = phosphate. B Displays in green curves the fitted surface only for the percentage of VC, which is the most explicative variable of the ordination (65%). Dashed lines enclose and separate the detected groups.

Table 3. Diversity patterns of macroinvertebrate communities in both groups of catchments identified in RDA ordination plot (Figure 3B). VC = catchment vegetation cover.

Vegetation cover	<70% VC		>70% VC		Response to increasing VC
Metric	**Mean**	**SD**	**Mean**	**SD**	
Structural					
Richness	14.3	2.1	20.1	6.2	Increase
Fisher's index	2.5	0.4	4.9	1.3	Increase
Evenness	0.7	0.1	0.7	0.1	Variable
Compositional					
% Scrappers	26.0	16.3	30.6	17.3	Increase
% Shredders	0.3	0.8	3.1	2.6	Increase

focus on considering larger spatial patterns that have more important consequences for stream conservation.

Macroinvertebrates diversity is more related to native VC at catchment and buffer scale than at a local scale. Many studies have documented the influence of upstream vegetation on stream ecological condition downstream, especially on water chemistry and habitat conditions [13,14,52]. For example, DO increases as the stream enters the forest [53]. Likewise, the similar relationships shown between water quality metrics and VC at catchment and buffer scale is not surprising given the strong longitudinal connectivity in a riverine landscape [13,54]. Riparian forests can influence water chemistry and habitat quality, which are directly linked to the macroinvertebrate metrics and help to minimize the effects of catchment deforestation on streams [4,42,55].

In this study, there were two particular sampling points in streams of catchments Zu4 and Zu8 (see Figure 1 and Table S1). Both had low values in water quality indices but a relatively high VC (60–68%). We assume that livestock and minor roads which are located parallel to the streams influence these metrics. For example, in [4] the authors found a stream affected by inputs of sediments and nutrients derived from roads and livestock, despite having a wide riparian buffer. We believe those elements are affecting the macroinvertebrate composition of very well forested catchments as discussed below.

Macroinvertebrate communities, indicators and vegetation cover

Analyzing the composition of macroinvertebrate communities of 23 Andean streams, we are able to identify direct relationships with the coverage of the catchment's native forest. The difference between macroinvertebrate assemblages increases when decreasing the percentage of native VC. We find a group with a more aggregated distribution when VC is large in the RDA plot. This pattern was also observed by [14] in a larger sample of streams in New Zealand. It supports the premise that deforestation is affecting the macroinvertebrate community assemblages in Ecuadorian Andean streams as in other Neotropical streams [4,21,56].

The difference between both groups of macroinvertebrate communities detected in the RDA is also revealed in the ANOSIM. Bücker et al. [10] found similar patterns in the San Francisco catchment having less variation at structural and compositional level and higher water quality metric values in forested than in deforested streams. Comparable results were also found in other Andean streams [11] and Amazon streams [21] of northern Ecuador. Nonetheless, among all metrics used here and mostly applied in similar studies, %5 Dominant taxa reveals a superior fitting to the regression slope despite it has not been used to reflect deforestation effects on streams before. In [36], authors denoted that high values of this metric indicate a community under environmental stress.

In terms of the FFG composition, we observe a direct response to increasing VC in both groups of macroinvertebrates (Table 3). In average, the abundance of shredders present in streams with > 70% VC in catchment is always higher than in less forested catchments. Shredders abundance is typically linked to leaf litter from riparian vegetation and it is sensitive to riparian zone impacts [43,57,58]. Shedders could be considered as an indicator of deforestation in Andean streams, despite proportion of shredders in macroinvertebrate communities in our catchments are always small (see Table 3). According to [4,5] this seems to be a general pattern in tropical streams. On average, scrapers abundance is slightly higher in more forested catchments in contrast to other Neotropical streams where their abundance is higher in less forested streams [4]. Additionally no association between scrappers and VC was detected.

Management and conservation

Allan et al. [13,47] have recommended that it is preferable to retain as much forest cover as possible in a catchment to preserve natural stream condition. However, in tropical regions like Ecuadorian Andes, deforestation has been increasing throughout the years affecting a large part of catchments along the Andes [29,59,60]. Deforestation has stopped in the upper Andean catchments because of topography and accessibility, preserving headwater forests and/or páramo [29]. However, this native vegetation in the headwaters is not enough to sustain healthy streams. In Andean countries, human activities are especially concentrated on midstream sections and valleys of catchments affecting water/habitat quality downstream and other rivers after the confluence. For example, within Guayas river basin in Ecuador there are several cities and in combination with intensive agricultural land use they are affecting the entire river network and its water quality [61]. Our results suggest that retaining over 70% native forest cover in Andean catchments now dominated by grasslands and keeping the riparian buffers, the water quality and stream biodiversity could be enhanced or at least preserved. However, this task is not so easy to achieve if there is not a strong regulation at national level.

The Ecuadorian constitution defines the right to water and promotes the protection, restoration and management of basins and water resources; but it does not mention how much forest needs to be retained in a catchment or a recommended width for riparian buffers. Best conservation actions that intend maintaining as much forest as possible have been focused on catchments for drinking water supply from initiatives of local governments (Municipalities) supported by non–governmental organizations, through the establishment of trust funds for catchment conservation. Also, local governments through Municipal Ordinances have decided to protect riparian strips with different buffer widths (15–30 m), depending if the stream is running in the city or outside the urban boundary. Riparian buffers could considerably reduce the effects of deforestation on benthic communities and water quality, but there is need for a proper management of riparian areas [4]. Unfortunately, Municipal Ordinances in Ecuador do not include the management of riparian buffers. Here, we assessed the effect of a 30 m buffer width on stream condition showing strong associations with the water quality metrics. Therefore, we also recommend protection and an effective management of native riparian forest buffers in Andean streams to assist headwater forests to sustain ecological condition of the streams.

We did not evaluate any management approaches of riparian areas, but we would like to recommend some strategies that governments can include in national regulations. We believe that the most important problems that managers will face in Andean streams are livestock density, residential areas and parallel roads along riparian zones which affect water quality and macroinvertebrate communities in forested catchments. Therefore, these problems must be addressed; for example, livestock exclusion jointly with a long–term research program to assess its effectiveness [62]; or an in–depth monitoring to further study road effects on stream and riparian vegetation [63]. Another important issue to be considered is the restoration of riparian vegetation, mainly because it enhances water quality, as our study indicates. Restoration positively affects habitat and macroinvertebrate community heterogeneity [48] and can be ecologically successful by using native plant species and avoiding the use of invasive plant species, which can affect structure and functionally of streams [64]. Finally, as mentioned by [4] we would like to emphasize that "riparian forest buffers are not a replacement for continuous forest, but can promote biodiversity and stream ecosystem integrity in tropical catchments affected by deforestation".

Supporting Information

Table S1 List of environmental variables considered in the 23 studied Andean streams. Data of water parameters are means obtained from several measurements in two sampling campaigns. Altitude was determined once at the sampling sites. Catchment area and percentage of vegetation cover (VC) at three spatial scales were calculated in ArcGIS 9.3 Geographic Information System software. T = Temperature, SC = Specific Conductance, DO = Dissolved Oxygen, NO_3^- = Nitrate, PO_4^{3-} = Phosphate, BOD_5 = Biochemical Oxygen Demand in 5 days. Turbidity in Nephelometric Turbidity Units or NTU and Total Coliforms (TC) as Colony Forming Units or CFU.

Table S2 Stepwise multiple linear regression analysis between macroinvertebrate metrics, land use and stream parameters for 23 Andean streams. Degrees of freedom for all regressions were 1, 20. AIC = Akaike Information Criterion used for selection of predictor variables. DO = Dissolved Oxygen, PO_4^{3-} = Phosphate, TC = Total Coliforms.

Acknowledgments

Special thanks are due to the personnel of the San Francisco Research Station, Nature and Culture International, Municipality of Loja and the Ministry of Environment for the provision of research facilities. Many thanks are due to C. Prieto to provide us additional information on stream parameters and macroinvertebrates of the Zamora catchment; and to L.M. Cun for his support with field sampling and stream measurements. The help of C. Romero, S. Aguilar and M. Villa with water analyses is appreciated. We would like to thank S. Bourret for proof reading, C. Espinosa for his assistance with R functions and routines, and P.F. Quintana–Ascencio and two anonymous reviewers for constructive comments on previous drafts of the manuscript.

Author Contributions

Conceived and designed the experiments: CIA HGF LB. Performed the experiments: CIA AL. Analyzed the data: CIA LB HH. Contributed reagents/materials/analysis tools: CIA HGF LB. Wrote the paper: CIA HH LB.

References

1. Laurance WF (1999) Reflections on the tropical deforestation crisis. Biol Conserv 91: 109–117.
2. Achard F, Eva HD, Stibig H-J, Mayaux P, Gallego J, et al. (2002) Determination of deforestation rates of the world's humid tropical forests. Science (80-) 297: 999–1002.
3. Lambin EF, Geist HJ, Lepers E (2003) Dynamics of l and-use and land -cover change in tropical regions. Annu Rev Environ Resour 28: 205–241.
4. Lorion CM, Kennedy BP (2009) Relationships between deforestation, riparian forest buffers and benthic macroinvertebrates in neotropical headwater streams. Freshw Biol 54: 165–180.
5. Wantzen KM, Wagner R (2006) Detritus processing by invertebrate shredders: a neotropical – temperate comparison. J North Am Benthol Soc 25: 216–232.
6. Iwata T, Nakano S, Inoue M (2003) Impacts of past riparian deforestation on stream communities in a tropical rain forest in Borneo. Ecol Appl 13: 461–473.
7. Ordóñez MV (2011) Influencia del uso del suelo y la cobertura vegetal natural en la integridad ecológica de los ríos altoandinos al noreste del Ecuador Universidad San Francisco de Quito.
8. Sierra R, Campos F, Chamberlin J (2002) Assessing biodiversity conservation priorities: ecosystem risk and representativeness in continental Ecuador. Landsc Urban Plan 59: 95–110.
9. Buytaert W, Cuesta-Camacho F, Tobón C (2011) Potential impacts of climate change on the environmental services of humid tropical alpine regions. Glob Ecol Biogeogr 20: 19–33.
10. Bücker A, Sondermann M, Frede H-G, Breuer L (2010) The influence of land-use on macroinvertebrate communities in montane tropical streams – a case study from Ecuador. Fundam Appl Limnol 177: 267–282.
11. Encalada AC, Calles J, Ferreira V, Canhoto CM, Graça MAS (2010) Riparian land use and the relationship between the benthos and litter decomposition in tropical montane streams. Freshw Biol 55: 1719–1733.
12. Crespo PJ, Feyen J, Buytaert W, Bücker A, Breuer L, et al. (2011) Identifying controls of the rainfall–runoff response of small catchments in the tropical Andes (Ecuador). J Hydrol 407: 164–174.
13. Allan JD (2004) Landscapes and riverscapes: the influence of land use on stream ecosystems. Annu Rev Ecol Evol Syst 35: 257–284.
14. Death RG, Collier KJ (2009) Measuring stream macroinvertebrate responses to gradients of vegetation cover: when is enough enough? Freshw Biol 55: 1447–1464.
15. Van Sickle J, Baker J, Herlihy A, Bayley P, Gregory S, et al. (2004) Projecting the biological condition of streams under alternative scenarios of human land use. Ecol Appl 14: 368–380.
16. Collier KJ, Hamer M (2012) The ecological condition of Waikato wadeable streams based on the Regional Ecological Monitoring of Streams (REMS) Programme. Waikato Regional Council, Technical Report, Waikato.
17. European Environment Agency (2012) European waters - assessment of status and pressures. EEA Report No 8. Copenhagen.
18. Environmental Protection Agency (2013) National Rivers and Streams Assessment 2008–2009: A Collaborative Survey. Draft February 28. Washington.
19. Perrie A, Morar S, Milne JR, Greenfield S (2012) River and stream water quality and ecology in the Wellington region: State and trends. Greater Wellington Regional Council, Publication No. GW/EMI-T-12/143, Wellington.
20. Parkyn SM, Davies-Colley RJ, Halliday NJ, Costley KJ, Croker GF (2003) Planted Riparian Buffer Zones in New Zealand: Do They Live Up to Expectations? Restor Ecol 11: 436–447.
21. Bojsen BH, Jacobsen D (2003) Effects of deforestation on macroinvertebrate diversity and assemblage structure in Ecuadorian Amazon streams. Arch für Hydrobiol 158: 317–342.

22. Miserendino ML, Casaux R, Archangelsky M, Di Prinzio CY, Brand C, et al. (2011) Assessing land-use effects on water quality, in-stream habitat, riparian ecosystems and biodiversity in Patagonian northwest streams. Sci Total Environ 409: 612–624.
23. Villamarín C, Rieradevall M, Paul MJ, Barbour MT, Prat N (2013) A tool to assess the ecological condition of tropical high Andean streams in Ecuador and Peru: The IMEERA index. Ecol Indic 29: 79–92.
24. Gergel SE, Turner MG, Miller JR, Melack JM, Stanley EH (2002) Landscape indicators of human impacts to riverine systems. Aquat Sci 64: 118–128.
25. Stewart JS, Wang L, Lyons J, Horwatich JA, Bannerman R (2001) Influences of watershed, riparian-cprridor and reach-scale characteristics on aquatic biota in agricultural watersheds. J Am Water Resour Assoc 37: 1475–1487.
26. Sutherland AB, Meyer JL, Gardiner EP (2002) Effects of land cover on sediment regime and fish assemblage structure in four southern Appalachian streams. Freshw Biol 47: 1791–1805.
27. Göttlicher D, Obregón A, Homeier J, Rollenbeck R, Nauss T, et al. (2009) Land-cover classification in the Andes of southern Ecuador using Landsat ETM+ data as a basis for SVAT modelling. Int J Remote Sens 30: 1867–1886.
28. Thies B, Meyer H, Nauss T, Bendix J (2014) Projecting land-use and land-cover changes in a tropical mountain forest of Southern Ecuador. J Land Use Sci 9: 1–33.
29. Torrachi S, Tapia MF, Escudero A, de la Cruz M (2013) Deforestación en una región montañosa megadiversa en los Andes: Dinámica del paisaje en el sur de Ecuador. In: de la Cruz M, Maestre FT, editors. Avances en el Análisis Espacial de Datos Ecológicos: Aspectos Metodológicos y Aplicados. Móstoles: ECESPA–Asociación Española de Ecología Terrestre. pp. 275–289.
30. Stein H, Springer M, Kohlmann B (2008) Comparison of two sampling methods for biomonitoring using aquatic macroinvertebrates in the Dos Novillos River, Costa Rica. Ecol Eng 34: 267–275.
31. De Cáceres M, Legendre P, Moretti M (2010) Improving indicator species analysis by combining groups of sites. Oikos 119: 1674–1684.
32. Sites RW, Willig MR, Linit MJ (2003) Macroecology of Aquatic Insects: A Quantitative Analysis of Taxonomic Richness and Composition in the Andes Mountains of Northern Ecuador. Biotropica 35: 226–239.
33. Jacobsen D (2004) Contrasting patterns in local and zonal family richness of stream invertebrates along an Andean altitudinal gradient. Freshw Biol 49: 1293–1305.
34. Tarras-Wahlberg N., Flachier A, Lane S., Sangfors O (2001) Environmental impacts and metal exposure of aquatic ecosystems in rivers contaminated by small scale gold mining: the Puyango River basin, southern Ecuador. Sci Total Environ 278: 239–261.
35. American Public Health Association (1995) Standard Methods for the Examination of Water and Wastewater. 19th ed. American Public Health Association, editor Washington DC.
36. Barbour MT, Gerritsen J, Snyder BD, Stribling JB (1999) Rapid Bioassessment Protocols for Use in Streams and Wadeable Rivers: Periphyton, Benthic Macroinvertebrates, and Fish. 2nd ed. Environmental Protection Agency.
37. Hauer FR, Resh VH (2007) Macroinvertebrates. In: Hauer FR, Lamberti GA, editors. Methods in Stream Ecology. San Diego CA: Academic Press. pp. 435–454.
38. Ríos-Touma B, Encalada AC, Prat Fornells N (2011) Macroinvertebrate Assemblages of an Andean High-Altitude Tropical Stream: The Importance of Season and Flow. Int Rev Hydrobiol 96: 667–685.
39. Jacobsen D, Encalada AC (1998) The Macroinvertebrate fauna of Ecuadorian high-land streams in the wet and dry season. Arch für Hydrobiol 142: 53–70.
40. Roldán Pérez G (2003) Bioindicación de la calidad del agua en Colombia. Uso del método BMWP/Col. 1st ed. Medellín: Editorial Universidad de Antioquia.
41. Fernádez HR, Domínguez E (2001) Guía para la determinación de los artrópodos bentónicos sudamericanos. 1st ed. Tucumán: Universidad Nacional de Tucumán.
42. Sponseller R, Benfield E, Valett H (2001) Relationships between land use, spatial scale and stream macroinvertebrate communities. Freshw Biol.
43. Cummins KW, Merritt RW, Andrade PCN (2005) The use of invertebrate functional groups to characterize ecosystem attributes in selected streams and rivers in south Brazil. Stud Neotrop Fauna Environ 40: 69–89.
44. Tomanova S, Goitia E, Helešic J (2006) Trophic levels and functional feeding groups of macroinvertebrates in neotropical streams. Hydrobiologia 556: 251–264.
45. Brown R (1970) A Water Quality Index – Do We Dare? Water Sew Work 11: 339–343.
46. Lenat DR (1988) Water quality assessment of streams using a qualitative collection method for benthic macroinvertebrates. J North Am Benthol Soc 7: 222–233.
47. Allan JD, Erickson DL, Fay J (1997) The influence of catchment land use on stream integrity across multiple spatial scales. Freshw Biol 37: 149–161.
48. Harding JS, Benfield EF, Bolstad P V, Helfman GS, Jones III EBD (1998) Stream biodiversity: The ghost of land use past. Proc Natl Acad Sci U S A 95: 14843–14847.
49. Legendre P, Gallagher E (2001) Ecologically meaningful transformations for ordination of species data. Oecologia 129: 271–280.
50. R Development Core Team (2011) R: A language and environment for statistical computing. R Foundation for Statistical Computing, Vienna. http://www.r-project.org/.
51. Bocard D, Gillet F, Legendre P (2011) Numerical Ecology with R. 1st ed. New York: Springer.
52. Harding JS, Claassen K, Evers N (2006) Can forest fragments reset physical and water quality conditions in agricultural catchments and act as refugia for forest stream invertebrates? Hydrobiologia 568: 391–402.
53. Suga CM, Tanaka MO (2012) Influence of a forest remnant on macroinvertebrate communities in a degraded tropical stream. Hydrobiologia 703: 203–213.
54. Shandas V, Alberti M (2009) Exploring the role of vegetation fragmentation on aquatic conditions: Linking upland with riparian areas in Puget Sound lowland streams. Landsc Urban Plan 90: 66–75.
55. Rios SL, Bailey RC (2006) Relationship between Riparian Vegetation and Stream Benthic Communities at Three Spatial Scales. Hydrobiologia 553: 153–160.
56. Couceiro SRM, Hamada N, Luz SLB, Forsberg BR, Pimentel TP (2006) Deforestation and sewage effects on aquatic macroinvertebrates in urban streams in Manaus, Amazonas, Brazil. Hydrobiologia 575: 271–284.
57. Lecerf A, Dobson M, Dang CK, Chauvet E (2005) Riparian plant species loss alters trophic dynamics in detritus-based stream ecosystems. Oecologia 146: 432–442.
58. Lecerf A, Richardson JS (2010) Litter decomposition can detect effects of high and moderate levels of forest disturbance on stream condition. For Ecol Manage 259: 2433–2443.
59. James LA, Marcus WA, Harden CP (2006) Human impacts on headwater fluvial systems in the northern and central Andes. Geomorphology 79: 249–263.
60. Molina A, Vanacker V, Balthazar V, Mora D, Govers G (2012) Complex land cover change, water and sediment yield in a degraded Andean environment. J Hydrol 472: 25–35.
61. Borbor-Córdova MJ, Boyer EW, McDowell WH, Hall CA (2006) Nitrogen and phosphorus budgets for a tropical watershed impacted by agricultural land use: Guayas, Ecuador. Biogeochemistry 79: 135–161.
62. Sarr D (2002) Riparian livestock exclosure research in the Western United States: A critique and some recommendations. Environ Manage 30: 516–526.
63. Jones JA, Swanson FJ, Wemple BC, Snyder KU (2000) Effects of roads on hydrology, geomorphology, and disturbance patches in stream networks. Conserv Biol 14: 76–85.
64. Hladyz S, Kajsa A, Giller PS, Woodward G (2011) Impacts of an aggressive riparian invader on community structure and ecosystem functioning in stream food webs. J Appl Ecol 48: 443–452.

Permissions

All chapters in this book were first published in PLOS ONE, by The Public Library of Science; hereby published with permission under the Creative Commons Attribution License or equivalent. Every chapter published in this book has been scrutinized by our experts. Their significance has been extensively debated. The topics covered herein carry significant findings which will fuel the growth of the discipline. They may even be implemented as practical applications or may be referred to as a beginning point for another development.

The contributors of this book come from diverse backgrounds, making this book a truly international effort. This book will bring forth new frontiers with its revolutionizing research information and detailed analysis of the nascent developments around the world.

We would like to thank all the contributing authors for lending their expertise to make the book truly unique. They have played a crucial role in the development of this book. Without their invaluable contributions this book wouldn't have been possible. They have made vital efforts to compile up to date information on the varied aspects of this subject to make this book a valuable addition to the collection of many professionals and students.

This book was conceptualized with the vision of imparting up-to-date information and advanced data in this field. To ensure the same, a matchless editorial board was set up. Every individual on the board went through rigorous rounds of assessment to prove their worth. After which they invested a large part of their time researching and compiling the most relevant data for our readers.

The editorial board has been involved in producing this book since its inception. They have spent rigorous hours researching and exploring the diverse topics which have resulted in the successful publishing of this book. They have passed on their knowledge of decades through this book. To expedite this challenging task, the publisher supported the team at every step. A small team of assistant editors was also appointed to further simplify the editing procedure and attain best results for the readers.

Apart from the editorial board, the designing team has also invested a significant amount of their time in understanding the subject and creating the most relevant covers. They scrutinized every image to scout for the most suitable representation of the subject and create an appropriate cover for the book.

The publishing team has been an ardent support to the editorial, designing and production team. Their endless efforts to recruit the best for this project, has resulted in the accomplishment of this book. They are a veteran in the field of academics and their pool of knowledge is as vast as their experience in printing. Their expertise and guidance has proved useful at every step. Their uncompromising quality standards have made this book an exceptional effort. Their encouragement from time to time has been an inspiration for everyone.

The publisher and the editorial board hope that this book will prove to be a valuable piece of knowledge for researchers, students, practitioners and scholars across the globe.

List of Contributors

David L. Jones
School of the Environment, Natural Resources & Geography, Bangor University, Bangor, Gwynedd, United Kingdom

Paul Simfukwe
School of the Environment, Natural Resources & Geography, Bangor University, Bangor, Gwynedd, United Kingdom

Paul W. Hill
School of the Environment, Natural Resources & Geography, Bangor University, Bangor, Gwynedd, United Kingdom

Robert T. E. Mills
Ecole Polytechnique Fédérale de Lausanne (EPFL), School of Architecture, Civil and Environmental Engineering (ENAC), Laboratory of Ecological Systems (ECOS), Lausanne, Switzerland

Bridget A. Emmett
Centre for Ecology and Hydrology, Environment Centre Wales, Bangor, Gwynedd, United Kingdom

Jiabo Chen
Department of Natural Resources, College of Environment and Natural Resources, Zhejiang University, Hangzhou, Zhejiang Province, China

Jun Lu
Department of Natural Resources, College of Environment and Natural Resources, Zhejiang University, Hangzhou, Zhejiang Province, China, Ministry of Education Key Lab of Environment Remediation and Ecological Health, Zhejiang University, Hangzhou, Zhejiang Province, China

Ian R. Waite
U.S. Geological Survey, Portland, Oregon, United States of America

Jonathan G. Kennen
U.S. Geological Survey, West Trenton, New Jersey, United States of America

Jason T. May
U.S. Geological Survey, Sacramento, California, United States of America

Larry R. Brown
U.S. Geological Survey, Sacramento, California, United States of America

Thomas F. Cuffney
U.S. Geological Survey, Raleigh, North Carolina, United States of America

Kimberly A. Jones
U.S. Geological Survey, Salt Lake City, Utah, United States of America

James L. Orlando
U.S. Geological Survey, Sacramento, California, United States of America

Jacqueline Loos
Institute of Ecology, Leuphana University, Lueneburg, Germany

Ine Dorresteijn
Institute of Ecology, Leuphana University, Lueneburg, Germany

Jan Hanspach
Institute of Ecology, Leuphana University, Lueneburg, Germany

Pascal Fust
Organic Agricultural Science Group, University Kassel, Witzenhausen, Germany

La´szló Rakosy
Department Taxonomy and Ecology, Babes-Bolay University, Cluj-Napoca, Romania

Joern Fischer
Institute of Ecology, Leuphana University, Lueneburg, Germany

Sara A. O. Cousins
Landscape Ecology, Department of Physical Geography and Quaternary Geology, Stockholm University, Stockholm, Sweden

Mitja Kaligaric
Department, Faculty of Natural Sciences and Mathematics, Maribor, Slovenia

Faculty of Agriculture and Life Sciences, University of Maribor, Pivola 10, Hoče, Slovenia

Branko Bakan
Department, Faculty of Natural Sciences and Mathematics, Maribor, Slovenia

Regina Lindborg
Landscape Ecology, Department of Physical Geography and Quaternary Geology, Stockholm University, Stockholm, Sweden

Anja Madelen Ödman and Pål Axel Olsson
Biodiversity, Department of Biology, Lund University, Lund, Sweden

D. Richard Cameron
The Nature Conservancy, San Francisco, California, United States of America

Jaymee Marty
The Nature Conservancy, San Francisco, California, United States of America

Robert F. Holland
Geobotanical Phenomenology, Auburn, California, United States of America

Guangju Zhao
Institute of Soil and Water Conservation, Northwest A&F University, Yangling, Shaanxi Province, China, Institute of Soil and Water Conservation, Chinese Academy of Sciences & Ministry of Water Resources, Yangling, Shaanxi Province, China

Xingmin Mu
Institute of Soil and Water Conservation, Northwest A&F University, Yangling, Shaanxi Province, China Institute of Soil and Water Conservation, Chinese Academy of Sciences & Ministry of Water Resources, Yangling, Shaanxi Province, China

Alex Strehmel
University of Kiel, Institute of Natural Resource Conservation, Department of Hydrology and Water Resources Management, Kiel, Germany

Peng Tian
College of Resource and Environment, Northwest A&F University, Yangling, Shaanxi Province, China

Paul T. Leisnham
Department of Environmental Science and Technology, University of Maryland, College Park, Maryland, United States of America

Shannon L. LaDeau
Cary Institute of Ecosystem Studies, Millbrook, New York, United States of America

Steven A. Juliano
School of Biological Sciences, Illinois State University, Normal, Illinois, United States of America

Maria João Santos
Center for Spatial and Textual Analysis, Spatial History Project, and Bill Lane Center for the American West, Stanford University, Stanford, California, United States of America

Terry Watt
Terry Watt Planning Associates, San Francisco, California, United States of America

Stephanie Pincetl
California Center for Sustainable Communities, Institute of the Environment University of California Los Angeles, Los Angeles, California, United States of America

Filipe Batista e Silva
Institute for Environment and Sustainability, Joint Research Centre, Ispra, Italy
Centro de Estudos de Geografia e Ordenamento do Territó rio, Porto, Portugal

Eric Koomen and Vasco Diog
Department of Spatial Economics/SPINlab, VU University Amsterdam, Amsterdam, The Netherlands

Carlo Lavalle
Institute for Environment and Sustainability, Joint Research Centre, Ispra, Italy

João M. B. Carreiras
Tropical Research Institute (IICT), Lisboa, Portugal
National Centre for Earth Observation (NCEO), Centre for Terrestrial Carbon Dynamics (CTCD), University of Sheffield,Sheffield, South Yorkshire, United Kingdom

Joshua Jones
Department of Geography and Earth Sciences, Aberystwyth University, Aberystwyth, Ceredigion, United Kingdom

Richard M. Lucas
Department of Geography and Earth Sciences, Aberystwyth University, Aberystwyth, Ceredigion, United Kingdom
Centre for Ecosystem Science, School of Biological, Earth and Environmental Sciences, University of New South Wales, Sydney, New South Wales, Australia

Cristina Gabriel
Tropical Research Institute (IICT), Lisboa, Portugal

Annamaria Bevivino
ENEA (Italian National Agency for New Technologies, Energy and Sustainable Economic Development) Casaccia Research Center, Technical Unit for Sustainable Development and Innovation of Agro-Industrial System, Rome, Italy

Patrizia Paganin
ENEA (Italian National Agency for New Technologies, Energy and Sustainable Economic Development) Casaccia Research Center, Technical Unit for Sustainable Development and Innovation of Agro-Industrial System, Rome, Italy

Giovanni Bacci
Consiglio per la Ricerca e la Sperimentazione in Agricoltura - Research Centre for the Soil-Plant System, Rome, Italy
Laboratory of Microbial and Molecular Evolution, Department of Biology, University of Florence, Florence, Italy

Alessandro Florio
Consiglio per la Ricerca e la Sperimentazione in Agricoltura - Research Centre for the Soil-Plant System, Rome, Italy

Maite Sampedro Pellicer
ENEA (Italian National Agency for New Technologies, Energy and Sustainable Economic Development) Casaccia Research Center, Technical Unit for Sustainable Development and Innovation of Agro-Industrial System, Rome, Italy

Maria Cristiana Papaleo
Laboratory of Microbial and Molecular Evolution, Department of Biology, University of Florence, Florence, Italy

Alessio Mengoni
Laboratory of Microbial and Molecular Evolution, Department of Biology, University of Florence, Florence, Italy

Luigi Ledda
Dipartimento di Agraria, University of Sassari, Sassari, Italy

Renato Fani
Laboratory of Microbial and Molecular Evolution, Department of Biology, University of Florence, Florence, Italy

Anna Benedetti
Consiglio per la Ricerca e la Sperimentazione in Agricoltura - Research Centre for the Soil-Plant System, Rome, Italy

Claudia Dalmastri
ENEA (Italian National Agency for New Technologies, Energy and Sustainable Economic Development) Casaccia Research Center, Technical Unit for Sustainable Development and Innovation of Agro-Industrial System, Rome, Italy

David M. Theobald
Conservation Science Partners, Inc., Fort Collins, Colorado, United States of America
Department of Fish, Wildlife, and Conservation Biology, Colorado State University, Fort Collins, Colorado, United States of America

Kabindra Adhikari
Department of Soil Science, University of Wisconsin2Madison, Madison, Wisconsin, United States of America

Alfred E. Hartemink
Department of Soil Science, University of Wisconsin2Madison, Madison, Wisconsin, United States of America

Budiman Minasny
Department of Environmental Sciences, The University of Sydney, Sydney, New South Wales, Australia

Rania Bou Kheir
Department of Agro-ecology, Aarhus University, Tjele, Denmark

Mtte B. Greve
Department of Agro-ecology, Aarhus University, Tjele, Denmark

Mogens H. Greve
Department of Agro-ecology, Aarhus University, Tjele, Denmark

Kai Riess, Franz Oberwinkler, Robert Bauer and Sigisfredo Garnica
Plant Evolutionary Ecology, Institute of Evolution and Ecology, University of Tü bingen, Tü bingen, Germany

Yaai Dang
State Key Laboratory of Soil Erosion and Dryland Farming on the Loess Plateau, Institute of Soil and Water Conservation, Northwest A&F University, Yangling, Shaanxi, China
International Center for Climate and Global Change Research, School of Forestry & Wildlife Sciences, Auburn University, Auburn, Alabama, United States of America

Wei Ren
International Center for Climate and Global Change Research, School of Forestry & Wildlife Sciences, Auburn University, Auburn, Alabama, United States of America

Bo Tao
International Center for Climate and Global Change Research, School of Forestry & Wildlife Sciences,

Auburn University, Auburn, Alabama, United States of America

Guangsheng Chen
International Center for Climate and Global Change Research, School of Forestry & Wildlife Sciences, Auburn University, Auburn, Alabama, United States of America

Chaoqun Lu
International Center for Climate and Global Change Research, School of Forestry & Wildlife Sciences, Auburn University, Auburn, Alabama, United States of America

Jia Yang
International Center for Climate and Global Change Research, School of Forestry & Wildlife Sciences, Auburn University, Auburn, Alabama, United States of America,

Shufen Pan
International Center for Climate and Global Change Research, School of Forestry & Wildlife Sciences, Auburn University, Auburn, Alabama, United States of America

Guodong Wang
College of Science, Northwest A&F University, Yangling, Shaanxi, China

Shiqing Li
1 State Key Laboratory of Soil Erosion and Dryland Farming on the Loess Plateau, Institute of Soil and Water Conservation, Northwest A&F University, Yangling, Shaanxi, China

Hanqin Tian
International Center for Climate and Global Change Research, School of Forestry & Wildlife Sciences, Auburn University, Auburn, Alabama, United States of America

Bernard W. T. Coetzee
Centre for Invasion Biology, Department of Botany and Zoology, Stellenbosch University, Stellenbosch, Western Cape, South Africa, School of Biological Sciences, Monash University, Melbourne, Victoria, Australia

Kevin J. Gaston
Environment and Sustainability Institute, University of Exeter, Penryn, Cornwall, United Kingdom

Steven L. Chown
School of Biological Sciences, Monash University, Melbourne, Victoria, Australia

Chelsea G. Himsworth
School of Population and Public Health, University of British Columbia, Vancouver, British Columbia, Canada
Animal Health Centre, British Columbia Ministry of Agriculture, Abbotsford, British Columbia, Canada

Kirbee L. Parsons
School of Population and Public Health, University of British Columbia, Vancouver, British Columbia, Canada

Alice Y. T. Feng
School of Population and Public Health, University of British Columbia, Vancouver, British Columbia, Canada

Thomas Kerr
British Columbia Centre for Excellence in HIV/AIDS, St. Paul's Hospital, Vancouver, British Columbia, Canada,
Department of Medicine, University of British Columbia, Vancouver, British Columbia, Canada

Claire M. Jardine
Department of Pathobiology, University of Guelph, Guelph, Ontario, Canada

David M. Patrick
School of Population and Public Health, University of British Columbia, Vancouver, British Columbia, Canada

Vanessa M. Adams
Northern Australia National Environmental Research Program Hub and Research Institute for the Environment and Livelihoods, Charles Darwin University, Darwin, NT, Australia
Australian Research Council Centre of Excellence for Coral Reef Studies, James Cook University, Townsville, QLD, Australia

Robert L. Pressey
Australian Research Council Centre of Excellence for Coral Reef Studies, James Cook University, Townsville, QLD, Australia

Carlos Iñiguez-Armijos
Departamento de Ciencias Naturales, Universidad Te´cnica Particular de Loja (UTPL), Secciόn de Ecología, Loja, Ecuador

Adrián Leiva
Departamento de Ciencias Naturales, Universidad Técnica Particular de Loja (UTPL), Sección de Ecología, Loja, Ecuador Hans-Georg Frede Institute for Landscape Ecology and Resources
Management (ILR), Research Centre for Bio Systems, Land Use and Nutrition (IFZ), Justus-Liebig-University Giessen, Giessen, Germany

Henrietta Hampel
Universidad de Cuenca, Quinta Balzaín, Cuenca, Ecuador

Lutz Breuer
Institute for Landscape Ecology and Resources Management (ILR), Research Centre for Bio Systems, Land Use and Nutrition (IFZ), Justus-Liebig-University Giessen, Giessen, Germany

Index

A
Agricultural Ecosystems, 150, 196
Agricultural Landscapes, 40-41, 49-50, 56, 66
Agricultural Management, 150
Anthropogenic Activities, 7, 90, 150
Anthropogenic Influences, 81, 213, 220

B
Benthic Indicators, 241
Bioassessment Metrics, 19, 37
Biochemical Analyses, 155
Bird Diversity Patterns, 53
Brazilian Amazon, 126, 149
Bulk Density in the Peat Areas, 181
Butterfly Diversity, 40-41, 44-45, 47-50, 56

C
California's Rangeland Ecosystems, 68
Chinese Ecosystem Research Network (CERN), 204
Classification and Regression Tree (CART), 20
Classification System, 101, 164-168, 171, 177, 179
Colourless Monilioid Cells, 191
Commercial Land, 94, 112-115, 117-124, 168, 170
Comparisons of Biodiversity, 212
Conservation, 8, 17, 22, 37-40, 44-45, 49-51, 54-60, 64-66, 68, 70-72, 74-80, 82, 88, 90-92, 103-111, 126, 128-130, 146-149, 162, 164-165, 167, 170-171, 173, 175, 177, 190, 192, 201-202, 206, 208, 211-214, 217, 219-222, 230-233, 239-242, 247-249
Cork-oak Forest (CO), 151-152, 161
Cultural-historic Landscapes, 40

D
Deforestation, 58-59, 126-130, 132, 134-139, 145-149, 201, 212, 219-222, 241-242, 245, 248-250
Digital Mapping, 178, 189
Dissolved Organic Carbon, 1, 6, 202

E
Economic Forecasts, 112, 124
Endophytic Sebacinales, 191-192, 195-197, 199
Equity Issues, 230, 239
Extensive Time-series Analysis, 126

F
Flow Accumulation, 179, 183

G
Global Atmospheric Carbon Storage, 201
Global Biosphere, 164

H
Haplotype Structure, 193
Healthy Andean Streams, 241
Herbaceous Plants, 191-192, 195, 197, 199
Heterogeneous Landscapes, 40, 51
Hydrologic Runoff Metrics, 23, 37

I
Infrastructure Planning, 108
Inner-city Neighborhood, 223-224
Institute of Terrestrial Ecology (ITE), 2
Intensive Agricultural Production, 7

L
Land Acquisition, 103-106, 109-111
Land Cover Change Dynamics, 126
Land Use History, 51, 53, 56, 58-59, 61-62, 104, 203
Land Use Mapping Errors, 120
Lithic Xerorthents, 155

M
Macroinvertebrate, 19-20, 22-27, 30, 32-34, 36-39, 241-243, 245-250
Macroinvertebrate Data, 19-20, 22-23, 37, 242, 245
Mediterranean Region, 150-151, 162
Meta-analysis, 101, 125, 211-213, 215, 217, 219-221
Mixed Methods Approach, 223
Model Parameterization, 204
Mosquitoes, 93-95, 99-102
Mountainous Watershed, 7
Multiple Regression Analysis (MRA), 7

N
National Monitoring Schemes, 1
Nomenclature, 23, 52, 117, 119, 124-125, 195
Norway Rat, 223, 229

P
Phylogenetic Analysis, 150, 160, 193, 200
Phylogenetic Diversity, 191-192, 195
Physico-chemical Water Quality, 10
Plant Diversity Patterns, 53, 55

Plant Phenotypes, 193
Plateau Region, 201, 206, 211
Political Systems, 51-52, 54-55
Protected Area Ecological Performance, 212
Public Conservation Management Data, 71

Q
Quantitative Findings, 226

R
Rangeland, 68-71, 74-79, 91, 165-166, 172, 177
Rangeland Habitats, 68-71, 74-75
Relative Difference (RD), 119
Remote Sensing Data, 126-127, 242
Riparian Variables, 21, 32, 246
River Water Quality, 7-8, 10, 12-17

S
Sandy Calcareous Grassland, 58
Seasonal Changes, 91, 150, 161, 191
Sediment Load, 80-84, 86-92
Serendipita Herbamans, 191, 193, 195-199
Sessile Species Diversity, 51
Socio-economic Factors, 7-8, 10, 12, 14-16, 219
Soil Bacterial Community, 150-151, 156, 159
Soil Quality Indicator, 1, 5
Soil Sampling, 1, 5-6, 60-61, 151, 155, 162
Spatial Scale, 19-20, 25, 30, 36, 101, 241-242, 245
Spatiaotemporal Patterns, 208

T
Taxonomic Composition, 156
Temporal Habitat Segregation, 93
Temporal Variation, 80, 89, 91, 211
The Cascade Rules, 231-232, 234-235, 239
The Redundancy Analysis, 14
Topographical Factors, 16
Topography, 7-8, 10, 12, 14-18, 40, 50, 52, 55, 60, 130, 148, 190, 248
Transmission Electron Microscopy (TEM), 191
Transylvania, 40-42, 44-45, 48-50
Typic Dystroxerepts, 155

W
Widespread Development Patterns, 164

Y
Yellow River Basin, 80-84, 87-88, 90-92